Algebraic Combinatorics
and Applications

Springer-Verlag Berlin Heidelberg GmbH

Anton Betten
Axel Kohnert
Reinhard Laue
Alfred Wassermann
Editors

Algebraic Combinatorics and Applications

Proceedings of the Euroconference
Algebraic Combinatorics and Applications (ALCOMA),
held in Gößweinstein, Germany,
September 12-19, 1999

Springer

Editors

Anton Betten
Axel Kohnert
Reinhard Laue
Alfred Wassermann

Lehrstuhl II für Mathematik
Universität Bayreuth
95440 Bayreuth
Germany

e-mail:
anton.betten@uni-bayreuth.de
kohnert@uni-bayreuth.de
reinhard.laue@uni-bayreuth.de
alfred.wassermann@uni-bayreuth.de

Library of Congress Cataloging-in-Publication Data

Euroconference Algebraic Combinatorics and Applications (1999 : Gössweinstein, Germany)
Algebraic combinatorics and applications : proceedings of the Euroconference Algebraic Combinatorics and Applications (ALCOMA), held in Gössweinstein, Germany, on September 12-19, 1999 / Anton Betten ... [et al.], editors.
p. cm.
Includes bibliographical references.
ISBN 978-3-540-41110-9
1. Combinatorial analysis--Congresses. I. Betten, Anton. II. Title.

QA164 .E84 1999
511'.6--dc21

00-054947

Mathematics Subject Classification (2000): 05-XX, 20-XX, 81-XX, 92-XX

ISBN 978-3-540-41110-9 ISBN 978-3-642-59448-9 (eBook)
DOI 10.1007/978-3-642-59448-9

http://www.springer.de

Originally published by Springer-Verlag Berlin Heidelberg New York 2001

Typeset by the authors using a Springer TEX macro package
Cover design: *design & production* GmbH, Heidelberg

SPIN 10723244 46/3142LK - 5 4 3 2 1 0 – Printed on acid-free paper

Preface

This volume contains articles based on talks given at the Euroconference Alcoma99, held in Gößweinstein in northern Bavaria, Germany, from 12 to 19 September 1999. The place is in the centre of a tourist region Fränkische Schweiz with a famous Basilica, only a few meters away from the conference centre. The weather was extremely fine, supporting the good atmosphere of the conference. There were more than 80 participants and 40 talks, including hour-long invited talks by Clausen, Dress, Foata, Fowler, Grabmeier, Gutman, Lascoux, Laue, Lulek, Morris, Paule, Strehl, Varmuza and Zimmermann. The conference brought together researchers from the different fields of algebraic combinatorics and its applications. The result was that pure mathematicians, applied mathematicians, computer scientists, physicists, and chemists, both senior and young researchers from different European countries communicated in stimulating discussions everywhere in the small village.

The main theme of the conference was group actions in various areas. There was an easy rule to follow for the selection of these areas: Just touch upon the items that are related to the scientific work of Professor A. Kerber. Indeed, many authors took the opportunity to dedicate their article to him on occasion of his 60th birthday. He had the vision to turn group actions into a tool which with the help of computers can solve practical problems by just pressing the right buttons. So, the long path of his research started with the representation theory of symmetric groups, Polya's and Redfield's combinatorial art of counting, culminating with the software products used in industry for chemical structure elucidation, contributions to solid state physics, research on error correcting codes, and on t-designs. The whole spectrum is present in these proceedings. Several articles highlight the questions from these areas which may be answered by adopting this algebraic point of view.

The Conference was sponsored by the European Community. The organizers who are also the editors would like to thank them for their financial support. Also thanks are due to the local Catholic Church for providing an excellent lecture hall and the tourist office for organizing lodging and social events. The editors especially thank the Springer Verlag, in particular Dr. Peters and Mrs. Allewelt, for publishing these proceedings and for their patience during the editing procedure.

Last but not least we want to thank all the referees for their invalueable help.

July, 2000

Anton Betten
Axel Kohnert
Reinhard Laue
Alfred Wassermann

Contents

Introduction

Algebraic methods and in particular groups play an important role in combinatorics. Computer algebra systems now allow us to handle the necessary computations and so applications to a wide range of subjects are now accessible. A feedback between theory and practice is needed for both sides to profit from this development. The Euroconference Alcoma99 aimed at bringing together researchers from algebraic combinatorics and its applications. An important role is played by the actions of finite groups on different sets of objects. Group theory appeals to a theoretician, since it is a beautiful theory, relying only on a few axioms but yielding strong theorems. Moreover, groups appear in many different areas in various ways, usually classifying objects like functions, graphs, molecules, spaces, and even groups themselves. Therefore, practical applications also rely on group theoretical methods.

The range of applications of group actions covers important aspects of Chemistry, Physics, Algebraic Combinatorics and Group Theory itself. There is a common basis and a big overlap in the methods used. This means that researchers in any of these areas are well prepared for an interdisciplinary exchange of ideas. The participants of the conference contributed articles from their discipline to a wider audience in this spirit. So, in particular for the invited talks an understandable presentation of the ideas from the disciplines was prefered to the latest news for specialists only. This volume thus may serve as an introduction into several nice and important areas of applications of group actions as well as providing the related theoretical background related. Some articles also refer to software which is available via the internet and which was demonstrated at the conference.

Symmetric Groups

As tableaux techniques are a cornerstone of the representation theory of the symmetric group, there are several papers studying these objects. The article by Blessenohl and Jöllenbeck studies a generalisation of the Robinson-Schensted algorithm. Clausen in his paper considers for which groups tableaux methods can be extended for studying representation theory. Also the paper by Golojuch et al., which has a physical background, shows the relevance of the Robinson-Schensted costruction.

The paper by Fiedler gives a new method for the computation of normal forms of covariant tensors, using methods dealing with the group ring of the symmetric group. Lascoux presents a new description of the Bruhat order for the affine symmetric group. Morris and Almestady apply the method introduced by Fischer for calculating the characters of group extensions to compute the Fischer matrices of the generalized symmetric group and one of its covering groups.

Number Theory

The paper by Andrews et al. shows the application of combinatorial methods for the solution of diophatine equations. Whereas the two papers by Dress and Luca are devoted to the study of binary recurrence sequences. In his paper Warnaar he gives an overview of the history of Bailey's Lemma.

Constructive Group Actions

There are several contributions dealing with the construction of combinatorial objects up to isomorphism. Here the isomorphism classes are group orbits and the stabilizer of an object is its automorphism group. The article of Laue contains a theory for constructing representatives from the isomorphism classes in a reasonable generality. New results on t-designs highlight the approach. In particular, the first simple 9-designs with small parameters are presented, double coset methods lead to the construction of 7-designs with trivial automorphism group. Also large numbers of isomorphism types are determined. Betten and Betten use tactical decompositions together with group actions to construct all proper linear spaces on 18 points. Delandtsheer presents a proof for an important part of the classification theorem of finite flag-transitive linear spaces, covering the case of an alternating group as the socle of the prescribed automorphism group. These linear spaces are in particular Steiner 2-designs. Sebille provides explicit examples of block-transitive but point-imprimitive 2-designs which prove the necessity of the exceptions in the classification theorem by Cameron and Praeger. Grabmeier and Lambe use completely different methods for their compuation of resolutions over finite p-groups. They are based on efficient data structures for the computation. This is also true for the methods used in in the article by Rehfinger et al. on methods for the computation of linear codes.

Applications to Chemistry and Biology

Graphs model interactions of atoms by edges between corresponding vertices. Thus graph theory was greatly fostered by problems from chemistry. In particular, constructing molecules with prescribed properties corresponds to finding representatives from orbits of a symmetric group acting on the set of atoms and the mathematical results are rated according to chemical relevance. There are contributions of Gutman on the energy of a graph, by Schocker on the intervals of degree sequences that are realizable by graphs, and by Fowler et al. on Fulleren graphs with results, conjectures and beautiful illustrations. A graphical approach to the elucidation of evolutionary patterns of species in biology is given by Dress et al.

Applications to Physics

In a series of papers Lulek and his students study the connection between the Heisenberg model and the group action. In his paper Lulek shows how to describe this connection. The paper by Golujuch et al. was already mentioned in the section dealing with the symmetric group, which show the strong interaction between physics of the Heisenberg model and methods first used in the study of the symmetric group. In the paper by Kuzma et al. the authors studies combinatorial methods for the Bethe Ansatz. The paper of Wal studies the combinatorial methods in the case of the Bethe Ansatz in a hexagon.

MacMahon's Partition Analysis V: Bijections, Recursions, and Magic Squares

George E. Andrews[1*], Peter Paule[2], Axel Riese[3**], and Volker Strehl[4***]

[1] Department of Mathematics
The Pennsylvania State University
University Park, PA 16802, USA
andrews@math.psu.edu

[2] Research Institute for Symbolic Computation
Johannes Kepler University Linz
A–4040 Linz, Austria
Peter.Paule@risc.uni-linz.ac.at

[3] Research Institute for Symbolic Computation
Johannes Kepler University Linz
A–4040 Linz, Austria
Axel.Riese@risc.uni-linz.ac.at

[4] Computer Science Institute-Informatik 8
Friedrich-Alexander-Universität Erlangen-Nürnberg
D-91058 Erlangen, Germany
strehl@informatik.uni-erlangen.de

Dedicated to Professor A. Kerber at the occasion of his 60th birthday

Abstract A significant portion of MacMahon's famous book "Combinatory Analysis" is devoted to the development of "Partition Analysis" as a computational method for solving problems in connection with linear homogeneous diophantine inequalities and equations, respectively. Nevertheless, MacMahon's ideas have not received due attention with the exception of work by Richard Stanley. A long range object of a series of articles is to change this situation by demonstrating the power of MacMahon's method in current combinatorial and partition-theoretic research. The renaissance of MacMahon's technique partly is due to the fact that it is ideally suited for being supplemented by modern computer algebra methods. In this paper we illustrate the use of Partition Analysis and of the corresponding package `Omega` by focusing on three different aspects of combinatorial work: the construction of bijections (for the Refined Lecture Hall Partition Theorem), exploitation of recursive patterns (for Cayley compositions), and finding nonnegative integer solutions of linear systems of diophantine equations (for magic squares of size 3).

* Partially supported by National Science Foundation Grant DMS-9870060.
** Supported by SFB-grant F1305 of the Austrian FWF.
*** Supported by a visiting researcher grant of the J. Kepler University Linz.

1 Introduction

The initial motive for the revival of MacMahon's Partition Analysis was the beautiful refinement of a classic result due to Euler – the number of partitions of N into distinct parts equals the number of partitions of N into odd parts [1, p. 5] – that was discovered by M. Bousquet-Mélou and K. Eriksson [7] only recently:

Theorem 1.1 ("Lecture Hall Partition Theorem"). *The number of partitions of N of the form $N = b_1 + b_2 + \cdots + b_n$ wherein*

$$\frac{b_n}{n} \geq \frac{b_{n-1}}{n-1} \geq \cdots \geq \frac{b_1}{1} \geq 0$$

equals the number of partitions of N into odd parts each $\leq 2n-1$.

Note that in lecture hall partitions some parts can be 0. For example, if $N = 13$ and $n = 3$ we have the 10 lecture hall partitions $0+0+13$, $0+1+12$, $0+2+11$, $0+3+10$, $1+2+10$, $0+4+9$, $1+3+9$, $0+5+8$, $1+4+8$, and $2+4+7$. On the other hand there are also 10 partitions of $N = 13$ into parts from $\{1,3,5\}$, namely $1^0 3^1 5^2$, $1^3 3^0 5^2$, $1^2 3^2 5^1$, $1^5 3^1 5^1$, $1^8 3^0 5^1$, $1^1 3^4 5^0$, $1^4 3^3 5^0$, $1^7 3^2 5^0$, $1^{10} 3^1 5^0$, and $1^{13} 3^0 5^0$.

The same authors also derived a further refinement of Euler's classic result; see the "Refined Lecture Hall Partition Theorem" (Theorem 2.13) below. Section 2 will be devoted to the construction of a bijective proof of it.

In [7] Bousquet-Mélou and Eriksson gave two different proofs of this theorem, one using Bott's formula for the affine Coxeter group $\tilde{C}_n$, and one of bijective-combinatorial nature. In [3] the first named author presented a proof following an entirely different approach. Namely, an approach which is based on the observation that MacMahon's Partition Analysis, surveyed in [12, Vol. 2, Sect. VIII, pp. 91–170], is perfectly tailored for theorems of this kind. In order to illustrate this point, recall the definition of MacMahon's Omega operator $\Omega_{\geq}$.

Definition 1.2. The operator $\Omega_{\geq}$ is given by

$$\underset{\geq}{\Omega} \sum_{s_1=-\infty}^{\infty} \cdots \sum_{s_r=-\infty}^{\infty} A_{s_1,\ldots,s_r} \lambda_1^{s_1} \cdots \lambda_r^{s_r} := \sum_{s_1=0}^{\infty} \cdots \sum_{s_r=0}^{\infty} A_{s_1,\ldots,s_r} ,$$

where the domain of the $A_{s_1,\ldots,s_r}$ is the field of rational functions over $\mathbb{C}$ in several complex variables and the λ_i are restricted to annuli of the form $1 - \epsilon < |\lambda_i| < 1 + \epsilon$.

Remark 1.3. It is convenient to treat everything involved analytically rather than formally because the method relies on unique Laurent series representations of a variety of rational functions. For a more detailed discussion of this aspect, the interested reader is referred to [4].

Let us now consider the instance $n = 3$ of Theorem 1.1. Obviously, the coefficient of q^N in

$$\frac{1}{(1-q)(1-q^3)(1-q^5)} \tag{1}$$

equals the number of partitions of N into odd parts each ≤ 5. On the other hand, the coefficient of q^N in

$$\underset{\geqq}{\Omega} \sum_{b_1,b_2,b_3 \geq 0} \lambda_1^{2b_3-3b_2} \lambda_2^{b_2-2b_1} q^{b_1+b_2+b_3} \tag{2}$$

gives exactly the number of the desired lecture hall partitions for n being fixed to 3. Note that due to the Omega operator $\Omega_{\geqq}$ only those partitions $b_1 + b_2 + b_3 = N$ are counted for which $2b_3 - 3b_2 \geq 0$ and $b_2 - 2b_1 \geq 0$. By geometric series expansion the triple sum can be brought into product form, which means that expression (2) can be rewritten as

$$\underset{\geqq}{\Omega} \frac{1}{(1-\lambda_1^2 q)(1-\frac{\lambda_2 q}{\lambda_1^3})(1-\frac{q}{\lambda_2^2})} \ . \tag{3}$$

where the factors in the denominator correspond to b_3, b_2, b_1 in this order.

Therefore all what remains for proving the Lecture Hall Partition Theorem for $n = 3$ is to show equality of the generating function expressions (1) and (3). For doing so, MacMahon introduced a catalogue of rules that describe the elimination of the λ-parameters involved. As an example, we state one of these rules in form of a lemma.

Lemma 1.4. *For any integer $s \geq 0$,*

$$\underset{\geqq}{\Omega} \frac{1}{(1-\lambda x)(1-\frac{y}{\lambda^s})} = \frac{1}{(1-x)(1-x^s y)} \ .$$

Proof. By geometric series expansion the left hand side equals

$$\underset{\geqq}{\Omega} \sum_{i,j \geq 0} \lambda^{i-sj} x^i y^j = \underset{\geqq}{\Omega} \sum_{j,k \geq 0} \lambda^k x^{sj+k} y^j \ ,$$

where the summation parameter i has been replaced by $sj + k$. But now $\Omega_{\geqq}$ sets λ to 1, which completes the proof. □

With this lemma in hand, the proof of "(1) = (3)" reduces to successive elimination of the $\Omega_{\geqq}$-parameters λ_1 and λ_2.

Proof (of the Lecture Hall Partition Theorem for $j = 3$). Split (3) additively into two parts by applying partial fraction decomposition $1/((1-\lambda_1 z)(1+\lambda_1 z)) = 1/(2(1-\lambda_1 z)) + 1/(2(1+\lambda_1 z))$ to the term $1/(1-\lambda_1^2 q)$. Then by using Lemma 1.4 eliminate from both summands the $\Omega_{\geqq}$-parameter λ_1. For the last step one observes that Lemma 1.4 can be applied again in order to eliminate λ_2; this way one arrives at (1). □

Already this particular example suggests that MacMahon's approach is an ideal candidate for being supplemented by modern computer algebra methods. But rather than implementing tables of rules – as, for example, the list of twelve fundamental evaluations given by MacMahon [12, Vol. II, pp. 102–103] – in [4, Theorem 2] we explain how this can be achieved "in one stroke" by a fairly general setting based on the "fundamental recurrence" for the $\Omega_{\geqq}$ operator. Using the procedures from the Mathematica package `Omega`, the problem of showing "(1) = (3)" is solved as follows:

We put the file `Omega.m` (together with the file `Readme.txt`) in the same directory in which we run our Mathematica session. After invoking Mathematica we load the package:

```
In[1]:= <<Omega.m
        Axel Riese's Omega implementation version 1.3 loaded
```

Now the proof of Theorem 1.1 for $n = 3$ can simply be done as follows. First we input the expression the $\Omega_{\geqq}$ operator acts on; see (3). Then we call the `OR`-procedure to eliminate the λ-variables $\lambda 1$ and $\lambda 2$:

```
In[2]:= f = 1 / ((1-λ1^2 q)(1-λ2 q/λ1^3)(1-q/λ2^2))
```

Out[2]=

$$\frac{1}{\left(1-\frac{\lambda 2\, q}{\lambda 1^3}\right)\left(1-\lambda 1^2 q\right)\left(1-\frac{q}{\lambda 2^2}\right)}$$

```
In[3]:= OR[f, λ1]
```

Out[3]=

$$\frac{1+\lambda 2\, q^3}{(1-q)\left(1-\frac{q}{\lambda 2^2}\right)\left(1-\lambda 2^2 q^5\right)}$$

```
In[4]:= OR[%, λ2]
```

Out[4]=

$$\frac{1}{(1-q)\left(1-q^3\right)\left(1-q^5\right)}$$

This proves the equality in question.

More information (theoretical background, usage, etc.) about the `Omega` package can be found in [4]. In this paper we focus on concrete applications concerning the construction of bijections, exploitation of recursive patterns, and finding nonnegative integer solutions of linear systems of diophantine equations.

In Sect. 2 we illustrate how the `Omega` package and, more generally, Partition Analysis can be used for the construction of a bijective proof, Proposition 2.10 and Theorem 2.15, for the Refined Lecture Hall Partition Theorem (Theorem 2.13), a refinement of Theorem 1.1 above. To this end we need another essential ingredient, namely an involution, defined in Sect. 2.5, that is equivalent to an involution discovered by Bousquet-Mélou and Eriksson in [8, Prop. 3.4].

In Sect. 3 we deal with a classical type of compositions that have been introduced and studied by A. Cayley [9]. Here the application of the `Omega` package leads to the discovery of a recursive pattern that enables to find a surprising generating function representation, Theorem 3.1, which encodes the solution to Cayley's problem.

Finally, in Sect. 4 we briefly describe how MacMahon's technique works for finding nonnegative integer solutions of systems of linear homogeneous diophantine equations. Instead of following MacMahon's table look-up approach, we again achieve elimination "in one stroke" by deriving a suitable analogue (Theorem 4.4) of the "fundamental recurrence" developed in [4, Theorem 2] for diophantine inequalities. We illustrate our Mathematica implementation by revisiting a section of MacMahon's book [12, Vol. 2, Sect. 407, p. 161].

2 A Lecture Hall Bijection

Following [3], for $n \geq 2$ let us define

$$f_n(y_1, \ldots, y_n) = \sum_{\frac{b_n}{n} \geq \frac{b_{n-1}}{n-1} \geq \cdots \geq \frac{b_1}{1} \geq 0} y_1^{b_1} y_2^{b_2} \cdots y_n^{b_n} . \tag{4}$$

We note that in [3] the notation $F_{n-1,0}(y_n, y_{n-1}, \ldots, y_1)$ is used instead of $f_n(y_1, \ldots, y_n)$.

The generating function version of Theorem 1.1 then is

$$f_n(q, \ldots, q) = \prod_{k=1}^{n} \frac{1}{1 - q^{2k-1}} \tag{5}$$

which was proved in [3] by using Partition Analysis. In order to give a flavor of the mechanism of MacMahon's method, we briefly sketch how this has been achieved.

By introducing parameters λ_1 up to λ_{n-1} where the jth lecture hall condition $(j-1)\, b_j \geq j\, b_{j-1}$ is represented by a factor $\lambda_{n-j+1}^{(j-1)\, b_j - j\, b_{j-1}}$, $(2 \leq j \leq n)$, the generating function expression (4), is encoded as an $\Omega_{\geq}$-expression:

$$f_n(y_1, \ldots, y_n) = \underset{\geq}{\Omega} \frac{1}{(1 - \lambda_1^{n-1} y_n)(1 - \frac{\lambda_2^{n-2}}{\lambda_1^{n}} y_{n-1})} \cdot \frac{1}{(1 - \frac{\lambda_3^{n-3}}{\lambda_2^{n-1}} y_{n-2}) \cdots (1 - \frac{\lambda_{n-1}}{\lambda_{n-2}^{3}} y_2)(1 - \frac{y_1}{\lambda_{n-1}^{2}})} .$$

For example, similar to the computation in the introduction, for $n = 2$ one has

$$\begin{aligned} f_2(y_1, y_2) &= \sum_{\frac{b_2}{2} \geq \frac{b_1}{1} \geq 0} y_1^{b_1} y_2^{b_2} \\ &= \underset{\geq}{\Omega} \sum_{b_1, b_2 \geq 0} \lambda_1^{b_2 - 2b_1} y_1^{b_1} y_2^{b_2} \\ &= \underset{\geq}{\Omega} \frac{1}{(1 - \lambda_1 y_2)(1 - \frac{y_1}{\lambda_1^2})} \\ &= \frac{1}{(1 - y_1 y_2^2)(1 - y_2)} . \end{aligned} \tag{6}$$

Since only one λ-variable is involved, elimination in the last line is elementary and can be done, for instance, by applying Lemma 1.4 with $s = 2$. This computation can be found also in MacMahon's book [12].

Setting y_1 and y_2 to q results in the desired product form $1/((1-q)(1-q^3))$, which proves instance $n = 2$ of (5).

The general situation turns out to be considerably more involved; nevertheless it can be handled by Partition Analysis as follows. The λ-elimination rules that are used for carrying out induction with respect to n, reveal that f_n must be of the form

$$f_n(y_1, \ldots, y_n) = \frac{P_n(y_1, \ldots, y_n)}{(1 - y_1 y_2^2 \cdots y_n^n)(1 - y_2^2 \cdots y_n^n) \cdots (1 - y_{n-1}^{n-1} y_n^n)(1 - y_n)} , \tag{7}$$

where $P_n(y_1, \ldots, y_n)$ is a polynomial in $y_1, \ldots, y_n$ with integer coefficients, which is defined by a recursive pattern. This recursive description then is used to prove that for the specialization $y_i = q$ the following factorization property [3, Lemma 1] holds:

$$p_n(q) := P_n(q, q, \ldots, q) - \prod_{k=2}^{n} \frac{1 - q^{kn - \binom{k}{2}}}{1 - q^{2k-1}} \quad (n \geq 2) . \tag{8}$$

Substituting this expression into the corresponding specialization $y_i = q$ of (7) gives the Lecture Hall Partition Theorem in the form of (5).

We want to remark that from representation (8) the polynomial structure of $p_n(q)$ is not entirely obvious. However, it is fully revealed by the following elementary bijection.

Definition 2.1. For fixed positive $n \in \mathbb{N}$ we define the following permutation of $\{1, \ldots, n\}$:

$$\sigma_n : j \mapsto \begin{cases} n - 2j + 1, & \text{if } 1 \leq 2j \leq n, \\ 2j - n, & \text{if } n < 2j \leq 2n . \end{cases}$$

Example 2.2. In order to illustrate the structure of the permutation σ_n we give explicit presentations for the first few n:

$$\sigma_1 = \begin{pmatrix} 1 \\ 1 \end{pmatrix}, \qquad \sigma_2 = \begin{pmatrix} 1\,2 \\ 1\,2 \end{pmatrix}, \qquad \sigma_3 = \begin{pmatrix} 1\,2\,3 \\ 2\,1\,3 \end{pmatrix},$$
$$\sigma_4 = \begin{pmatrix} 1\,2\,3\,4 \\ 3\,1\,2\,4 \end{pmatrix}, \quad \sigma_5 = \begin{pmatrix} 1\,2\,3\,4\,5 \\ 4\,2\,1\,3\,5 \end{pmatrix} \quad . \qquad \square$$

We will need the following property of σ_n.

Lemma 2.3. *For $1 \leq j \leq n$ we have*

$$j \cdot (2n - 2j + 1) = \binom{n+1}{2} - \binom{\sigma_n(j)}{2} .$$

Proof. By elementary computation,

$$\begin{aligned} j \cdot (2n - 2j + 1) &= \begin{cases} \frac{(n-(n-2j+1)+1)(n+(n-2j+1))}{2}, & \text{if } 1 \leq 2j \leq n, \\ \frac{(n+(2j-n))(n-(2j-n)+1)}{2}, & \text{if } n < 2j \leq 2n \end{cases} \\ &= \begin{cases} \binom{n+1}{2} - \binom{n-2j+1}{2}, & \text{if } 1 \leq 2j \leq n, \\ \binom{n+1}{2} - \binom{2j-n}{2}, & \text{if } n < 2j \leq 2n \end{cases} \\ &= \binom{n+1}{2} - \binom{\sigma_n(j)}{2} . \end{aligned} \qquad \square$$

Now it is easy to see that $p_n(q)$ is indeed a polynomial. Namely, since $kn - \binom{k}{2} = \binom{n+1}{2} - \binom{n+1-k}{2}$, by Lemma 2.3 we obtain

$$\begin{aligned} \left\{kn - \binom{k}{2} : 1 \leq k \leq n\right\} &= \left\{\binom{n+1}{2} - \binom{\sigma_n(k)}{2} : 1 \leq k \leq n\right\} \\ &= \{(n-k+1)(2k-1) : 1 \leq k \leq n\} . \end{aligned}$$

In other words, we have

$$\begin{aligned} p_n(q) &= \frac{1-q}{1-q^n} \cdot \prod_{k=1}^{n} \frac{1-q^{kn-\binom{k}{2}}}{1-q^{2k-1}} = \frac{1-q}{1-q^n} \cdot \prod_{k=1}^{n} \frac{1-q^{(n-k+1)(2k-1)}}{1-q^{2k-1}} \\ &= \prod_{k=2}^{n} \frac{1-q^{(n-k+1)(2k-1)}}{1-q^{2k-1}} \quad ; \end{aligned}$$

in particular, this gives

$$p_2(q) = 1, \quad p_3(q) = 1 + q^3, \quad p_4(q) = (1+q^5)(1+q^3+q^6)$$

and in general for $n \geq 3$,

$$\begin{aligned} p_n(q) &= (1 + q^{2n-3}) \\ &\quad \cdot (1 + q^{2n-5} + q^{2(2n-5)}) \cdots (1 + q^3 + q^{2\cdot 3} + \cdots + q^{(n-2)\cdot 3}) \ . \end{aligned}$$

After having recalled the Partition Analysis approach to the Lecture Hall Partition Theorem (Theorem 1.1) we devote the remaining part of this section to the question how Partition Analysis can help in the task of constructing a bijective proof for the Refined Lecture Hall Partition Theorem (Theorem 2.13) which is a substantially stronger result.

However, equipped with the `Omega` package, in Sect. 2.1 we will first consider lecture hall bijections for special cases. This study then leads to an algebraic representation of lecture hall partitions (Proposition 2.10) that will be proved in Sect. 2.2. In Sect. 2.3 this parametrization is used for a further refinement of the problem (Theorem 2.15) which extends to the case of the Refined Lecture Hall Partition Theorem (Theorem 2.13). The task of finding a bijective proof of this theorem then is reduced to the task of finding bijections σ_n (derived from Definition 2.1) and τ_n. The first task is solved in Sect. 2.4 where we use a crucial permutation that has been suggested by the use of the `Omega` package. The second task is solved in Sect. 2.5 by using a fundamental involution that was also discovered by Bousquet-Mélou and Eriksson, but who had used it in a different direction. Finally, combining the mappings σ_n and τ_n results in the desired refined lecture hall bijection.

2.1 Lecture Hall Bijections for Special Cases

Before we construct a general lecture hall bijection for Theorem 1.1 and its refined version, Theorem 2.13 below, we examine what Partition Analysis can do for us in various special cases.

The key to making constructive use of Partition Analysis is based on the fact that it delivers a parametrized representation of all nonnegative solutions of a given system of linear homogeneous diophantine inequalities.

Let us consider the case $n = 2$. Starting from (6),

$$f_2(y_1, y_2) = \sum_{\frac{b_2}{2} \geq \frac{b_1}{1} \geq 0} y_1^{b_1} y_2^{b_2} = \frac{1}{(1 - y_1 y_2^2)(1 - y_2)} ,$$

geometric series expansion,

$$\frac{1}{(1 - y_1 y_2^2)(1 - y_2)} = \sum_{\alpha_1, \alpha_2 \geq 0} y_1^{\alpha_1} y_2^{2\alpha_1 + \alpha_2} ,$$

reveals that the diophantine solution set

$$\mathcal{L}_2 := \left\{ \langle b_1, b_2 \rangle \in \mathbb{N}^2 : \tfrac{b_2}{2} \geq \tfrac{b_1}{1} \geq 0 \right\}$$

is identical with the set

$$\mathcal{L}_2^* := \{ \langle \alpha_1, 2\alpha_1 + \alpha_2 \rangle : \langle \alpha_1, \alpha_2 \rangle \in \mathbb{N}^2 \} ; \tag{9}$$

i.e.,

$$\mathcal{L}_2 = \mathcal{L}_2^* . \tag{10}$$

In other words, given a listing algorithm for $\mathbb{N}^2$, $\mathcal{L}_2$ can be constructed via the bijection

$$\omega_2 : \mathbb{N}^2 \to \mathcal{L}_2 \; : \; \langle \alpha_1, \alpha_2 \rangle \mapsto \langle \alpha_1, 2\alpha_1 + \alpha_2 \rangle \; ;$$

i.e.,

$$\mathcal{L}_2 = \omega_2(\mathbb{N}^2) = \mathcal{L}_2^* \; .$$

From now on the following mappings on partitions will play a significant rôle; it will be convenient to consider them as linear functionals on the vector space $\mathbb{Q}^n$.

Definition 2.4. We define the following three linear functionals on $\mathbb{Q}^n$:

$$|\langle x_1, x_2, \ldots, x_n \rangle| := x_1 + x_2 + \cdots + x_n \; ,$$
$$\|\langle x_1, x_2, \ldots, x_n \rangle\| := (2n-1) \cdot x_1 + (2n-3) \cdot x_2 + \cdots + 1 \cdot x_n \; ,$$

and

$$|||\langle x_1, x_2, \ldots, x_n \rangle||| := x_n - x_{n-1} + x_{n-2} - x_{n-3} + \cdots \pm x_1 \; .$$

We need also a couple of definitions for partitions with odd parts.

Definition 2.5 (Parametrizing partitions with odd parts). Let $\mathcal{O}_n$ denote the set of partitions with odd parts, all parts $\leq 2n-1$. For each $a = \langle a_1, a_2, \ldots, a_n \rangle \in \mathbb{N}^n$ let $\Psi_n(a)$ denote the partition where parts with size $2j-1$ occur with multiplicity a_{n-j+1}, i.e.,

$$\Psi_n(a) = 1^{a_n} 3^{a_{n-1}} \ldots (2j-1)^{a_{n-j+1}} \ldots (2n-1)^{a_1} \; .$$

If $\mathcal{O}_n(N)$ denotes the partitions from $\mathcal{O}_n$ with sum of parts equal to N, then

$$\mathcal{O}_n(N) = \Psi_n\{a \in \mathbb{N}^n : \|a\| = N\} \; .$$

If $\mathcal{O}_n(N,K)$ denotes the partitions from $\mathcal{O}_n$ with K parts and with sum of parts equal to N, then

$$\mathcal{O}_n(N,K) = \Psi_n\{a \in \mathbb{N}^n : \|a\| = N, |a| = K\} \; .$$

Considering the case $n = 2$ first, we denote for fixed N the corresponding set of lecture hall partitions by

$$\mathcal{L}_2(N) := \{\langle b_1, b_2 \rangle \in \mathcal{L}_2 : |\langle b_1, b_2 \rangle| = N\} \; . \tag{11}$$

But from (10) we know that

$$\mathcal{L}_2(N) = \mathcal{L}_2^*(N)$$

where

$$\mathcal{L}_2^*(N) := \{\langle \alpha_1, 2\alpha_1 + \alpha_2 \rangle \in \mathcal{L}_2^* : |\langle \alpha_1, 2\alpha_1 + \alpha_2 \rangle| = N\} \; .$$

Now, since we need to map any lecture hall partition onto a partition into odd parts, we only need to rewrite N, the sum of the b_i, as a linear combination of the corresponding parameters α_i; namely as

$$N = b_1 + b_2 = \alpha_1 + (2\alpha_1 + \alpha_2) = 3\alpha_1 + \alpha_2 \ ,$$

and we are immediately led to the bijection

$$\gamma_2 : \ \mathcal{L}_2^*(N) \to \mathcal{O}_2(N) \ : \ \langle \alpha_1, 2\alpha_1 + \alpha_2 \rangle \mapsto 1^{\alpha_2} 3^{\alpha_1} \ .$$

Before following the same approach for the higher cases, we introduce notation for the underlying lecture hall partition sets.

Definition 2.6. For integers $N, K \geq 0$ and $n \geq 1$, let

$$\begin{aligned} \mathcal{L}_n &:= \left\{ \langle b_1, b_2, \ldots, b_n \rangle \in \mathbb{N}^n : \tfrac{b_n}{n} \geq \tfrac{b_{n-1}}{n-1} \geq \cdots \geq \tfrac{b_1}{1} \geq 0 \right\} \ , \\ \mathcal{L}_n(N) &:= \{ b \in \mathcal{L}_n : |b| = N \} \ , \end{aligned}$$

and

$$\mathcal{L}_n(N, K) := \{ b \in \mathcal{L}_n : |b| = N, \ |||b||| = K \} \ .$$

We remark that we will need $\mathcal{O}_n(N, K)$ and $\mathcal{L}_n(N, K)$ for the Refined Lecture Hall Partition Theorem in Sect. 2.3.

For the case $n = 3$ we start out again by computing the required parameter representation with help of the `Omega` package:

$$\begin{aligned} f_3(y_1, y_2, y_3) &= \underset{\geq}{\Omega} \frac{1}{(1 - \lambda_1^2 y_3)(1 - \frac{\lambda_2^3}{\lambda_1^3} y_2)(1 - \frac{y_1}{\lambda_2^2})} \\ &= \frac{P_3(y_1, y_2, y_3)}{(1 - y_1 y_2^2 y_3^3)(1 - y_2^2 y_3^3)(1 - y_3)} \ , \end{aligned} \tag{12}$$

with

$$P_3(y_1, y_2, y_3) = 1 + y_2 y_3^2 \ . \tag{13}$$

Analogously to above, from (12) and (13) we can read off the desired parameter representation; namely

$$\mathcal{L}_3 = \mathcal{L}_3^* \ , \tag{14}$$

where

$$\begin{aligned} \mathcal{L}_3^* := \{ &\langle \alpha_1, 2\alpha_1 + 2\alpha_2 + r_2, 3\alpha_1 + 3\alpha_2 + \alpha_3 + 2r_2 \rangle : \\ &\langle \alpha_1, \alpha_2, \alpha_3 \rangle \in \mathbb{N}^3 \text{ and } r_2 \in \{0, 1\} \} \ . \end{aligned} \tag{15}$$

In other words, given a listing algorithm for $\mathbb{N}^3 \times \{0, 1\}$, $\mathcal{L}_3$ can be constructed via the bijection

$$\begin{aligned} \omega_3 : \ &\mathbb{N}^3 \times \{0, 1\} \to \mathcal{L}_3 \\ &\langle \alpha_1, \alpha_2, \alpha_3, r_2 \rangle \mapsto \langle \alpha_1, 2\alpha_1 + 2\alpha_2 + r_2, 3\alpha_1 + 3\alpha_2 + \alpha_3 + 2r_2 \rangle \ ; \end{aligned}$$

i.e.,

$$\mathcal{L}_3 = \omega_3(\mathbb{N}^3 \times \{0,1\}) = \mathcal{L}_3^* \ .$$

Therefore from (14) we learn that

$$\mathcal{L}_3(N) = \mathcal{L}_3^*(N)$$

where

$$\begin{aligned}\mathcal{L}_3^*(N) := \{\langle \alpha_1, 2\alpha_1 + 2\alpha_2 + r_2, 3\alpha_1 + 3\alpha_2 + \alpha_3 + 2r_2\rangle \in \mathcal{L}_3^* :\\ \alpha_1 + (2\alpha_1 + 2\alpha_2 + r_2) + (3\alpha_1 + 3\alpha_2 + \alpha_3 + 2r_2) = N\} \ .\end{aligned}$$

This time we can rewrite the b_i-sum N as follows,

$$\begin{aligned} N &= b_1 + b_2 + b_3 \\ &= \alpha_1 + (2\alpha_1 + 2\alpha_2 + r_2) + (3\alpha_1 + 3\alpha_2 + \alpha_3 + 2r_2) \\ &= \alpha_3 + 3(2\alpha_1 + r_2) + 5\alpha_2 \ , \end{aligned}$$

which leads us directly to the bijection

$$\begin{aligned} \gamma_3 : \ & \mathcal{L}_3^*(N) \to \mathcal{O}_3(N) \\ & \langle \alpha_1, 2\alpha_1 + 2\alpha_2 + r_2, 3\alpha_1 + 3\alpha_2 + \alpha_3 + 2r_2\rangle \mapsto 1^{\alpha_3} 3^{2\alpha_1 + r_2} 5^{\alpha_2} \ . \end{aligned} \tag{16}$$

For the case $n = 4$ things work analogously. For instance, one computes with the `Omega` package

$$f_4(y_1, y_2, y_3, y_4) = \frac{P_4(y_1, y_2, y_3, y_4)}{(1 - y_1 y_2^2 y_3^3 y_4^4)(1 - y_2^2 y_3^3 y_4^4)(1 - y_3^3 y_4^4)(1 - y_4)}$$

with

$$P_4(y_1, y_2, y_3, y_4) = 1 + y_3 y_4^2 + y_3^2 y_4^3 + y_2 y_3^2 y_4^3 + y_2 y_3^3 y_4^4 + y_2 y_3^4 y_4^6 \ .$$

However proceeding in this manner, the computation of the parametrized representation $\mathcal{L}_n^*$ of $\mathcal{L}_n$ that is needed for the construction of a lecture hall bijection, i.e., a bijection between $\mathcal{L}_n^*$ and $\mathcal{O}_n$, gets more and more involved. Nevertheless, guided by Partition Analysis this task can be accomplished by looking at the structure from an algebraic perspective as shown in the next section.

2.2 The Algebraic Structure of Lecture Hall Partitions

In view of the rational function representation (7) the parametrized representation $\mathcal{L}_n^*$ of $\mathcal{L}_n$ for arbitrary n has to combine two different aspects: the contribution emerging from the nice pattern of the denominator product, and the contribution of the numerator P_n. The latter one is much more involved since the polynomials P_n are growing rapidly with respect to the number of monomials involved; see Proposition 2.10 together with (17). This growth is also made explicit by the sufficiently complicated recursive scheme spelled out in [3]; however, we don't make use of this scheme in the algebraic approach explained below.

Before stating the main result, Proposition 2.10, of this section, it is necessary to introduce a couple of definitions.

Definition 2.7. For $n \in \mathbb{N}, n \geq 1$ we define sets of integer vectors

$$\mathcal{I}_n := \{\langle r_1, r_2, \ldots, r_n \rangle : 0 \leq r_j < j \ (1 \leq j \leq n)\} \ ,$$

and

$$\mathcal{I}_n^0 := \{r \in \mathcal{I}_n : r_n = 0\} \ .$$

The letter $\mathcal{I}$ is a mnemonic for *inversion vectors* as used in the combinatorial study of permutations, although we will not use this significance.

Definition 2.8. For $n \in \mathbb{N}, n \geq 1$ we define the n linearly independent vectors

$$\delta^j := \langle \underbrace{0, \ldots, 0}_{j-1}, j, \underbrace{0, \ldots, 0}_{n-j} \rangle \delta^n := \langle \underbrace{0, \ldots, 0}_{n-1}, 1 \rangle \ ,$$

and

$$\mathcal{D}_n^0 := \sum_{1 \leq j \leq n} \mathbb{N}\, \delta^j$$

which is the free $\mathbb{N}$-semimodule generated by the δ^j.

By the division property of the integers, each vector $a = \langle a_1, a_2, \ldots, a_n \rangle \in \mathbb{N}^n$ has a unique presentation

$$\begin{aligned} a &= \langle \alpha_1 \cdot 1 + r_1, \alpha_2 \cdot 2 + r_2, \ldots, \alpha_{n-1} \cdot (n-1) + r_{n-1}, \alpha_n \cdot 1 + r_n \rangle \\ &= \langle r_1, \ldots, r_n \rangle + \sum_{1 \leq j \leq n} \alpha_j\, \delta^j \end{aligned}$$

where $\langle r_1, \ldots, r_n \rangle \in \mathcal{I}_n^0$ and $\langle \alpha_1, \ldots, \alpha_n \rangle \in \mathbb{N}^n$, namely $\alpha_j = \mathtt{quot}(a_j, j)$ and $r_j = \mathtt{rem}(a_j, j)$ for $1 \leq j < n$, and $r_n = 0, \alpha_n = a_n$. In short,

$$\mathbb{N}^n = \mathcal{I}_n^0 \oplus \mathcal{D}_n^0 \ .$$

A parametrized representation of $\mathcal{L}_n$, the set of lecture hall partitions with n parts, is provided by the bijective mapping

$$\omega \ : \ \mathbb{N}^n \to \mathcal{L}_n \ : \ a = \langle a_1, \ldots, a_n \rangle \mapsto \langle b_1, \ldots, b_n \rangle$$

where

$$b_1 := a_1, \quad b_j := a_j + \left\lceil \frac{j \cdot b_{j-1}}{j-1} \right\rceil \ (1 < j \leq n) \ .$$

This parametrization is straight-forward from the inequality constraints between the b_i. For reasons of legibility we omit indexing ω by n since its definition is essentially independent of n.

Note that ω is not linear in general, but certain linearity properties of ω can be exhibited by choosing a second set of basis vectors. Besides the linearly independent vectors δ^j $(1 \leq j \leq n)$ introduced above we define another set of independent vectors $\epsilon^1, \epsilon^2, \ldots, \epsilon^n \in \mathbb{N}^n$ by applying ω to the δ^j.

Definition 2.9. For $n \in \mathbb{N}, n \geq 1$ we define the n linearly independent vectors

$$\epsilon^j := \omega(\delta^j) = \langle\underbrace{0, \ldots, 0}_{j-1}, j, j+1, \ldots, n\rangle \; (1 \leq j < n) \quad \text{and} \quad e^n := \omega(\delta^n) = \delta^n \;,$$

and

$$\mathcal{E}_n^0 := \sum_{1 \leq j \leq n} \mathbb{N}\, \epsilon^j$$

which is the free $\mathbb{N}$-semimodule generated by the ϵ^j.

It is easily checked that for any $a \in \mathbb{N}^n$ and $1 \leq j \leq n$

$$\omega(a + \delta^j) = \omega(a) + \epsilon^j$$

and hence in general

$$\omega(a + \textstyle\sum_j \alpha_j \, \delta^j) = \omega(a) + \textstyle\sum_j \alpha_j \, \epsilon^j \;.$$

In particular we have that ω is an isomorphism of semimodules

$$\omega \;:\; \mathcal{D}_n^0 \to \mathcal{E}_n^0 \;.$$

We summarize in form of a proposition.

Proposition 2.10. *For $n \in \mathbb{N}, n \geq 1$ we have the semilinear presentation*

$$\mathbb{N}^n = \mathcal{I}_n^0 \oplus \mathcal{D}_n^0 \;.$$

Moreover, if we put

$$\mathcal{R}_n^0 := \omega(\mathcal{I}_n^0)$$

we arrive at a semilinear presentation of $\mathcal{L}_n$, namely

$$\mathcal{L}_n = \omega(\mathbb{N}^n) = \omega(\mathcal{I}_n^0) \oplus \omega(\mathcal{D}_n^0) = \mathcal{R}_n^0 \oplus \mathcal{E}_n^0$$

where ω is an isomorphism between the semimodules $\mathcal{D}_n^0$ and $\mathcal{E}_n^0$.

Example 2.11. Let $n = 2$ and $a = \langle a_1, a_2 \rangle \in \mathbb{N}^2$. The $\mathcal{I}_2^0 \oplus \mathcal{D}_2^0$-decomposition in this case is trivial, namely $a = r + \alpha \in \mathcal{I}_2^0 \oplus \mathcal{D}_2^0$ with

$$r = \langle 0, 0 \rangle \quad \text{and} \quad \alpha = \langle \alpha_1, \alpha_2 \rangle = \alpha_1 \delta^1 + \alpha_2 \delta^2 \quad (= a) \ .$$

Then

$$\begin{aligned} \omega(a) &= \omega(r) + \omega(\alpha) = \langle 0, 0 \rangle + (\alpha_1 \epsilon^1 + \alpha_2 \epsilon^2) = \alpha_1 \langle 1, 2 \rangle + \alpha_2 \langle 0, 1 \rangle \\ &= \langle \alpha_1, 2\alpha_1 + \alpha_2 \rangle \ , \end{aligned}$$

in accordance with the definition (9) of $\mathcal{L}_2^*$. □

Example 2.12. Let $n = 3$ and $a = \langle a_1, a_2, a_3 \rangle \in \mathbb{N}^3$. By Euclidean division one computes

$$a = \langle \alpha_1 \cdot 1 + 0, \alpha_2 \cdot 2 + r_2, \alpha_3 \cdot 1 + 0 \rangle \quad \text{where } r_2 \in \{0, 1\} \ .$$

This means, the $\mathcal{I}_3^0 \oplus \mathcal{D}_3^0$-decomposition $a = r + \alpha \in \mathcal{I}_3^0 \oplus \mathcal{D}_3^0$ comes with

$$r = \langle 0, r_2, 0 \rangle \ \text{ and } \ \alpha = \langle \alpha_1, 2\alpha_2, \alpha_3 \rangle = \alpha_1 \delta^1 + \alpha_2 \delta^2 + \alpha_3 \delta^3 \ .$$

We obtain,

$$\begin{aligned} \omega(a) &= \omega(r) + \omega(\alpha) = \langle 0, r_2, 2r_2 \rangle + (\alpha_1 \epsilon^1 + \alpha_2 \epsilon^2 + \alpha_3 \epsilon^3) \\ &= \langle 0, r_2, 2r_2 \rangle + \alpha_1 \langle 1, 2, 3 \rangle + \alpha_2 \langle 0, 2, 3 \rangle + \alpha_3 \langle 0, 0, 1 \rangle \\ &= \langle \alpha_1, 2\alpha_1 + 2\alpha_2 + r_2, 3\alpha_1 + 3\alpha_2 + \alpha_3 + 2r_2 \rangle \ , \end{aligned}$$

in accordance with the definition (15) of $\mathcal{L}_3^*$. □

We conclude this section by the remark that the algebraic decomposition $\mathcal{L}_n = \mathcal{R}_n^0 \oplus \mathcal{E}_n^0$ is linked to the generating function presentation (7) in an obvious way. Namely, $\mathcal{R}_n^0$ corresponds to the numerator polynomial P_n, whereas $\mathcal{E}_n^0$ reflects the structure of the denominator product. More precisely, using the convention $y^a = y_1^{a_1} y_2^{a_2} \dots y_n^{a_n}$ whenever $a = \langle a_1, \dots, a_n \rangle \in \mathbb{N}^n$ we have

$$P_n(y_1, \dots, y_n) = \sum_{R \in \mathcal{R}_n^0} y^R \tag{17}$$

and

$$(1 - y_1 y_2^2 \cdots y_n^n)(1 - y_2^2 y_3^3 \cdots y_n^n) \cdots (1 - y_{n-1}^{n-1} y_n^n)(1 - y_n) = \prod_{j=1}^{n} \left(1 - y^{\epsilon^j}\right) \ .$$

2.3 The Refined Lecture Hall Partition Theorem

In the special cases $n = 2$ and $n = 3$ the mappings γ_2 and γ_3 have an important additional property. To illustrate this, let us fix two positive integers N and K.

Take $1^{\alpha_2}3^{\alpha_1} \in \mathcal{O}_2(N,K)$, i.e., $\|\langle \alpha_1, \alpha_2 \rangle\| = N$ and $|\langle \alpha_1, \alpha_2 \rangle| = K$. Then for $\langle b_1, b_2 \rangle := \gamma_2^{-1}(1^{\alpha_2}3^{\alpha_1}) \in \mathcal{L}_2(N)$ we observe also that

$$|||\langle b_1, b_2 \rangle||| = |||\langle \alpha_1, 2\alpha_1 + \alpha_2 \rangle||| = |\langle \alpha_1, \alpha_2 \rangle| = K \ .$$

This means, γ_2 is not only a bijection between $\mathcal{L}_2(N)$ and $\mathcal{O}_2(N)$ but also between $\mathcal{L}_2(N,K)$ and $\mathcal{O}_2(N,K)$.

Similarly, let $1^{\alpha_3}3^{2\alpha_1+r_2}5^{\alpha_2} \in \mathcal{O}_3(N,K)$ with $r_2 \in \{0,1\}$, i.e.,

$$\|\langle \alpha_2, 2\alpha_1 + r_2, \alpha_3 \rangle\| = N \quad \text{and} \quad |\langle \alpha_2, 2\alpha_1 + r_2, \alpha_3 \rangle| = K \ .$$

Then for $\langle b_1, b_2, b_3 \rangle := \gamma_3^{-1}(1^{\alpha_3}3^{2\alpha_1+r_2}5^{\alpha_2}) \in \mathcal{L}_3(N)$ we observe also that

$$\begin{aligned} |||\langle b_1, b_2, b_3 \rangle||| &= |||\langle \alpha_1, 2\alpha_1 + 2\alpha_2 + r_2, 3\alpha_1 + 3\alpha_2 + \alpha_3 + 2r_2 \rangle||| \\ &= |\langle \alpha_2, 2\alpha_1 + r_2, \alpha_3 \rangle| = K. \end{aligned}$$

This means, γ_3 is not only a bijection between $\mathcal{L}_3(N)$ and $\mathcal{O}_3(N)$ but also between $\mathcal{L}_3(N,K)$ and $\mathcal{O}_3(N,K)$.

The observation concerning this extra property with respect to the parameter K has already been made by Bousquet-Mélou and Eriksson. More precisely, they refined their theorem accordingly as follows [7].

Theorem 2.13 ("Refined Lecture Hall Partition Theorem"). *The number of partitions of N of the form $|\langle b_1, \ldots, b_n \rangle| = N$ wherein*

$$\frac{b_n}{n} \geq \frac{b_{n-1}}{n-1} \geq \cdots \geq \frac{b_1}{1} \geq 0$$

and

$$|||\langle b_1, \ldots, b_n \rangle||| = K \tag{18}$$

equals the number of partitions of N into exactly K odd parts each $\leq 2n-1$. In short,

$$\#\mathcal{L}_n(N,K) = \#\mathcal{O}_n(N,K) \ .$$

Example 2.14. From the partition sets listed after Theorem 1.1 we see, for instance, that

$$\mathcal{L}(13,5) = \{\langle 2,4,7 \rangle, \langle 1,4,8 \rangle, \langle 0,4,9 \rangle\}$$

and

$$\mathcal{O}(13,5) = \{1^1 3^4 5^0, 1^2 3^2 5^1, 1^3 3^0 5^2\} \ .$$ □

Our goal for the rest of the paper is to construct a lecture hall bijection Λ_n that takes also condition (18) into account. More precisely, for arbitrary nonnegative integers N and K we will construct a map

$$\Lambda_n \ : \ \mathcal{O}_n(N,K) \to \mathcal{L}_n(N,K)$$

that is a bijection.

We already know that both $\mathcal{O}_n$ and $\mathcal{L}_n$ are parametrized by $\mathbb{N}^n$. Hence, given

$$1^{a_n} 3^{a_{n-1}} \dots (2n-1)^{a_1} \in \mathcal{O}_n(N,K) \ ,$$

an obvious first step is to apply Ψ_n^{-1}; i.e.,

$$a := \langle a_1, a_2, \dots, a_n \rangle = \Psi_n^{-1}(1^{a_n} 3^{a_{n-1}} \dots (2n-1)^{a_1}) \in \mathbb{N}^n \ . \tag{19}$$

Next, by Euclidean division on the parts, we decompose a according to $\mathbb{N}^n = \mathcal{I}_n^0 \oplus \mathcal{D}_n^0$; see the first statement of Proposition 2.10. In other words, we compute r and α such that

$$a = r + \alpha \tag{20}$$

where

$$r = \langle r_1, \dots, r_n \rangle \in \mathcal{I}_n^0$$

and

$$\alpha = \langle \alpha_1, 2\alpha_2, \dots, (n-1)\,\alpha_{n-1}, \alpha_n \rangle = \sum_{j=1}^{n} \alpha_j \delta^j \in \mathcal{D}_n^0 \ .$$

Finally, in view of $\mathcal{L}_n = \omega(\mathbb{N}^n) = \mathcal{R}_n^0 \oplus \mathcal{E}_n^0$, the second part of Proposition 2.10, we need bijections

$$\tau_n \ : \ \mathcal{I}_n^0 \to \mathcal{R}_n^0 \quad \text{and} \quad \sigma_n \ : \ \mathcal{D}_n^0 \to \mathcal{E}_n^0$$

that respect the various functional relations with respect to N and K. More precisely, for any $r \in \mathcal{I}_n^0$ we must have

$$|r| = |||\tau_n(r)||| \quad \text{and} \quad \|r\| = |\tau_n(r)| \ , \tag{21}$$

and for any $\alpha \in \mathcal{D}_n^0$ we need

$$|\alpha| = |||\sigma_n(\alpha)||| \quad \text{and} \quad \|\alpha\| = |\sigma_n(\alpha)| \ . \tag{22}$$

Because then we would achieve our goal as follows:

For a as in (19), i.e., $\|a\| = N$ and $|a| = K$, and such that $a = r + \alpha$ as in (20) we have,

$$\tau_n(r) + \sigma_n(\alpha) \in \mathcal{R}_n^0 \oplus \mathcal{E}_n^0 = \mathcal{L}_n$$

with

$$|\tau_n(r) + \sigma_n(\alpha)| = |\tau_n(r)| + |\sigma_n(\alpha)| = \|r\| + \|\alpha\| = \|r + \alpha\| = \|a\| = N$$

and

$$|||\tau_n(r) + \sigma_n(\alpha)||| = |||\tau_n(r)||| + |||\sigma_n(\alpha)||| = |r| + |\alpha| = |r + \alpha| = |a| = K \ .$$

We summarize these considerations in form of a theorem.

Theorem 2.15. *Let $n \in \mathbb{N}$, $n \geq 1$. Suppose we have bijections*

$$\tau_n \; : \; \mathcal{I}_n^0 \to \mathcal{R}_n^0 \quad \textit{and} \quad \sigma_n \; : \; \mathcal{D}_n^0 \to \mathcal{E}_n^0$$

satisfying (21) *and* (22). *Let Γ_n be the bijection defined as*

$$\Gamma_n \; : \; \mathbb{N}^n = \mathcal{I}_n^0 \oplus \mathcal{D}_n^0 \to \mathcal{L}_n = \mathcal{R}_n^0 \oplus \mathcal{E}_n^0 \; : \; r + \alpha \mapsto \tau_n(r) + \sigma_n(\alpha) \; .$$

Then the map

$$\Lambda_n := \Gamma_n \circ \Psi_n^{-1} \; : \; \mathcal{O}_n(N,K) \to \mathcal{L}_n(N,K) \tag{23}$$

is a bijection.

2.4 The Bijection σ_n

The construction of the map σ_n is suggested by Partition Analysis. More precisely, we generalize the pattern that emerges from the special cases $n = 2$ and $n = 3$ as follows.

Definition 2.16. For $n \in \mathbb{N}$, $n \geq 1$, we define the linear transformation

$$\sigma_n \; : \; \mathbb{Q}^n \to \mathbb{Q}^n$$

on the basis vectors by setting

$$\sigma_n(\delta^j) := \epsilon^{\sigma_n(j)} \quad (1 \leq j \leq n) \; ,$$

where the σ_n on the right hand side is the permutation from Definition 2.1.

Since the meaning will be always clear from the context, for the linear transformation we use the same symbol σ_n as for the corresponding permutation.

Lemma 2.17. *The linear transformation σ_n provides a semimodule isomorphism between $\mathcal{D}_n^0$ and $\mathcal{E}_n^0$, i.e.,*

$$\sigma_n \; : \; \mathcal{D}_n^0 \to \mathcal{E}_n^0 \; ,$$

which satisfies the conditions (22).

Proof. The first part of the statement is obvious. Concerning property (22) we have,

$$\|\delta^j\| = j \cdot (2n - 2j + 1) = \binom{n+1}{2} - \binom{\sigma_n(j)}{2} = |\epsilon^{\sigma_n(j)}| \quad (1 \leq j \leq n) \; ,$$

$$|\delta^j| = \left\{ \begin{matrix} j, & \text{if } 1 \leq j < n, \\ 1, & \text{if } j = n \end{matrix} \right\} = |||\epsilon^{\sigma_n(j)}||| \quad (1 \leq j \leq n)$$

as a consequence of Lemma 2.3. By linearity it follows that for any $m = \sum_{j=1}^n m_j \, \delta^j \in \mathcal{D}_n^0$,

$$|\sigma_n(m)| = \|m\| \quad \text{and} \quad |||\sigma_n(m)||| = |m| \; . \qquad \square$$

We conclude this section by re-examining the special cases $n = 2$ and $n = 3$.

Example 2.18. For $n = 2$ the maps τ_2 and σ_2 are nothing but the corresponding ω-mappings. (Note that $\mathcal{I}_2^0 = \{\langle 0,0\rangle\}$ and $\sigma_2 = \left(\begin{smallmatrix}1&2\\1&2\end{smallmatrix}\right)$, the trivial permutation of $\{1,2\}$.) □

Example 2.19. In the case $n = 3$ the map τ_3 can again be chosen as the corresponding ω-mapping. But now the isomorphism σ_3 is non-trivial since $\sigma_3 = \left(\begin{smallmatrix}1&2&3\\2&1&3\end{smallmatrix}\right)$ as a permutation of $\{1,2,3\}$. Suppose we are given

$$a = \langle a_1, a_2, a_3\rangle = \Psi_3^{-1}(1^{a_3}3^{a_2}5^{a_1}) \in \mathbb{N}^3$$

with $\mathcal{I}_3^0 \oplus \mathcal{D}_3^0$- decomposition

$$a = r + \alpha = \langle 0, r_2, 0\rangle + \langle \alpha_2, 2\alpha_1, \alpha_3\rangle = \langle 0, r_2, 0\rangle + \alpha_2\delta^1 + \alpha_1\delta^2 + \alpha_3\delta^3 \ .$$

Then

$$\begin{aligned} \Gamma_3(a) &= \tau_3(r) + \sigma_3(\alpha) = \omega(r) + \alpha_2\epsilon^2 + \alpha_1\epsilon^1 + \alpha_3\epsilon^3 \\ &= \langle 0, r_2, 2r_2\rangle + \alpha_2\langle 0,2,3\rangle + \alpha_1\langle 1,2,3\rangle + \alpha_3\langle 0,0,1\rangle \\ &= \langle \alpha_1, 2\alpha_1 + 2\alpha_2 + r_2, 3\alpha_1 + 3\alpha_2 + \alpha_3 + 2r_2\rangle \ , \end{aligned}$$

which corresponds to the mapping γ_3; see (16). □

Finally, let us consider an example for $n = 4$. Now $\sigma_4 = \left(\begin{smallmatrix}1&2&3&4\\2&1&3&4\end{smallmatrix}\right)$ as a permutation of $\{1,2,3,4\}$. Take, for instance, $1^1 3^0 5^1 7^1 \in \mathcal{O}_4(13,3)$ and

$$a = \langle 1,1,0,1\rangle = \Psi_4^{-1}(1^1 3^0 5^1 7^1) \in \mathbb{N}^4.$$

As in the case $n = 3$ let us choose again $\tau_4 = \omega$. Consider the $\mathcal{I}_4^0 \oplus \mathcal{D}_4^0$-decomposition

$$a = r + \alpha = \langle 0,1,0,0\rangle + 1\cdot\delta^1 + 0\cdot\delta^2 + 0\cdot\delta^3 + 1\cdot\delta^4 \ .$$

Then we obtain

$$\begin{aligned} \Gamma_4(a) &= \tau_4(r) + \sigma_4(\alpha) = \omega(r) + 1\cdot\epsilon^3 + 0\cdot\epsilon^1 + 0\cdot\epsilon^2 + 1\cdot\epsilon^4 \\ &= \langle 0,1,2,3\rangle + 1\cdot\langle 0,0,3,4\rangle + 1\cdot\langle 0,0,0,1\rangle \\ &= \langle 0,1,2,3\rangle + \langle 0,0,3,5\rangle = \langle 0,1,5,8\rangle \ ; \end{aligned}$$

but $\langle 0,1,5,8\rangle \in \mathcal{L}_4(14,4)$! In other words, if we choose $\tau_4 = \omega$ then at least one of the functional properties we need with respect to N and K is violated. In this particular example we have

$$|r| = 1, \ |||\omega(r)||| = 2 \quad \text{and} \quad \|r\| = 5, \ |\omega(r)| = 6 \ .$$

Hence in general we need some extra effort to construct the bijection τ_n in a suitable manner. This is accomplished by an involutive approach described in the next section.

2.5 The Bijection τ_n

A Local Involution. We consider again the ω-mapping form Sect. 2.2 which affords a parametrization of lecture hall partitions:

$$\omega \; : \; \mathbb{N}^n \to \mathbb{N}^n \; : \; r = \langle r_1, r_2, \ldots, r_n \rangle \mapsto \langle R_1, R_2, \ldots, R_n \rangle = \omega(r) =: R$$

where

$$R_1 = r_1, \; R_k = r_k + \left\lceil \frac{k}{k-1} \, R_{k-1} \right\rceil \; (1 < k \leq n) \; .$$

Now the equation relating R_k and R_{k-1} can be written as

$$(R_k - r_k)(k-1) = kR_{k-1} + \varepsilon_{k-1}$$

where $0 \leq \varepsilon_{k-1} < k-1$. Yet another way of writing this equation is

$$\Delta_k = r_k(k-1) + \varepsilon_{k-1}$$

where we put[1]

$$\Delta_k := (k-1)R_k - kR_{k-1} \; .$$

Division by $k-1$ allows to recover (r_k, ε_{k-1}) from Δ_k:

$$r_k = \mathtt{quot}(\Delta_k, k-1), \; \varepsilon_{k-1} = \mathtt{rem}(\Delta_k, k-1) = \mathtt{rem}(-R_{k-1}, k-1) \; .$$

Note that ε_{k-1} depends only on R_{k-1}, not on R_k.

We will now say that the sequence R is *reduced at position* k, or k-*reduced*, if $0 \leq r_k < k$. Obviously we have the equivalence

$$0 \leq r_k < k \Delta_k < k(k-1)r_k = \mathtt{rem}(R_k + \varepsilon_{k-1}, k) \; . \tag{24}$$

Now note that the relation between R_{k-1} and R_k can also be seen "from the right", i.e., one can write

$$R_{k-1} = \left\lfloor \frac{k-1}{k} \, R_k \right\rfloor - s_{k-1}$$

where $s_{k-1} \geq 0$. This can be rewritten as

$$kR_{k-1} + ks_{k-1} = (k-1)R_k - \delta_k$$

where $0 \leq \delta_k < k$, or as

$$\Delta_k = s_{k-1}k + \delta_k \; .$$

It follows that the k-reducibility condition (24) transforms into another equivalent statement:

$$0 \leq s_{k-1} < k-1 \; .$$

---

[1] Note: the Lecture Hall Condition is equivalent to saying "$\Delta_k \geq 0$" for $1 \leq k \leq n$. The condition for $k = 1$ is void, of course.

Again, the pair (s_{k-1}, δ_k) can be recovered from Δ_k:

$$s_{k-1} = \texttt{quot}(\Delta_k, k),\ \delta_k = \texttt{rem}(\Delta_k, k) = \texttt{rem}(-R_k, k)\ .$$

This shows that δ_k only depends on R_k, not on R_{k-1}.

Now assume that the sequence R is both k-reduced and $(k+1)$-reduced; we will say k^+*-reduced* for short. This means that we have equations

$$\begin{aligned} \Delta_k &= r_k(k-1) + \varepsilon_{k-1}, && \text{where } 0 \le r_k < k \\ \Delta_{k+1} &= s_k(k+1) + \delta_{k+1}, && \text{where } 0 \le s_k < k\ . \end{aligned}$$

This suggests the following: since the conditions put on r_k and s_k are precisely the same, and since ε_{k-1} depends only on R_{k-1} and δ_{k+1} depends only on R_{k+1}, we may simply exchange the rôles of r_k and s_k – thus producing a new k^+-reduced sequence R' which differs from R only in position k, namely, we define

$$R'_k := R_k - r_k + s_k, \quad R' = \langle R_1, \ldots, R_{k-1}, R'_k, R_{k+1}, \ldots, R_n \rangle$$

and we have

$$\begin{aligned} R_k &= r_k + \left\lceil \frac{k}{k-1} R_{k-1} \right\rceil = \left\lfloor \frac{k}{k+1} R_{k+1} \right\rfloor - s_k\ , \\ R'_k &= s_k + \left\lceil \frac{k}{k-1} R_{k-1} \right\rceil = \left\lfloor \frac{k}{k+1} R_{k+1} \right\rfloor - r_k \end{aligned}$$

and an immediate consequence of this is

$$R_k + R'_k = \left\lceil \frac{k}{k-1} R_{k-1} \right\rceil + \left\lfloor \frac{k}{k+1} R_{k+1} \right\rfloor\ . \tag{25}$$

It is clear from the exchange argument that this mapping $\phi_k : R \mapsto R'$ is an involution on the set of k^+-reduced sequences for any f

Global Involution for Reduced Sequences. A sequence $R = \langle R_1, R_2, \ldots, R_n \rangle$ is *reduced* if it is k-reduced for all $k \le n$. Note that this is equivalent to saying that $R \in \mathcal{R}_n$ where

$$\mathcal{R}_n := \omega(\mathcal{I}_n)\ .$$

For reduced sequences the involutive procedure $\phi_k : R \mapsto R'$ at position k, as described in the previous section, can be simultaneously executed[2] for all positions $k = n - 2j - 1$, $(0 \le j < \lfloor n/2 \rfloor)$. Each of the ϕ_k keeps all the R_{n-2i}, $(0 \le i < \lfloor n/2 \rfloor)$ fixed, so that these actions commute with each other. We denote by $R \mapsto \Phi_n(R)$ this simultaneous involution on $\mathcal{R}_n$.

[2] In order to deal with the case $k = 1$ consistently one has to put formally $R_0 = 0$.

From the fact that the action of Φ_n keeps all the R_{n-2i} $(0 \le i < \lfloor n/2 \rfloor)$ fixed it follows that

$$|R| + |||R||| = |\Phi_n(R)| + |||\Phi_n(R)|||$$

or equivalently

$$|||R - \Phi_n(R)||| = |\Phi_n(R) - R| \ . \tag{26}$$

The main property of Φ_n w.r.t. reduced sequences is

$$|||R + \Phi_n(R)||| = \left\lceil \frac{n+1}{n} R_n \right\rceil \ . \tag{27}$$

The *proof* uses the result (25) from the previous section and a simple rearrangement of terms:

$$\begin{aligned}
&|||R + \Phi_n(R)||| \\
&\quad = 2\sum_{k\ge 0} R_{n-2k} - \sum_{k\ge 0} (R_{n-2k-1} + R'_{n-2k-1}) \\
&\quad = 2\sum_{k\ge 0} R_{n-2k} - \sum_{k\ge 0} \left(\left\lceil \frac{n-2k-1}{n-2k-2} R_{n-2k-2} \right\rceil + \left\lfloor \frac{n-2k-1}{n-2k} R_{n-2k} \right\rfloor \right) \\
&\quad = 2\sum_{k\ge 0} R_{n-2k} - \sum_{k>0} \left(\left\lceil \frac{n-2k+1}{n-2k} R_{n-2k} \right\rceil + \left\lfloor \frac{n-2k-1}{n-2k} R_{n-2k} \right\rfloor \right) \\
&\qquad - \left\lfloor \frac{n-1}{n} R_n \right\rfloor \\
&\quad = 2\sum_{k\ge 0} R_{n-2k} - \sum_{k>0} 2R_{n-2k} - \left\lfloor \frac{n-1}{n} R_n \right\rfloor \\
&\quad = 2R_n - \left\lfloor \frac{n-1}{n} R_n \right\rfloor = \left\lceil \frac{n+1}{n} R_n \right\rceil \ .
\end{aligned}$$

The Extension Step. Let $r \in \mathcal{I}_n$, $r' = \langle r, r_{n+1} \rangle \in \mathcal{I}_{n+1}$ and let $S = \langle S_1, \ldots, S_n \rangle \in \mathcal{R}_n$ such that

$$|r| = |||S||| \quad \text{and} \quad ||r|| = |S| \ .$$

Then for $S' := \langle \Phi_n(S), r_{n+1} + \lceil \frac{n+1}{n} S_n \rceil \rangle \in \mathcal{R}_{n+1}$:

$$|r'| = |||S'||| \quad \text{and} \quad ||r'|| = |S'| \ .$$

The sequence S' belongs indeed to $\mathcal{R}_{n+1}$: from $S \in \mathcal{R}_n$ we have $\Phi_n(S) \in \mathcal{R}_n$ because Φ_n is an involution on $\mathcal{R}_n$. Reducibility at position $n+1$ follows from the fact that $\Phi_n(S)$ has the same last element as S, namely S_n. By the same argument: if $r' \in \mathcal{I}^0_{n+1}$, i.e., if $r_{n+1} = 0$, then $S' \in \mathcal{R}^0_{n+1}$.

The *proof* of the asserted properties of S' is by simple verification, using the properties (26) and (27) mentioned in the previous section.

$$\begin{aligned} |||S'||| &= r_{n+1} + \left\lceil \frac{n+1}{n} S_n \right\rceil - |||\Phi_n(S)||| \\ &= r_{n+1} + \left\lceil \frac{n+1}{n} S_n \right\rceil + |||S||| - \left\lceil \frac{n+1}{n} S_n \right\rceil \\ &= r_{n+1} + |r| = |r'| \ , \end{aligned}$$

$$\begin{aligned} |S'| &= |\Phi_n(S)| + r_{n+1} + \left\lceil \frac{n+1}{n} S_n \right\rceil \\ &= |S| + |||S - \Phi_n(S)||| + r_{n+1} + |||S + \Phi_n(S)||| \\ &= |S| + 2|||S||| + r_{n+1} \\ &= \|r\| + 2|r| + r_{n+1} = \|r'\| \ . \end{aligned}$$

We will use the extension step in the following obvious way.

Iterative Construction of τ_n.

Suppose that $\tau_n : \mathcal{I}_n \to \mathcal{R}_n$ is a bijection that satisfies

$$|r| = |||\tau_n(r)||| \quad \text{and} \quad \|r\| = |\tau_n(r)|$$

then, writing $S = \tau_n(r)$,

$$\tau_{n+1} \ : \ \langle r, r_{n+1} \rangle \mapsto \left\langle \Phi_n(S), r_{n+1} + \left\lceil \tfrac{n+1}{n} S_n \right\rceil \right\rangle$$

is a bijection $\tau_{n+1} \ : \ \mathcal{I}_{n+1} \to \mathcal{R}_{n+1}$ that satisfies

$$|r'| = |||\tau_{n+1}(r')||| \quad \text{and} \quad \|r'\| = |\tau_{n+1}(r')| \ .$$

Since the existence of such a bijection can easily be checked for small values of n, it follows that such a bijection $\tau_n \ : \ \mathcal{I}_n \to \mathcal{R}_n$ and $\tau_n \ : \ \mathcal{I}_n^0 \to \mathcal{R}_n^0$ exists for all $n \in \mathbb{N}$.

In particular, we can construct a specific sequence τ_n by starting with $\tau_2 = \omega$ and $\tau_3 = \omega$.

Example 2.20. Choosing $\tau_3 = \omega$ one can easily check that the iterative construction then gives

$$\begin{aligned} \tau_4 \ : \ & \mathcal{I}_4 \to \mathcal{R}_4 \\ & \langle r_1, r_2, r_3, r_4 \rangle \mapsto \\ & \quad \left\langle r_1, 2r_1 + \left\lfloor \tfrac{2r_2+r_3}{3} \right\rfloor, 2r_2 + r_3, 2r_2 + r_3 + r_4 + \left\lceil \tfrac{2r_2+r_3}{3} \right\rceil \right\rangle \ . \end{aligned}$$ □

Example 2.21. Consider

$$\mathcal{O}_4(13,3) = \{1^1 3^0 5^1 7^1, 1^0 3^2 5^0, 1^0 3^1 5^2 7^0\}$$

and

$$\mathcal{L}_4(13,3) = \{\langle 0,0,5,8\rangle, \langle 0,1,5,7\rangle, \langle 1,2,4,6\rangle\} \ .$$

The corresponding bijective map Λ_4 between these two sets details as follows:

$$\Psi_4^{-1}(1^1 3^0 5^1 7^1) = \langle 1,1,0,1\rangle = r + \alpha \in \mathcal{I}_4^0 \oplus \mathcal{D}_4^0$$

where

$$r = \langle 0,1,0,0\rangle \quad \text{and} \quad \alpha = 1\cdot\delta^1 + 0\cdot\delta^2 + 0\cdot\delta^3 + 1\cdot\delta^4 \ ;$$

this gives

$$\Gamma_4(\langle 1,1,0,1\rangle) = \tau_4(r) + \sigma_4(\alpha) = \langle 0,0,2,3\rangle + \langle 0,0,3,5\rangle = \langle 0,0,5,8\rangle \ .$$

The second entry:

$$\Psi_4^{-1}(1^0 3^2 5^0 7^1) = \langle 1,0,2,0\rangle = r + \alpha \in \mathcal{I}_4^0 \oplus \mathcal{D}_4^0$$

where

$$r = \langle 0,0,2,0\rangle \quad \text{and} \quad \alpha = 1\cdot\delta^1 + 0\cdot\delta^2 + 0\cdot\delta^3 + 1\cdot\delta^4 \ ;$$

this gives

$$\Gamma_4(\langle 1,0,2,0\rangle) = \tau_4(r) + \sigma_4(\alpha) = \langle 0,1,2,3\rangle + \langle 0,0,3,4\rangle = \langle 0,1,5,7\rangle \ .$$

The third entry:

$$\Psi_4^{-1}(1^0 3^1 5^2 7^0) = \langle 0,2,1,0\rangle = r + \alpha \in \mathcal{I}_4^0 \oplus \mathcal{D}_4^0$$

where

$$r = \langle 0,0,1,0\rangle \quad \text{and} \quad \alpha = 0\cdot\delta^1 + 1\cdot\delta^2 + 0\cdot\delta^3 + 0\cdot\delta^4 \ ;$$

this gives

$$\Gamma_4(\langle 0,2,1,0\rangle) = \tau_4(r) + \sigma_4(\alpha) = \langle 0,0,1,2\rangle + \langle 1,2,3,4\rangle = \langle 1,2,4,6\rangle \ . \qquad \square$$

We conclude this section by the remark that all our bijections can be reverted easily.

2.6 Concluding Remarks

With the Benefit of Hindsight. If we take variables $y_1, y_2, \ldots, y_n$ and define as in Sect. 2.2 for $r = \langle r_1, r_2, \ldots, r_n\rangle \in \mathbb{N}^n$ the monomial

$$y^r := y_1^{r_1} y_2^{r_2} \cdots y_n^{r_n}$$

then we have from the semilinear representation of lecture hall partitions

$$f_n(y_1, y_2, \ldots, y_n) = \sum\{y^b : b \in \mathcal{L}_n = \mathcal{R}_n^0 \oplus \mathcal{E}_n^0\}$$
$$= \frac{\sum\{y^R : R \in \mathcal{R}_n^0\}}{\prod_{1 \le j \le n}(1 - y^{\epsilon^j})} .$$

Multiplying the numerator and the denominator of this last fraction by $1 - y_n = 1 - y^{\omega(n\,\delta^n)}$ leads (thanks to a finite geometric series) to

$$f_n(y_1, y_2, \ldots, y_n) = \frac{\sum\{y^R : R \in \mathcal{R}_n\}}{\prod_{1 \le j < n}(1 - y^{\epsilon^j}) \cdot (1 - y^{\omega(n\,\delta^n)})}$$
$$= \sum\{y^b : b \in \mathcal{L}_n = \mathcal{R}_n \oplus \mathcal{E}_n\}$$

where $\mathcal{E}_n = \omega(\mathcal{D}_n)$, and where $\mathcal{D}_n$ is the free semimodule generated by $\delta^1, \ldots, \delta^{n-1}$ and $n\,\delta^n = \langle 0, \ldots, 0, n \rangle (= \omega(n\,\delta^n))$. Obviously: $\mathbb{N}^n = \mathcal{I}_n \oplus \mathcal{D}_n$. This avoids the special treatment of the last basis vector.

In other words, we could have used the semilinear presentation $\mathcal{L}_n = \mathcal{R}_n \oplus \mathcal{E}_n$ instead of $\mathcal{L}_n = \mathcal{R}_n^0 \oplus \mathcal{E}_n^0$ – and everything would have gone through equally well. In particular, the essential properties of the mapping λ_n are available for both $\lambda_n : \mathcal{I}_n \to \mathcal{R}_n$, and $\lambda_n : \mathcal{I}_n^0 \to \mathcal{R}_n^0$.

The above use of the presentation $\mathcal{L}_n = \mathcal{R}_n^0 \oplus \mathcal{E}_n^0$ was motivated by what Ω-calculus (or better: its implementation) had suggested. Automatic simplification led to cancelling the factor $1 - y_n^n$, which made things less homogeneous.

Other Bijections. The involutory approach we have used for the construction of τ_n is essentially equivalent to an involution discovered by Bousquet-Mélou and Eriksson in [8, Prop. 3.4]. However, they applied this tool in a different direction; in addition, our presentation differs very much from that in [8].

We also want to remark that the limiting case $n \to \infty$ of the Refined Lecture Hall Partition Theorem (Theorem 2.13) finds a much more direct bijective treatment. In fact, before the finite version in form of Theorem 2.13 had been discovered, in 1994 C. Bessenrodt [6, Prop. 2.2] described a very elegant bijection between the underlying sets of this limiting case. To this end Bessenrodt uses 2-modular Young diagrams in order to formulate a new version of a classic bijection due to Sylvester. Another variant of Sylvester's bijection was given by D. Kim and A. Yee [11] in 1999; they essentially describe the inverse of Bessenrodt's map. This gives rise to the following problem.

Problem 2.22. Is there any bijective proof of Theorem 2.13 that in the limit $n \to \infty$ converges to Bessenrodt's bijection?

Being based on an iterative use of the involution Φ_n, our bijection does not have an infinite version. More generally one can ask the following.

Problem 2.23. Is there any bijective proof of Theorem 2.13 without using the involution Φ_n?

It might well be possible that there is a simpler bijection in case the refinement condition is dropped. So we conclude by raising another problem.

Problem 2.24. Is there any simpler lecture hall bijection for the version of Theorem 1.1, i.e., in case the refinement condition (18) is dropped?

Note added in proof: A.E. Yee [18] developed a bijective approach which is different to our bijection and which seems to solve Problem 2.22 and Problem 2.23.

3 Cayley Compositions

3.1 Introduction

In [9], A. Cayley poses and solves the following problem:

> "It is required to find the number of [compositions] into a given number of parts, such that the first part is unity, and that no part is greater than twice the preceding part.
> Commencing to form the [compositions] in question, these are (read vertically):

$$\begin{array}{c|cc|cccccc|} 1 & 1 & 1 & 1 & 1 & 1 & 1 & 1 & 1 \\ & 1 & 2 & 1 & 1 & 2 & 2 & 2 & 2 \\ & & & 1 & 2 & 1 & 2 & 3 & 4 \end{array} \quad \&c. \ldots \text{"}$$

We shall call such compositions, Cayley compositions. Let us define $c_j(n)$ to be the total number of Cayley compositions ending in n and having j parts. Thus from Cayley's enumeration we see that $c_1(1) = 1$, $c_2(1) = c_2(2) = 1$, $c_3(1) = c_3(2) = 2$, $c_3(3) = c_3(4) = 1$.

Clearly the Cayley composition with j parts and largest last part is $1, 2, 4, \ldots, 2^{j-1}$. So if we define a Cayley polynomial as

$$C_j(q) = \sum_{n \geq 0} c_j(n)\, q^n \ ,$$

then $C_j(q)$ has degree 2^{j-1}. Returning to Cayley enumeration, we see that

$$\begin{aligned} C_1(q) &= q \ , \\ C_2(q) &= q + q^2 \ , \\ C_3(q) &= 2q + 2q^2 + q^3 + q^4 \ . \end{aligned}$$

Cayley's Theorem. *The number of [Cayley compositions with j parts] is equal to the number of partitions of $2^{j-1}-1$ into the parts $1, 1', 2, 4, \ldots, 2^{j-2}$. Or again, it is equal to twice the sum of the number of partitions of $0, 1, 2, \ldots, 2^{j-2}-1$ respectively into the parts $1, 1', 2, 4, \ldots, 2^{j-3}$ (where the number of partitions of 0 counts for 1).*

Cayley closes [9] with this example:

"... the partitions of $0, 1, 2, 3$, &c. with the parts $1, 1', 2, \ldots$ are

$$\begin{aligned}&(\cdot)\ ,\\ &1, 1'\ ,\\ &1+1, 1+1', 1'+1', 2\ ,\\ &1+1+1, 1+1+1', 1+1'+1', 1'+1'+1', 2+1, 2+1'\ ,\end{aligned}$$

the numbers of which are $1, 2, 4, 6$. Hence, by the first part of the theorem, the number of 3-partitions is 6, and by the second part of the theorem, the number of 4-partitions is

$$2(1+2+4+6) = 26\ ."$$

Cayley's proof of his theorem is quite elegant, efficient and elementary. Our object here is not to improve on Cayley. Rather we wish to show that a direct application of the Partition Analysis paradigm developed by P.A. MacMahon [12] (and subsequently implemented in Mathematica [4]) allows one to obtain easily:

Theorem 3.1. *For $j \geq 2$:*

$$C_j(q) = \sum_{h=1}^{j-2} \frac{b_{j-h-1}\,(-1)^{h-1}\,q^{2^h-1}}{(1-q)(1-q^2)(1-q^4)\cdots(1-q^{2^{h-1}})} + \frac{(-1)^j\,q^{2^{j-1}-1}\,(1-q^{2^{j-1}})}{(1-q)(1-q^2)(1-q^4)\cdots(1-q^{2^{j-2}})}\ , \tag{28}$$

where b_n is the coefficient of q^{2^n-1} in the power series expansion of

$$\frac{1}{1-q}\prod_{m=0}^{\infty}\frac{1}{1-q^{2^m}}\ . \tag{29}$$

It hardly needs to be pointed out that (28) is a surprising representation of a polynomial. Indeed the right-hand side does not look like a polynomial at all. However when $j=3$, we note $b_1 = 2$ and

$$\frac{2q}{1-q} - \frac{q^3\,(1-q^4)}{(1-q)(1-q^2)} = \frac{2q-q^3-q^5}{1-q} = 2q+2q^2+q^3+q^4 = C_3(q)\ .$$

From Theorem 3.1, Cayley's Theorem follows as a natural corollary.

In Sect. 3.2, we shall apply Partition Analysis to Cayley compositions. This will yield Theorem 3.1 quite directly. The short Sect. 3.3 will derive Cayley's Theorem from Theorem 3.1. In Sect. 3.4 we briefly describe some generalizations and relations to other work.

3.2 Partition Analysis and Cayley Compositions

The following is the only strictly Partition Analysis identity that is required:

$$\underset{\geqq}{\Omega} \frac{\lambda}{(1-\lambda^2 A)(1-B/\lambda)} = \frac{1+B}{(1-A)(1-AB^2)} . \tag{30}$$

While (30) is not in MacMahon's fundamental list [12, p.102], it is easily proved:

$$\begin{aligned}
\underset{\geqq}{\Omega} \frac{\lambda}{(1-\lambda^2 A)(1-B/\lambda)} &= \underset{\geqq}{\Omega} \sum_{r,s\geq 0} A^r B^s \lambda^{2r+1-s} \\
&= \sum_{r\geq 0} \sum_{s=0}^{2r+1} A^r B^s \\
&= \sum_{r\geq 0} \frac{A^r(1-B^{2r+2})}{1-B} \\
&= \frac{1}{(1-A)(1-B)} - \frac{B^2}{(1-AB^2)(1-B)} \\
&= \frac{(1-AB^2)-B^2(1-A)}{(1-A)(1-B)(1-AB^2)} \\
&= \frac{1+B}{(1-A)(1-AB^2)} .
\end{aligned}$$

We remark that applying the `Omega` package would give (30) in one stroke.

Let us now consider a $j+1$ variable generating function for Cayley compositions:

$$\begin{aligned}
p_j(x_0,x_1,\ldots,x_j) &:= \sum_{\substack{n1,n2,\ldots,n_j\geq 1 \\ n1\leq 2, n_{i+1}\leq 2n_i}} x_0^1 x_1^{n_1} \cdots x_j^{n_j} \\
&= \underset{\geqq}{\Omega} \sum_{n1,\ldots,n_j\geq 1} x_0^1 x_1^{n_1} \cdots x_j^{n_j} \lambda_1^{2-n_1} \lambda_2^{2n_1-n_2} \cdots \lambda_j^{2n_{j-1}-n_j} \\
&= \underset{\geqq}{\Omega} \frac{x_0 x_1 \cdots x_j \lambda_1 \lambda_2 \cdots \lambda_j}{(1-\frac{\lambda_2^2 x_1}{\lambda_1})(1-\frac{\lambda_3^2 x_2}{\lambda_2})\cdots(1-\frac{\lambda_j^2 x_{j-1}}{\lambda_{j-1}})(1-\frac{x_j}{\lambda_j})} \\
&= \underset{\geqq}{\Omega} \frac{x_0 x_1 \cdots x_j \lambda_1 \lambda_2 \cdots \lambda_{j-1}}{(1-\frac{\lambda_2^2 x_1}{\lambda_1})\cdots(1-\frac{\lambda_{j-1}^2 x_{j-2}}{\lambda_{j-2}})} \frac{(1+x_j)}{(1-\frac{x_{j-1}}{\lambda_{j-1}})(1-\frac{x_{j-1}x_j^2}{\lambda_{j-1}})}
\end{aligned}$$

(by applying (30) to λ_j)

$$= \underset{\geq}{\Omega} \frac{x_0 \cdots x_j \lambda_1 \cdots \lambda_{j-1}}{(1 - \frac{\lambda_2^2 x_1}{\lambda_1}) \cdots (1 - \frac{\lambda_{j-1}^2 x_{j-2}}{\lambda_{j-2}})} \frac{(1+x_j)}{(x_{j-1} - x_{j-1}x_j^2)}$$

$$\cdot \left(\frac{x_{j-1}}{1 - \frac{x_{j-1}}{\lambda_{j-1}}} - \frac{x_{j-1}x_j^2}{1 - \frac{x_{j-1}x_j^2}{\lambda_{j-1}}} \right)$$

$$= \frac{x_j}{1-x_j} \underset{\geq}{\Omega} \frac{x_0 \cdots x_{j-1} \lambda_1 \cdots \lambda_{j-1}}{(1 - \frac{\lambda_2^2 x_1}{\lambda_1}) \cdots (1 - \frac{\lambda_{j-1}^2 x_{j-2}}{\lambda_{j-2}})} \left(\frac{1}{1 - \frac{x_{j-1}}{\lambda_{j-1}}} - \frac{x_j^2}{1 - \frac{x_{j-1}x_j^2}{\lambda_{j-1}}} \right)$$

$$= \frac{x_j}{1-x_j} \left(p_{j-1}(x_0, \ldots, x_{j-1}) - p_{j-1}(x_0, \ldots, x_{j-2}, x_{j-1}x_j^2) \right) . \quad (31)$$

We now note that for $j \geq 2$

$$C_j(q) = p_{j-1}(1, 1, \ldots, 1, q) .$$

So (31) implies the recurrence

$$C_j(q) = \frac{q}{1-q} \left(C_{j-1}(1) - C_{j-1}(q^2) \right) . \quad (32)$$

It is interesting to note that once the recurrence (32) has been found, here by using Ω-calculus, it can be also proved by straight-forward combinatorial reasoning.

Combinatorial proof of (32). Comparing the coefficients of q^n on both sides of (32) after shifting $j \to j+1$, we see that (32) is equivalent to

$$C_j(1) - \sum_{l=1}^{m-1} c_j(l) = \begin{cases} c_{j+1}(2m), & \text{if } n = 2m, \\ c_{j+1}(2m-1), & \text{if } n = 2m-1 . \end{cases}$$

First, suppose that $n = 2m$ and let

$$\{\langle 1, n_2, \ldots, n_j, 2m \rangle\}$$

be the set of Cayley compositions with $j+1$ parts ending in $2m$. Its cardinality is $c_{j+1}(2m)$. Each part of a Cayley composition is less or equal twice the preceding part. Hence, if we omit the last entry $2m$ from all these tuples, the resulting set

$$\{\langle 1, n_2, \ldots, n_j \rangle\}$$

is running through all Cayley compositions with j parts ending in elements $n_j \geq m$. The cardinality of this set is exactly $C_j(1) - \sum_{l=1}^{m-1} c_j(l)$. The case $n = 2m-1$ is analogous. □

Now let us iterate recurrence (32) which implies something very close to (28) namely

$$\begin{aligned} C_j(q) &= \sum_{h=1}^{j-2} \frac{C_{j-1-h}(1)\,(-1)^{h-1}\,q^{2^h-1}}{(1-q)(1-q^2)(1-q^4)\cdots(1-q^{2^{h-1}})} \\ &\quad + \frac{(-1)^j\,q^{2^{j-1}-1}(C_1(1)-C_1(q^{2^{j-1}}))}{(1-q)(1-q^2)(1-q^4)\cdots(1-q^{2^{j-2}})} \\ &= \sum_{h=1}^{j-2} \frac{C_{j-1-h}(1)\,(-1)^{h-1}\,q^{2^h-1}}{(1-q)(1-q^2)(1-q^4)\cdots(1-q^{2^{h-1}})} \\ &\quad + \frac{(-1)^j\,q^{2^{j-1}-1}(1-q^{2^{j-1}})}{(1-q)(1-q^2)(1-q^4)\cdots(1-q^{2^{j-2}})} \,. \end{aligned}$$

Consequently

$$\begin{aligned} q^{2^{j-1}}C_j(q^{-1}) &= -\sum_{h=1}^{j-2} \frac{C_{j-1-h}(1)\,q^{2^{j-1}}}{(1-q)(1-q^2)(1-q^4)\cdots(1-q^{2^{h-1}})} \\ &\quad + \frac{1}{(1-q)(1-q^2)(1-q^4)\cdots(1-q^{2^{j-2}})} \,. \end{aligned} \tag{33}$$

Now we observe the magic of (33). The $C_j(q)$ have degree 2^{j-1} and the lowest power of q appearing is q^1. Consequently $q^{2^{j-1}}C_j(q^{-1})$ is a polynomial of degree $2^{j-1}-1$. Now let us examine the right-hand side of (33) as an analytic function of q with $|q|<1$ (even though we know a priori that it is a polynomial of degree $2^{j-1}-1$). The terms in the sum all have $q^{2^{j-1}}$ as the lowest power of q appearing. They, therefore, contribute nothing to this polynomial; i.e. they must be cancelled out by the tail of the expansion of

$$\frac{1}{(1-q)(1-q^2)(1-q^4)\cdots(1-q^{2^{j-2}})} \,.$$

Hence $q^{2^{j-1}}C_j(q^{-1})$ is the polynomial made up of the first 2^{j-1} terms of the power series expansion of

$$\prod_{n=0}^{\infty} \frac{1}{1-q^{2^n}} \,.$$

Therefore $C_j(1)$ is the coefficient of $q^{2^{j-1}-1}$ in

$$\frac{1}{1-q} \prod_{n=0}^{\infty} \frac{1}{1-q^{2^n}} \,,$$

and this completes the proof of Theorem 3.1.

3.3 Cayley's Theorem

The first assertion in Cayley's Theorem is equivalent to the last sentence in Sect. 3.2. The second assertion is equivalent to the statement that the sum of the first 2^{j-1} coefficients in

$$\frac{1}{1-q}\prod_{n=0}^{\infty}\frac{1}{1-q^{2^n}}$$

equals the coefficient of q^{2^j-1} in the same series. To see this we note that if

$$F(q)=\frac{1}{1-q}\prod_{n=0}^{\infty}\frac{1}{1-q^{2^n}} \ ,$$

then

$$F(q)=\frac{1+q}{1-q}\,F(q^2)=(1+2q+2q^2+2q^3+\cdots)\,F(q^2)$$

and comparison of the coefficients of q^{2^j-1} on both sides of this identity is the second assertion in Cayley's Theorem.

3.4 Generalizations and Observations

The entire development so far can be generalized by essentially replacing "2" by "k" throughout where $k>1$ is an integer. In doing so, one replaces (30) by

$$\underset{\geqslant}{\Omega}\frac{\lambda^{k-1}}{(1-\lambda^k A)(1-B/\lambda)}=\frac{1+B+\cdots+B^{k-1}}{(1-A)(1-AB^k)} \ ,$$

and (32) by

$$C_j(k;q)=\frac{q}{1-q}\left(C_{j-1}(k;1)-C_{j-1}(k;q^k)\right) \ . \tag{34}$$

The rest of Sect. 3.2 can be generalized accordingly. For example, $C_3(3;q)$ is

$$3q+3q^2+3q^3+2q^4+2q^5+2q^6+q^7+q^8+q^9 \ ,$$

and the coefficient of q^{9-1} in

$$\begin{aligned}\frac{1}{1-q}\prod_{n=0}^{\infty}\frac{1}{1-q^{3^n}}\\ =1+2q+3q^2+5q^3+7q^4+9q^5+12q^6+15q^7+18q^8+23q^9+\cdots\end{aligned}$$

is $18=C_3(3;1)$.

It should be emphasized that a combinatorial proof of (32) (or more generally (34)) is quite straight-forward. The point here is that Partition

Analysis reveals these recurrences without any combinatorial reasoning on the part of the investigator.

It should be pointed out that H. Minc [13,14] in his work on groupoids studied enumeration problems that are essentially equivalent to Cayley compositions. In a subsequent paper [2], inspired by [14], it was shown that (in our notation)

$$\sum_{j=0}^{\infty} p_j(q,q,\ldots,q) = \frac{q}{1+\sum_{j=1}^{\infty} \frac{(-1)^j q^{2^{j+1}-j-1}}{(1-q)(1-q^3)(1-q^7)\cdots(1-q^{2^j-1})}} .$$

This is the generating function for all Cayley compositions classified according to the number being composed (not largest summand).

4 Linear Homogeneous Diophantine Equations

The fundamental step in our construction of a bijective proof for the Refined Lecture Hall Partition Theorem (Theorem 2.13) was the computation of a parametrized representation of lecture hall partitions; see Sect. 2.2. This was achieved by extrapolation from the first special cases that have been computed by applying the `Omega` package to $\Omega_{\geqq}$-expressions which encode generating functions whose summation parameters satisfy constraints in form of linear homogeneous diophantine inequalities.

In this concluding section we want to explain briefly that generating functions involving constraints in form of linear homogeneous diophantine *equations* can be handled in an analogous fashion as already observed by MacMahon. This does not come as entire surprise since any equation is equivalent to two inequalities. However, for various reasons, e.g., from efficiency point-of-view, it pays off indeed to have a closer look at this aspect of MacMahon's method.

To this end we follow MacMahon and consider:

Definition 4.1. The operator $\Omega_=$ is given by

$$\underset{=}{\Omega} \sum_{s_1=-\infty}^{\infty} \cdots \sum_{s_r=-\infty}^{\infty} A_{s_1,\ldots,s_r} \lambda_1^{s_1} \cdots \lambda_r^{s_r} := A_{0,\ldots,0} .$$

This means, all nontrivial power-products in the λ's are killed by the $\Omega_=$ operator which, alternatively, can be viewed as a constant term operator.

As already pointed out by MacMahon [12, Vol. 2, Sect. VIII, p. 104], this operator is related to $\Omega_{\geqq}$, for instance, as follows:

$$\underset{=}{\Omega} F(\lambda) = \underset{\geqq}{\Omega} F(\lambda) + \underset{\geqq}{\Omega} F(1/\lambda) - F(1) .$$

In other words, the rules for the $\Omega_{\geqq}$ operator in principle would be sufficient to carry out elimination of λ-variables from $\Omega_=$-expressions. However, it turns out that the use of special $\Omega_=$-rules that are tailored in the spirit of Lemma 1.4 is much more convenient – especially with respect to efficiency of computer algebra implementation. Despite having developed his theory long time before the age of computers, this was exactly the program carried out by MacMahon in his book. There he presents a collection of such rules, for instance

$$\underset{=}{\Omega}\frac{1}{(1-\lambda^2x)\left(1-\frac{y}{\lambda}\right)\left(1-\frac{z}{\lambda}\right)}=\frac{1+xyz}{(1-xy^2)(1-xz^2)}\ ; \tag{35}$$

see [12, Vol. 2, Sect. 351, p. 105]. The proofs of many of these rules are quite elementary but in case of several λ-variables, elimination can be much more cumbersome.

In Sect. 4.2 we present a very general elimination mechanism, Theorem 4.4. As an application we will have a look at magic squares of size 3. But before doing so, we discuss two elementary examples.

4.1 Introductory Examples

We had mentioned that Partition Analysis has not received due attention with the exception of work by R. Stanley. For instance, in the pioneering paper [15] containing his proof of the Anand-Dumir-Gupta conjecture, Stanley makes essential use of an $\Omega_=$-method that MacMahon describes as "The Method of Elliott"; see [12, Vol. 2, Sects. 358 and 359]. Stanley's interest in the problem of solving linear homogeneous equations for nonnegative integers is also reflected by his book [16] that contains many further references to this problem area. An additional reference is the chapter on rational generating function in Stanley's textbook [17].

Example 4.2. We illustrate the use of rule (35) by choosing an example from [16, Ex. 3.5]: find all nonnegative integer solutions $\langle a_1, a_2, a_3\rangle \in \mathbb{N}^3$ of

$$a_1+a_2-2a_3=0\ . \tag{36}$$

First we encode the corresponding generating function as an $\Omega_=$ expression,

$$\begin{aligned}\sum_{a_1+a_2-2a_3=0} x_1^{a_1}x_2^{a_2}x_3^{a_3} &= \underset{=}{\Omega}\sum_{a_1,a_2,a_3\geq 0}\lambda^{a_1+a_2-2a_3}x_1^{a_1}x_2^{a_2}x_3^{a_3}\\ &= \underset{=}{\Omega}\frac{1}{(1-\lambda x_1)(1-\lambda x_2)(1-\frac{x_3}{\lambda^2})}\ .\end{aligned}$$

Now due to

$$\underset{=}{\Omega}F(\lambda)=\underset{=}{\Omega}F(1/\lambda)\ ,$$

rule (35) gives

$$\underset{=}{\Omega}\frac{1}{(1-\lambda x_1)(1-\lambda x_2)(1-\frac{x_3}{\lambda^2})}=\frac{1+x_1x_2x_3}{(1-x_1^2x_3)(1-x_2^2x_3)}\ .$$

By geometric series expansion we obtain the desired parametrized representation of the solution set of (36), namely

$$\langle a_1, a_2, a_3\rangle = \{n_1\langle 2,0,1\rangle + n_2\langle 0,2,1\rangle + r\langle 1,1,1\rangle : \langle n_1, n_2\rangle \in \mathbb{N}^2, r \in \{0,1\}\} .$$

This means,

$$\{\langle 2,0,1\rangle, \langle 0,2,1\rangle, \langle 1,1,1\rangle\}$$

is the set of *fundamental solutions*, whereas

$$\{\langle 2,0,1\rangle, \langle 0,2,1\rangle\}$$

is called the set of *completely fundamental solutions*; note that $2\langle 1,1,1\rangle = \langle 2,0,1\rangle + \langle 0,2,1\rangle$. This terminology, together with corresponding ring and module theoretic considerations, traces back to Hilbert's syzygy theorem [10]; for further information consult, e.g., [15], [16] or [17]. □

In order to treat also such $\Omega_=$-problems in a purely automatic fashion, we developed a procedure that has also been implemented in Mathematica.

Example 4.3. We illustrate its use by taking another example from [16, Ex. 5.14] (see also [17, Ch. 4, Example 4.6.15]): find all nonnegative integer solutions $\langle a_1, a_2, a_3, a_4\rangle \in \mathbb{N}^4$ of

$$a_1 + a_2 - a_3 - a_4 = 0 \ . \tag{37}$$

Encoding the corresponding generating function as an $\Omega_=$-expression results in

$$\sum_{a_1+a_2-a_3-a_4=0} x_1^{a_1} x_2^{a_2} x_3^{a_3} x_4^{a_4} = \underset{=}{\Omega} \sum_{a_1,a_2,a_3,a_4 \geq 0} \lambda^{a_1+a_2-a_3-a_4} x_1^{a_1} x_2^{a_2} x_3^{a_3} x_4^{a_4}$$
$$= \underset{=}{\Omega} \frac{1}{(1-\lambda x_1)(1-\lambda x_2)(1-\frac{x_3}{\lambda})(1-\frac{x_4}{\lambda})} .$$

The λ-elimination rule that is needed for this situation is also to find in MacMahon's book; see [12, Vol. 2, Sect. 351, p. 105]. Nevertheless, this time we will apply our procedure:

```
In[2]:= ?OEqR

        OEqR[expr, z] applies the OmegaEq operator to expr
        eliminating the variable z.

In[3]:= f = 1 / ((1-x1 λ)(1-x2 λ)(1-x3/λ)(1-x4/λ))

Out[3]=
```

$$\frac{1}{(1-\lambda\,\mathrm{x1})\,(1-\lambda\,\mathrm{x2})\,\left(1-\frac{\mathrm{x3}}{\lambda}\right)\,\left(1-\frac{\mathrm{x4}}{\lambda}\right)}$$

```
In[4]:= OEqR[f, λ]

Out[4]=
```

$$\frac{1 - \mathrm{x1\,x2\,x3\,x4}}{(1 - \mathrm{x1\,x3})\,(1 - \mathrm{x2\,x3})\,(1 - \mathrm{x1\,x4})\,(1 - \mathrm{x2\,x4})}$$

By geometric series expansion we again obtain the desired parametrized representation of the solution set of (37), but this time – due to the *minus sign* in the numerator polynomial – the representation in this form is not of the same type as in the previous example. Namely, despite the fact that the fundamental solutions again are immediate from the factors of the denominator, namely

$$\{\langle 1,0,1,0\rangle, \langle 0,1,1,0\rangle, \langle 1,0,0,1\rangle, \langle 0,1,0,1\rangle\} \ ,$$

the numerator monomial gives rise to the syzygy

$$\langle 1,0,1,0\rangle + \langle 0,1,0,1\rangle = \langle 0,1,1,0\rangle + \langle 1,0,0,1\rangle \ .$$

Instead of writing the syzygy additively, its multiplicative version reads as

$$x_1x_2x_3x_4 = (x_1x_3)(x_2x_4) = (x_2x_3)(x_1x_4) \ . \qquad \square$$

One of MacMahon's interests was to use Partition Analysis for discovering syzygetic relations. Below we will give an example in connection with magic squares.

4.2 The Fundamental Recurrence

For automatic elimination of λ-variables with respect to $\Omega_=$ we use essentially the same method as described in [4]. The only difference consists in the replacement of the "fundamental recurrence" [4, Theorem 2] by the corresponding result for the $\Omega_=$ operator which we formulate as Theorem 4.4.

Before we state our result, we must recall the homogeneous symmetric functions, denoted by $h_j(x_1, x_2, \ldots, x_n)$, which are given by

$$\sum_{j=0}^{\infty} h_j(x_1, x_2, \ldots, x_n)\, t^j = \frac{1}{(1 - tx_1)(1 - tx_2)\cdots(1 - tx_n)} \ .$$

Theorem 4.4 ("Fundamental Recurrence"). *For n and m positive integers and a any integer,*

$$\underset{=}{\Omega} \frac{\lambda^a}{(1 - A_1\lambda)(1 - A_2\lambda)\cdots(1 - A_n\lambda)(1 - \frac{B_1}{\lambda})(1 - \frac{B_2}{\lambda})\cdots(1 - \frac{B_m}{\lambda})} = \frac{P_{n,m,a}(A_1, \ldots, A_n; B_1, \ldots, B_m)}{\prod_{i=1}^{n}\prod_{j=1}^{m}(1 - A_iB_j)} \ ,$$

where for $n > 1$,

$$\begin{aligned}
&P_{n,m,a}(A_1,\ldots,A_n;B_1,\ldots,B_m)\\
&\quad=\frac{1}{A_n-A_{n-1}}\\
&\qquad\cdot\Big\{A_n\prod_{j=1}^{m}(1-A_{n-1}B_j)\,P_{n-1,m,a}(A_1,\ldots,A_{n-2},A_n;B_1,\ldots,B_m)\\
&\qquad\quad-A_{n-1}\prod_{j=1}^{m}(1-A_nB_j)\,P_{n-1,m,a}(A_1,\ldots,A_{n-2},A_{n-1};B_1,\ldots,B_m)\Big\}
\end{aligned}$$

and for $n = 1$,

$$\begin{aligned}
&P_{1,m,a}(A_1;B_1,\ldots,B_m)\\
&\quad=\begin{cases}A_1^{-a}, & \text{if } a\le 0,\\ A_1^{-a}-\prod_{j=1}^{m}(1-A_1B_j)\sum_{j=0}^{a-1}A_1^{j-a}h_j(B_1,\ldots,B_m), & \text{if } a>0\ .\end{cases}
\end{aligned}$$

Proof. The proof of the recurrence is exactly as the proof in the $\Omega_{\geq}$-case [4, Theorem 2]. We again use the convenient identity

$$\frac{1}{(1-A_n\lambda)(1-A_{n-1}\lambda)}=\frac{1}{A_n-A_{n-1}}\Big(\frac{A_n}{1-A_n\lambda}-\frac{A_{n-1}}{1-A_{n-1}\lambda}\Big)\ ,$$

and the rest follows as before.

The $n = 1$ case splits into two cases as before:
Case $a \leq 0$.

$$\begin{aligned}
&\underset{=}{\Omega}\frac{\lambda^a}{(1-A_1\lambda)(1-\frac{B_1}{\lambda})(1-\frac{B_2}{\lambda})\cdots(1-\frac{B_m}{\lambda})}\\
&\quad=\underset{=}{\Omega}\sum_{h=0}^{\infty}\sum_{n_1,\ldots,n_m\ge 0}A_1^hB_1^{n_1}\cdots B_m^{n_m}\lambda^{a+h-n_1-\cdots-n_m}\\
&\quad=\sum_{n_1,\ldots,n_m\ge 0}A_1^{n_1+\cdots+n_m-a}B_1^{n_1}\cdots B_m^{n_m}\\
&\quad=\frac{A_1^{-a}}{(1-A_1B_1)(1-A_1B_2)\cdots(1-A_1B_m)}\ ,
\end{aligned}$$

which means that when $a \leq 0$

$$P_{1,m,a}(A_1;B_1,\ldots,B_m)=A_1^{-a}\ .$$

Case $a > 0$.

$$\begin{aligned}
&\underset{=}{\Omega}\frac{\lambda^a}{(1-A_1\lambda)(1-\frac{B_1}{\lambda})(1-\frac{B_2}{\lambda})\cdots(1-\frac{B_m}{\lambda})} \\
&= \underset{=}{\Omega}\sum_{h=0}^{\infty}\sum_{n_1,\ldots,n_m\geq 0} A_1^h B_1^{n_1}\cdots B_m^{n_m}\lambda^{a+h-n_1-\cdots-n_m} \\
&= \sum_{\substack{n_1,\ldots,n_m\geq 0\\ n_1+\cdots+n_m\geq a}} A_1^{n_1+\cdots+n_m-a}B_1^{n_1}\cdots B_m^{n_m} \\
&= \left(\sum_{n_1,\ldots,n_m\geq 0} - \sum_{\substack{n_1,\ldots,n_m\geq 0\\ n_1+\cdots+n_m<a}}\right) A_1^{n_1+\cdots+n_m-a}B_1^{n_1}\cdots B_m^{n_m} \\
&= \frac{A_1^{-a}}{(1-A_1B_1)(1-A_1B_2)\cdots(1-A_1B_m)} - \sum_{j=0}^{a-1} A_1^{j-a}h_j(B_1,\ldots,B_m)\ ,
\end{aligned}$$

so for $a > 0$

$$P_{1,m,a}(A_1;B_1,\ldots,B_m) = A_1^{-a} - \prod_{j=1}^{m}(1-A_1B_j)\sum_{j=0}^{a-1}A_1^{j-a}h_j(B_1,\ldots,B_m)\ ,$$

as desired. □

Example 4.5. Starting with Sect. 406, in his book [12, Vol. 2, p. 149] MacMahon studied magic squares of size 3. More precisely, given an arbitrary integer a_{10} he is interested to find all nonnegative solutions a_1 up to a_9 of the corresponding eight equations

$$\begin{aligned}
a_1 + a_2 + a_3 = a_4 + a_5 + a_6 = a_7 + a_8 + a_9 = a_{10}\ ,\\
a_1 + a_4 + a_7 = a_2 + a_5 + a_8 = a_3 + a_6 + a_9 = a_{10}\ ,\\
a_1 + a_5 + a_9 = a_3 + a_5 + a_7 = a_{10}\ .
\end{aligned} \tag{38}$$

MacMahon then states that the corresponding generating function

$$f = \sum_{a_1,\ldots,a_{10}} x_1^{a_1}x_2^{a_2}x_3^{a_3}x_4^{a_4}x_5^{a_5}x_6^{a_6}x_7^{a_7}x_8^{a_8}x_9^{a_9}x_{10}^{a_{10}}\ ,$$

where the nonnegative integers a_i satisfy the equations in (38), turns into the $\Omega_=$-expression

$$\begin{aligned}
f = \underset{=}{\Omega}\,&\frac{1}{(1-\lambda_1\lambda_4\lambda_7x_1)(1-\lambda_1\lambda_5x_2)(1-\lambda_1\lambda_6\lambda_8x_3)(1-\lambda_2\lambda_4x_4)}\\
&\cdot\frac{1}{(1-\lambda_2\lambda_5\lambda_7\lambda_8x_5)(1-\lambda_2\lambda_6x_6)(1-\lambda_3\lambda_4\lambda_8x_7)}\\
&\cdot\frac{1}{(1-\lambda_3\lambda_5x_8)(1-\lambda_3\lambda_6\lambda_7x_9)(1-x_{10}/(\lambda_1\lambda_2\lambda_3\lambda_4\lambda_5\lambda_6\lambda_7\lambda_8))}\ .
\end{aligned}$$

We eliminate λ_1 up to λ_6 with our package as follows:

```
In[2]:= f = 1/((1-λ1 λ4 λ7 x1) (1-λ1 λ5 x2) (1-λ1 λ6 λ8 x3) *
              (1-λ2 λ4 x4) (1-λ2 λ5 λ7 λ8 x5) (1-λ2 λ6 x6) *
              (1-λ3 λ4 λ8 x7) (1-λ3 λ5 x8) (1-λ3 λ6 λ7 x9) *
              (1-x10/(λ1 λ2 λ3 λ4 λ5 λ6 λ7 λ8)));

In[3]:= g = OEqR[OEqR[OEqR[OEqR[OEqR[f,λ1],λ6],λ5],λ2],λ3]

Out[3]=
```

$$-(-1+\mathrm{x1}\,\mathrm{x10}^3\,\mathrm{x2}\,\mathrm{x3}\,\mathrm{x4}\,\mathrm{x5}\,\mathrm{x6}\,\mathrm{x7}\,\mathrm{x8}\,\mathrm{x9})\,/\,\Big((1-\lambda 8^2\,\mathrm{x10}\,\mathrm{x3}\,\mathrm{x5}\,\mathrm{x7})\left(1-\tfrac{\mathrm{x10}\,\mathrm{x2}\,\mathrm{x6}\,\mathrm{x7}}{\lambda 7}\right)\left(1-\tfrac{\mathrm{x10}\,\mathrm{x3}\,\mathrm{x4}\,\mathrm{x8}}{\lambda 7}\right)\left(1-\tfrac{\mathrm{x1}\,\mathrm{x10}\,\mathrm{x6}\,\mathrm{x8}}{\lambda 8}\right)\left(1-\tfrac{\mathrm{x10}\,\mathrm{x2}\,\mathrm{x4}\,\mathrm{x9}}{\lambda 8}\right)(1-\lambda 7^2\,\mathrm{x1}\,\mathrm{x10}\,\mathrm{x5}\,\mathrm{x9})\Big)$$

Note that λ_7 and λ_8 correspond to the last two equations in (38). So setting $\lambda_7 = \lambda_8 = 1$ corresponds to dropping the conditions on the diagonals. Furthermore, if we also set $x_i = x$ for $0 \le i \le 9$ and $x_{10} = y$ we obtain

```
In[4]:= g /. {λ7->1,λ8->1,x1->x,x2->x,x3->x,x4->x,x5->x,x6->x,
             x7->x,x8->x,x9->x,x10->y}

Out[4]=
```

$$-\frac{-1+x^9\,y^3}{(1-x^3\,y)^6}$$

which is the result obtained by MacMahon [12, Vol. 2, Sect. 407, p. 161] who concludes his (hand) computation by the remark, "In this the coefficient of $x^{3n}y^n$ represents the number of squares such that in each column and in each row the sum of the numbers is n. It has the value $3\binom{n+3}{4} + \binom{n+2}{2}$." □

4.3 Concluding Remarks

At the end of his Ω-treatment of the syzygetic theory of magic square enumeration MacMahon writes [12, Vol. 2, Sect. 409, p. 164], "There is no theoretical difficulty in dealing with the squares of higher orders, but even in the case $n = 4$ there is practical difficulty in handling the $\Omega_=$ generating function."

With the Mathematica implementation of the Fundamental Recurrence, Theorem 4.4 above, it was our hope to be able to treat at least the cases $n = 4$ and $n = 5$ in purely automatic fashion. But it turned out that the problems Mathematica had with the rational function arithmetics were too involved. In order to overcome these computational difficulties we decided to take a different approach which is essentially based on partial fraction decomposition and which proceeds iteratively.

We already mentioned that Stanley [15] had used the $\Omega_=$-method in a way that MacMahon describes as "The Method of Elliott"; see [12, Vol. 2,

Sects. 358 and 359]. Also this algorithm proceeds iteratively with basic steps being partial fraction decompositions of the type

$$\frac{1}{(1-x\lambda^A)(1-\frac{y}{\lambda^B})} = \frac{1}{1-xy\lambda^{A-B}}\left(\frac{1}{1-x\lambda^A} + \frac{1}{1-\frac{y}{\lambda^B}} - 1\right) ,$$

where A and B are positive integers.

Our new $\Omega_=$-algorithm [5] is a variation of this Elliott iteration but uses a different partial fraction decomposition for the fundamental steps. Computations show that concerning efficiency this new strategy is far superior to the method of Elliott and also considerably faster than the implementation based on the Fundamental Recurrence. Moreover, we adapted this new method also to the $\Omega_{\geqq}$-situation where we achieved a similar speed-up. Finally we want to mention that using this new approach the computation of the generating function for general magic squares of order 4 has been reduced to a basic problem of computer algebra, namely to the task of adding 254 rational functions, i.e., to simplify them to a single one. So far we were not able to accomplish this in Mathematica.

References

1. G.E. Andrews, *The Theory of Partitions*, Encyclopedia of Mathematics and Its Applications, Vol. 2, G.-C. Rota ed., Addison-Wesley, Reading, 1976. (Reissued: Cambridge University Press, Cambridge, 1985.)
2. G.E. Andrews, *The Rogers-Ramanujan reciprocal and Minc's partition function*, Pacific J. Math. **95** (1981), 251–256.
3. G.E. Andrews, *MacMahon's partition analysis I: The lecture hall partition theorem*, in "Mathematical essays in honor of Gian-Carlo Rota's 65th birthday" (B.E. Sagan et al., eds.), Prog. Math. **161**, Birkhäuser, Boston, 1998, pp. 1–22.
4. G.E. Andrews, P. Paule and A. Riese, *MacMahon's partition analysis III: The Omega package*, (to appear).
5. G.E. Andrews, P. Paule and A. Riese, *MacMahon's partition analysis VI: The fast algorithm*, (in preparation).
6. C. Bessenrodt, *A bijection for Lebesgue's identity in the spirit of Sylvester*, Discrete Math. **132** (1994), 1–10.
7. M. Bousquet-Mélou and K. Eriksson, *Lecture hall partitions*, Ramanujan J. **1** (1997), 101–111.
8. M. Bousquet-Mélou and K. Eriksson, *Lecture hall partitions II*, Ramanujan J. **1** (1997), 165–187.
9. A. Cayley, *On a problem in the partition of numbers*, Philosophical Mag. **13** (1857), 245–248 (reprinted: The Coll. Math. Papers of A. Cayley, Vol. III, Cambridge University Press, Cambridge, 1890, pp. 247–249).
10. D. Hilbert, *Über die Theorie der algebraischen Formen*, Math. Ann. **36** (1890), 473–534.
11. D. Kim and A.J. Yee, *A note on partitions into distinct parts and odd parts*, Ramanujan J. **3** (1999), 227–231.
12. P.A. MacMahon, *Combinatory Analysis*, 2 vols., Cambridge University Press, Cambridge, 1915–16 (reprinted: Chelsea, New York, 1960).

13. H. Minc, *The free commutative entropic logarithmetic*, Proc. Roy. Soc. Edinburgh **65** (1959), 177–192.
14. H. Minc, *A problem in partitions: Enumeration of elements of a given degree in the free commutative entropic groupoid*, Proc. Edinburgh Math. Soc. **11** (1959), 223–224.
15. R.P. Stanley, *Linear homogeneous diophantine equations and magic labelings of graphs*, Duke Math. J. **40** (1973), 607–632.
16. R.P. Stanley, *Combinatorics and Commutative Algebra*, Birkhäuser, Boston, 1983.
17. R.P. Stanley, *Enumerative Combinatorics - Volume 1*, Wadsworth, Monterey, California, 1986.
18. A.J. Yee, *On Combinatorics of Lecture Hall Partitions*, (to appear).

Note on the Proper Linear Spaces on 18 Points

A. Betten[1] and D. Betten[2]

[1] Fakultät für Mathematik und Physik
Universität Bayreuth
95440 Bayreuth
Anton.Betten@uni-bayreuth.de
[2] Mathematisches Seminar der Universität Kiel
Ludewig-Meyn-Str. 4
24098 Kiel
betten@math.uni-kiel.de

Abstract In [4] we constructed and enumerated all proper linear spaces on 17 points using the so-called TDO-method. This method is also strong enough for the construction and enumeration of all proper linear spaces on 18 points. In the present note we list the results. We get 2 412 890 proper linear spaces on 18 points.

AMS subject classification: 05B25, 05B30, 51E99

1 Introduction

A linear space on a set of v points is a set of subsets called lines such that each line contains at least two points and any two points are contained in exactly one line (cf. [1]). A linear space is called proper if all lines have at least three and at most $v-1$ points. All proper linear spaces on at most 17 points have been classified, [9], [12], [4], see also [11]. In the present note, we extend this classification to the collection of proper linear spaces on 18 points. The number of linear spaces on v points, subsequently denoted by PLIN(v) grows very fast with v. The numbers PLIN(v) for $2 \le v \le 18$ are

v	2	3	4	5	6	7	8	9	10	11	12	13	14	15	16	17	18
PLIN(v)	0	0	0	0	0	1	0	1	1	1	3	7	1	119	398	161 925	2 412 890

In our computer programs for the construction and analysis of finite geometries we make extensive use of the well known notion of a *tactical decomposition* (TD) of an incidence matrix, see [10]. An important tactical decomposition, which we call TDA, is induced by the orbits of the automorphism group, [5]. Another tactical decomposition is defined by a successive ordering process which we call TDO. It is canonical and may be calculated very quickly, [8], [2], [3]. Each tactical decomposition may be described by its *TD-scheme*, and these schemes are good invariants for the structure of geometries. In order to construct geometries we go in the opposite direction: we start with a given initial parameter set and refine these parameters step by step. See for instance [3], where we proceeded up to parameter depth 2.

Continuing this refining process we will eventually reach a TDO-scheme (or a set of TDO-schemes). From these schemes we then generate the related geometries. Here, a TDO-scheme may or may not be realizable, and if it is, it may produce several geometries.

Using this method, we constructed in [4] all proper linear spaces on 17 points. By the same method it is even possible to construct also the proper linear spaces on 18 points. In the present note we list the results in this case and point out some special spaces and situations.

2 Table of Results

In this section, we present the classification of proper linear spaces on 18 points. We refer to [4] for all definitions and notions used in the following.

Table 1 displays the numbers of proper linear spaces for the various line cases, i. e. distributions of block lengths. For instance $(4^{18}\,3^{15})$ means that we look for a linear space which has 18 blocks of length 4 and 15 blocks of length 3. Similar as in [4] we compute for each line case the set of possible tactical decompositions which may arise as TDO of linear spaces of that type. We took only those line cases into the table where the parameter calculation produced at least one TDO-scheme. In the first three columns we give the number of the case, the line case and the number of TDO-schemes one gets in this case. Since many line cases do not lead to TDO-schemes we get gaps in the (original) numbering.

We call a TDO-scheme *discrete* if all classes of points and all classes of blocks have only one element. In this case the TDO-scheme coincides with the incidence matrix of the space and nothing more is to be done. Only those TDO-schemes which are non-discrete have to be handed to the generator program. The number of non-discrete TDO-schemes for each line case is in column 4. For instance the 740 TDO-schemes in line case no. 15 split into 60 non-discrete TDOs and $740 - 60 = 680$ discrete TDOs.

In column 5 of the table we list the number of linear spaces which are constructed from all non-discrete TDO-schemes in the corresponding line case. In column 6 we give the total number (up to isomorphism) of linear spaces for the respective line case, i. e. the sum of the number of discrete TDO-schemes and the number of geometries constructed from all non-discrete TDO-schemes:

$$\text{number of geometries} = \text{column } 6 = \text{column } 3 - \text{column } 4 + \text{column } 5 \ .$$

Table1. Proper linear spaces on 18 points by their line type

no.	line case	# TDO	# TDO .n.d.	# GEO .n.d.	# GEO total
6	$(4^5\,3^{41})$	158 419	87	2 488	160 820
7	$(4^6\,3^{39})$	1 139 617	198	2 657	1 142 076
8	$(4^7\,3^{37})$	407 499	249	115 796	523 046
9	$(4^8\,3^{35})$	166 614	73	379	166 920
10	$(4^9\,3^{33})$	170 060	293	2 171	171 938
11	$(4^{10}\,3^{31})$	124 848	75	288	125 061
12	$(4^{11}\,3^{29})$	48 164	236	428	48 356
13	$(4^{12}\,3^{27})$	14 336	148	170	14 358
14	$(4^{13}\,3^{25})$	3 623	165	207	3 665
15	$(4^{14}\,3^{23})$	740	60	66	746
16	$(4^{15}\,3^{21})$	273	97	164	340
17	$(4^{16}\,3^{19})$	18	8	6	16
18	$(4^{17}\,3^{17})$	5	5	2	2
19	$(4^{18}\,3^{15})$	3	3	2	2
21	$(4^{20}\,3^{11})$	2	2	0	0
22	$(4^{21}\,3^{9})$	1	1	0	0
32	$(5^3\,4^5\,3^{31})$	813	27	84	870
33	$(5^3\,4^6\,3^{29})$	1 413	57	52	1 408
34	$(5^3\,4^7\,3^{27})$	1 713	132	194	1 775
35	$(5^3\,4^8\,3^{25})$	118	4	3	117
36	$(5^3\,4^9\,3^{23})$	160	74	70	156
37	$(5^3\,4^{10}\,3^{21})$	35	6	1	30
38	$(5^3\,4^{11}\,3^{19})$	14	10	5	9
39	$(5^3\,4^{12}\,3^{17})$	2	1	0	1
40	$(5^3\,4^{13}\,3^{15})$	6	6	2	2
41	$(5^3\,4^{14}\,3^{13})$	1	1	1	1
42	$(5^3\,4^{15}\,3^{11})$	4	4	1	1
43	$(5^3\,4^{16}\,3^{9})$	1	1	1	1
54	$(5^6\,4^6\,3^{19})$	1	1	0	0
55	$(5^6\,4^7\,3^{17})$	6	6	1	1
57	$(5^6\,4^9\,3^{13})$	6	6	4	4
63	$(5^6\,4^{15}\,3^{1})$	1	1	0	0
73	$(5^9\,4^9\,3^{3})$	1	1	1	1
85	$(6^1\,4^3\,3^{40})$	1	1	469	469
86	$(6^1\,4^4\,3^{38})$	2	2	29 622	29 622
87	$(6^1\,4^5\,3^{36})$	7 506	38	264	7 732
88	$(6^1\,4^6\,3^{34})$	10 112	78	394	10 428

continued on next page

continued from previous page

no.	line case	# TDO	# TDO .n.d.	# GEO .n.d.	# GEO total
89	$(6^1\,4^7\,3^{32})$	351	2	16	365
90	$(6^1\,4^8\,3^{30})$	907	30	184	1 061
91	$(6^1\,4^9\,3^{28})$	330	9	40	361
92	$(6^1\,4^{10}\,3^{26})$	557	39	67	585
93	$(6^1\,4^{11}\,3^{24})$	32	3	33	62
94	$(6^1\,4^{12}\,3^{22})$	57	24	94	127
95	$(6^1\,4^{13}\,3^{20})$	1	1	1	1
96	$(6^1\,4^{14}\,3^{18})$	2	1	1	2
100	$(6^1\,4^{18}\,3^{10})$	1	1	0	0
102	$(6^1\,4^{20}\,3^{6})$	2	2	0	0
110	$(6^1\,5^3\,4^4\,3^{28})$	146	44	115	217
111	$(6^1\,5^3\,4^5\,3^{26})$	8	7	10	11
112	$(6^1\,5^3\,4^6\,3^{24})$	25	13	12	24
114	$(6^1\,5^3\,4^8\,3^{20})$	1	1	0	0
116	$(6^1\,5^3\,4^{10}\,3^{16})$	1	1	0	0
118	$(6^1\,5^3\,4^{12}\,3^{12})$	7	7	6	6
129	$(6^1\,5^6\,4^4\,3^{18})$	5	5	2	2
157	$(6^2\,4^5\,3^{31})$	30	1	0	29
159	$(6^2\,4^7\,3^{27})$	4	3	3	4
163	$(6^2\,4^{11}\,3^{19})$	5	5	0	0
207	$(6^3\,3^{36})$	1	1	12	12
217	$(6^3\,4^{10}\,3^{16})$	1	1	1	1
276	$(7^1\,4^7\,3^{30})$	10	4	17	23
277	$(7^1\,4^8\,3^{28})$	1	1	6	6
278	$(7^1\,4^9\,3^{26})$	13	1	4	16
280	$(7^1\,4^{11}\,3^{22})$	3	3	1	1
total:		2 258 639	2 367	156 618	2 412 890

In Table 2, we list for each line case the distribution of automorphism group orders for the geometries which come from the non-discrete TDO-schemes. Of course the linear spaces which correspond to the discrete TDO-schemes all have a trivial automorphism group.

Table2. The distribution of automorphism group orders

no.	TDO.n.d.:	distr. of aut. group orders
6	2488:	$1^{1778}, 2^{630}, 3^{74}, 6^{6}$
7	2657:	$1^{1479}, 2^{936}, 3^{234}, 6^{8}$
8	115796:	$1^{113673}, 2^{2012}, 3^{75}, 4^{24}, 6^{12}$
9	379:	$1^{167}, 2^{122}, 3^{86}, 6^{2}, 21^{2}$
10	2171:	$1^{904}, 2^{1056}, 3^{113}, 4^{57}, 6^{29}, 12^{6}, 18^{4}, 36^{1}, 54^{1}$
11	288:	$1^{24}, 2^{194}, 3^{66}, 6^{4}$
12	428:	$1^{35}, 2^{348}, 3^{33}, 4^{6}, 6^{4}, 12^{2}$
13	170:	$1^{9}, 2^{151}, 3^{7}, 6^{2}, 18^{1}$
14	207:	$1^{30}, 2^{173}, 3^{3}, 6^{1}$
15	66:	$1^{2}, 2^{46}, 3^{5}, 4^{12}, 6^{1}$
16	164:	$1^{21}, 2^{99}, 3^{21}, 4^{4}, 6^{13}, 8^{2}, 12^{2}, 24^{2}$
17	6:	2^{6}
18	2:	2^{2}
19	2:	$6^{1}, 18^{1}$
32	84:	$1^{6}, 2^{66}, 3^{8}, 6^{4}$
33	52:	$2^{22}, 3^{28}, 6^{2}$
34	194:	$1^{3}, 2^{173}, 3^{6}, 4^{8}, 6^{4}$
35	3:	$2^{1}, 3^{2}$
36	70:	$2^{39}, 3^{6}, 4^{11}, 6^{10}, 12^{2}, 18^{1}, 36^{1}$
37	1:	2^{1}
38	5:	$2^{4}, 6^{1}$
40	2:	$6^{1}, 12^{1}$
41	1:	6^{1}
42	1:	20^{1}
43	1:	2^{1}
55	1:	4^{1}
57	4:	$3^{1}, 6^{1}, 12^{1}, 36^{1}$
73	1:	108^{1}
85	469:	$1^{264}, 2^{122}, 3^{7}, 4^{57}, 6^{4}, 8^{6}, 9^{1}, 12^{3}, 18^{1}, 24^{3}, 72^{1}$
86	29622:	$1^{27971}, 2^{1386}, 3^{37}, 4^{194}, 6^{9}, 8^{23}, 12^{1}, 24^{1}$
87	264:	$1^{48}, 2^{216}$
88	394:	$1^{78}, 2^{266}, 3^{28}, 4^{18}, 8^{4}$
89	16:	$1^{8}, 3^{8}$
90	184:	$1^{58}, 2^{113}, 3^{3}, 4^{3}, 6^{6}, 12^{1}$
91	40:	$1^{4}, 2^{22}, 4^{14}$
92	67:	$1^{3}, 2^{60}, 4^{4}$
93	33:	$2^{6}, 4^{13}, 6^{3}, 8^{9}, 24^{2}$

continued on next page

continued from previous page

no.	TDO.n.d.:	distr. of aut. group orders
94	94:	$1^6, 2^{44}, 3^{13}, 4^6, 6^9, 8^7, 12^2, 18^2, 24^2, 72^3$
95	1:	3^1
96	1:	3^1
110	115:	$1^3, 2^{74}, 3^4, 4^{22}, 6^1, 8^{10}, 12^1$
111	10:	2^{10}
112	12:	$1^1, 2^6, 6^5$
118	6:	$2^1, 6^1, 8^1, 12^1, 24^1, 72^1$
129	2:	8^2
159	3:	2^3
207	12:	$8^1, 16^1, 24^3, 48^1, 72^1, 144^1, 240^1, 432^1, 648^1, 1296^1$
217	1:	20^1
276	17:	$1^6, 2^{10}, 6^1$
277	6:	$1^2, 3^3, 21^1$
278	4:	$1^2, 2^1, 6^1$
280	1:	2^1

3 Some Special Situations and Geometries

3.1 The Largest Case

Case no. 7 yields the largest number of geometries. Here one can see the advantage of the TDO-method. Among the 1 139 617 TDO-schemes only 198 are non-discrete and need further attention. They generate only 2 657 linear spaces. So, the overwhelming part of this case consists of discrete TDO-schemes, and after having constructed the TDO-schemes, nearly all the work is done.

Because of the high number we did not save all geometries. We put into a file those geometries which came from a non-discrete TDO-scheme, the other geometries which correspond to discrete TDO-schemes were only counted. We use the following notation for the geometries saved in [6]: The first number is the number of the line case, the second number denotes the TDO-scheme and the third number is the number of the geometry generated by this TDO-scheme. For instance 10-151-288 is the linear space no. 288 which is generated starting from TDO-scheme no. 151 belonging to line case no. 10. Of course this numbering is not canonical but depends on the algorithm which has been used.

3.2 A Highly Productive TDO-Scheme

Very remarkable is case no. 8, especially the TDO-scheme no. 249 (the last one). This special TDO-scheme generates 114 672 proper linear spaces, i. e. nearly as much as all 249 non-discrete TDO-schemes together. We present this TDO-scheme here. First in the form that it came out from the program calculating the TDO-schemes, and then in the permutated form which we took for the generating process.

Finally we display one of the 114 672 geometries which were generated from this specific TDO-scheme (cf. Fig. 1).

TDO-scheme:

	1	6	1	12	6	6	12
3	1	2	0	4	0	0	0
2	0	3	1	0	3	0	0
1	1	0	1	0	0	6	0
12	0	1	0	2	1	1	3

Transposed and permuted TDO-scheme:

	1	12	2	3
6	1	2	0	0
12	0	3	0	0
6	0	2	1	1
6	0	2	1	0
12	0	2	0	1
1	1	0	0	3
1	1	0	2	0

This TDO-scheme is remarkable in another respect: It was the only one where the generation of the geometries proved to be really hard. We had to find a good permutation for the scheme, had to transpose it and we took another generation program (generating by respecting some canonical ordering). In addition, the tests we took in between had to be well chosen.

Remark: The choice of the permutation of the TDO-scheme and the choice of the intermediate tests may affect the computing time enormously. It seems a crucial problem to find good criteria for these choices. Otherwise one has to carry out long experiments to get a suitable conditioning for the computer run.

Special attention was also needed for the following TDO-schemes: 9-1, 10-151, 10-293 and 16-97. Line cases 85 and 86 were done by transposing and using the order preserving program.

There is a second TDO-scheme which produced many linear spaces, namely TDO-scheme 86-1. We display this scheme together with one of its geometries in Fig. 2. We choose the geometry 86-1-24786, which has the largest automorphism group (order 24).

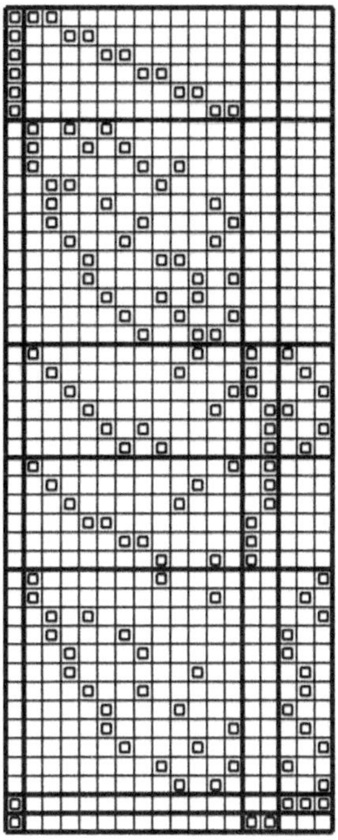

Figure1. One of the 114 672 geometries of TDO-scheme 8-249

3.3 The 12 Latin Squares of Order 6

Line case 207 leads to the well known 12 Latin squares of order 6, see for instance [7] and [11, p. 99]. Compare also [2] where we have explained how a Latin square may be viewed as a linear space: We label the rows and columns and digits by $r_1, \ldots, r_6$ and $c_1, \ldots c_6$ and $d_1, \ldots, d_6$. The Latin Squares of order 6 are the linear spaces of line type $(6^3, 3^{36})$ where the three 6-lines are special: They are $\{r_1, \ldots, r_6\}$, $\{c_1, \ldots, c_6\}$ and $\{d_1, \ldots, d_6\}$. The 36 3-lines correspond to the entries of the Latin Square. Any automorphism either permutes the three 6-lines or not. If it fixes all three, we call the automorphism inner.

These 12 squares have rather large automorphism groups. Using the TDA-schemes of the corresponding linear spaces, we get the orbits on the entries of the Latin Square from the orbits on the 3-blocks in the space.

One Latin Square (cf. Fig. 3, using $d_i = i$ for $i = 1, \ldots, 6$) has special properties. The automorphism group has exactly two orbits on the entries. The diagonal elements of the square form one orbit (indicated by small circles), and the other orbit is formed by all off-diagonal elements. The automorphim group is of order 240, generated by the three permutations α, β and τ. Here,

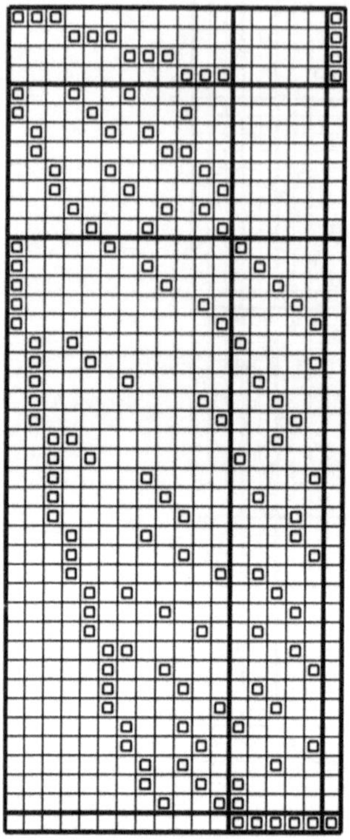

TDO-scheme:

	12	5	1
4	3	0	1
8	3	0	0
30	2	1	0
1	0	5	1

Figure2. A geometry for TDO-scheme 86-1 (with $|Aut| = 24$)

α and β generate the subgroup of inner automorphisms isomorphic to Sym_5. This group is extended by τ, the mapping which interchanges row i with column i ($i = 1, \ldots, 6$). Recall that there are two non-equivalent subgroups $\mathrm{Sym}_5 \leq \mathrm{Sym}_6$. One of them acts transitively on the six elements and the other one is the stabilizer of Sym_6 on one element. It is remarkable that this special Latin square displays both actions. The group Sym_5 of inner automorphisms acts transitively on the six rows $r_1, r_2, \ldots, r_6$ (and also on the six columns). The other action with a fixed point happens on the six digits: the digit $1 = d_1$ is fixed and Sym_5 acts on the five other digits $d_2, d_3, \ldots, d_6$ in the natural way. Formally, we have the following isomorphism

$$\begin{aligned} \mathrm{Sym}_{\{r_1,\ldots,r_6\}} \simeq \mathrm{Sym}_6 \geq \mathrm{Sym}_5 &\rightarrow \mathrm{Sym}_{\{d_2,\ldots,d_6\}}, \\ (r_1\, r_4\, r_6\, r_3\, r_2) = \alpha_1 &\mapsto \alpha_3 = (d_2\, d_4\, d_3\, d_5\, d_6), \\ (r_1\, r_5\, r_3\, r_2) = \beta_1 &\mapsto \beta_3 = (d_2\, d_5\, d_4\, d_6) \end{aligned}$$

between these two types of groups Sym_5.

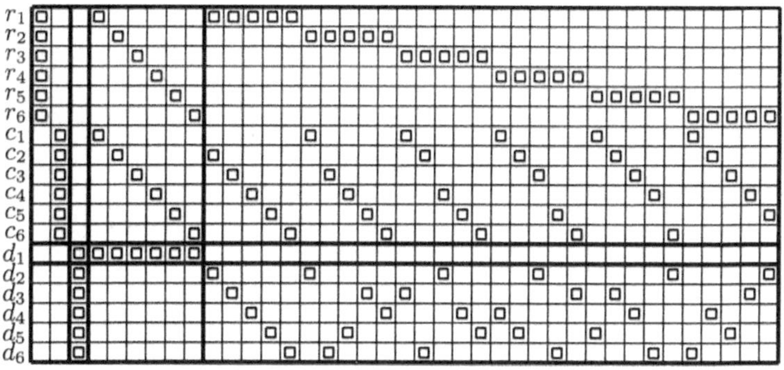

1	2	3	4	5	6
2	1	6	5	3	4
3	6	1	2	4	5
4	5	2	1	6	3
5	3	4	6	1	2
6	4	5	3	2	1

$\mathrm{Aut} = \langle \alpha, \beta, \tau \rangle,$

$\alpha = \alpha_1 \alpha_2 \alpha_3,$

$\beta = \beta_1 \beta_2 \beta_3,$

$\alpha_1 = (r_1\, r_4\, r_6\, r_3\, r_2) \quad \alpha_2 = (c_1\, c_4\, c_6\, c_3\, c_2) \quad \alpha_3 = (d_2\, d_4\, d_3\, d_5\, d_6),$

$\beta_1 = (r_1\, r_5\, r_3\, r_2) \quad \beta_2 = (c_1\, c_5\, c_3\, c_2) \quad \beta_3 = (d_2\, d_5\, d_4\, d_6),$

$\tau = (r_1\, c_1)(r_2\, c_2)(r_3\, c_3)(r_4\, c_4)(r_5\, c_5)(r_6\, c_6)$

Figure3. The TDA-decomposition of the special Latin Square

3.4 Some Geometries with a Rather Large Group

Besides the Latin squares there are 26 linear spaces in the list which have automorphism group order ≥ 20. Here are the numbers of these spaces together with the group order (in brackets):

9-1-108(21),	42-4-1(20),	86-1-24786(24),	94-24-14(24),
9-1-5(21),	57-5-1(36),	93-1-14(24),	118-7-1(24),
10-151-263(36),	73-1-1(108),	93-3-12(24),	118-7-2(72),
10-151-288(54),	85-1-372(24),	94-1-8(72),	217-1-1(20),
16-97-7(24),	85-1-464(24),	94-1-13(72),	277-1-6(21).
16-97-8(24),	85-1-468(72),	94-1-14(72),	
36-32-13(36),	85-1-469(24),	94-24-2(24),	

Among these spaces those species might be of special interest, where the automorphism group has only few orbits on points (or on blocks or on flags). There are 4 spaces with exactly two point orbits. Let us illustrate these 4 spaces. We display in each case the TDO-scheme, the TDA-scheme and the incidence matrix with its TDA. In addition, generators for the automorphism group acting on the points $1, \ldots, 18$ are shown (labelled by capital letters $A, B, C, \ldots$). All groups are soluble and a presentation for the group is given. This presentation is adapted to a composition series, i. e. we have

$$1 \trianglelefteq \langle A \rangle \trianglelefteq \langle A, B \rangle \trianglelefteq \langle A, B, C \rangle \trianglelefteq \ldots \trianglelefteq \mathrm{Aut}\mathcal{S}$$

where $\mathcal{S}$ is the linear space in question and successive factors of this series have prime order. The order of the group is thus the product of the indices, i. e. the product of the exponents in the first column of the presentation. We identified the groups in the table of small groups contained in the algebra software package GAP [13] (Version 4). This catalogue is based on the work of H. U. Besche, B. Eick and E. O'Brien. We denote the n-th group of order m in this catalogue by $m\#_{\mathrm{GAP}}n$.

The Linear Space 010-151-288 The automorphism group has order 54 and is isomorphic to $54\#_{\mathrm{GAP}}5$. The TDA-scheme is isomorphic to the TDO-scheme, i. e. the automorphism group induces no refinement.

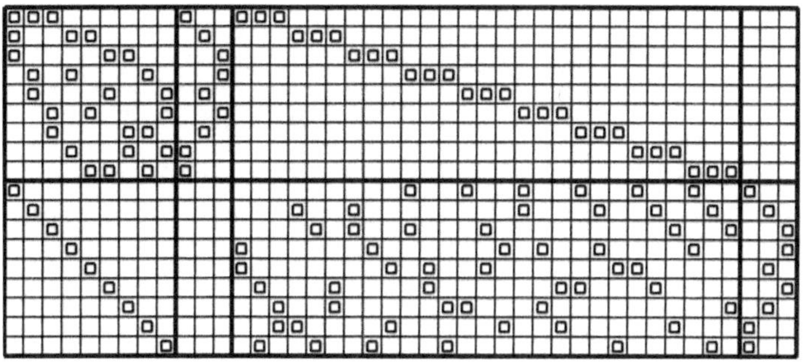

TDA-scheme and the TDO-scheme:

	9	3	27	3
9	3	1	3	0
9	1	0	6	1

$A^3 = id,$

$B^3 = id, A^B = A,$

$C^2 = id, A^C = A^2, B^C = B^2,$

$D^3 = id, A^D = A,\ B^D = AB, C^D = C$

$A = (1\,8\,9)(2\,5\,7)(3\,6\,4)(10\,18\,17)(11\,16\,14)(12\,13\,15)$

$B = (2\,7\,5)(3\,6\,4)(10\,12\,11)(13\,16\,18)(14\,17\,15)$

$C = (4\,6)(5\,7)(8\,9)(11\,12)(13\,14)(15\,16)(17\,18)$

$D = (1\,2\,3)(4\,9\,7)(5\,6\,8)(11\,14\,16)(12\,13\,15)$

The Linear Space 073-1-1 The automorphism group has order 108 (isomorphic to $108\#_{\mathrm{GAP}}17$) and TDO-scheme and TDA-scheme are isomorphic.

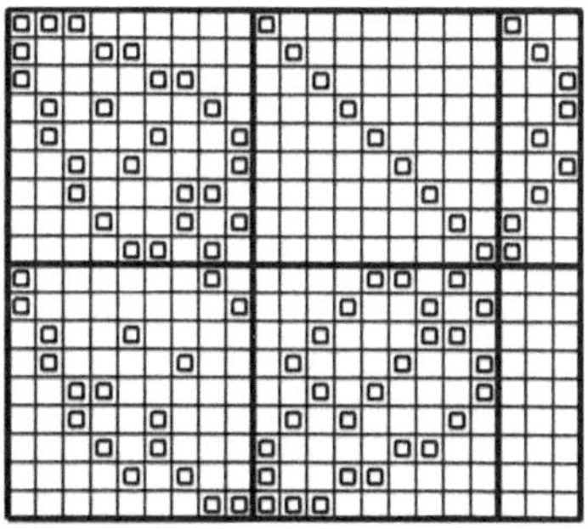

TDA-scheme and the TDO-scheme:

	9	9	3
9	3	1	1
9	2	3	0

$A^3 = id,$

$B^3 = id, A^B = A,$

$C^2 = id, A^C = A^2, B^C = B^2,$

$D^3 = id, A^D = A,\ B^D = AB, C^D = C,$

$E^2 = id, A^E = A,\ B^E = B^2,\ C^E = C, D^E = D^2$

$$A = (1\,8\,9)(2\,5\,7)(3\,6\,4)(10\,11\,18)(12\,13\,17)(14\,16\,15)$$
$$B = (2\,7\,5)(3\,6\,4)(10\,15\,12)(11\,14\,13)(16\,17\,18)$$
$$C = (4\,6)(5\,7)(8\,9)(10\,11)(12\,14)(13\,15)(16\,17)$$
$$D = (1\,2\,3)(4\,9\,7)(5\,6\,8)(12\,17\,13)(14\,16\,15)$$
$$E = (2\,3)(4\,7)(5\,6)(12\,15)(13\,14)(16\,17)$$

The Linear Space 085-1-468 The automorphism group has order 72 (isomorphic to $72\#_{\mathrm{GAP}}44$). The TDO-scheme and the TDA-scheme are isomorphic.

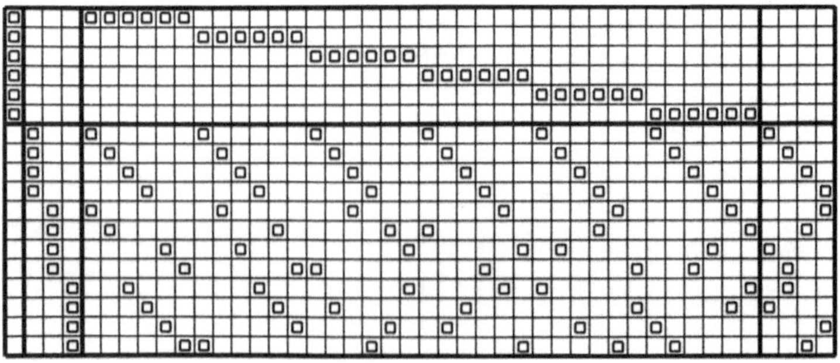

TDA-scheme and the TDO-scheme:

	1	3	36	4
6	1	0	6	0
12	0	1	6	1

$A^2 = id,$
$B^2 = id, A^B = A,$
$C^3 = id, A^C = A, B^C = B,$
$D^2 = id, A^D = A, B^D = B,\ C^D = C^2,$
$E^3 = id, A^E = B, B^E = AB, C^E = C,\ D^D = D$

$A = (2\,4)(3\,5)(7\,8)(9\,10)(11\,12)(13\,14)(15\,16)(17\,18)$
$B = (1\,6)(3\,5)(7\,10)(8\,9)(11\,13)(12\,14)(15\,18)(16\,17)$
$C = (7\,13\,16)(8\,14\,15)(9\,12\,18)(10\,11\,17)$
$D = (1\,6)(2\,4)(3\,5)(11\,17)(12\,18)(13\,16)(14\,15)$
$E = (1\,2\,3)(4\,5\,6)(7\,8\,9)(12\,13\,14)(15\,18\,16)$

The Linear Space 094-24-2 The automorphism group is isomorphic to Sym_4 of order 24 ($\simeq 24\#_{\mathrm{GAP}}12$). The TDA-scheme is a bit finer than the TDO-scheme. In the natural action on four points, the generators can be identified as $A = (1\,2)(3\,4)$, $B = (1\,3)(2\,4)$, $C = (1\,2\,4)$ and $D = (1\,2)$, for example.

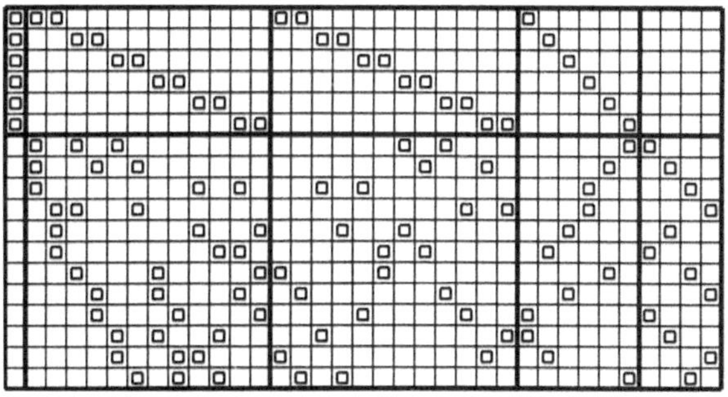

TDO-scheme:

	1	12	18	4
6	1	2	3	0
12	0	3	3	1

TDA-scheme:

	1	12	12	6	4
6	1	2	2	1	0
12	0	3	2	1	1

$A^2 = id,$
$B^2 = id, A^B = A,$
$C^3 = id, A^C = B, B^C = AB,$
$D^2 = id, A^D = A, B^D = AB, C^D = C^2$

$$A = (2\,5)(3\,6)(7\,11)(8\,12)(9\,10)(13\,17)(14\,18)(15\,16)$$
$$B = (1\,4)(3\,6)(7\,14)(8\,13)(9\,16)(10\,15)(11\,18)(12\,17)$$
$$C = (1\,2\,6)(3\,4\,5)(7\,15\,12)(8\,14\,9)(10\,13\,11)(16\,17\,18)$$
$$D = (2\,6)(3\,5)(7\,12)(8\,11)(9\,10)(13\,14)(17\,18)$$

References

1. L. M. Batten and A. Beutelspacher: The theory of finite linear spaces. Cambridge University Press, Cambridge 1993.
2. A. Betten and D. Betten: Regular linear spaces. *Beiträge Algebra Geom.* **38** (1997), 111–124.
3. A. Betten and D. Betten: Linear spaces with at most 12 points. *J. of Combinatorial Designs* **7** (1999), 119–145.
4. A. Betten and D. Betten: The Proper Linear Spaces on 17 Points. *Discrete Applied Mathematics* **95** (1999), 83–108.
5. A. Betten and D. Betten: Tactical decompositions and some configurations v_4. *J. of Geom.* **66** (1999), 27–41.
6. A. Betten and D. Betten: Addendum to *Note on the Proper Linear Spaces on 18 Points* `http://www.mathe2.uni-bayreuth.de/betten/PUB/pub_proper18.html`
7. D. Betten: Die 12 lateinischen Quadrate der Ordnung 12. *Mitt. Math. Sem. Giessen* **163** (1984), 181–188.
8. D. Betten and M. Braun: A tactical decomposition for incidence structures. *Ann. Disc. Math.* **52** (1992), 37–43.
9. A. E. Brouwer: The linear spaces on 15 points. *Ars Combin.* **12** (1981), 3–35.
10. P. Dembowski: *Finite geometries.* Classics in Mathematics. Springer-Verlag, Berlin, 1997. Reprint of the 1968 original.
11. C. Colbourn, J. Dinitz: CRC Handbook of Combinatorial Designs, CRC press, Boca Raton, New York, London, Tokyo, 1996.
12. G. Heathcote: Linear spaces on 16 points. *J. Combin. Des.* **1** (1993), 359–378.
13. The GAP Group: *GAP — Groups, Algorithms, and Programming*, Version 4.1; Aachen, St. Andrews, 1999. (http://www-gap.dcs.st-and.ac.uk/~gap)

Variation über ein Thema von Knuth, Robinson, Schensted und Schützenberger

Dieter Blessenohl and Armin Jöllenbeck*

Mathematisches Seminar der Christian-Albrechts-Universität zu Kiel, Ludewig-Meyn-Str. 4, 24098 Kiel, Germany

Abstract The crucial point of many approaches to representation theory of symmetric groups is the algorithm of Robinson and Schensted. Roughly speaking, this is a sorting algorithm with documentation. From the theorem of Schützenberger we take the idea for a description of the Robinson-Schensted correspondence without using the documentation part of the algorithm, which therefore plays no rôle in our approach. On the other hand we analyze the Knuth relations more thoroughly than usual. By means of a slight generalization of the sorting part of the algorithm we get another associative product on the free monoid X^* over a countable alphabet X. The canonical map of X^* onto the plactic monoid is also a homomorphism with respect to this product.

Adalbert Kerber zum 60. Geburtstag

1 Einführung

Der Robinson-Schensted-Algorithmus ist das kombinatorische Herz der Charaktertheorie der symmetrischen Gruppen – zumindest bei dem in [1] vorgestellten Aufbau dieser Theorie. Bei diesem Zugang mit Hilfe eines nichtkommutativen Überbaus aus geeigneten Bialgebren ist der Algorithmus von Robinson und Schensted auch der einzige, der benötigt wird. Abweichend vom üblichen Vorgehen werden dabei über den Satz von Robinson und Schensted hinaus allerdings auch die Sätze von Knuth und Schützenberger benutzt. Wir stellen deshalb hier eine Version vor, bei der alle für [1] notwendigen Ergebnisse in durchsichtiger Form erreicht werden. Der Robinson-Schensted-Algorithmus ist, kurz gesagt, ein Sortieralgorithmus mit Dokumentation. Wir verzichten auf die dokumentierende Hälfte und analysieren stattdessen genauer die Knuth-Relationen, wodurch sich am Ende die Robinson-Schensted-Bijektion in augenfälliger Weise ergibt.

Für jedes $n \in \mathbb{N}_0$ sei $\underline{n} := \{1, 2, \dots, n\}$, insbesondere $\underline{0} := \emptyset$. Für jede Menge X sei $X^n := X^{\underline{n}} := \{f \mid f : \underline{n} \to X \text{ Abbildung}\}$ die Menge der

* Dieser Artikel entstand im Rahmen des DFG-Projekts BL 488/1-1.

n-Tupel über X. Mit S_n bezeichnen wir die Teilmenge der Bijektionen in $(\underline{n})^n \subseteq \mathbb{N}^n$, die *symmetrische Gruppe* auf $\underline{n}$. Für alle $i \in \underline{n-1}$ sei $\tau_i \in \mathrm{S}_n$ die *Transposition* $(i, i+1)$. Für die Verknüpfung „Nacheinander von Abbildungen“ benutzen wir das Symbol $\circ$. Schließlich sei $\mathbb{N}^* := \bigcup_{k \geq 0} \mathbb{N}^k$. Die Elemente von $\mathbb{N}^*$ nennen wir Worte über $\mathbb{N}$. Ist $q \in \mathbb{N}^k$, so nennen wir $|q| := k$ die Länge von q und schreiben q_i für das Bild von i unter der Abbildung q. Durch die *Konkatenation* $qr := q_1 \dots q_n r_1 \dots r_m$, $q \in \mathbb{N}^n$, $r \in \mathbb{N}^m$, wird $\mathbb{N}^*$ zu einem freien Monoid mit neutralem Element $\emptyset$, frei erzeugt von $\mathbb{N}$ $(= \mathbb{N}^1)$. Ist $p = p_1 \dots p_k \in \mathbb{N}^k$ und $p_1 + \dots + p_k = n$, so heißt p eine *Zerlegung* von n $(p \models n)$. Gilt außerdem $p_1 \geq p_2 \geq \dots \geq p_k$, so heißt p *Partition* von n $(p \vdash n)$. Für jede Partition $p = p_1 \dots p_k$ heißt

$$\mathrm{R}(p) := \{(i,j) \mid i,j \in \mathbb{N},\ 1 \leq i \leq k,\ 1 \leq j \leq p_i\} \subseteq \mathbb{N} \times \mathbb{N}$$

der Rahmen zu p. Üblicherweise versteht man unter einem Standardtableau der Gestalt p eine Abbildung $\tau : \mathrm{R}(p) \to \mathbb{N}$ mit der Eigenschaft, monoton wachsend in den Zeilen und streng monoton wachsend in den Spalten von $\mathrm{R}(p)$ zu sein. Abweichend davon nennen wir das „von links unten nach rechts oben herausgelesene Wort“ ein *Standardtableau* zu p, und bezeichnen mit $\mathrm{ST}^p \subseteq \mathbb{N}^*$ die Menge aller solcher Standardtableaux. Man überlegt sich leicht, daß sich aus dem herausgelesenen Wort die Partition p eindeutig bestimmen läßt. Schließlich nennen wir die Elemente von $\mathrm{SYT}^p := \mathrm{ST}^p \cap \mathrm{S}_n$ *Standard-Young-Tableaux* zu p.

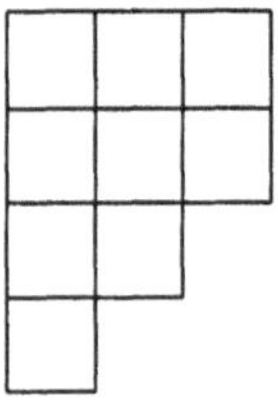

R(3321)

1	1	2
2	3	3
4	4	
5		

$5\,44\,233\,112 \in \mathrm{ST}^{3321}$

Für alle $a, b \in \mathbb{N}$ bezeichnen wir mit

$$\langle a, b \rangle := \{c \in \mathbb{N} \mid \min(a,b) \leq c \leq \max(a,b)\}$$

das „Intervall der natürlichen Zahlen zwischen a und b“. Für alle $\pi \in \mathrm{S}_n$ und $a, b \in \underline{n}$ sei

$$\langle a, b \rangle_\pi := \langle a\pi^{-1}, b\pi^{-1} \rangle \pi \quad ,$$

der Abschnitt in der „Bildzeile“ von π zwischen a und b.

Zunächst beschreiben wir den Algorithmus mitsamt seinen Hauptergebnissen, wobei wir uns auf Permutationen beschränken und auf alle Beweise verzichten. Sei π eine Permutation von $\underline{n}$. Der Algorithmus läuft in n Schritten ab. Im i-ten Zwischenschritt sind die Zahlen $1\pi, 2\pi, \dots, i\pi$ zu einem Standardtableau verarbeitet. Dann ist $(i+1)\pi$ in die „1. Zeile“ dieses Tableau einzuordnen, und zwar ans Ende, falls $(i+1)\pi$ größer ist als alle Einträge in der 1. Zeile, sonst an die Stelle der kleinsten Zahl in der 1. Zeile größer als $(i+1)\pi$. Diese wird dann nach dem gleichen Verfahren in der 2. Zeile einsortiert usw. Nach n Schritten ist ein Standard-Young-Tableau entstanden. Wir illustrieren dies an einem Beispiel: $\pi = 453281796 \in \mathrm{S}_9$.

$$\begin{array}{ccccccccc}
4 & 45 & 35 & 25 & 258 & 158 & 157 & 1579 & 1569 \\
 & & 4 & 3 & 3 & 2 & 28 & 28 & 27 \\
 & & & 4 & 4 & 3 & 3 & 3 & 38 \\
 & & & & & 4 & 4 & 4 & 4
\end{array}$$

Entstanden ist der Rahmen R(4221) und das Standardtableau $4\,38\,27\,1569 \in \mathrm{SYT}^{4221}$. Allgemein erhält man eine Partition p von n und ein Element von SYT^p, welches wir $\mathrm{P}(\pi)$ nennen. Dieser Algorihmus wird durch ein weiteres Tableau der gleichen Gestalt dokumentiert. Im i-ten Schritt bekommt das neu hinzugenommene Feld die Nr. i, im Beispiel also:

$$\begin{array}{ccccccccc}
1 & 12 & 12 & 12 & 125 & 125 & 125 & 1258 & 1258 \\
 & & 3 & 3 & 3 & 3 & 37 & 37 & 37 \\
 & & & 4 & 4 & 4 & 4 & 4 & 49 \\
 & & & & & 6 & 6 & 6 & 6
\end{array}$$

Man erhält ein Standardtableau $\mathrm{Q}(\pi) = 6\,49\,37\,1258$. Insgesamt haben wir so eine Abbildung definiert:

$$\mathrm{RS}: \quad \mathrm{S}_n \to \bigcup_{p \vdash n} \mathrm{SYT}^p \times \mathrm{SYT}^p \quad , \quad \pi \mapsto (\mathrm{P}(\pi), \mathrm{Q}(\pi)) \quad .$$

Für die folgenden Resultate verweisen wir auf [2].

Satz 1.1 (Robinson–Schensted). RS *ist eine Bijektion.*

Satz 1.2 (Schützenberger). *Für alle* $\pi \in \mathrm{S}_n$ *gilt:*

$$\mathrm{RS}(\pi^{-1}) = (\mathrm{Q}(\pi), \mathrm{P}(\pi)) \quad .$$

Korollar 1.3. *Es gilt:*

$$\left| \left\{ \pi \mid \pi \in \mathrm{S}_n, \pi^2 = \mathrm{id}_{\underline{n}} \right\} \right| = \left| \bigcup_{p \vdash n} \mathrm{SYT}^p \right| \quad .$$

Seien $\pi, \rho \in \mathrm{S}_n$. Wir nennen π einen *Links-Knuth-Nachbarn* von ρ, in Zeichen $\pi \;{}_{\mathrm{K}}\!\smile\; \rho$, wenn es ein $j \in \underline{n-1}$ gibt mit

(i) $(j, j+1) \circ \pi = \rho \quad ,$

(ii) $(j-1)\pi \in \langle j\pi, (j+1)\pi \rangle$ oder $(j+2)\pi \in \langle j\pi, (j+1)\pi \rangle \quad .$

Z.B. ist $2341 \;{}_{\mathrm{K}}\!\smile\; 2314 \;{}_{\mathrm{K}}\!\smile\; 2134$.

Satz 1.4 (Knuth). *Für alle* $\pi, \rho \in \mathrm{S}_n$ *ist genau dann* $\mathrm{P}(\pi) = \mathrm{P}(\rho)$, *wenn es* $\pi_1, \ldots \pi_k \in \mathrm{S}_n$ *gibt mit* $\pi = \pi_1 \;{}_{\mathrm{K}}\!\smile \cdots {}_{\mathrm{K}}\!\smile\; \pi_k = \rho$.

Bemerkung 1.5. *Für alle* $\pi \in \mathrm{S}_n$ *ist* $\mathrm{P}(\pi) = \mathrm{P}(\mathrm{P}(\pi))$.

2 Eine Verknüpfung

Wir definieren $\mathbb{N}^*_{\text{mon}} \subseteq \mathbb{N}^*$ durch $\emptyset \in \mathbb{N}^*_{\text{mon}}$ und

$$w = w_1 w_2 \cdots w_k \in \mathbb{N}^*_{\text{mon}} \quad :\Longleftrightarrow \quad w_1 \leq w_2 \leq \cdots \leq w_k \quad .$$

Ist $q = q^{(1)} \cdots q^{(m)}$ mit $q^{(1)}, \dots, q^{(m)} \in \mathbb{N}^*_{\text{mon}}$ und dabei m minimal gewählt, so sind $q^{(1)}, \dots, q^{(m)}$ eindeutig bestimmt. Wir nennen $q^{(1)}, \dots, q^{(m)}$ die *monotonen Bausteine* und m die *monotone Länge* von q. Zum Beispiel ist für $q = 112\,1345\,444$ die monotone Länge gleich 3. Die monotonen Bausteine sind $q^{(1)} = 112$, $q^{(2)} = 1345$ und $q^{(3)} = 444$. Wir definieren Abbildungen

$$\lhd, \blacktriangleleft : \mathbb{N}^*_{\text{mon}} \times \mathbb{N}^* \to \mathbb{N}^*$$

folgendermaßen: Sei $w = w_1 \cdots w_k \in \mathbb{N}^*_{\text{mon}}$. Wir setzen

$$w \lhd \emptyset := w \quad \text{und} \quad w \blacktriangleleft \emptyset := \emptyset \quad .$$

Für alle $x \in \mathbb{N}$ setzen wir

$$\left.\begin{array}{l} w \lhd x := wx \\ w \blacktriangleleft x := \emptyset \end{array}\right. , \quad \text{falls} \quad w = \emptyset \text{ oder } w_k \leq x \text{ ist.}$$

Ist $k > 0$ und $x < w_k$, so sei $i := \min\{j \in \underline{k} \mid x < w_j\}$ und

$$\begin{aligned} w \lhd x &:= w_1 \cdots w_{i-1} x w_{i+1} \cdots w_k \quad , \\ w \blacktriangleleft x &:= w_i \quad . \end{aligned}$$

Schließlich setzen wir

$$\begin{aligned} w \lhd x_1 \cdots x_l &:= (w \lhd x_1) \lhd x_2 \cdots x_l \quad , \\ w \blacktriangleleft x_1 \cdots x_l &:= (w \blacktriangleleft x_1)((w \lhd x_1) \blacktriangleleft x_2 \cdots x_l) \quad . \end{aligned}$$

Bemerkung 2.1. *Für alle* $w \in \mathbb{N}^*_{\text{mon}}$, $x \in \mathbb{N}$ *und* $q, u, v \in \mathbb{N}^*$ *gilt*

$$\begin{aligned} w \lhd uv &= (w \lhd u) \lhd v \quad , \\ w \blacktriangleleft uv &= (w \blacktriangleleft u)((w \lhd u) \blacktriangleleft v) \quad , \end{aligned}$$

Wir definieren rekursiv eine Verknüpfung $\diamond$ auf $\mathbb{N}^*$ folgendermaßen:
Sei $q \in \mathbb{N}^*$ und $q = q^{(1)} \cdots q^{(m)}$ die Zerlegung von q in monotone Bausteine. Wir setzen

$$q \diamond \emptyset := q \quad .$$

Für alle $x \in \mathbb{N}$ setzen wir

$$q \diamond x := x \quad , \text{falls } m = 0 \quad ,$$

$$q \diamond x := (q^{(1)} \cdots q^{(m-1)} \diamond (q^{(m)} \blacktriangleleft x))(q^{(m)} \lhd x) \quad \text{falls } m > 0.$$

Schließlich setzen wir

$$q \diamond x_1 \cdots x_l := (q \diamond x_1) \diamond x_2 \cdots x_l \quad .$$

Z.B. ist $123\,12 \diamond 211 = 123\,122 \diamond 11 = 3\,122\,112 \diamond 1 = 3\,1222\,111$.

Bemerkung 2.2. *Für alle* $q, u, v, r \in \mathbb{N}^*$ *gilt*

$$q \diamond uv = (q \diamond u) \diamond v \quad ,$$

$$q \diamond r = (q^{(1)} \cdots q^{(m-1)} \diamond (q^{(m)} \blacktriangleleft r))(q^{(m)} \lhd r) \text{ falls } q \neq \emptyset \quad .$$

Die Verknüpfung $\diamond$ ist die Verallgemeinerung der sortierenden Hälfte des Robinson-Schensted-Algorithmus auf beliebige Worte, anstelle von Standardtableaux, wobei wir uns allerdings besonders für $\emptyset \diamond r$ interessieren. In der üblichen Weise zeigt man:

Satz 2.3. *Ist* w *ein Standardtableau und* $q \in \mathbb{N}^*$, *so ist* $w \diamond q$ *ein Standardtableau. Insbesondere ist* $\emptyset \diamond q$ *ein Standardtableau für alle* $q \in \mathbb{N}^*$. *Ist* q *ein Standardtableau, so ist* $\emptyset \diamond q = q$.

3 Knuth-Relationen

Wir erklären die Links-Knuth-Nachbarschaft ${}_{K}\!\smile$ auf $\mathbb{N}^3$ wie folgt:
Für alle $a, b, c \in \mathbb{N}$ mit $a \leq b \leq c$ setzen wir

$$acb \;{}_{K}\!\smile\; cab, \text{ falls } b < c, \quad \text{und} \quad bac \;{}_{K}\!\smile\; bca, \text{ falls } a < b \quad .$$

Außerdem sei ${}_{K}\!\smile$ symmetrisch. Für alle $q, r \in \mathbb{N}^*$ setzen wir ferner:

$$q \;{}_{K}\!\smile\; r \quad :\Longleftrightarrow \quad \text{es gibt } u, v \in \mathbb{N}^* \text{ und } abc, a'b'c' \in \mathbb{N}^3 \text{ mit}$$
$$abc \;{}_{K}\!\smile\; a'b'c' \quad \text{und} \quad q = uabcv, \; r = ua'b'c'v \quad .$$

Schließlich definieren wir die Links-Knuth-Äquivalenz ${}_{K}\!\sim$ auf $\mathbb{N}^*$ als die transitive und reflexive Hülle der Links-Knuth-Nachbarschaft. Für jedes $q \in \mathbb{N}^*$ heißt ${}_{K}[\,q\,] := \{r \in \mathbb{N}^* \mid q \;{}_{K}\!\sim\; r\}$ die Links-Knuth-Klasse von q. Z.B. sind die Äquivalenzklassen von ${}_{K}\!\sim$ auf den Worten vom Inhalt 111, 21, 12 und 3:

$$\{123\}, \{132, 312\}, \{213, 231\}, \{321\},$$
$$\{112\}, \{121, 211\},$$
$$\{122\}, \{212, 221\},$$
$$\{111\}.$$

Wir bemerken noch, daß auf $\mathbb{N}^0 \cup \mathbb{N}^1 \cup \mathbb{N}^2$ die Links-Knuth-Äquivalenz die Gleichheit ist.

Da ${}_{K}\!\sim$ nach Definition eine Kongruenzrelation bzgl. der Konkatenation auf $\mathbb{N}^*$ ist, wird auf der Menge $\mathbb{N}^*/{}_{K}\!\sim$ der Links-Knuth-Klassen eine assoziative Verknüpfung induziert – das plaktische Monoid [3]. Die folgenden Sätze zeigen u.a., daß ${}_{K}\!\sim$ auch eine Kongruenzrelation bzgl. $\diamond$ ist (Schlußbemerkung iii).

Satz 3.1. *Für alle* $q, r \in \mathbb{N}^*$ *ist* $qr \;{}_{K}\!\sim\; q \diamond r$ *und insbesondere* $r \;{}_{K}\!\sim\; \emptyset \diamond r$.

Satz 3.2. *Für alle* $w, q, r \in \mathbb{N}^*$ *mit* $q \;{}_{\mathrm{K}}\!\!\sim r$ *gilt* $w \diamond q = w \diamond r$.

Korollar 3.3. *Für alle* $q, r \in \mathbb{N}^*$ *gilt:* $q \;{}_{\mathrm{K}}\!\!\sim r \iff \emptyset \diamond q = \emptyset \diamond r$.

Der Zusammenhang von Links-Knuth-Nachbarschaft und Robinson-Schensted-Algorithmus zeigt sich besonders deutlich im Beweis des folgenden Spezialfalls von 3.1.

Hilfssatz 3.4. *Für alle* $w \in \mathbb{N}^*_{\mathrm{mon}}$ *und* $x \in \mathbb{N}$ *ist* $wx \;{}_{\mathrm{K}}\!\!\sim w \diamond x$.

Beweis. Für $w = \emptyset$ ist nichts zu zeigen. Sei also $w = w_1 \dots w_k \in \mathbb{N}^k$ und $k \geq 1$. Ist $w_k \leq x$, so ist $wx = w \diamond x$ und ebenfalls nichts zu zeigen. Sei also $x < w_k$ und $i \in \underline{k}$ so gewählt, daß $w_{i-1} \leq x < w_i$ ist. Wegen $x < w_i \leq w_{i+1} \leq \dots \leq w_k$ ist

$$\begin{aligned} w_i \dots w_{k-1} w_k x \;{}_{\mathrm{K}}\!\!\sim w_i \dots w_{k-1} x w_k \;{}_{\mathrm{K}}\!\!\sim w_i \dots w_{k-2} x w_{k-1} w_k \\ {}_{\mathrm{K}}\!\!\sim \dots {}_{\mathrm{K}}\!\!\sim w_i x w_{i+1} \dots w_k \quad . \end{aligned}$$

Wegen $w_1 \leq \dots \leq w_{i-1} \leq x < w_i$ ist

$$\begin{aligned} w_1 \dots w_i x \;{}_{\mathrm{K}}\!\!\sim w_1 \dots w_i w_{i-1} x \;{}_{\mathrm{K}}\!\!\sim w_1 \dots w_i w_{i-2} x \\ {}_{\mathrm{K}}\!\!\sim \dots {}_{\mathrm{K}}\!\!\sim w_i w_1 \dots w_{i-1} x \quad . \end{aligned}$$

Da ${}_{\mathrm{K}}\!\!\sim$ eine Kongruenzrelation ist, folgt

$$wx \;{}_{\mathrm{K}}\!\!\sim w_i w_1 \dots w_{i-1} x w_{i+1} \dots w_k = w \diamond x \quad .$$

□

Satz 3.1 folgt mit Induktion nach der monotonen Länge von q und der Länge von r mittels 3.4. In 3.2 kann man sich wegen 2.2 auf den Fall $q, r \in \mathbb{N}^3$ und $q \;{}_{\mathrm{K}}\!\smile r$ beschränken. Dann folgt der Satz mit Hilfe einer Induktion nach der monotonen Länge von w.

Bemerkung 3.5. *Nach 2.3, 3.1 und 3.3 liegt in jeder Links-Knuth-Klasse genau ein Standardtableau.*

Nach 3.1 und 3.2 ist für alle $w, q, r \in \mathbb{N}^*$

$$(w \diamond q) \diamond r = w \diamond qr = w \diamond (q \diamond r) \quad ,$$

d.h. $\diamond$ ist eine assoziative Verknüpfung.

Offenbar ist S_n (als Teilmenge von $\mathbb{N}^*$) eine Vereinigung von Links-Knuth-Klassen. Für alle $\pi, \rho \in \mathrm{S}_n$ setzen wir

$$\begin{aligned} \pi \smile_{\mathrm{K}} \rho \quad &:\iff \quad \pi^{-1} \;{}_{\mathrm{K}}\!\smile \rho^{-1} \quad , \\ \pi \sim_{\mathrm{K}} \rho \quad &:\iff \quad \pi^{-1} \;{}_{\mathrm{K}}\!\!\sim \rho^{-1} \end{aligned}$$

und nennen $\smile_{\mathrm{K}}$ die Rechts-Knuth-Nachbarschaft und $\sim_{\mathrm{K}}$ die Rechts-Knuth-Äquivalenz auf S_n. Für jedes $\pi \in \mathrm{S}_n$ heißt $[\pi]_{\mathrm{K}} := \{\rho \in \mathrm{S}_n \mid \pi \sim_{\mathrm{K}} \rho\}$ die Rechts-Knuth-Klasse von π.

Die Links-Knuth-Nachbarschaft für Permutationen ist in der Einführung beschrieben. Durch Invertieren folgt daraus:

$$\pi \smile_{\mathrm{K}} \rho \iff \text{es gibt ein } i \in \underline{n-1} \text{ mit } \rho = \pi \circ \tau_i$$
$$\text{und } \{(i-1),(i+2)\} \cap \langle i, i+1\rangle_\pi \neq \emptyset \quad .$$

Z.B. ist $2314 \smile_{\mathrm{K}} 1324 \smile_{\mathrm{K}} 1423$.

Satz 3.6. *Für alle* $p \vdash n$ *ist* SYT^p *eine Rechts-Knuth-Klasse.*

Daß SYT^p gegen Rechts-Knuth-Nachbarschaft abgeschlossen ist, sieht man leicht. Durch Induktion nach n folgt andererseits, daß SYT^p in einer Rechts-Knuth-Klasse enthalten ist.

4 Teppiche

Wir definieren $\rho \in \mathrm{S}_n$ durch $i\rho := n - i + 1$ für $1 \leq i \leq n$. Es ist

$$\rho = n\,(n-1)\,(n-2)\ldots 2\,1$$

eine Involution in S_n. Für alle $j \in \underline{n-1}$ ist

$$\rho \circ \tau_j = \tau_{n-j+1} \circ \rho \quad .$$

Für alle $\pi, \phi, \psi \in \mathrm{S}_n$ gilt:

$$\pi \;{}_{\mathrm{K}}\!\smile \phi \iff \pi \circ \rho \;{}_{\mathrm{K}}\!\smile \phi \circ \rho \iff \rho \circ \pi \;{}_{\mathrm{K}}\!\smile \rho \circ \phi \quad ,$$

$$\pi \smile_{\mathrm{K}} \psi \iff \rho \circ \pi \smile_{\mathrm{K}} \rho \circ \psi \iff \pi \circ \rho \smile_{\mathrm{K}} \psi \circ \rho$$

Die beiden folgenden Lemmata enthalten die entscheidenen Beziehungen zwischen Rechts- und Links-Knuth-Relationen.

Lemma 4.1. *Seien* $\pi \in \mathrm{S}_n$ *und* $i, j \in \underline{n-1}$ *mit*

$$\pi \circ \tau_i \smile_{\mathrm{K}} \pi \;{}_{\mathrm{K}}\!\smile \tau_j \circ \pi \quad .$$

Dann gibt es ein $\sigma \in \mathrm{S}_n$ *mit*

$$\pi \circ \tau_i \;{}_{\mathrm{K}}\!\smile \sigma \smile_{\mathrm{K}} \tau_j \circ \pi \quad .$$

Wir illustrieren dies mit:

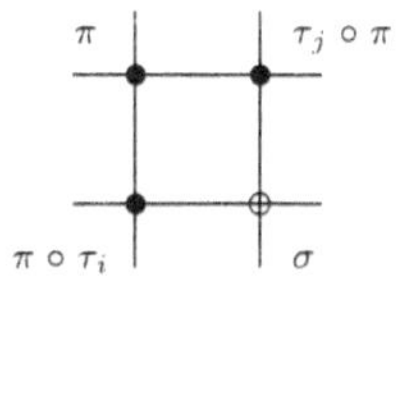

Beweis. Sei

$$\pi = \begin{pmatrix} \cdots & j-1 & j & j+1 & j+2 & \cdots \\ \cdots & a & b & c & d & \cdots \end{pmatrix} \quad .$$

Nach Voraussetzung gilt:

$$a \in \langle b, c\rangle \quad \text{oder} \quad d \in \langle b, c\rangle$$

und

$$i-1 \in \langle i, i+1\rangle_\pi \quad \text{oder} \quad i+2 \in \langle i, i+1\rangle_\pi \quad .$$

Indem man gegebenenfalls π durch $\rho\circ\pi$ oder $\pi\circ\rho$ oder $\rho\circ\pi\circ\rho$ ersetzt, kann man o.B.d.A. annehmen

$$a \in \langle b, c\rangle \text{ und } i-1 \in \langle i, i+1\rangle_\pi \quad .$$

1. *Fall:* $|\{i, i+1\} \cap \{a, b, c\}| \leq 1$. Dann folgt $a\tau_i \in \langle b\tau_i, c\tau_i\rangle$. also $\pi \circ \tau_i \;{}_{\mathrm{K}}\!\smile\; \tau_j \circ (\pi \circ \tau_i)$. Wäre nun $i-1 \notin \langle i, i+1\rangle_{\tau_j\circ\pi}$, so wäre $\{b, c\} = \{i-1, i+1\}$ und daher $a = i$, ein Widerspruch zur Voraussetzung. Also ist $(\tau_j \circ \pi) \circ \tau_i \smile_{\mathrm{K}} \tau_j \circ \pi$.
2. *Fall:* $\{i, i+1\} \subseteq \{a, b, c\}$. Dann ist $\{i, i+1\} = \{a, c\}$ und $b = i-1$. Wegen $a \in \langle b, c\rangle$ folgt $c = i+1$, $a = i$ und damit

$$\pi \circ \tau_i \;{}_{\mathrm{K}}\!\smile\; \tau_{j-1} \circ (\pi \circ \tau_i) = (\tau_j \circ \pi) \circ \tau_{i-1} \smile_{\mathrm{K}} \tau_j \circ \pi \quad .$$

□

Lemma 4.2. *Seien $\pi, \rho, \rho^* \in S_n$ mit*

$$\pi \smile_{\mathrm{K}} \rho \quad , \quad \pi \smile_{\mathrm{K}} \rho^* \quad \textit{und} \quad \rho \;{}_{\mathrm{K}}\!\sim\; \rho^* \quad .$$

Dann ist

$$\rho = \rho^* \quad .$$

Beweis. Sei $\rho = \pi \circ \tau_i$, $\rho^* = \pi \circ \tau_l$. Angenommen, es ist $i \neq l$. Nach Voraussetzung gibt es $\rho_1, \ldots, \rho_k \in S_n$ mit

$$\rho = \rho_1 \;{}_{\mathrm{K}}\!\smile\; \rho_2 \;{}_{\mathrm{K}}\!\smile\; \cdots \;{}_{\mathrm{K}}\!\smile\; \rho_k = \rho^* \quad ,$$

d.h. es gibt $j_2, \ldots, j_k \in \underline{n-1}$ mit

$$\rho^* = \tau_{j_k} \circ \cdots \circ \tau_{j_2} \circ \rho \quad ,$$

also

$$\pi \circ \tau_l \circ \tau_i = \tau_{j_k} \circ \cdots \circ \tau_{j_2} \circ \pi \quad .$$

Wegen $i \neq l$ gibt es ein $m \in \underline{n-1}$, so daß m und $m+1$ in den Bildzeilen von π und $\pi \circ \tau_l \circ \tau_i$ in verschiedener Reihenfolge stehen. Die Reihenfolge von m und $m+1$ bleibt aber unter Links-Knuth-Vertauschungen ungeändert, ein Widerspruch. □

Wir nennen zwei Rechts-Knuth-Klassen A und B benachbart, wenn es $\alpha \in A$ und $\beta \in B$ gibt mit $\alpha \,{}_{\mathrm{K}}\!\smile \beta$. Analog nennen wir Links-Knuth-Klassen C und D benachbart, wenn es Rechts-Knuth-Nachbarn $\gamma \in C$ und $\delta \in D$ gibt.

Satz 4.3. *Die Links-Knuth-Nachbarschaft stiftet zwischen je zwei benachbarten Rechts-Knuth-Klassen eine Bijektion, d.h., sind A, B benachbarte Rechts-Knuth-Klassen, so gibt es zu jedem $\pi \in A$ genau ein $\pi' \in B$ mit $\pi \,{}_{\mathrm{K}}\!\smile \pi'$ und umgekehrt.*

Beweis. Seien $\alpha \in A$ und $\beta \in B$ mit $\alpha \,{}_{\mathrm{K}}\!\smile \beta$ und sei $\pi \in A$. Dann gibt es $\alpha_0, \ldots, \alpha_k \in A$ mit

$$\alpha = \alpha_0 \,{}_{\mathrm{K}}\!\smile \alpha_1 \,{}_{\mathrm{K}}\!\smile \cdots \,{}_{\mathrm{K}}\!\smile \alpha_k = \pi \quad .$$

Eine k-fache Anwendung von 4.1 liefert $\beta_0, \ldots, \beta_k \in B$ mit

$$\beta = \beta_0 \,{}_{\mathrm{K}}\!\smile \beta_1 \,{}_{\mathrm{K}}\!\smile \cdots \,{}_{\mathrm{K}}\!\smile \beta_k$$

und $\beta_i \smile_{\mathrm{K}} \alpha_i$ für $0 \leq i \leq k$, insbesondere $\pi \smile_{\mathrm{K}} \beta_k$. Aus 4.2 folgt, daß es höchstens ein $\pi' \in B$ geben kann mit $\pi \smile_{\mathrm{K}} \pi'$. □

Durch Invertieren erhält man die analoge Aussage für Links-Knuth-Klassen.

Satz 4.4. *Für alle $p \vdash n$ enthält* SYT^p *eine Involution.*

7	8	9
4	5	6
2	3	
1		

$\longrightarrow$

1	3	6
2	5	9
4	8	
7		

1 23 456 789 $\qquad$ 7 48 259 136 = (17)(24)(38)(69)

Sei $p \vdash n$. Wir nennen

$$\mathcal{T}^p := \{\pi \in \mathrm{S}_n \mid \emptyset \diamond \pi \in \mathrm{SYT}^p\}$$

den Teppich zu p. Nach 3.3 ist $\mathcal{T}^p$ eine Vereinigung von Links-Knuth-Klassen. Nach 3.5 gilt $|A \cap \mathrm{SYT}^p| = 1$ für jede Links-Knuth-Klasse $A \subseteq \mathcal{T}^p$. Nach 3.6 ist SYT^p eine Rechts-Knuth-Klasse, also $(\mathrm{SYT}^p)^{-1}$ eine Links-Knuth-Klasse, die wegen 4.4 in $\mathcal{T}^p$ enthalten ist. Wegen 4.3 haben alle in $\mathcal{T}^p$ enthaltenen Links-Knuth-Klassen die gleiche Mächtigkeit, nämlich $|(\mathrm{SYT}^p)^{-1}| = |\mathrm{SYT}^p|$. Durch Invertieren werden aus den in $\mathcal{T}^p$ enthaltenen Links-Knuth-Klassen die Rechts-Knuth-Klassen der Elemente von $(\mathrm{SYT}^p)^{-1}$. Eine mehrfache Anwendung von 4.3 zeigt, daß $\mathcal{T}^p$ Vereinigung von Rechts-Knuth-Klassen ist. Aus 4.3 folgt weiter, daß jede Rechts-Knuth-Klasse $A \subseteq \mathcal{T}^p$ und jede Links-Knuth-Klasse $B \subseteq \mathcal{T}^p$ einen nichtleeren Schnitt haben. Da $|A| = |\mathrm{SYT}^p|$und dies auch die Anzahl der in $\mathcal{T}^p$ enthaltenen Links-Knuth-Klassen ist, folgt $|A \cap B| = 1$.

Korollar 4.5. *Für jedes* $\pi \in \mathcal{T}^p$ *seien* $\mathrm{T}(\pi)$ *und* $\mathrm{T}^-(\pi)$ *definiert durch*

$$\{\mathrm{T}(\pi)\} :={}_\mathrm{K}[\,\pi\,] \cap \mathrm{SYT}^p \quad \textit{und} \quad \{\mathrm{T}^-(\pi)\} := [\,\pi\,]_\mathrm{K} \cap (\mathrm{SYT}^p)^{-1} \quad .$$

Dann ist die Abbildung

$$\pi \mapsto (\mathrm{T}(\pi), \mathrm{T}^-(\pi))$$

eine Bijektion von $\mathcal{T}^p$ *auf* $\mathrm{SYT}^p \times (\mathrm{SYT}^p)^{-1}$. *Außerdem ist*

$$\mathrm{T}(\pi) = \emptyset \diamond \pi \quad \textit{und} \quad \mathrm{T}^-(\pi) = \mathrm{T}(\pi^{-1})^{-1} \quad .$$

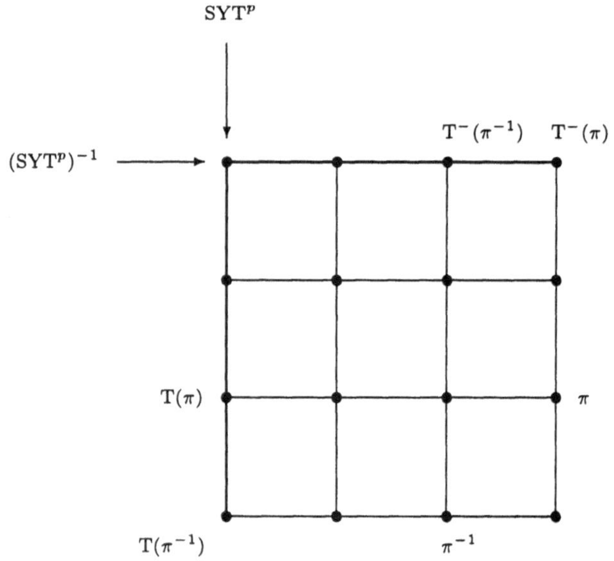

5 Schlußbemerkungen

(i) Man macht sich leicht klar, daß die Abbildung P aus der Einführung mit T übereinstimmt, d.h. daß $\emptyset \diamond \pi$ das Standard-Young-Tableau $\mathrm{P}(\pi)$ zu π ist. Daß andererseits $\mathrm{Q}(\pi) = \mathrm{T}^-(\pi)^{-1} = \mathrm{T}(\pi^{-1}) = \mathrm{P}(\pi^{-1})$ ist, haben wir nicht bewiesen. Dies ist der Satz von Schützenberger, der gerade besagt, daß $\emptyset \diamond \pi^{-1}$ den Algorithmus, der zu $\emptyset \diamond \pi$ führt, dokumentiert – und vice versa. Die Robinson-Schensted-Bijektion

$$\mathrm{RS} : \mathrm{S}_n \to \bigcup_{p \vdash n} (\mathrm{SYT}^p \times \mathrm{SYT}^p)$$

erscheint in unserer Version also in der Form

$$\pi \mapsto (\mathrm{T}(\pi), \mathrm{T}(\pi^{-1})) = (\emptyset \diamond \pi, \emptyset \diamond \pi^{-1}).$$

(ii) Ist $A \subseteq \mathrm{S}_n$ eine Links-Knuth-Klasse, so ist A^{-1} eine Rechts-Knuth-Klasse, die im gleichen Teppich wie A enthalten ist. Wie oben gezeigt ist $|A \cap A^{-1}| = 1$. Wegen $(A \cap A^{-1})^{-1} = A^{-1} \cap A$ ist das einzige Element von $A \cap A^{-1}$ eine Involution. Die Involutionen bilden daher ein gemeinsames Repräsentantensystem für die Links- und für die Rechts-Knuth-Klassen in S_n, woraus insbesondere 1.3 folgt.

(iii) Wegen 3.1 ist ${}_{\mathrm{K}}\!\sim$ nicht nur eine Kongruenzrelation bzgl. Konkatenation sondern auch bzgl. $\diamond$ und $\mathbb{N}^*/_{\mathrm{K}}\!\sim$ mit der von $\diamond$ induzierten Verknüpfung das plaktische Monoid. Die von $\mathbb{N}$ $(= \mathbb{N}^1)$ zusammen mit $\emptyset$ erzeugte Halbgruppe von $(\mathbb{N}^*, \diamond)$ besteht genau aus den Standardtableaux und ist nach 3.5 ein Repräsentantensystem für $\mathbb{N}^*/_{\mathrm{K}}\!\sim$, insbesondere isomorph zum plaktischen Monoid. Nach 3.1 ist dabei $\emptyset$ neutrales Element in dieser Halbgruppe.

(iv) Für alle $p \vdash n$ bezeichne p' die zu p konjugierte Partition. Man kann sich leicht überlegen, daß $\emptyset \diamond (\rho \circ \pi) \in \mathrm{ST}^{p'}$ ist für alle $\pi \in \mathrm{ST}^p$. Genauer gesagt ist $\emptyset \diamond (\rho \circ \pi)$ dasjenige Standardtableau, das aus π durch „Spiegeln an der Hauptdiagonalen“ entsteht wie im folgenden Beispiel:

1	2	6
3	7	9
4	8	
5		

$\pi = 5\,48\,379\,126$

1	3	4	5
2	7	8	
6	9		

$\emptyset \diamond (\rho \circ \pi) = 69\,278\,1345$

Wegen $(\mathcal{T}^p)^{-1} = \mathcal{T}^p$ folgt insbesondere, da $\rho \circ \mathcal{T}^p$ ein Teppich ist,

$$\rho \circ \mathcal{T}^p = \mathcal{T}^{p'} = (\mathcal{T}^{p'})^{-1} = \mathcal{T}^p \circ \rho$$

und daher

$$\rho \circ \mathcal{T}^p \circ \rho = \mathcal{T}^p \quad .$$

(v) Die Teppiche der S_4:

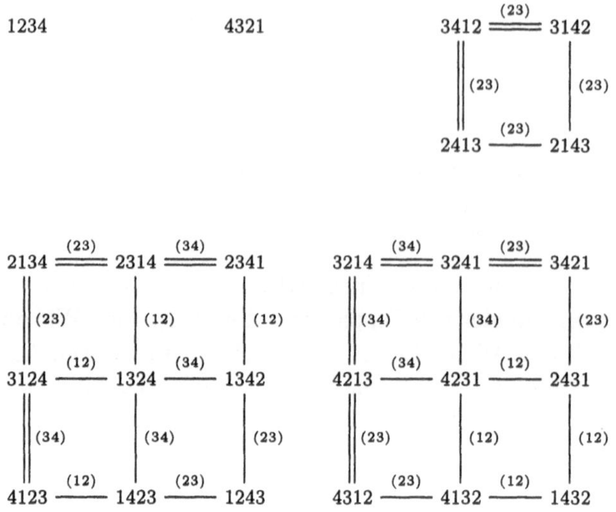

Dabei markieren Doppelstriche die Klassen SYT^p bzw. $(\mathrm{SYT}^p)^{-1}$.

(vi) Eine ausführlichere Darstellung dieser Variation des Robinson-Schensted-Algorithmus enthalten die in Vorbereitung befindlichen Lecture Notes „Nichtkommutative Charaktere der symmetrischen Gruppen“ der beiden Autoren.

Literatur

1. A. Jöllenbeck: Nichtkommutative Charaktertheorie. Dissertation Kiel 1998. Bayreuther Mathematische Schriften, p. 1-41, Heft 56, 1999.
2. D. E. Knuth: The art of computer programming. Vol. 3: Sorting and searching. Addison-Wesley Series in Computer Science and Information Processing. Reading, Mass. 1974.
3. M. Lothaire: Combinatorics on words (2nd ed.). Encyclopedia of Mathematics and Its Applications 17. Cambridge University Press 1997.

Elements of a General Algebraic Theory of Standard Tableaux

Michael Clausen[1]

Universität Bonn, Institut für Informatik V,
Römerstr. 164, D-53117 Bonn, Germany
clausen@cs.uni-bonn.de

Abstract We introduce the class of groups having a multiplicity-free character graph (MC-groups). This class includes both the class of symmetric and the class of solvable groups. We show that every MC-group has an S_n-like representation theory with generalized standard tableaux as main combinatorial objects. As an illustration, we give a detailed description of the irreducible representations of supersolvable groups.

1 Introduction

The ordinary representation theory of the symmetric group S_n has always been an interplay between group theory, linear algebra, and combinatorics. Here are three examples illustrating this interplay:

- irreducible characters of S_n corresponding to partitions of n,
- irreducible representations of $\mathbb{C}S_n$ corresponding to standard tableaux,
- permutations of S_n corresponding to standard bitableaux.

After a brief review of this classical case in Section 3, we introduce the class of MC-groups in Section 4. Symmetric groups as well as all solvable groups belong to this class. We show that there are analogues of standard tableaux for every MC-group G. Every standard tableau defines a primitive idempotent in the group algebra $\mathbb{C}G$. This leads to a direct sum decomposition of $\mathbb{C}G$ into minimal left ideals and to canonical bases in these ideals that are indexed by standard tableaux.

This general theory of standard tableaux fits in a particular nice way to the class of supersolvable groups, see Section 5. In contrast to the symmetric group case we have an action of a supersolvable group G on the set of standard tableaux (w.r.t. a chief series of G) such that the G-orbits are characterized by their common 'shape' or type. This allows to specify a full set $D_1, \ldots, D_h$ of (induced monomial) irreducible representations of $\mathbb{C}G$ by triples (U_i, L_i, λ_i), where U_i is a subgroup of G, L_i is a transversal of the left cosets of U_i in G, and λ_i is a linear character of U_i such that D_i is the induced monomial representation $\lambda_i \uparrow_{L_i} G$.

Section 6 points to papers that on this theoretical basis have solved several computational problems very efficiently. Among these problems are the construction of a full set of irreducible representations of a supersolvable group, the design of fast Fourier transforms of those groups, the efficient computation of the irreducible characters of p-groups, and DFT-based collection in pc-presented groups.

2 Basic Definitions and Facts

To make this paper essentially self-contained, this section recalls some basics from representation theory. For more details and proofs we refer to Serre's book [13].

The set $\mathbb{C}G := \{a | a : G \to \mathbb{C}\}$ of all complex-valued functions on the finite group G becomes a $\mathbb{C}$-space under pointwise addition and scalar multiplication. It becomes even an associative algebra, the so-called group algebra, by defining the product of a and b in $\mathbb{C}G$ via the following convolution formula:

$$(ab)(x) := \sum_{g \in G} a(g) b(g^{-1}x),$$

for all $x \in G$. An element $a \in \mathbb{C}G$ is typically written as a formal sum $a = \sum_{g \in G} a_g g$, where $a_g := a(g)$. With this convention, the convolution formula reads as follows:

$$\Big(\sum_{g \in G} a_g g\Big)\Big(\sum_{h \in G} b_h h\Big) = \sum_{x \in G}\Big(\sum_{g \in G} a_g b_{g^{-1}x}\Big)x.$$

A d-dimensional representation of $\mathbb{C}G$ is an algebra morphism $D : \mathbb{C}G \to \mathbb{C}^{d \times d}$. D is called irreducible, if no linear subspace of $\mathbb{C}^d$ (except 0 and $\mathbb{C}^d$) is invariant under all $D(g)$, $g \in G$. It can be shown that D is irreducible iff it is surjective. Two representations D and D', both of degree d, are equivalent ($D \sim D'$), iff there is an invertible matrix $X \in \mathrm{GL}(d, \mathbb{C})$ such that $D'(g) = XD(g)X^{-1}$, for all $g \in G$.

Theorem 2.1 (Wedderburn's Structure Theorem). *Let G denote a finite group. Then its complex group algebra $\mathbb{C}G$ is isomorphic to an algebra of block-diagonal matrices:*

$$D : \mathbb{C}G \longrightarrow \oplus_{k=1}^{h} \mathbb{C}^{d_k \times d_k}.$$

Every such isomorphism D, a so-called discrete Fourier transform of $\mathbb{C}G$, is the direct sum of a full set of pairwise inequivalent irreducible representations $D_1, \ldots, D_h$. Furthermore, the number h of blocks equals the number of conjugacy classes of G.

In turn, every full set $D_1, \ldots, D_h$ defines via $D := \oplus D_k$ a DFT of $\mathbb{C}G$. According to Maschke's Theorem, every representation F of $\mathbb{C}G$ is equivalent to a direct sum of irreducible representations F_i: $F \sim F_1 \oplus \ldots \oplus F_s$. Furthermore, by the Krull-Remak-Schmidt Theorem, the multiplicity

$$\langle F | D_k \rangle := |\{j \le s | F_j \sim D_k\}|$$

with which the irreducible representation D_k occurs in F, is well-defined. As we will see in a moment, this multiplicity can be computed using characters.

The character χ of a representation F of $\mathbb{C}G$ is a function $\chi : G \to \mathbb{C}$ defined by $\chi(g) := \mathrm{Trace}(F(g))$, for $g \in G$. Characters are constant on conjugacy classes and two representations are equivalent iff their characters are equal. Characters corresponding to irreducible representations are called irreducible characters. A character is called linear iff it corresponds to a representation of degree 1. By $\mathrm{Irr}(G)$ we denote the set of all irreducible characters of G. The space $CF(G, \mathbb{C})$ of all complex-valued class functions on G becomes an inner product space by $\langle\chi|\psi\rangle := |G|^{-1} \sum_{g\in G} \chi(g)\overline{\psi(g)}$.

Theorem 2.2. *For a finite group G with h conjugacy classes the following is true:*

(1) $\mathrm{Irr}(G) = \{\chi_1, \ldots, \chi_h\}$ *is an orthonormal basis of* $CF(G, \mathbb{C})$.
(2) *Let F and D_k be representations of $\mathbb{C}G$ with characters χ and χ_k, respectively. If D_k is irreducible, then the multiplicity $\langle D_k|F\rangle$ with which D_k occurs in F equals $\langle\chi_k|\chi\rangle$.*
(3) *If $e_k := e_{\chi_k} := \frac{\chi_k(1)}{|G|} \sum_{g\in G} \chi_k(g^{-1})g$, then $e_1, \ldots, e_h$ are a basis of the center of $\mathbb{C}G$. Moreover, $1 = e_1 + \ldots + e_h$ and $e_k e_j = \delta_{kj} e_k$. (The e_k are called central primitive idempotents in $\mathbb{C}G$.)*
(4) *If M is a left $\mathbb{C}G$-module affording the representation F with character χ, then $M = \oplus_{k=1}^{h} e_k M$ (isotypic decomposition). If M_k is a simple module affording the character χ_k, then $e_k M$, the isotypic component of type χ_k, is isomorphic to the direct sum of $\langle\chi_k|\chi\rangle$ copies of M_k. Every simple submodule of M affording the character χ_k is contained in $e_k M$.*

There are two reciprocal techniques to construct new representations: restriction and induction. We begin with the trivial process of restriction. Let H be a subgroup of G. Then $\mathbb{C}H$ can be viewed as a subalgebra of $\mathbb{C}G$. If D is a representation of $\mathbb{C}G$, its restriction to $\mathbb{C}H$ is a representation of $\mathbb{C}H$ denoted by $D \downarrow H =: F$. In turn, D is called an extension of F. Analogously, $\chi{\downarrow}H$ denotes the restriction of the character χ of G to H.

Of special importance is the process of induction, where a representation of $\mathbb{C}G$ is constructed from a representation of $\mathbb{C}H$, where H is a subgroup of G. In terms of modules, the construction is straightforward: let L be a left ideal in $\mathbb{C}H$. Then $\mathbb{C}GL$ is a left ideal in $\mathbb{C}G$ and with the left coset decomposition $G = \sqcup_{i=1}^{r} g_i H$ one obtains the decomposition $\mathbb{C}GL = \oplus_{i=1}^{r} g_i L$ as a $\mathbb{C}$-space. In particular, $\mathbb{C}GL$ has dimension $[G : H] \cdot \dim L$. The left $\mathbb{C}G$-module $\mathbb{C}GL$ is said to be induced by L. A look at the corresponding matrix representations leads to the following definition.

Definition 2.3. Let H be a subgroup of the group G, $T := (g_1, \ldots, g_r)$ a transversal of the left cosets of H in G and let F be a representation of $\mathbb{C}H$ of degree f. The induced representation $F{\uparrow}_T G$ of $\mathbb{C}G$ of degree $f \cdot r$ is defined by

$$F{\uparrow}_T G(x) := (\dot{F}(g_i^{-1} x g_j))_{1\le i,j\le r} \in (\mathbb{C}^{f\times f})^{r\times r},$$

for $x \in G$, where $\dot{F}(y) := F(y)$ if $y \in H$, and $\dot{F}(y)$ is the f-square all zero matrix, if $y \in G \setminus H$.

It is easily checked that this defines a representation of $\mathbb{C}G$. Taking different transversals gives possibly different, but equivalent representations. Thus in non-critical situations we sometimes write $F{\uparrow}G$ instead of $F{\uparrow}_T G$. Note that $F{\uparrow}_T G(x)$, $x \in G$, is a block matrix, with exactly one non-zero block in each block row and in each block column. In particular, if F is of degree 1, then, for all $x \in G$, the matrix $F{\uparrow}_T G(x)$ is monomial. (Recall that a matrix is called monomial iff it has exactly one non-zero entry in each row and in each column. A representation D of $\mathbb{C}G$ is said to be monomial iff $D(g)$ is monomial, for all $g \in G$.) A group G is called an M-group if every irreducible representation is equivalent to a monomial one. Below we will give an alternative proof to the well-known fact that supersolvable groups are M-groups. There is a close connection between restriction and induction. A more precise statement reads as follows.

Theorem 2.4 (Frobenius Reciprocity Theorem). *Let H be a subgroup of G. Furthermore, let F and D be irreducible representations of $\mathbb{C}H$ and $\mathbb{C}G$, respectively. Then the multiplicity $\langle F|D{\downarrow}H\rangle$ of F in $D{\downarrow}H$ equals the multiplicity $\langle D|F{\uparrow}G\rangle$ of D in $F{\uparrow}G$: $\langle F|D{\downarrow}H\rangle = \langle D|F{\uparrow}G\rangle$.*

If N is a normal subgroup of G and F a representation of $\mathbb{C}N$, then for $g \in G$ we define a new representation F^g of $\mathbb{C}N$ by $F^g(n) := F(g^{-1}ng)$ for all $n \in N$. F and F^g are called conjugate representations. As $\{F(n)|n \in N\} = \{F^g(n)|n \in N\}$, F is irreducible iff F^g is irreducible, and G acts on $\mathrm{Irr}(N)$ by conjugation via $g\chi := (N \ni n \mapsto \chi(g^{-1}ng))$. We need the following special case of Clifford theory.

Theorem 2.5 (Clifford's Theorem). *Let N be a normal subgroup in G of prime index p, and let F be an irreducible representation of $\mathbb{C}N$. For a fixed $g \in G \setminus N$, let T denote the transversal $(1, g, g^2, \ldots, g^{p-1})$ of the cosets of N in G. Then exactly one of the following two cases applies.*

(1) *All F^{g^i} are equivalent. Then there are exactly p irreducible representations $D_0, \ldots, D_{p-1}$ of $\mathbb{C}G$ extending F. The D_k are pairwise inequivalent and satisfy $F{\uparrow}G \sim D_0 \oplus \ldots \oplus D_{p-1}$. Moreover, if $\chi^0, \chi^1, \ldots, \chi^{p-1}$ are the linear characters of the cyclic group G/N in a suitable order, we have $D_k = \chi^k \otimes D_0$ for all k, i.e., $D_k(x) = \chi^k(xN)D_0(x)$, for all $x \in G$.*
(2) *The representations F^{g^i} are pairwise inequivalent. In this case, the induced representation $F{\uparrow}G$ is irreducible.*

3 Standard Tableaux for Symmetric Groups

The basic combinatorial objects in the representation theory of the symmetric group S_n are the partitions of n and the standard tableaux. Recall that a partition α of n (for short: $\alpha \vdash n$) is a non-increasing sequence of positive integers summing up to n. (In the sequel we will identify the partition $\alpha = (\alpha_1, \ldots, \alpha_r)$ with its Young diagram, which is the set $\cup_{i=1}^{r}\{(i,j) | 1 \leq j \leq \alpha_i\}$. Such a diagram is illustrated by α_i left-justified cells in row i.) Let $X(\mathcal{C}_n)$ denote the poset of all partitions of m, $m \leq n$, ordered by inclusion. The unusual notation $X(\mathcal{C}_n)$ for the first n levels of the Young lattice will become clear in a moment. A standard tableau of shape (or type) α, α a partition of n, is obtained by filling the numbers $1, 2, \ldots, n$ into the n cells such that the entries from left to right in each row and down each column are increasing. Thus a standard tableau of shape α encodes a path in $X(\mathcal{C}_n)$ starting by the partition (1) and ending up with α. Fig. 1 shows a standard tableau as a sequence of partitions ($\equiv$ characters).

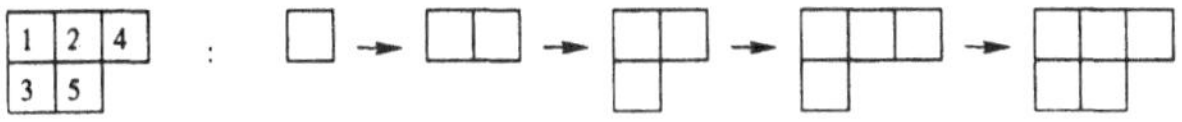

Figure1. A standard tableau of type $(3, 2)$.

Why do partitions and standard tableaux play such a role in the representation theory of S_n? First of all, the conjugacy class of a permutation σ in S_n is characterized by the cycle structure of σ. Hence there is a natural bijection between partitions of n and conjugacy classes of S_n. According to Wedderburn's Theorem, the number of partitions of n equals the number of equivalence classes of irreducible representations of S_n, which also coincides with the number of irreducible characters of S_n. In fact, the irreducible characters of S_n are indexed by the partitions. The deeper reason why standard tableaux are the right combinatorial objects for describing the irreducible representations of S_n is provided by the following theorem.

Theorem 3.1 (Young's Branching Rule). *Let α be a partition of n. Then the irreducible character χ_α of S_n restricted to S_{n-1} splits as follows:*

$$\chi_\alpha \downarrow S_{n-1} = \sum_{\beta \subset \alpha} \chi_\beta,$$

where the sum runs over all partitions β of $n-1$ that are contained in α. In particular, $\chi_\alpha \downarrow S_{n-1}$ is multiplicity-free.

Thus the branching behaviour of the irreducible characters corresponding to the elements of the subgroup chain

$$\mathcal{C}_n := (S_n \supset S_{n-1} \supset \ldots \supset S_1 = \{1\})$$

is described by the Young lattice $X(\mathcal{C}_n)$. As the partitions parametrize the irreducible characters of the symmetric groups, the Young lattice will also be called the character graph $X(\mathcal{C}_n)$ of the chain $\mathcal{C}_n$. Fig. 2 illustrates $X(\mathcal{C}_4)$.

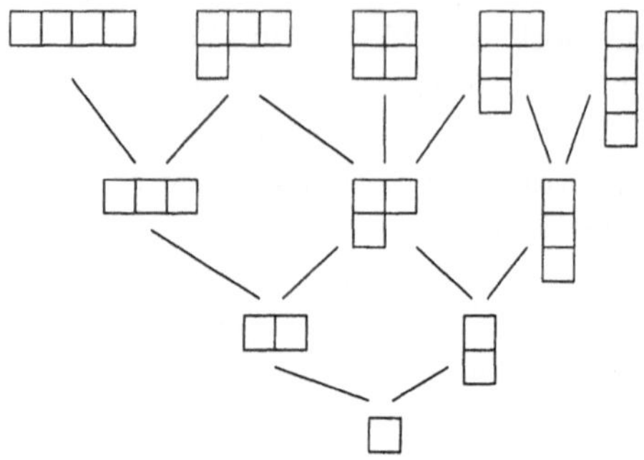

Figure2. Character graph $X(\mathcal{C}_4)$.

Let $\mathrm{ST}(\alpha)$ denote the set of all standard tableaux of type α. The elements in $\mathrm{ST}(\alpha) \times \mathrm{ST}(\alpha) =: \mathrm{SBT}(\alpha)$ are called standard bitableaux of type α. As a consequence of Young's Rule, the degree $\chi_\alpha(1)$ of the irreducible character χ_α equals the number of all standard tableaux of shape (or type) α. In combination with Wedderburn's Theorem we obtain the following well-known result.

Theorem 3.2. *The number of permutations in S_n equals the number of standard bitableaux with entries $1, 2, \ldots, n$:*

$$|S_n| = \sum_{\alpha \vdash n} |\mathrm{SBT}(\alpha)| = \sum_{\alpha \vdash n} |\mathrm{ST}(\alpha)|^2.$$

The Robinson-Schensted correspondence describes in a purely combinatorial way a bijection between permutations and standard bitableaux. This bijection has an number of remarkable properties, see, e.g., [9,14]. For algebraic variants we refer to [4,8,11].

Now all what we need to extend the notion of standard tableaux to a larger class of groups is a subgroup chain like $\mathcal{C}_n$ and a multiplicity-free branching behaviour of the characters. This will be our next topic.

4 Standard Tableaux for MC-Groups

Let G be a finite group. Throughout this section,

$$\mathcal{C} = (G = G_n \supset G_{n-1} \supset \ldots \supset G_0 = \{1\})$$

denotes a chain of subgroups of G. The character graph $X(\mathcal{C})$ of G is an edge-weighted graph. Its set of nodes is the union $\cup_{i=0}^n \mathrm{Irr}(G_i)$. (The elements of $\mathrm{Irr}(G_i)$ will be called the nodes of level i.) All edges in this graph connect at most nodes of adjacent levels. More precisely, $\chi_i \in \mathrm{Irr}(G_i)$ is connected to $\chi_{i-1} \in \mathrm{Irr}(G_{i-1})$ by an edge iff its weight, defined by $\langle \chi_i | \chi_{i-1} \rangle := \langle \chi_i {\downarrow} G_{i-1} | \chi_{i-1} \rangle$ is positive. We will call $X(\mathcal{C})$ multiplicity-free iff all (non-zero) weights are equal to 1. In this case, every irreducible character of any G_i remains multiplicity-free when restricted to G_{i-1}.

Definition 4.1. If the finite group G has a chain $\mathcal{C}$ of subgroups such that the corresponding character graph $X(\mathcal{C})$ is multiplicity-free, then G is called an MC-group (w.r.t. $\mathcal{C}$). The class of MC-groups will be denoted by $\mathcal{MC}$.

The symmetric group S_n is an MC-group w.r.t. $\mathcal{C}_n$, see the last section. In the sequel, G will always denote an MC-group w.r.t. $\mathcal{C}$.

The set of all paths in $X(\mathcal{C})$ from level 0 to $\chi \in \mathrm{Irr}(G)$ will be denoted by $\mathrm{ST}_\mathcal{C}(\chi)$, thus

$$\mathrm{ST}_\mathcal{C}(\chi) := \{(\chi_0, \dots, \chi_n = \chi) \mid \chi_i \in \mathrm{Irr}(G_i),\ \langle \chi_i {\downarrow} G_{i-1} | \chi_{i-1} \rangle = 1, \forall i \geq 1\}.$$

Elements in $\mathrm{ST}_\mathcal{C}(\chi)$ are called $\mathcal{C}$-standard tableaux of type χ, whereas elements in $\mathrm{SBT}_\mathcal{C}(\chi) := \mathrm{ST}_\mathcal{C}(\chi) \times \mathrm{ST}_\mathcal{C}(\chi)$ will be called $\mathcal{C}$-standard bitableaux of type χ. By induction, one easily proves that $\chi(1) = |\mathrm{ST}_\mathcal{C}(\chi)|$. Thus we obtain the following generalization of Theorem 3.2.

Theorem 4.2. *If G is an MC-group w.r.t. $\mathcal{C}$ then the order of G equals the number of $\mathcal{C}$-standard bitableaux:*

$$|G| = \sum_{\chi \in \mathrm{Irr}(G)} |\mathrm{SBT}_\mathcal{C}(\chi)| = \sum_{\chi \in \mathrm{Irr}(G)} |\mathrm{ST}_\mathcal{C}(\chi)|^2.$$

We are going to describe an algebraic variant of this fact. To this end, we define for every $\chi \in \mathrm{Irr}(G)$ and every $w = (\chi_0, \dots, \chi_n = \chi) \in \mathrm{ST}_\mathcal{C}(\chi)$ the element

$$e(w) := e_{\chi_0} \cdot e_{\chi_1} \cdots e_{\chi_n}.$$

As each e_{χ_i} is in the center of $\mathbb{C}G_i$, the above product defining $e(w)$ does not depend on the ordering of the factors. Furthermore, by Theorem 2.2, $e(w)$ is non-zero and idempotent. In fact, $e(w)$ is even primitive, see Janusz [10].

Theorem 4.3. *If G is an MC-group w.r.t. $\mathcal{C}$, then the following holds.*

(1) *For every irreducible character χ of G and every $\mathcal{C}$-standard tableau w of type χ, the element $e(w)$ is a primitive idempotent affording the character χ. ($e(w)$ will be called the primitive idempotent of the standard tableau w.)*

(2) *The $\mathcal{C}$-standard tableaux w of G induce a direct sum decomposition of $\mathbb{C}G$ into minimal left ideals $\mathbb{C}Ge(w)$:*

$$\mathbb{C}G = \bigoplus_{\chi \in \mathrm{Irr}(G)} \bigoplus_{w \in \mathrm{ST}_\mathcal{C}(\chi)} \mathbb{C}Ge(w).$$

(3) *The $\mathcal{C}$-standard bitableaux (v, w) of G induce a direct sum decomposition of $\mathbb{C}G$ into one-dimensional linear subspaces $e(v)\mathbb{C}Ge(w)$:*

$$\mathbb{C}G = \bigoplus_{\chi \in \mathrm{Irr}(G)} \bigoplus_{(v,w) \in \mathrm{ST}_\mathcal{C}(\chi)^2} e(v)\mathbb{C}Ge(w).$$

Proof. (1). $e_\chi = (\prod_{i<n} 1)e_\chi = \prod_{i<n}(\sum_{\chi_i \in \mathrm{Irr}(G_i)} e_{\chi_i})e_\chi = \sum_{w \in \mathrm{ST}_{\mathcal{C}}(\chi)} e(w)$. Thus e_χ is the sum of $\chi(1)$ pairwise orthogonal idempotents $e(w)$; in particular, all $e(w)$ are primitive.

(2). Obvious by (1).

(3). Let $M = \mathbb{C}Ge(w)$ and $v = (\chi_0, \ldots, \chi_n) \in \mathrm{ST}_{\mathcal{C}}(\chi)$. Then the factors of $e(v)$ cause an isotypic filtration of M:

$$e(v)M = e_{\chi_0} \cdot (e_{\chi_1} \cdot (\ldots \cdot (e_{\chi_{n-1}} M) \ldots)).$$

As $\langle \chi_i \downarrow G_{i-1} | \chi_{i-1} \rangle = 1$ (by assumption), $e(v)M$ is a simple $\mathbb{C}G_0$-module, hence one-dimensional. □

Open Problem 1. For which $g \in G$ is $e(v)ge(w)$ non-zero? How does G act on the lines $e(v)\mathbb{C}Ge(w)$? (For partial answers see the next section.)

Theorem 4.4. *The class $\mathcal{MC}$ of all MC-groups satisfies the following closure properties.*

(1) *If G and H are MC-groups, then their direct product is an MC-group as well.*
(2) *Every epimorphic image of an MC-group is an MC-group.*
(3) *Let N be a normal subgroup of G such that G/N is abelian. If N is an MC-group, then G is an MC-group, too. In particular, every solvable group G is an MC-group w.r.t. any composition series of G.*

Proof. (1). Let G and H be MC-groups with respect to $(G_n \supset \ldots \supset G_0)$ and $(H_m \supset \ldots \supset H_0)$, respectively. Consider the subgroup chain

$$G_n \times H_m \supset G_n \times H_{m-1} \supset \ldots \supset G_n \times H_0 \supset G_{n-1} \times H_0 \supset \ldots \supset G_0 \times H_0.$$

As $\mathrm{Irr}(G_i \times H_j) \equiv \mathrm{Irr}(G_i) \otimes \mathrm{Irr}(H_j)$, one easily checks that $G \times H$ is an MC-group w.r.t. the above chain.

(2). Let G be an MC-group w.r.t. $G_n \supset \ldots \supset G_0$ and let N be a normal subgroup of G. We claim that G/N is an MC-group with respect to the chain $G_nN/N \supseteq \ldots \supseteq G_0N/N$. If $\overline{F}$ is a representation of some $\mathbb{C}(H/N)$, H a subgroup of G containing N, then $F(h) := \overline{F}(hN)$ is well-defined and yields a representation of $\mathbb{C}H$. As the sets of operators $F[H]$ and $\overline{F}[H/N]$ are equal, F is irreducible iff the same is true for $\overline{F}$. Furthermore, if $\overline{F}$ is an irreducible representation of $\mathbb{C}(G_iN/N)$, then $\overline{F}\downarrow(G_{i-1}N/N)$ is multiplicity-free, since $F\downarrow G_{i-1}$ is multiplicity-free.

(3). This follows by Clifford theory. □

Open Problem 2. Describe the class of MC-groups.

5 Standard Tableaux for Supersolvable Groups

In this section we will study the class of supersolvable groups for which the above general technique fits very well. Recall that a finite group G is called supersolvable iff there exists a chain

$$\mathcal{C} = (G = G_n \supset G_{n-1} \supset \ldots \supset G_1 \supset G_0 = \{1\})$$

such that each G_i is a normal subgroup in G and all indices $[G_i : G_{i-1}] =: p_i$ are prime. Thus $\mathcal{C}$ is a chief series of G with chief factors G_i/G_{i-1} of prime order. For example, all nilpotent groups (especially all groups of prime power order) are supersolvable. Before describing the irreducible representations of supersolvable groups, we need some preparations.

The exponent e of a finite group G is the least common multiple of the orders of all $g \in G$. A representation D of $\mathbb{C}G$ is called e-monomial iff for all $g \in G$ the matrix $D(g)$ is monomial and all non-zero entries in $D(g)$ are eth roots of unity. Now we can state the main result of this section.

Theorem 5.1. *Let G be a supersolvable group of exponent e, and $\chi \in \mathrm{Irr}(G)$. Then the following holds.*

(1) *The character graph $X(\mathcal{C})$ of any chief series $\mathcal{C}$ of G is multiplicity-free.*
(2) *G acts transitively on the set $\mathrm{ST}_{\mathcal{C}}(\chi)$ of all $\mathcal{C}$-standard tableaux of type χ by $g(\chi_0, \ldots, \chi_n = \chi) := (g\chi_0, \ldots, g\chi_n)$.*
(3) *For all $g \in G$ and all $w \in \mathrm{ST}_{\mathcal{C}}(\chi)$ we have $ge(w)g^{-1} = e(gw)$.*
(4) *Let U be the stabilizer of $w \in \mathrm{ST}_{\mathcal{C}}(\chi)$ and L a transversal of the left cosets of U in G. Then $\{\ell e(w) | \ell \in L\}$ is a $\mathbb{C}$-basis of $\mathbb{C}Ge(w)$ and the corresponding representation D of $\mathbb{C}G$ is e-monomial. More precisely, if the 1-dimensional $\mathbb{C}U$-module $e(w)\mathbb{C}Ge(w)$ affords the linear character λ, then $e(w)$ equals the central primitive idempotent corresponding to λ: $e(w) = e_\lambda$, and $D = \lambda\uparrow_L G$.*

Proof. (1). This follows by Clifford's Theorem.

(2). By Clifford's Theorem, G acts transitively on the irreducible constituents of $\chi \downarrow G_{n-1}$. Observing that G_{n-1} acts trivially on $\mathrm{Irr}(G_{n-1})$, an induction on n yields our claim.

(3). This follows from $ge_{\chi_i}g^{-1} = e_{g\chi_i}$, for all $\chi_i \in \mathrm{Irr}(G_i)$.

(4). Let $M := \mathbb{C}Ge(w)$. By (2) and (3), G acts transitively on the set of lines $\{e(v)M | v \in \mathrm{ST}_{\mathcal{C}}(\chi)\}$ according to $ge(v)M = e(gv)gM = e(gv)M$. Choosing any nonzero vector $x_v \in e(v)M$ yields a basis $(x_v)_{v \in \mathrm{ST}_{\mathcal{C}}(\chi)}$ of M, and, by (2), the corresponding matrix representation is monomial. Now we choose the x_v in such a way that the non-zero entries in the representation matrices of the group elements are all eth roots of unity. To this end, let $U \leq G$ denote the stabilizer of $w \in \mathrm{ST}_{\mathcal{C}}(\chi)$ and L a left coset transversal of U in G. As for $g \in G$, $0 \neq ge(w) = e(gw)ge(w) \in e(gw)M$ and G acts transitively on $\mathrm{ST}_{\mathcal{C}}(\chi)$, the set $\{ge(w) | g \in L\}$ is a $\mathbb{C}$-basis of M. As U stabilizes the line

$e(w)M = \mathbb{C}e(w)$, there exists a linear character λ of U such that $ue(w) = \lambda(u)e(w)$, for all $u \in U$. Now let $e_\lambda = |U|^{-1}\sum_{u\in U}\lambda(u^{-1})u \in \mathbb{C}U$ denote the central primitive idempotent corresponding to λ. Then $e_\lambda e(w) = e(w) = e(w)e_\lambda$, and hence $\mathbb{C}Ge(w) \leq \mathbb{C}Ge_\lambda$. Comparing dimensions we see that even $\mathbb{C}Ge(w) = \mathbb{C}Ge_\lambda$. Thus $e_\lambda = ae(w)$, for some $a \in \mathbb{C}G$. But then, $e(w) = e_\lambda e(w) = ae(w)e(w) = ae(w) = e_\lambda$. □

Note that the last theorem not only shows that supersolvable groups are M-groups, but it also yields full information about the monomial irreducible representations.

Supersolvable groups are often described by a consistent power-commutator presentation. The following example shows such a pc-presentation of a group G_{128} of order 128. In the presentation, trivial commutator relations are omitted.

$$G_{128} = \Big\langle g_7, g_6, g_5, g_4, g_3, g_2, g_1 \;\Big|\; g_1^2 = g_2^2 = g_4^2 = g_5^2 = g_6^2 = 1, g_3^2 = g_1, g_7^2 = g_4,$$
$$[g_2, g_6] = [g_2, g_7] = [g_3, g_4] = [g_3, g_5] = [g_3, g_6] = g_1, [g_3, g_7] = g_2,$$
$$[g_4, g_5] = g_2 \cdot g_1, [g_4, g_6] = g_3 \cdot g_1, [g_5, g_7] = g_3, [g_6, g_7] = g_5 \Big\rangle .$$

Every such presentation defines a chief series $\mathcal{C}$; the (normal) subgroup G_i of G in $\mathcal{C}$ is generated by $g_1, \ldots, g_i$. In our example we obtain the chief series $\mathcal{C} := (G_7 \supset \ldots \supset G_0)$. The corresponding character graph $X(\mathcal{C})$ is shown in Fig. 3.

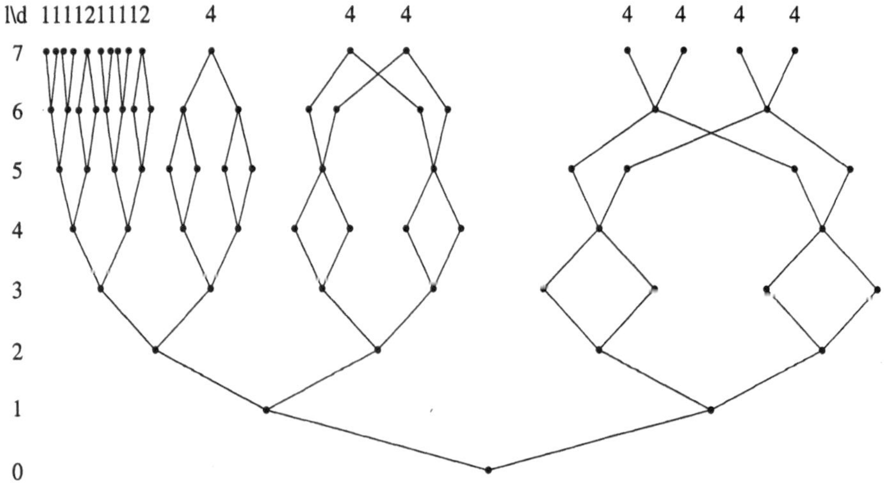

Figure3. Character graph of G_{128}.

The numbers on the left of the character graph denote the levels of the graph, whereas the numbers in the first row denote the degrees of the irreducible characters of G_{128}.

Open Problem 3. Are there Robinson-Schensted like bijections for supersolvable groups given by consistent pc-presentations?

6 Final Remarks

At least implicitly, the above framework has been the basis for designing various efficient algorithms for supersolvable groups. Roughly speaking, character graphs and generalized standard tableaux help to organize efficient computations.

In [2], we have designed a near-optimal algorithm for constructing monomial DFTs for supersolvable groups. As input, a consistent pc-presentation of a supersolvable group G is expected. This presentation defines a chief series $\mathcal{C}$ of G. The algorithm is based on Clifford theory and constructs bottom-up a $\mathcal{C}$-adapted DFT of $\mathbb{C}G$, which turns out to be e-monomial, where e denotes the exponent of G. Of great advantage is the fact that no field arithmetic is needed. One has only to work with integers modulo e. Furthermore, no group operations are required at all, for all information about G is already contained in the pc-presentation. Altogether, the running time of the algorithm is essentially proportional to the length of the output. As has been shown by Baum [1], such $\mathcal{C}$-adapted monomial DFTs are well-suited for designing fast Fourier transforms of supersolvable groups.

The fast DFT-generation algorithm has been used as a subroutine to solve other computational problems. Thümmel [15] has designed an algorithm that computes from a pc-presented p-group G in time $O(p \cdot h \cdot |G|)$ its h conjugacy classes as well as the character table of G. Omrani and Shokrollahi [12] have combined the fast DFT-generation algorithm with Galois theory to construct a full set of irreducible representations of a supersolvable group G over those finite fields K, for which the group algebra KG is non-split semisimple. In [7] we propose DFT-based collection in supersolvable groups. For a thorough treatment of these computational problems we refer to [3,5,7].

Acknowledgement. I would like to thank Meinard Müller for carefully reading the manuscript.

References

1. Baum, U.: Existence and efficient construction of fast Fourier transforms for supersolvable groups. Computational Complexity, **1** (1991), 235–256.
2. Baum, U., Clausen, M.: Computing irreducible representations of supersolvable groups. Mathematics of Computation, Volume **63**, Number 207 (1994) 351–359.
3. Bürgisser, P., Clausen, M., Shokrollahi, M.A.: Algebraic Complexity Theory. Grundlehren der mathematischen Wissenschaften, Volume **315**, Springer Verlag, Berlin, 1997.
4. Clausen, M.: Multivariate polynomials, standard tableaux, and representations of symmetric groups. J. Symbolic Computation, **11** (1991), 483–522.

5. Clausen, M., Baum, U.: Fast Fourier Transforms. BI-Wissenschaftsverlag, Mannheim, 1993.
6. Clausen, M., Baum, U.: Ein kombinatorischer Zugang zur Darstellungstheorie überauflösbarer Gruppen. Bayreuther Mathematische Schriften, **44** (1993), 99–107.
7. Clausen, M., Müller, M.: A fast program generator of fast Fourier transforms. AAECC-13, Honolulu, Nov. 1999. To appear in LNCS.
8. Désarménien, J, Kung, J.P.S., Rota, G.-C.: Invariant theory, Young bitableaux and combinatorics. Advances in Mathematics **27** (1978), 63–92.
9. Greene, C.: An extension of Schensted's theorem. Advances in Mathematics **14** (1974), 254–265.
10. Janusz, G.J.: Primitive idempotents in group algebras. Proc. Am. Math. Soc. **17** (1966), 520–523.
11. Kerber, A.: Algebraic Combinatorics via Finite Group Actions. BI-Wissenschaftsverlag, 1991.
12. Omrani, A., Shokrollahi, M.A.: Computing irreducible representations of supersolvable groups over small finite fields. Mathematics of Computation, Volume **66**, Number 218
13. Serre, J.P.: Linear Representations of Finite Groups. Graduate Texts in Mathematics, Springer, 1986.
14. Thomas, G.P.: On Schensted's construction and the multiplication of Schur functions. Advances in Mathematics **30** (1978), 8–32.
15. Thümmel, A.: Computing character tables of p-groups. pp. 150–154 in Proceedings ISSAC'96, Zürich, Switzerland.

Finite Flag-transitive Linear Spaces with Alternating Socle

Anne Delandtsheer

Université Libre de Bruxelles
Faculté des Sciences Appliquées, CP 165/11
Avenue F.D. Roosevelt, 50
B-1050 Bruxelles
Belgium
Email : adelandt@ulb.ac.be

Abstract This paper is a contribution to the classification of all pairs (S, G) where S is a finite non-trivial linear space and G is a flag-transitive automorphism group of S. We prove that if $Alt(n) \trianglelefteq G \leq Aut\ Alt(n)$ with $n > 4$, then $S = PG(3,2)$ and $G \cong Alt(7)$ or $G \cong Alt(8) \cong PSL\ (4,2)$.
Keywords : linear space, 2-design, flag-transitivity.

1 Introduction

A **linear space** S is an incidence structure of points and lines such that any two points are incident with exactly one line, any point is incident with at least two lines and any line with at least two points. S is said to be **trivial** if every line is incident with exactly two points. S is **finite** if it has a finite number v of points.

A **flag** of S is a pair (x, L) where x is a point and L is a line incident with x. If S is non-trivial and admits an automorphism group acting transitively on its flags, then all lines are incident with the same number k of points, so that S is a 2-$(v, k, 1)$ design (with $2 < k < v$). In the late 1980's, F.Buekenhout, A. Delandtsheer, J.Doyen, P.D.Kleidman, M.W.Liebeck et J.Saxl combined their efforts in order to classify all pairs (S, G) where S is a finite non-trivial linear space and $G \leq$ Aut S is a flag-transitive automorphism group of S. This classification was announced in [4], with the project of the six authors to publish its proof in book form. It appears now that this proof will be published piecemeal. The present paper, as well as [3] , [6], [8] and [10], is one of the pieces.

The basic result, proved in [3], is the following

Reduction theorem. *If G is a flag-transitive group of automorphisms of a finite non-trivial linear space S, then one of the following occurs:*

(a) *G is almost simple, i.e. G has a non-abelian simple normal subgroup N such that* $N \trianglelefteq G \leq Aut\ N$;

(b) *G is of affine type, i.e. the set of points of S carries the structure of an affine space* $AG(n,p)$ *(p prime) invariant under G, and G contains the translation group T of* $AG(n,p)$.

Case *(a)* splits up into four subcases :

(a.1) *N is an alternating group Alt* (n) *with* $n > 4$,

(a.2) *N is a sporadic simple group,*

(a.3) *N is a Lie-Chevalley finite simple group,*

(a.4) *N is a finite simple group of exceptional type.*

The purpose of the present paper is to handle case (a.1). Our result, already presented at the First International Conference on Finite Geometry and Combinatorics in Deinze in 1986, is the following :

Theorem. *Let S be a finite non-trivial linear space having an automorphism group G which acts flag-transitively on S.*
If $Alt(n) \trianglelefteq G \leq Aut\ Alt(n)$ *with* $n > 4$, *then* $S = PG(3,2)$ *and* $G \cong Alt(7)$ *or* $G \cong Alt\ (8) \cong PSL\ (4,2)$.

Note that in any trivial linear space, flag-transitivity is equivalent to 2-transitivity on points, so that the problem reduces to the well-known classification of finite 2-transitive permutation groups.

2. Useful lemmas

For convenience, we list here a few lemmas that will be needed in the proof of our theorem.

Lemma 1. *If b is the total number of lines of a* $2-(v,k,1)$ *design with* $2 < k < v$ *and if r is the number of lines through a point, then*

(i) $r = (v-1)/(k-1) \geq k$

(ii) $b = v(v-1)/k(k-1) \geq v$

(iii) $v \geq k^2 - k + 1$

(iv) $r > \sqrt{v}$

Proof. (i),(ii) and (iii) are well-known arithmetical properties of 2-$(v,k,1)$ designs.
Using $r \geq k$ twice, we get

$$k^2 - k + 1 \leq r(k-1) + 1 = v < r^2,$$

which proves (iv).

From now on, we shall always assume that S is a $2-(v,k,1)$ design as before and that G is a flag-transitive automorphism group of S. The following result is fundamental :

Lemma 2. (Higman and McLaughlin [7]). *G acts primitively on the points of S , i.e. the stabilizer G_x of any point x of S is a maximal subgroup of G of index v.*

Lemma 3. (Buekenhout, Delandtsheer and Doyen [3]).

(i) *r divides $(|G_x|, v-1)$,*

(ii) *If Δ is a union of orbits of G_x on the set of points of S (with $x \notin \Delta$), then every line through x intersects Δ in a constant number of points k' and $rk' = |\Delta|$.*

If m is an integer and p a prime number, we will denote by $\mathrm{cont}(p,m)$ the largest integer n such that p^n divides m.

Lemma 4. ([3]) *If p is a prime divisor of r, then $\mathrm{cont}(p,|G_x|) = \mathrm{cont}(p,|G|)$ for any point x of S.*

3. Proof of the theorem

By hypothesis S is a finite non-trivial linear space with line size $k > 2$, G is an automorphism group of S which acts flag-transitively on S and $Alt(n) \trianglelefteq G \leq Aut\ Alt(n)$ with $n > 4$. In other words, G is a symmetric or alternating group, except if $n = 6$ and $G \cong M_{10}$, $PGL(2,9)$ or $P\Gamma L(2,9)$. These three exceptions will be handled in part 1 of the proof, while part 2 will handle the general case.

PART 1.

The group G is $M_{10}, PGL(2,9)$ or $P\Gamma L(2,9)$. The first two of these groups have order 720 and, using the Atlas [5], it is easy to check that their maximal subgroups have indices 2,10,36 or 45 ; $P\Gamma L(2,9)$ has order 1440 and maximal subgroups of indices 2, 36 or 45.

From the assumptions on S and from Lemma 2, we conclude that the number v of points of S must be 2,10,36 or 45. Using Lemma 1, it is easy to show that v cannot be 2 nor 10, and that if $v = 36$, then $k = 6$. But a linear space on 36 points and line size 6 would be an affine plane of order 6, whose nonexistence is well known [2]. Hence $v = 45$. But v is the index of G_x in G, so that G_x must have order 16 or 32. Hence $(|G_x|, v-1) = 4$, which implies $r \leq 4$ by Lemma 3 (i). This contradicts Lemma 1 (iv) and ends Part 1 of the proof.

PART 2.

In what follows, we fix a point x of S. The stabilizer G_x is a maximal subgroup of $G \cong Alt(n)$ or $Sym(n)$, and G_x acts both on the point-set of S and on the set $\Omega_n = \{1, 2, \cdots, n\}$. We distinguish three cases, according as the action of G_x on Ω_n is

(1) not transitive,
(2) transitive but not primitive, or
(3) primitive.

Case 1: G_x is not transitive on Ω_n.

Since G_x acting on Ω_n is not transitive, it has at least two orbits. Since G acts primitively on the point-set of S, the subgroup G_x is maximal in G, so that in its action on Ω_n, G_x is necessarily the full stabilizer in $G = Alt(n)$ or $Sym(n)$ of a subset X of Ω_n, of size $s < n/2$. Indeed, the case where G_x has two orbits of equal size $n/2$ on Ω_n is ruled out by the maximality of G_x in G, because the stabilizer in G of this partition of Ω_n would be between G_x and G.

The subgroup G_x fixes exactly one point in S (by the flag-transitivity of G) and stabilizes exactly one s-subset in Ω_n, so that we may identify the point x of S with the unique s-subset of Ω_n stabilized by G_x. Conversely each s-subset of Ω_n corresponds in this way to a point of S, because G acts transitively on all the s-subsets of Ω_n. Since we will repeatedly switch from points in S to the corresponding s-subsets in Ω_n, any point of S will be denoted by a lower case letter while the corresponding s-subset in Ω_n will be denoted by the corresponding capital letter. Note that $s \geq 2$. Indeed $s = 1$ would imply $v = n$, and so, since $n \geq 5$, G would act 3-transitively on the point-set of S, forcing the line size to be 2, which contradicts the non-triviality of S.

Two points y, y' of S are in the same orbit under G_x if and only if the corresponding s-subsets Y, Y' of Ω_n intersect X in the same number of points. Thus G acting on S has rank $s + 1$, each G_x-orbit O_i on the point-set of S corresponding to a possible size $i \in \{0, 1, \cdots, s\}$ of the intersection of X with an s-subset of Ω_n.

Fix a line L through x. Since G is flag-transitive on S, L must intersect every orbit O_i by Lemma 3 (ii). Let $y_i \in O_i \cap L$ for some $i < s$. The subgroup G_{xy_i} leaves the flag (x, L) invariant and, in its action on Ω_n, stabilizes the subsets X, Y_i and their complements X^c and Y_i^c. More precisely, G_{xy_i} is the full stabilizer in $Alt(n)$ or $Sym(n)$ of each of the four subsets $X \cap Y_i, X \cap Y_i^c, X^c \cap Y_i$ and $X^c \cap Y_i^c$. Hence G_{xy_i} acts on $X^c \cap Y_i$ as $Sym(s - i)$ and acts on $X^c \cap Y_i^c$ at least as $Alt(n - 2s + i)$.

Consider a subset Y_{s-1} of Ω_n intersecting X in $s - 1$ points. The group $G_{XY_{s-1}}$ has two orbits on the s-subsets of X^c, namely those which contain the unique element e of $Y_{s-1} \cap X^c$ and those which do not contain e. Consider now a subset Y_{s-2} of Ω_n intersecting X in $s - 2$ elements and such that the corresponding point y_{s-2} is collinear with x and y_{s-1} in S. We deduce that $G_{y_{s-2}}$ can map an s-subset of X^c which contains e onto an s-subset of X^c which does not contain e. It follows that the group $G_{XY_{s-2}}$ does not fix e. Hence, if L denotes the line passing through x, y_{s-1} and y_{s-2}, the group G_{xL}

(which contains $G_{xy_{s-1}}$ and $G_{xy_{s-2}}$) must act transitively on the s-subsets of X^c, and so also on O_0. Since all the lines of S through x have a point in O_0, we conclude that x is incident with only one line, contradicting the fact that S has more than one line.

Case 2 : G_x acts transitively but not primitively on Ω_n.

Remember first that, by the flag-transitivity of G on S, the stabilizer G_x of any point x of S fixes no other point than x in S. Hence there is a canonical bijection α from the set of points of S onto the set of stabilizers of points of S, which can in turn be identified with the left cosets of G_x in G.

The hypothesis of case 2 implies that G_x acting on Ω_n stabilizes some partition P of Ω_n into t classes of size s, with $t \geq 2$, $s \geq 2$ and $s\,t = n$. For short, any partition of Ω_n into t classes of size s will be called a $t \times s$ partition. Because of the maximality of G_x as a subgroup of G, G_x acting on Ω_n is the full stabilizer of P in G. Since $G \cong Alt(n)$ or $Sym(n)$, G_x therefore contains all the even permutations of Ω_n preserving the partition P. We will now prove that there is no other nontrivial partition of Ω_n whose full stabilizer in G is G_x.

Suppose to the contrary that G_x preserves both a $t_1 \times s_1$ partition P_1 and a $t_2 \times s_2$ partition P_2 of Ω_n with $t_1, t_2, s_1, s_2 \geq 2$ and $s_1 t_1 = s_2 t_2 = n$. We will denote by C^i_a the class of a in the partition P_i. Suppose that there is an element $b \in C^1_a \cap C^2_a$, with $b \neq a$. If C^2_a is not contained in C^1_a, there is an element $c \in C^2_a \backslash C^1_a$. The tricycle (a, b, c) (i.e. the even permutation of Ω_n having (a, b, c) as a cycle and fixing all the other elements of Ω_n) belongs to G and preserves P_2 but not P_1 , a contradiction. Hence if $b \in C^1_a \cap C^2_a$ with $b \neq a$, then $C^2_a \subseteq C^1_a$. Similarly $C^1_a \subseteq C^2_a$. We conclude that if $|C^1_a \cap C^2_a| \geq 2$, then $C^1_a = C^2_a$. Suppose now that $C^1_a \cap C^2_a = \{a\}$ and that $|C^1_a| =: s_1 \geq 3$. Let $b \in C^2_a \backslash C^1_a$ and $c, d \in C^1_b$. Then the tricycle (b, c, d) preserves C^1_b and P_1, but not C^2_b since $C^2_b \neq C^1_b$, so that $C^2_b \cap C^1_b = \{b\}$. This element of G_x should preserve P_2. Since it fixes a, it stabilizes $C^2_a = C^2_b$, but it also maps b onto c, with $c \notin C^2_b$ (otherwise $c \in C^1_b \cap C^2_b$ would force $C^1_b = C^2_b$, contradicting the fact that $a \in C^2_b \backslash C^1_b$). We conclude that if $C^1_a \cap C^2_a = \{a\}$, then $|C^1_a| = |C^2_a| = 2$, and so both P_1 and P_2 are $t \times 2$ partitions, where $t = n/2$.

If $t \geq 3$, let $b \in C^2_a \backslash C^1_a$ and let $c \notin C^1_a \cup C^2_a$. Then the permutation having a cycle of length 2 in C^1_b and C^1_c and fixing all the points of $\Omega_n \backslash (C^1_b \cup C^1_c)$ is even and preserves P_1. Since it fixes a, it must stabilize $C^2_a = C^2_b = \{a, b\}$, a contradiction because b is not fixed. Hence $s = t = 2$, and so $n = 4$, contradicting the hypothesis of our theorem.

Therefore, for any point x in S, its stabilizer G_x preserves exactly one $t \times s$ partition of Ω_n. This defines an injection β from $\{G_x; x \in S\} = \{gG_x; g \in G\}$

into the set $\mathcal{P}$ of all $t \times s$ partitions of Ω_n. This map β is onto, because $G \geq Alt(n)$ acts transitively on $\mathcal{P}$. Hence we get a canonical bijection $\alpha\beta$ from the point-set of S onto $\mathcal{P}$, allowing us to identify the points of S with the $t \times s$ partitions of Ω_n. For the sake of shortness we shall use the same symbol to denote a point of S and the corresponding $t \times s$ partition of Ω_n. We shall also use freely such expressions as "the $t \times s$ partitions belonging to a line of S".

The general idea of the argument in case 2 is the following. Fix a point x of S and consider the orbits under G_x of the $t \times s$ partitions other than x. By the flag-transitivity assumption, every line through x must intersect each of these orbits. But if two partitions x and y differ "very little "in Ω_n, their common stabilizer G_{xy} will be "almost as big "as G_x, while the stabilizer G_{xy} of two points in the linear space S must be "significantly smaller"than G_x, because G_{xy} stabilizes a line, whereas G_x acts transitively on the lines through x.

By a j-**cycle** we mean a j-tuple up to a cyclic reordering of its j components. Two points x and y of S, i.e. two $t \times s$ partitions of Ω_n, will be called j-**cyclic** with respect to each other if there is a suitable labeling $C_0, C_1, \cdots, C_{t-1}$ of the t classes of x such that y has exactly $t-j$ classes $C_j, \cdots, C_{t-1}$ in common with x and each class $C_i (i = 0, \cdots, j-1)$ contains an element c_i such that the j classes of y which differ from those of x are the subsets $(C_i - \{c_i\}) \cup \{c_{i+1}\}$, where the indices are computed modulo j. In what follows, we fix a point x and we define the **cycle** of y (with respect to x) to be the cycle $(C_0, C_1, \cdots, C_{j-1})$. If $s \geq 3$, the cycle $(c_0, c_1, \cdots, c_{j-1})$ determines y uniquely from x and is uniquely determined by y ; the elements c_i will be called the **cyclic elements** of y. If $s = 2$, there are two such "opposite"cycles $(c_0, c_1, \cdots, c_{j-1})$ and $(c'_{j-1}, c'_{j-2}, \cdots, c'_0)$ where $C_i = \{c_i, c'_i\}$, and this pair of cycles of elements determines y uniquely from x and is uniquely determined by y.
We will decompose our proof into seven steps.

Step 1. *Let x and y_j be two $j-$cyclic partitions with cycle $(C_0, C_1, \cdots, C_{j-1})$, admitting $(c_0, c_1, \cdots, c_{j-1})$ as a cycle of elements (with $c_i \in C_i$). There is a permutation $g \in G$ preserving x and y, fixing $C_j \cup C_{j+1} \cup \cdots \cup C_t$ elementwise and having $(c_0, c_1, \cdots, c_{j-1})$ as a cycle.*

Proof .

Let us write $C_i = \{c_i, c_i^1, c_i^2, \cdots, c_i^{s-1}\}$ for every $i = 0, \cdots, j-1$. If j is odd, we choose a permutation g consisting of the s cycles $(c_0, c_1, \cdots, c_{j-1})$ and $(c_0^l, \cdots, c_{j-1}^l)$ for $l = 1, \cdots, s-1$ of length j, so that g is an even permutation of Ω_n, and so $g \in G$. By construction, g preserves x and y_j and is as announced. If j is even, we choose a permutation g having two nontrivial cycles : the cycle $(c_0, \cdots, c_{j-1})$ of length j and the cycle

$(c_0^1, c_1^1, \cdots, c_{j-1}^1, c_0^2, c_1^2, \cdots, c_{j-1}^2, \cdots, c_0^{s-1}, c_1^{s-1}, \cdots, c_{j-1}^{s-1})$ of length $(s-1)j$. Since g has exactly two cycles of even length, g is an even permutation of Ω_n, and so $g \in G$. By construction, g satisfies all our requirements.

Step 2. *For any line L of S through x, G_{xL} acts on the t classes of x as $Sym(t)$.*

Proof. G_x being transitive on the lines through x, the line L must contain at least one point in each G_x-orbit of points. Hence, for every $j = 2, \cdots, t$, the line L contains at least one j-cyclic partition y_j with respect to x. It follows from Step 1 that there is an element of $G_{xy_j} \leq G_{xL}$ which stabilizes $t-j$ classes of x and permutes cyclically the other j classes. We see by taking $j = t$ that G_{xL} acting on the t classes of x contains a t-cycle, and so is transitive on this set of t classes. For $j = t-1$, G_{xL} contains an element preserving one class of x and permuting cyclically the other $t-1$ classes. Hence G_{xL} is 2-transitive on this set of t classes. Repeating this argument for every j going from t to 2, we conclude that G_{xL} acts t-transitively (i.e. as $Sym\,(t)$) on the t classes of x.

Step 3. *For any line L of S through x and for any two classes C_0, C_1 of the $t \times s$ partition x, the number of $t \times s$ partitions belonging to L, which are 2-cyclic with respect to x and whose cycle is precisely (C_0, C_1), is a constant $m_2 \geq 1$ independent of L and of (C_0, C_1). Moreover, m_2 divides s^2.*

Proof. Let y be a 2-cyclic $t \times s$ partition with respect to x, with cycle (C_0, C_1), and let L be the line through x and y. For this choice of L and of the cycle (C_0, C_1), the number m_2 in the above statement is not zero. We know by Step 2 that G_{xL} acts 2-transitively on the t classes of x. Hence, for any given line L through x, m_2 is independent of the cycle (C_0, C_1). On the other hand, G_x acts transitively on the lines through x, and so m_2 is also independent of L.

Now, if we count in two ways the number of $t \times s$ partitions which are 2-cyclic with respect to x and whose cycle is precisely (C_0, C_1), we get

$$rm_2 = \begin{cases} s^2 & \text{if } s \geq 3 \\ s^2/2 = 2 & \text{if } s = 2. \end{cases}$$

This proves Step 3.

Step 4. $s \geq 3$.

Proof. Suppose that $s = 2$. Consider two classes $C_0 = \{c_0, c_0'\}$ and $C_1 = \{c_1, c_1'\}$ of x. Since there are only two other ways of partitioning the set $\{c_0, c_0', c_1, c_1'\}$ into two classes of size 2, there are only two cyclic partitions y and z with cycle (C_0, C_1) with respect to x. By the flag-transitivity, any line through x must contain at least one of them, and so there are at most two lines through x. It follows that $v \leq 3$, contradicting the fact that S is non-trivial. Hence $s \geq 3$ and the 2-cyclic partitions are uniquely determined by their pairs of cyclic elements.

Step 5. $m_2 = 1$, $r = s^2$ and $t \leq 3$.

Proof. Suppose that $m_2 \geq 2$. Then there are two distinct $t \times s$ partitions y and z which are 2-cyclic with respect to x with cycle (C_0, C_1) and which belong to L. The orbit under G_{xz} of the pair of cyclic elements (y_0, y_1) of y contains at least $(s-1)^2$ such pairs of elements. Each of these $(s-1)^2$ pairs of elements defines a 2−cyclic partition with respect to x with cycle (C_0, C_1). Since $s \geq 3$, this yields $(s-1)^2$ distinct 2-cyclic partitions with cycle (C_0, C_1) belonging to the orbit of y under G_{xz}, hence belonging to the line L (because y is a point of the line L containing the two fixed points x and z). Therefore, by Step 3,

$$(s-1)^2 \leq m_2 \quad | \quad s^2.$$

Since $s \geq 3$ (by Step 4) , it follows that $m_2 = s^2$, so that L is the unique line of S through x, a contradiction. Hence $m_2 = 1$, and so the number r of lines through a point is s^2.

If $t \geq 4$, then L contains a 3-cyclic $t \times s$ partition y with cycle (C_0, C_1, C_2) and a 3-cyclic $t \times s$ partition z with cycle (C_1, C_2, C_3) (with respect to x). An argument similar to the one above shows that the number m_3 of 3-cyclic $t \times s$ partitions on L with cycle (C_0, C_1, C_2) satisfies

$$s(s-1)^2 \leq m_3 \quad | \quad s^3,$$

so that $m_3 = s^3$. From $rm_3 = s^3$, we deduce that $r = 1$, a contradiction.

Step 6. $t \neq 3$.

Proof. Suppose that $t = 3$. By Step 3 , L contains a 2-cyclic partition y with cyclic elements $y_0 \in C_0$ and $y_1 \in C_1$ and a 2-cyclic partition z with cyclic elements $z_1 \in C_1$ and $z_2 \in C_2$. Since $Alt(n)$ contains a permutation of the elements of Ω_n which stabilizes the partitions x and z but maps y_0 onto any other element of C_0 and y_1 onto any element of C_1 distinct from z_1, we get an element of $G_{xz} \leq G_{xL}$ mapping $y \in L$ onto at least $s(s-1)$ 2-cyclic partitions with cycle (C_0, C_1). Hence $m_2 \geq 6$, contradicting Step 5.

Step 7. *The final contradiction for Case 2.*

Proof. We know by Steps 5 and 6 that $t = 2$, so that $\Omega_n = C_0 \cup C_1$. If $s = 3$, then all $2 \times s$ partitions other than x in L are 2-cyclic with cycle (C_0, C_1). Since the number of points on L is at least 3, there are at least two such partitions belonging to L, so that $m_2 \geq 2$, contradicting Step 5.

We shall call **partition of the second type** with respect to x any partition having one class intersecting C_0 in $s-2$ elements (and C_1 in 2 elements). There are $\binom{s}{2}^2 = s^2(s-1)^2/4$ of them if $s \geq 5$, and $s^2(s-1)^2/8$ if $s = 4$. Since $r = s^2$ by Step 5, it follows from the flag-transitivity of G that each line through x contains precisely $(s-1)^2/4$ of them if $s \geq 5$ and $(s-1)^2/8$ if $s = 4$. But $9/8$ is not an integer, and so $s \geq 5$.

Hence, by the flag-transitivity of G on S, each line through x contains a 2-cyclic partition y and $(s-1)^2/4$ partitions of the second type with respect to x. Therefore the stabilizer G_{xy} of x and y, which stabilizes the line through x and y, must have an orbit of length $\leq (s-1)^2/4$ on the partitions of the second type. However, the $\binom{s}{2}^2$ partitions of the second type with respect to x split into three orbits under G_{xy}, whose respective lengths are $(s-1)^2, (s-1)\binom{s-1}{2}$ and $\binom{s-1}{2}^2$, a contradiction.

Case 3: G_x is primitive on Ω_n.

Note first that if G is $Sym(n)$, then G_x cannot be $Alt(n)$ because $[G : G_x] = v > 2$. We will reduce the problem to a few small values of n.

If the number r of lines through a point is even, then, by Lemma 4, the group G_x contains a Sylow 2-subgroup of $G \cong Alt(n)$ or $Sym(n)$ with $n \geq 5$. Hence G_x contains a subgroup acting transitively on 4 elements of Ω_n and fixing all the other elements, so that $n \leq 8$ by a theorem of Marggraf (see [12, Theorem 13.5]).

If r is odd, let p be any prime divisor of r, so that p divides the order of G_x because G is flag-transitive. By Lemma 4, G_x contains a Sylow p-subgroup of G, and so G_x acting on Ω_n contains an even permutation having exactly one cycle of length p and $n-p$ fixed points. By a result of Jordan [9] (see [12, Theorem 13.9]), the primitivity of G_x on Ω_n implies that $n - p \leq 2$. Since $n-2 \leq p \leq n$, p^2 does not divide $|G|$, and so p^2 does not divide r. It follows that r is either a prime, namely $n-2, n-1$ or n, or the product of two twin primes, namely $(n-2)n$. On the other hand, $r^2 > v$ by Lemma 1, and $v \geq \lfloor (n+1)/2 \rfloor !/2$ by a result of Bochert [1] (see [12, Theorem 14.2]). From the inequality

$$\lfloor (n+1)/2 \rfloor !/2 < r^2$$

and the above conditions on r, we readily deduce that $n = 5, 6, 7, 8$ or 13. Finally, using Sims' table [11] of primitive groups of degree ≤ 20, together with the arithmetical conditions $n \geq 5$, $[G : G_x] = v < r^2$, $k \leq r | (v-1, |G_x|)$, $(v-1)/r = k - 1 \geq 2$, it is straightforward to check that the only possibilities are $n = 7$ or 8, $v = 15, r = 7, k = 3$, corresponding to the groups $G \cong Alt(7)$ or $Alt(8) \cong PSL(4,2)$ in their usual 2-transitive action on the points of S=PG(3,2).

REFERENCES

1. A. Bochert, *Uber die Zahl verschiedener Werte, die eine Funktion gegebener Buchstaben durch Vertauschung derselben erlangen kann*, Math. Ann. **33** (1889), 584-590.
2. R.H. Bruck and H.J. Ryser, *The nonexistence of finite projective planes*, Canad. J. Math. **1** (1949), 88-93.
3. F. Buekenhout, A. Delandtsheer and J. Doyen, *Finite linear spaces with flag-transitive groups*, J. Combin. Theory Ser. A **49** (1988), 268-293.
4. F.Buekenhout, A. Delandtsheer, J. Doyen, P.D. Kleidman, M.W.Liebeck et J.Saxl, *Linear spaces with flag-transitive automorphism groups* , Geom. Dedicata 36 (1990), 89-94.
5. J.H. Conway, R.T. Curtis, S.P. Norton, R.A. Parker and R.A. Wilson, *Atlas of finite groups*, Clarendon Press, Oxford, 1985.
6. A. Delandtsheer, *Flag-transitive finite simple groups*, Arch. Math. 47 (1986), 195-400.
7. D.G. Higman and J.E. McLaughlin, *Geometric ABA groups*, Illinois J. Math. 6, 382-397.
8. P.B. Kleidman, *The finite flag-transitive linear spaces with an exceptional automorphism group*, to appear
9. C. Jordan, *Sur la limite de transitivité des groupes non alternés*, Bull. Soc. Math. France **1** (1873), 40-71.
10. M.W. Liebeck, *The classification of finite linear spaces with flag-transitive automorphism groups of affine type*, J. Combin. Theory Ser. A **84** (1998), 196-235.
11. C.C. Sims, *Computational methods in the study of permutation groups*, Computational problems in abstract algebra, J. Leech (editor), Pergamon Press, Oxford, 1970, pp. 169-183.
12. H. Wielandt, *Finite permutation groups*, Academic Press, New York, 1964.

A Characterization of Certain Binary Recurrence Sequences

Andreas Dress[1] and Florian Luca[2]

[1] Mathematics Department
Bielefeld University
Postfach 10 01 31
33 501 Bielefeld
Germany
dress@mathematik.uni-bielefeld.de
[2] Mathematics Department
Bielefeld University
Postfach 10 01 31
33 501 Bielefeld
Germany
fluca@mathematik.uni-bielefeld.de

Abstract In this note, we show that if $(A_n)_{n\geq 0}$ is a sequence of integers such that $(|A_n|)_{n\geq 0}$ diverges to infinity and

$$\limsup_{n\to\infty} \frac{|A_n^2 - A_{n+1}A_{n-1}|}{\sqrt{|A_n|}} < \frac{1}{\sqrt{2}}$$

holds, then $(A_n)_{n\geq 0}$ is binary recurrent from some n on. An application regarding quadratic Pisot numbers is also presented.

AMS *Mathematics Subject Classification*: 11P21, 11K06

1 Introduction

Let $(A_n)_{n\geq 0}$ be a sequence of integers satisfying the following two assumptions:

(i) $\lim_{n\to\infty} |A_n| = \infty$.

(ii)

$$\limsup_{n\to\infty} \frac{|A_n^2 - A_{n+1}A_{n-1}|}{\sqrt{|A_n|}} < \frac{1}{\sqrt{2}}. \tag{1}$$

It seems natural to ask whether one can characterize all sequences of integers $(A_n)_{n\geq 0}$ satisfying (i) and (ii) above.

We begin by pointing out some families of sequences of integers $(A_n)_{n\geq 0}$ that satisfy Conditions (i) and (ii).

Assume that r and s are two integers and that

$$A_{n+2} = rA_{n+1} + sA_n \tag{2}$$

holds for all $n \geq 0$. If $s = 0$, then $(A_n)_{n\geq 1}$ is just a geometrical progression of ratio r. In this case, the quantity $|A_n^2 - A_{n+1}A_{n-1}|$ vanishes for all $n \geq 2$, and condition (i) is always satisfied provided that $A_1 \neq 0$ and $|r| > 1$.

Assume now that $(r, s) = (2, -1), (-2, -1)$. In this case, there exist two constants a and b such that

$$A_n = a + nb \qquad \text{for all } n \geq 0,$$

or

$$A_n = (-1)^n(a + nb) \qquad \text{for all } n \geq 0$$

according to whether $r = 2$ or $r = -2$. One can now check that in this case

$$|A_n^2 - A_{n+1}A_{n-1}| = b^2$$

holds for all $n \geq 1$, hence (ii) is fullfiled. Condition (i) is fulfilled if and only if $b \neq 0$.

Assume now that r and s satisfy $r^2 + 4s > 0$ and $rs \neq 0$. Let α and β be the two roots of the equation

$$x^2 - rx - s = 0.$$

Notice that $|\alpha|$ and $|\beta|$ are distinct. Assume that $|\alpha| > |\beta|$. Clearly, $|\alpha| > 1$. It is well known then that there exist two constants a and b such that

$$A_n = a\alpha^n + b\beta^n \qquad \text{for all } n \geq 0. \tag{3}$$

We may assume $ab \neq 0$: Otherwise, Formula (3) and the fact that A_n is an integer for all n, implies that A_n is a geometrical progression of integer ratio which is a case already treated. Using Formula (3), one can show that

$$|A_n^2 - A_{n+1}A_{n-1}| = |s|^{n-1}|ab(\alpha - \beta)^2|$$

must hold. Consequently, Condition (ii) implies $|s| \leq |\alpha|^{1/2}$, or $|\beta| \leq |\alpha|^{-1/2}$.

Notice that, in fact, the above inequalities are strict and are, therefore, also equivalent with Condition (ii): Indeed, if $|s| = |\alpha|^{1/2}$, it follows that $|\alpha| = s^2$ is an integer. Hence, $\beta = r - \alpha$ is an integer too. Since $|\alpha| > 1$ and $|\beta| = |\alpha|^{-1/2} < 1$, it follows that $\beta = 0$, therefore $s = 0$, which is impossible.

More precisely, a straight-forward computation shows that $|\beta| < |\alpha|^{-1/2}$ holds if and only if r and s satisfy – in addition to the conditions $rs \neq 0$ and $r^2 + 4s > 0$ – the inequality

$$s^2 - \frac{1}{s} < |r|.$$

In the above discussion, we have assumed that $(A_n)_{n\geq 0}$ is binary recurrent, and we found necessary and sufficient conditions for (i) and (ii) to hold.

Our main result is that every sequence of integers satisfying (i) and (ii) must be one of the above, at least from some n on. That is, we have the following:

Theorem 1.1. *If* $(A_n)_{n\geq 0}$ *satisfies* (i) *and* (ii), *then there exist* $n_0 \in \mathbf{N}$ *and two integers* r *and* s *such that*

$$A_{n+2} = rA_{n+1} + sA_n \qquad \text{for all } n \geq n_0,$$

and either one of the following holds:

1) $s = 0$;
2) $(r,\ s) = (2,\ -1),\ (-2,\ -1)$;
3) $rs \neq 0,\ r^2 + 4s > 0$, *and* $s^2 - 1/s < |r|$.

We also present the following:

Corollary 1.2. *For every real number* x, *let* $||x|| = \min(|x - n| : n \in \mathbf{Z})$ *be the distance from* x *to the nearest integer. Assume that* α *is a real number with* $|\alpha| > 1$ *such that there exists* $n_0 \in \mathbf{N}$ *with*

$$||\alpha^n|| < \frac{1}{4\sqrt{2}|\alpha|^{3/2}} \cdot |\alpha|^{-n/2} \qquad \text{for all } n \geq n_0. \tag{4}$$

Then either one of the following holds:

1. $\alpha \in \mathbf{Z}$.

2. α *is a real irrational quadratic number and if one denotes its conjugate by* β, *then* $|\beta| < |\alpha|^{-1/2}$.

Conversely, if α *is a real number with* $|\alpha| > 1$ *which satisfies either Condition* 1 *or Condition* 2 *above, then there exists some* $n_0 \in \mathbf{N}$ *such that inequality* (4) *holds.*

The above condition $|\alpha| > 1$ is meant to rule out trivialities (notice that if $0 < |\alpha| \leq 1$, then formula (4) always holds for some $n_0 \in \mathbf{N}$).

In the above corollary, the constant

$$\frac{1}{4\sqrt{2}|\alpha|^{3/2}}$$

can be replaced by

$$\frac{1}{\sqrt{2}(2 + |\alpha|^{3/2} + |\alpha|^{-3/2})} - \epsilon$$

for any arbitrary $\epsilon > 0$.

The referee pointed out that M. Mignotte (see [2]) showed that if $\alpha > 1$ is a real number such that $||\alpha^n||$ is small enough for all integer values of $n \geq 0$, then α is an algebraic number. In particular, Corollaries 1 or 2 from [2] imply that if α satisfies inequalities (4) for all positive integers $n \geq 0$, then α is an algebraic number. However, it is unclear to us if the results of [2] imply our Corollary.

Remark 1.3. One may ask if our theorem remains true when one replaces Condition (ii) with the weaker requirement that the lim sup appearing at (ii) is finite. Unfortunately, this is not so. To see why, let $P \in \mathbf{Z}[X]$ be an irreducible polynomial of the form

$$P(X) := X^3 + a_1x^2 + a_2X + a_3.$$

Let α, β and γ be the roots of P. Assume that P satisfies the following conditions

1. $|a_3| = 1$;
2. $|\alpha| > 1$;
3. $|\beta| = |\gamma| < 1$.

An example of such a polynomial is $X^3 - 3X^2 - 2X - 1$.

Define $A_n := \alpha^n + \beta^n + \gamma^n$. It is well known that A_n is an integer for all $n \in \mathbf{N}$. Moreover, it is clear that there is no $n_0 \in \mathbf{N}$ such that $(A_n)_{n \geq n_0}$ is binary recurrent. Notice, however, that $(A_n)_{n \geq 0}$ is ternary recurrent.

A straight-forward computation shows that

$$A_n^2 - A_{n-1}A_{n+1} = (\alpha\beta)^{n-1}(\alpha-\beta)^2 + (\alpha\gamma)^{n-1}(\alpha-\gamma)^2 + (\beta\gamma)^{n-1}(\beta-\gamma)^2.$$

From Conditions 1-3 above, it follows that β and γ are complex conjugates, and

$$\rho := |\beta| = |\gamma| = \frac{1}{\sqrt{|\alpha|}}.$$

Let

$$\zeta = \frac{\beta}{\rho}.$$

With these notations, one has

$$|A_n^2 \quad A_{n-1}A_{n+1}| = \\ |\alpha|^{n/2}\left|\zeta^{n-1}\frac{(\alpha-\beta)^2}{|\alpha|\beta} + \overline{\zeta}^{n-1}\frac{(\alpha-\gamma)^2}{|\alpha|\gamma} + \left(\frac{\beta\gamma}{\sqrt{|\alpha|}}\right)^{n-1}\frac{(\beta-\gamma)^2}{\sqrt{|\alpha|}}\right|.$$

It is now clear that

$$\limsup_{n\to\infty} \frac{|A_n^2 - A_{n-1}A_{n+1}|}{\sqrt{|A_n|}} = \limsup_{n\to\infty} \left|\zeta^{n-1}\frac{(\alpha-\beta)^2}{|\alpha|\beta} + \overline{\zeta}^{n-1}\frac{(\alpha-\gamma)^2}{|\alpha|\gamma}\right| < \infty.$$

We mention that in our previous work [1] we characterized all sequences of integers $(A_n)_{n\geq 0}$ with $\lim_{n\to\infty}|A_n| = \infty$ for which $|A_n^2 - A_{n+1}A_{n-1}|$ is bounded. The results of [1] are clearly an immediate consequence of the present work.

2 Proofs

Proof of the theorem.

Condition (i), implies that for every positive constant C_1 there exists some $n_0 \in \mathbf{N}$ with $|A_n| \geq C_1$ for $n \geq n_0$. In what follows, we shall use this fact for several values of C_1. We also use the fact that there exists some ϵ with $0 < \epsilon < 1/\sqrt{2}$ such that

$$|A_n^2 - A_{n+1}A_{n-1}| < \left(\frac{1}{\sqrt{2}} - \epsilon\right)|A_n|^{1/2} \tag{5}$$

holds for n large enough. Since we are not interested in what happens for small values of n, we assume from now on that inequality (5) holds for all $n \geq 0$.

We proceed in several steps.

1. *There exists some $n_0 \in \mathbf{N}$ such that either A_n has constant sign for $n \geq n_0$ or $(-1)^n A_n$ has constant sign for $n \geq n_0$.*

It suffices to show that both sequences $(A_{2n})_{n\geq 0}$ and $(A_{2n+1})_{n\geq 0}$ have constant signs for n large enough. Assume n_0 is such that $|A_n| > 0$ holds for all $n > n_0$. If either one of the above two sequences does not have constant sign for $n > n_0$, it follows that there exists some $n > n_0$ such that $A_nA_{n+2} < 0$. Then,

$$|A_n| < A_{n+1}^2 + |A_nA_{n+2}| = |A_{n+1}^2 - A_nA_{n+2}| < \frac{1}{\sqrt{2}}|A_n|^{1/2} < |A_n|^{1/2},$$

which is a contradiction. □

Notice that if $(A_n)_{n\geq 0}$ satisfies (i) and (ii), then both $(-A_n)_{n\geq 0}$ and $((-1)^n A_n)_{n\geq 0}$ satisfy (i) and (ii). Moreover, if $(A_n)_{n\geq 0}$ satisfies the conclusion of the theorem, then both $(-A_n)_{n\geq 0}$ and $((-1)^n A_n)_{n\geq 0}$ satisfy the conclusion of the theorem as well: Indeed, if $(A_n)_{n\geq n_0}$ satisfies recurrence (2), then $(-A_n)_{n\geq n_0}$ satisfies also recurrence (2), while $((-1)^n A_n)_{n\geq n_0}$ satisfies recurrence (2) with r replaced by $-r$.

By **1**, it follows that we may assume that $A_n > 0$ holds for all n.

2. *$A_{n+1} > A_n$ holds for all n.*

We first show that $A_{n+1} \neq A_n$ must hold for all n. Indeed, if $A_{n+1} = A_n$ for some n, then

$$A_n|A_{n+1} - A_{n+2}| = |A_{n+1}^2 - A_nA_{n+2}| < \frac{1}{\sqrt{2}}A_n^{1/2} < A_n^{1/2}.$$

Since $A_n > 0$, this implies $A_{n+2} = A_{n+1}$. By induction, it follows that $A_m = A_n$ for all $m > n$ which contradicts (i). Now assume that $A_{n+1} < A_n$. Then,

$$A_{n+2}A_n \leq A_{n+1}^2 + |A_{n+1}^2 - A_nA_{n+2}| < A_{n+1}^2 + \frac{1}{\sqrt{2}}A_{n+1}^{1/2} < A_{n+1}^2 + A_{n+1} =$$

$$A_{n+1}(A_{n+1}+1) \leq A_{n+1}A_n,$$

and, hence,

$$A_{n+2} < A_{n+1}.$$

By induction, this implies $0 < A_{m+1} < A_m$ for all $m \geq n$ which is impossible. □

Next, denote

$$\delta_n := A_n^2 - A_{n+1}A_{n-1}. \tag{6}$$

3. *If*

$$\liminf_{n\to\infty} \frac{A_{n+1}-A_n}{\sqrt{A_{n+1}}} < 1,$$

then there exists $n_1 \in \mathbf{N}$ *such that* $(A_n)_{n\geq n_1}$ *is an arithmetical progression.*

Assume that there exists ϵ_1 such that $0 < \epsilon_1 < 1$ and

$$A_{n+1} - A_n < (1-\epsilon_1)A_{n+1}^{1/2},$$

holds for infinitely many $n \in \mathbf{N}$. Choose n_1 such that

$$A_n > \left(\frac{2}{\epsilon_1} - 1\right)^2$$

or – equivalently –

$$2 - \epsilon_1 < \epsilon_1 A_n^{1/2}$$

holds for all $n > n_1$. Assume that

$$A_{n+1} - A_n =: d,$$

for some $n > n_1$, where $d < (1-\epsilon_1)A_{n+1}^{1/2}$. It follows that

$$\begin{aligned} A_{n+2}A_n &= A_{n+1}^2 \pm |A_{n+1}^2 - A_nA_{n+2}| = (A_n+d)^2 \pm |\delta_{n+1}| \\ &= A_n^2 + 2A_nd + d^2 \pm |\delta_{n+1}|. \end{aligned}$$

Hence,

$$A_{n+2} = A_n + 2d + \left(\frac{d^2 \pm |\delta_{n+1}|}{A_n}\right). \tag{7}$$

Notice that

$$\frac{d^2 \pm |\delta_{n+1}|}{A_n}$$

is an integer and that

$$0 \leq \frac{d^2 + |\delta_{n+1}|}{A_n} < \frac{(1-\epsilon_1)^2 A_{n+1} + A_{n+1}^{1/2}}{A_n} <$$

$$\frac{(1-\epsilon_1)(A_n+d) + A_{n+1}^{1/2}}{A_n} = |1-\epsilon_1| + \frac{(1-\epsilon_1)d + A_{n+1}^{1/2}}{A_n} =$$

$$|1-\epsilon_1| + \frac{(d+A_{n+1}^{1/2})-\epsilon_1 d}{A_n} < |1-\epsilon_1| + \frac{((1-\epsilon_1)A_{n+1}^{1/2}+A_{n+1}^{1/2})-\epsilon_1 d}{A_n} =$$

$$|1-\epsilon_1| + \frac{(2-\epsilon_1)A_{n+1}^{1/2}-\epsilon_1 d}{A_n} = |1-\epsilon_1| + \frac{\epsilon_1 A_{n+1}^{1/2}A_{n+1}^{1/2}-\epsilon_1 d}{A_n} =$$

$$1-\epsilon_1+\epsilon_1\frac{A_{n+1}-d}{A_n} = 1.$$

Hence, this argument together with formula (7) show that

$$A_{n+2} = A_n + 2d = A_{n+1} + d.$$

By induction, one can show that $A_{m+1} = A_m + d$ for all $m \geq n$. Hence, $(A_m)_{m\geq n}$ is an arithmetical progression. □

From now on, we assume that for every $\epsilon_1 > 0$ there exists a positive integer n_1 with

$$A_{n+1} - A_n \geq (1-\epsilon_1)A_{n+1}^{1/2} \qquad \text{for all } n \geq n_1. \tag{8}$$

4. *The sequence $(A_{n+1}/A_n)_{n\geq 0}$ is convergent to a limit $\alpha > 1$. Moreover, there exists $n_1 \in \mathbf{N}$ such that*

$$\left|\frac{A_{n+1}}{A_n} - \alpha\right| \leq \frac{1}{\sqrt{2}A_n} \qquad \text{holds for all } n \geq n_1. \tag{9}$$

By formula (8) for $\epsilon_1 = \sqrt{2}\epsilon$, it follows that one may choose $n_1 \in \mathbf{N}$ such that

$$A_{n+1} - A_n > (1-\sqrt{2}\epsilon)A_{n+1}^{1/2}$$

holds for all $n > n_1$. Then, for $n > n_1$ one has

$$\left|\frac{A_{n+1}}{A_n} - \frac{A_{n+2}}{A_{n+1}}\right| = \frac{|A_{n+1}^2 - A_nA_{n+2}|}{A_nA_{n+1}} = \frac{|\delta_{n+1}|}{A_nA_{n+1}} =$$

$$\frac{|\delta_{n+1}|}{A_{n+1}-A_n}\left(\frac{1}{A_n} - \frac{1}{A_{n+1}}\right) < \frac{1}{\sqrt{2}}\left(\frac{1}{A_n} - \frac{1}{A_{n+1}}\right). \tag{10}$$

The last inequality above follows from the fact that

$$A_{n+1} - A_n > \sqrt{2}\left(\frac{1}{\sqrt{2}} - \epsilon\right)A_{n+1}^{1/2} > \sqrt{2}|\delta_{n+1}|.$$

Summing up inequalities (10) for n, $n+1$, ..., $n+m-1$, we get

$$\left|\frac{A_{n+1}}{A_n} - \frac{A_{n+m+1}}{A_{n+m}}\right| \leq \sum_{i=1}^{m}\left|\frac{A_{n+i}}{A_{n+i-1}} - \frac{A_{n+i+1}}{A_{n+i}}\right| <$$

$$\frac{1}{\sqrt{2}}\sum_{i=1}^{m}\Big(\frac{1}{A_{n+i-1}}-\frac{1}{A_{n+i}}\Big)<\frac{1}{\sqrt{2}A_n}, \tag{11}$$

for all $m \geq 1$. From inequality (11), it follows that the sequence $(A_{n+1}/A_n)_{n\geq 0}$ is a Cauchy sequence and, hence, convergent. Passing to the limit with m in formula (11), we get inequality (9).

Clearly, $A_{n+1} > A_n$ implies $\alpha \geq 1$. To see that $\alpha > 1$ must hold, notice that inequality (9) with $\alpha = 1$ implies

$$A_{n+1} - A_n \leq \frac{1}{\sqrt{2}}.$$

But this forces $A_{n+1} = A_n$ for all $n \geq n_1$, which is impossible. □

In what follows, we will show that, in fact, A_{n+1}/A_n is much closer to α than is suggested by inequality (9).

5. *There exists a constant C such that*

$$\Big|\frac{A_{n+1}}{A_n}-\alpha\Big| < \frac{C}{A_n^{3/2}} \qquad \text{for all } n. \tag{12}$$

Start by choosing $\epsilon_1 > 0$ such that $\alpha - \epsilon_1 > 1$. Choose $n_2 > n_1$ such that

$$A_n > \frac{1}{\sqrt{2}\epsilon_1}$$

or – equivalently –

$$\epsilon_1 > \frac{1}{\sqrt{2}A_n}$$

holds for all $n > n_2$. By inequality (9), it follows that

$$(\alpha - \epsilon_1)A_n < A_{n+1} < (\alpha + \epsilon_1)A_n$$

holds for all $n > n_2$. In particular, it follows that there exist two constants C_1 and C_2 such that

$$C_1(\alpha - \epsilon_1)^n < A_n < C_2(\alpha + \epsilon_1)^n$$

holds for all n. Now choose $n_3 > n_2$ so that

$$n^2 < \sqrt{2}\alpha A_n$$

holds for all $n > n_3$. By inequality (9), it follows that

$$\Big|\frac{A_{n+1}}{A_n}-\alpha\Big| \leq \frac{1}{\sqrt{2}A_n} < \frac{\alpha}{n^2}$$

or – equivalently –

$$\alpha\Big(1-\frac{1}{n^2}\Big)A_n < A_{n+1} < \alpha\Big(1+\frac{1}{n^2}\Big)A_n$$

holds for all $n > n_3$. Since the infinite products

$$\prod_{n\geq 2}\left(1-\frac{1}{n^2}\right) \quad \text{and} \quad \prod_{n\geq 2}\left(1+\frac{1}{n^2}\right)$$

both converge to positive real numbers, it follows that there exist two constants C_3 and C_4 such that

$$C_3\alpha^n < A_n < C_4\alpha^n$$

holds for all n.

We now investigate again the expression

$$\left|\frac{A_{n+2}}{A_{n+1}} - \frac{A_{n+1}}{A_n}\right|.$$

We have

$$\left|\frac{A_{n+2}}{A_{n+1}} - \frac{A_{n+1}}{A_n}\right| = \frac{|\delta_{n+1}|}{A_nA_{n+1}} < \frac{1}{\sqrt{2}A_nA_{n+1}^{1/2}} < \frac{C_5}{\alpha^{3n/2}} \qquad \text{for } n > n_0, \tag{13}$$

where C_5 is a constant. Applying inequality (13) for $n,\ n+1,\ \ldots,\ n+m-1$, and summing up all those inequalities we get

$$\left|\frac{A_{n+m+1}}{A_{n+m}} - \frac{A_{n+1}}{A_n}\right| \leq \sum_{k=n}^{n+m-1}\left|\frac{A_{k+2}}{A_{k+1}} - \frac{A_{k+1}}{A_k}\right| <$$

$$\frac{C_5}{\alpha^{3n/2}}\sum_{k=0}^{m-1}\frac{1}{\alpha^{3k/2}} < \frac{C_6}{\alpha^{3n/2}} < \frac{C_7}{A_n^{3/2}}. \tag{14}$$

Here,

$$C_6 = \frac{C_5\alpha^{3/2}}{\alpha^{3/2}-1}$$

and

$$C_7 = C_6C_4^{3/2}.$$

One may now pass to the limit with m in inequality (14) to conclude that inequality (12) holds with any $C > C_7$. □

The arguments employed at Steps 1-5 remain valid, with small modifications, when one replaces condition (ii) with the requirement that the lim sup appearing at condition (ii) is finite. In what follows, we use the full strength of condition (ii) to prove that:

6. *There exists $n_2 \in \mathbf{N}$ such that $(\delta_n)_{n\geq n_2}$ is a geometrical progression.*

Using the fact that $\delta_n = A_n^2 - A_{n+1}A_{n-1}$, one immediately sees that

$$\delta_n^2 - \delta_{n+1}\delta_{n-1} \equiv 0 \pmod{A_n}$$

holds. In particular, the number

$$\frac{\delta_n^2 - \delta_{n+1}\delta_{n-1}}{A_n}$$

is an integer for all n. We show that it is zero. In order to achieve this, it suffices to show that it is smaller than 1 in absolute value. Notice that

$$\frac{|\delta_n^2 - \delta_{n+1}\delta_{n-1}|}{A_n} \leq \frac{\delta_n^2 + |\delta_{n+1}\delta_{n-1}|}{A_n} < \Big(\frac{1}{\sqrt{2}} - \epsilon\Big)^2 \Big(1 + \frac{\sqrt{A_{n+1}A_{n-1}}}{A_n}\Big). \quad (15)$$

Since A_{n+1}/A_{n-1} is convergent, it follows that the sequence

$$1 + \frac{\sqrt{A_{n+1}A_{n-1}}}{A_n}$$

is convergent and its limit is 2. From inequality (15), it follows that

$$\frac{|\delta_n^2 - \delta_{n+1}\delta_{n-1}|}{A_n} < 1,$$

holds for n enough large. Hence, there exists $n_4 \in \mathbf{N}$ such that

$$\delta_n^2 = \delta_{n+1}\delta_{n-1}$$

holds for all $n > n_4$. □

Notice that if $\delta_n = 0$ for some $n > n_4$, it then follows that $\delta_m = 0$ for all $m > n_4$. In particular, this implies that $(A_m)_{m \geq n_4}$ is a geometrical progression. Since $(A_m)_{m \geq n_4}$ is a geometrical progression of integers, it follows that its ratio is an integer as well. Hence, $(A_n)_{n \geq 0}$ satisfies the conclusion of the theorem. We now assume that $\delta_n \neq 0$ for any $n > n_4$. Denote by $-s := \delta_{n+1}/\delta_n$ the ratio of the geometrical progression $(\delta_n)_{n \geq n_4}$.

7. *α is irrational.*

Assume that $\alpha = p/q$ is rational. By Step 5, it follows that

$$\left|\frac{A_{n+1}}{A_n} - \frac{p}{q}\right| < \frac{C}{A_n^{3/2}}$$

holds for some constant C and for all n. Hence,

$$|qA_{n+1} - pA_n| < \frac{Cq}{A_n^{1/2}}$$

holds for all n. If n_5 is such that $A_n > (Cq)^2$ holds for all $n > n_5$, we get that

$$|qA_{n+1} - pA_n| < 1,$$

or $A_{n+1} = \alpha A_n$. Hence, $(A_n)_{n \geq n_5}$ is a geometrical progression of ratio α. This forces $\delta_n = 0$ for all $n > n_5$, which is a contradiction.

Hence, α is irrational. □

Denote

$$d_n := \gcd(A_n,\ A_{n+1}).$$

8. *$d_n d_{n+1} < A_{n+1}^{1/2}$ holds for all $n > n_4$.*

This follows by noticing that $d_n d_{n+1}$ divides A_{n+1}^2 as well as $A_n A_{n+2}$ and, hence, $\delta_{n+1} = A_{n+1}^2 - A_n A_{n+2}$. Since $\delta_n \neq 0$ for $n > n_4$, it follows that

$$d_n d_{n+1} \le |\delta_{n+1}| < A_{n+1}^{1/2}.$$ □

9. *The final step.*

Recall that

$$\delta_n = A_n^2 - A_{n+1}A_{n-1}$$

and

$$\delta_{n+1} = A_{n+1}^2 - A_n A_{n+2}.$$

Multiplying the first relation with δ_{n+1} and the second with δ_n, we get that

$$\delta_{n+1}A_n^2 - \delta_{n+1}A_{n+1}A_{n-1} = \delta_n A_{n+1}^2 - \delta_n A_n A_{n+2},$$

or

$$A_n(\delta_{n+1}A_n + \delta_n A_{n+2}) = A_{n+1}(\delta_n A_{n+1} + \delta_{n+1}A_{n-1}).$$

Assume that $n > n_4$. Dividing both sides of the above equation by d_n, we get

$$\frac{A_n}{d_n}(\delta_{n+1}A_n + \delta_n A_{n+2}) = \frac{A_{n+1}}{d_n}(\delta_n A_{n+1} + \delta_{n+1}A_{n-1}).$$

It now follows that A_{n+1}/d_n divides $\delta_{n+1}A_n + \delta_n A_{n+2}$. Hence, there exists an integer u_n such that

$$\delta_n A_{n+2} + \delta_{n+1}A_n = \frac{u_n}{d_n}A_{n+1},$$

or

$$A_{n+2} - \frac{u_n}{d_n\delta_n}A_{n+1} + \frac{\delta_{n+1}}{\delta_n}A_n = 0.$$

Since $\delta_{n+1}/\delta_n = -s$ for all $n > n_4$, it follows that

$$A_{n+2} - \frac{u_n}{d_n\delta_n}A_{n+1} - sA_n = 0 \tag{16}$$

or – equivalently –

$$\frac{u_n}{d_n\delta_n} = \frac{A_{n+2} - sA_n}{A_{n+1}} =: r_n.$$

It remains to show that r_n is constant from some n on. Since A_{n+1}/A_n is convergent, it follows that $|r_n|$ is bounded.

Next, let us notice that equation (16) can be rewritten as

$$\frac{A_{n+2}}{A_n} - r_n \frac{A_{n+1}}{A_n} - s = 0. \tag{17}$$

From equation (17), it follows that

$$\left|\alpha^2 - r_n\alpha - s\right| = \left|\left(\frac{A_{n+2}}{A_n} - \alpha^2\right) + r_n\left(\frac{A_{n+1}}{A_n} - \alpha\right)\right| =$$

$$\left|\left(\frac{A_{n+2}}{A_{n+1}} - \alpha\right)\frac{A_{n+1}}{A_n} + \alpha\left(\frac{A_{n+1}}{A_n} - \alpha\right) + r_n\left(\frac{A_{n+1}}{A_n} - \alpha\right)\right|. \tag{18}$$

From Step 5, the fact that $|r_n|$ is bounded, and equation (18), it follows that there exists a constant C_8 such that

$$|\alpha^2 - r_n\alpha - s| < \frac{C_8}{A_n^{3/2}}. \tag{19}$$

Applying inequality (19) for n and $n+1$, it follows that

$$|r_n - r_{n+1}| \leq \frac{1}{\alpha}(|\alpha^2 - r_n\alpha - s| + |\alpha^2 - r_{n+1}\alpha - s|) < \frac{2C_8}{\alpha A_n^{3/2}},$$

or

$$|r_{n+1} - r_n| < \frac{C_9}{A_n^{3/2}}, \tag{20}$$

where $C_9 = 2C_8/\alpha$.

Notice now that $d_n d_{n+1}\delta_{n+1} = (-s)d_n d_{n+1}\delta_n$ is a common multiple of both the denominator of r_n and of r_{n+1}. Hence,

$$|d_n d_{n+1}\delta_{n+1}(r_n - r_{n+1})|$$

is an integer. From inequality (20) and Step 8, we get

$$|d_n d_{n+1}\delta_{n+1}(r_n - r_{n+1})| < \frac{C_9(d_n d_{n+1})|\delta_{n+1}|}{A_n^{3/2}} <$$

$$C_9\frac{A_{n+1}}{A_n^{3/2}} = C_9\left(\frac{A_{n+1}}{A_n}\right)\cdot\frac{1}{A_n^{1/2}}, \tag{21}$$

and the right hand side of inequality (21) tends to zero when n tends to infinity. Hence, there exists $n_6 \in \mathbf{N}$ such that

$$|d_n d_{n+1}\delta_{n+1}(r_n - r_{n+1})| < 1$$

holds for all $n > n_6$. This implies that $r_{n+1} = r_n$ holds for all $n > n_6$. Hence, $r_n = r$ for all $n \geq n_6$ and

$$A_{n+2} = rA_{n+1} + sA_n$$

holds for all $n > n_6$. It remains to show that r is, in fact, an integer. But this follows easily from the general theory of binary recurrence sequences.

Hence, $(A_n)_{n \geq n_6}$ is indeed binary recurrent. The remaining of the assertions of the theorem follow from the arguments from the beginning of the paper.

The theorem is therefore completely proved.

The Proof of the Corollary.

Let A_n be the closest integer to α^n. Write

$$A_n = \alpha^n + \zeta_n,$$

where

$$|\zeta_n| < \frac{1}{4\sqrt{2}} |\alpha|^{-(n+3)/2}. \tag{22}$$

Clearly, $(A_n)_{n \geq 0}$ satisfies condition (i). One may now use inequality (22) to conclude that

$$\limsup_{n \to \infty} \frac{|A_n^2 - A_{n+1}A_{n-1}|}{\sqrt{|A_n|}} \leq \frac{2|\alpha|^{-3/2} + 1 + |\alpha|^{-5/2}}{4\sqrt{2}} < \frac{1}{\sqrt{2}}.$$

Hence, $(A_n)_{n \geq 0}$ satisfies condition (ii) as well. The assertions of the corollary follow now immediately from the theorem.

3 Acknowledgements

We thank the referee for pointing out to us Mignotte's paper [2].

Work by the second author was done while he was visiting Bielefeld. He would like to thank the Mathematics Department in Bielefeld for its hospitality during the period when this paper was written and the Alexander von Humbold Foundation for support.

We would also like to thank Friederike Jordan as well as Anton Betten and the editorial staff of these Proceedings for their technical help in preparing the final version of this manuscript.

References

1. A. Dress, F. Luca, Unbounded Sequences $(A_n)_{n \geq 0}$ with $A_{n+1}A_{n-1} - A_n^2$ Bounded are of Fibonacci Type, *in these Proceedings.*
2. M. Mignotte, A characterization of integers, *Amer. Math. Monthly* **84** (1977), 278-281.

Unbounded Integer Sequences $(A_n)_{n\geq 0}$ with $A_{n+1}A_{n-1} - A_n^2$ Bounded are of Fibonacci Type

Andreas Dress[1] and Florian Luca[2]

[1] Mathematics Department
Bielefeld University
Postfach 10 01 31
33 501 Bielefeld
Germany
dress@mathematik.uni-bielefeld.de

[2] Mathematics Department
Bielefeld University
Postfach 10 01 31
33 501 Bielefeld
Germany
fluca@mathematik.uni-bielefeld.de

Abstract It is well known that the Fibonacci sequence $(F_n)_{n\geq 0}$ defined by $F_0 = 0$, $F_1 = 1$, and $F_{n+2} = F_{n+1} + F_n$ for all $n \geq 0$ satisfies the condition $F_n^2 - F_{n+1}F_{n-1} = (-1)^n$ for all $n \in \mathbf{N}$. In this note, we show that – somehow conversely – if $(A_n)_{n\geq 0}$ is a sequence of integers such that $(|A_n|)_{n\geq 0}$ diverges to infinity and $|A_n^2 - A_{n+1}A_{n-1}|$ remains bounded, then $(A_n)_{n\geq 0}$ is binary recurrent from some n on. An application to real quadratic units is also presented.

Key words: real quadratic units, binary recurrent sequences, Fibonacci sequences.

1 Introduction

Let $(A_n)_{n\geq 0}$ be a sequence of integers satisfying the following two assumptions:

(i) $\lim_{n\to\infty} |A_n| = \infty$.

(ii) There exists a constant C such that $|A_n^2 - A_{n+1}A_{n-1}| < C$ for all $n \geq 1$.

It seems natural to ask whether one can characterize all sequences of integers $(A_n)_{n\geq 0}$ satisfying (i) and (ii) above.
Notice that the following three families of sequences satisfy the above two conditions:

(a) A sequence $(A_n)_{n\geq 0}$ such that either $(A_n)_{n\geq 0}$ or $((-1)^n A_n)_{n\geq 0}$ is an arithmetic progression with integer step size r, $r \neq 0$.

(b) Geometrical progressions of integer ratio r, $|r| > 1$.

(c) Non-zero binary recurrence sequences whose characteristic equation has roots that are real irrational units.

Recall that a binary recurrence sequence $(A_n)_{n\geq 0}$ is a sequence such that

$$A_{n+2} = rA_{n+1} + sA_n \qquad \text{for } n \geq 0, \tag{1}$$

where r and s are some integers, and its characteristic equation is

$$x^2 - rx - s = 0. \tag{2}$$

Note that

$$A_{n+1}^2 - A_{n+2}A_n = s(A_n^2 - A_{n+1}A_{n-1})$$

then holds for all $n \geq 0$. Hence, a sequence satisfying (i), (ii), and (c) is a sequence satisfying recurrence (1) for some integers $r \neq 0$ and s such that $r^2 + 4s > 0$ and $|s| = 1$. If one denotes by α and β the two roots of the characteristic equation (2), then $|\alpha\beta| = 1$ and

$$A_n = a\alpha^n + b\beta^n \qquad \text{for } n \geq 0, \tag{3}$$

where a and b are easily computed in terms of A_0, A_1, α, and β, and neither of them can vanish.

We leave it to the reader to check that any sequence belonging to the families (a)-(c) above satisfies (i) and (ii).

A natural question therefore is whether there are any other sequences satisfying (i) and (ii). Our main result is:

Theorem 1.1. *If* $(A_n)_{n\geq 0}$ *satisfies* (i) *and* (ii), *then there exists some* $n_0 \in \mathbf{N}$ *such that* $(A_n)_{n\geq n_0}$ *is a sequence of type* (a), (b), *or* (c).

In other words, our Theorem shows that if $(A_n)_{n\geq 0}$ is a sequence of integers satisfying (i) and (ii), then $(A_n)_{n\geq 0}$ can differ from a sequence of type (a), (b), or (c) in at most finitely many terms.

Remark 1.2. For any sequence $(A_n)_{n\geq 0}$ and for any positive integer k, let

$$D_{n,k} := \begin{vmatrix} A_n & A_{n+1} & \dots & A_{n+k} \\ A_{n+1} & A_{n+2} & \dots & A_{n+k+1} \\ \dots & \dots & \dots & \dots \\ A_{n+k} & A_{n+k+1} & \dots & A_{n+2k} \end{vmatrix}.$$

Notice that $D_{n,1} = A_nA_{n+2} - A_{n+1}^2$. Notice also that if $(A_n)_{n\geq 0}$ is binary recurrent, then $D_{n,2} = 0$ holds for all $n \geq 0$. Thus, our theorem simply asserts that if $(A_n)_{n\geq 0}$ is a sequence of integers such that $\lim_{n\to\infty} |A_n| = \infty$ and $|D_{n,1}| < C$ holds for all $n \geq 0$, then there exists $n_0 \in \mathbf{N}$ with $D_{n,2} = 0$ for all $n \geq n_0$.

It would be interesting to investigate whether a more general statement holds. That is, we propose the following

Question 1.3. Let $k > 1$ be a positive integer and let $(A_n)_{n\geq 0}$ be a sequence of integers such that $\lim_{n\to\infty} |A_n| = \infty$. Is it then true that if $|D_{n,k}| < C$ holds for all $n \in \mathbf{N}$, then there exists $n_0 \in \mathbf{N}$ with $D_{n,k+1} = 0$ for all $n > n_0$?

We also present the following

Corollary 1.4. *For every real number x, let $||x|| = \min(|x-n| : n \in \mathbf{Z})$ be the distance from x to the nearest integer. Assume that b is a real number with $|b| > 1$ such that there exist a non-zero real number a and a constant C with*

$$||ab^n|| < C|b|^{-n} \qquad \text{for all } n \geq 0. \tag{4}$$

Then either one of the following holds:

1. $b \in \mathbf{Z}$, $a \in \mathbf{Q}$, and $ab^m \in \mathbf{Z}$ for some integer $m \geq 0$.

2. b is a real irrational quadratic unit and a is an element in $\mathbf{Q}[b]$ with $a(b - b^{-1}) \in \mathbf{Z}$.

Conversely, if either $b \in \mathbf{Z}^*$ or b is a real irrational quadratic unit b, then inequality (4) holds for $a = 1$ with some constant C.

The above condition $|b| > 1$ is meant to rule out trivialities (notice that if $0 < |b| \leq 1$ and a is any real number, then formula (4) holds for some C).

The referee pointed out that M. Mignotte (see [3]) showed that if a and b are real numbers with $b > 1$ such that $||ab^n||$ is small enough for all integer values of $n \geq 0$, then b is an algebraic number. In particular, Corollaries 1 or 2 from [3] imply that if a and b satisfy inequalities (4) for all positive integers $n \geq 0$, then b is an algebraic number. However, it is unclear to us if the results of [3] imply our Corollary 4.

The first part of Corollary 4 holds if one replaces the function $||x||$ by $\{x\}$ where $\{x\}$ is the classical fractional part $x - \lfloor x \rfloor$ of x. The second part of Corollary 4 fails always in this setting. That is,

Corollary 1.5. *If b is a nonzero real number such that*

$$\{b^n\} < C|b|^{-n} \qquad \text{for } n \geq 0, \tag{5}$$

where C is some constant, then $|b| < 1$ or $b \in \mathbf{Z}$.

2 Proofs

Proof of the Theorem.

Without loss of generality, we assume $C > 1$ and $A_n \neq 0$ for all $n \in \mathbf{N}$. By condition (i), it follows that for every C_1 there exists some $n_0 \in \mathbf{N}$ such that $|A_n| \geq C_1$ for $n \geq n_0$. In what follows, we shall use this fact for several values of C_1. We proceed in several steps.

1. *There exists some $n_0 \in \mathbf{N}$ such that either A_n has constant sign for $n \geq n_0$ or $(-1)^n A_n$ has constant sign for $n \geq n_0$.*

It suffices to show that both sequences $(A_{2n})_{n\geq 0}$ and $(A_{2n+1})_{n\geq 0}$ have constant signs for n large enough. Assume n_0 is such that $|A_n| > C$ holds for all $n > n_0$. If either one of the above two sequences does not have constant sign for $n > n_0$, it follows that there exists some $n > n_0$ such that $A_n A_{n+2} < 0$. Then,

$$C^2 < A_{n+1}^2 < A_{n+1}^2 + |A_n A_{n+2}| = |A_{n+1}^2 - A_n A_{n+2}| < C,$$

which is a contradiction.

Notice that if $(A_n)_{n\geq 0}$ satisfies (i) and (ii), then both $(-A_n)_{n\geq 0}$ and $((-1)^n A_n)_{n\geq 0}$ satisfy (i) and (ii). Moreover, if $(A_n)_{n\geq 0}$ satisfies one of the properties (a)-(c), then both $(-A_n)_{n\geq 0}$ and $((-1)^n A_n)_{n\geq 0}$ satisfy the same property (a)-(c) as $(A_n)_{n\geq 0}$. By **1**, it follows that we may assume that $A_n > 0$ holds for all $n > n_0$.

2. *There exists some n_0 such that $A_{n+1} > A_n$ holds for all $n > n_0$.*

Assume that $A_n > C$ holds for all $n \geq n_0$. We first show that $A_{n+1} \neq A_n$ must hold for all $n > n_0$. Indeed, if $A_{n+1} = A_n$ for some $n > n_0$, then

$$A_{n+1}|A_{n+1} - A_{n+2}| = |A_{n+1}^2 - A_n A_{n+2}| < C.$$

Since $A_{n+1} > C$, it follows that $A_{n+2} = A_{n+1}$. By induction, it follows that $A_m = A_n$ for all $m > n$ which contradicts (i). Now assume that $A_{n+1} < A_n$. Then,

$$A_{n+2}A_n \leq A_{n+1}^2 + |A_{n+1}^2 - A_n A_{n+2}| < A_{n+1}^2 + C,$$

or

$$A_{n+2} < \frac{A_{n+1}^2 + C}{A_n} \leq \frac{A_{n+1}^2 + C}{A_{n+1} + 1} < A_{n+1},$$

because $A_{n+1} > C$. By induction, it follows that $A_{m+1} < A_m$ for all $m \geq n$. Since $A_m > C$ for all $m \geq n$, it follows that the sequence is finite. This contradiction shows that $(A_n)_{n\geq n_0}$ is strictly increasing.

Denote

$$\delta_n := A_n^2 - A_{n+1}A_{n-1}. \tag{6}$$

3. *If* $\liminf_{n\to\infty} |\delta_n| = 0$, *then there exists some $n_0 \in \mathbf{N}$ such that $(A_n)_{n\geq n_0}$ is a sequence of type* (b).

Choose $n_0 \in \mathbf{N}$ with $\delta_{n_0} = 0$ such that $A_n > C^2$ holds for all $n > n_0$. We show that $\delta_n = 0$ for $n > n_0$. Assume that this is not so. Let $n \geq n_0$ be such that $\delta_n = 0$ but $\delta_{n+1} \neq 0$. Since $\delta_n = 0$, it follows that $A_{n+1} \mid A_n^2$. Since

$$\delta_{n+1} = A_{n+1}^2 - A_n A_{n+2}$$

and $\delta_{n+1} \neq 0$, it follows that

$$0 \neq \delta_{n+1}^2 = (A_{n+1}^2 - A_n A_{n+2})^2 \equiv A_n^2 A_{n+2}^2 \equiv 0 \pmod{A_{n+1}}. \tag{7}$$

By formula (7), it follows that

$$C^2 > \delta_{n+1}^2 > A_{n+1}, \tag{8}$$

which is a contradiction. Hence, $\delta_{n+1} = 0$. It now follows that $\delta_n = 0$ for all $n \geq n_0$, therefore $(A_n)_{n\geq n_0}$ is a sequence of type (b).

From now on, we assume that there exists n_0 such that $|\delta_n| > 0$ for $n > n_0$.

4. $\gcd(A_n, A_{n+1}) < C$ *for all positive integers* n.

This follows by noticing that $\gcd(A_n, A_{n+1}) \mid \delta_n$. Hence,

$$\gcd(A_n, A_{n+1}) \leq |\delta_n| < C.$$

5. *If* $\liminf_{n\to\infty}(A_{n+1} - A_n) < \infty$, *then there exists* $n_0 \in \mathbf{N}$ *such that* $(A_n)_{n\geq n_0}$ *is a sequence of type* (a).

Assume that

$$\liminf_{n\to\infty}(A_{n+1} - A_n) < C_1,$$

for some constant C_1. Choose n_0 such that $A_n > C_1^2 + C$ for all $n > n_0$. Assume that

$$A_{n+1} - A_n = d,$$

for some $n > n_0$, where $d < C_1$. Since $A_{n+1}^2 > C^2 > C$, it follows that

$$A_{n+2}A_n = A_{n+1}^2 \pm |A_{n+1}^2 - A_nA_{n+2}| = (A_n + d)^2 \pm \delta_n = A_n^2 + 2A_nd + d^2 \pm \delta_n.$$

Hence,

$$A_{n+2} = A_n + 2d + \left(\frac{d^2 \pm \delta_n}{A_n}\right).$$

Since

$$|d^2 \pm \delta_n| < d^2 + C < C_1^2 + C < A_n \qquad \text{for } n \geq n_0$$

and

$$\frac{d^2 \pm \delta_n}{A_n} \in \mathbf{Z},$$

it follows that

$$\frac{d^2 \pm \delta_n}{A_n} = 0$$

and

$$A_{n+2} = A_n + 2d = A_{n+1} + d.$$

By induction, one can show that $A_{m+1} = A_m + d$ for all $m \geq n$. Hence, $(A_n)_{n\geq 0}$ is a sequence of type (a) for n large enough.

From now on, we assume

$$\lim_{n\to\infty}(A_{n+1} - A_n) = \infty. \tag{9}$$

6. *The sequence $(A_{n+1}/A_n)_{n\geq 0}$ is convergent to a limit α. Moreover, for every $C_1 > 0$, there exists $n_0 \in \mathbf{N}$ such that*

$$\left|\frac{A_{n+1}}{A_n} - \alpha\right| \leq \frac{1}{C_1 A_n} \tag{10}$$

holds for all $n \geq n_0$.

Choose $n_0 \in \mathbf{N}$ such that $A_{n+1} - A_n > CC_1$ holds for all $n > n_0$. Then,

$$\left|\frac{A_{n+1}}{A_n} - \frac{A_{n+2}}{A_{n+1}}\right| = \frac{|A_{n+1}^2 - A_n A_{n+2}|}{A_n A_{n+1}} < \frac{C}{A_n A_{n+1}} =$$

$$\frac{C}{A_{n+1} - A_n}\left(\frac{1}{A_n} - \frac{1}{A_{n+1}}\right) < \frac{1}{C_1}\left(\frac{1}{A_n} - \frac{1}{A_{n+1}}\right). \tag{11}$$

Summing up inequalities (11) for n, $n+1$, ..., $n+m-1$, we get

$$\left|\frac{A_{n+1}}{A_n} - \frac{A_{n+m+1}}{A_{n+m}}\right| \leq \sum_{i=1}^{m} \left|\frac{A_{n+i}}{A_{n+i-1}} - \frac{A_{n+i+1}}{A_{n+i}}\right| <$$

$$\frac{1}{C_1}\sum_{i=1}^{m}\left(\frac{1}{A_{n+i-1}} - \frac{1}{A_{n+i}}\right) < \frac{1}{C_1 A_n}, \tag{12}$$

for all $m \geq 1$. From inequality (12), it follows that the sequence $(A_{n+1}/A_n)_{n\geq 0}$ is a Cauchy sequence and, hence, convergent. Passing to the limit with m in formula (12) we get inequality (10).

7. *$\alpha > 1$ and α is irrational.*

Clearly, $A_{n+1}/A_n > 1$, implies $\alpha \geq 1$. Assume now that $\alpha = a/b$ is rational. Choose $C_1 > b$. Then, for n enough large, inequality (10) implies

$$\frac{|bA_{n+1} - aA_n|}{bA_n} = \left|\frac{A_{n+1}}{A_n} - \frac{a}{b}\right| \leq \frac{1}{C_1 A_n} < \frac{1}{bA_n}.$$

Hence, $bA_{n+1} = aA_n$ and therefore $\delta_n = 0$ for n large enough which contradicts our assumption.

8. *There exists n_0 such that $(A_n)_{n\geq n_0}$ is a sequence of type* (c).

Let n_0 be such that $|\delta_n| > 0$ for $n \geq n_0$. Recall that

$$\delta_n = A_n^2 - A_{n-1}A_{n+1}$$

and

$$\delta_{n+1} = A_{n+1}^2 - A_n A_{n+2}.$$

Hence,

$$\delta_{n+1}(A_n^2 - A_{n-1}A_{n+1}) = \delta_n(A_{n+1}^2 - A_n A_{n+2}),$$

or

$$A_n(\delta_{n+1}A_n + \delta_n A_{n+2}) = A_{n+1}(\delta_n A_{n+1} + \delta_{n+1}A_{n-1}) \tag{13}$$

holds for all n. Let $d_n := \gcd(A_n, A_{n+1})$. From equation (13), it follows that

$$\frac{A_{n+1}}{d_n} v_n = \delta_{n+1} A_n + \delta_n A_{n+2}, \tag{14}$$

where v_n is a divisor of $\delta_n A_{n+1} + \delta_{n+1} A_{n-1}$. Notice that v_n is bounded. Indeed, this follows because

$$v_n = d_n \left(\delta_{n+1} \frac{A_n}{A_{n+1}} + \delta_n \frac{A_{n+2}}{A_{n+1}} \right),$$

where d_n divides δ_n, $|\delta_n|$ and $|\delta_{n+1}|$ are all bounded by C, and A_{n+1}/A_n is convergent. Since $|d_n|$, $|\delta_n|$ and $|v_n|$ can take only finitely many values, it follows that there exist integers d, δ, δ', and v such that $d_n = d$, $\delta_n = \delta$, $\delta_{n+1} = \delta'$, and $v_n = v$ for infinitely many $n \in \mathbf{N}$. Dividing by A_n and passing to the limit in equation (14) with respect to those n, we get

$$\alpha \frac{v}{d} = \delta' + \delta \alpha^2$$

or

$$\alpha^2 + \frac{\delta'}{\delta} = \alpha \frac{v}{d\delta}.$$

Hence, α is quadratic. We now show that δ_n/δ_{n+1} is constant from some n on. Indeed, by the previous argument, it follows that $\delta_n/\delta_{n+1} = \delta/\delta'$ for infinitely many n. Assume that there exist infinitely many $n \in \mathbf{N}$ with $\delta_n/\delta_{n+1} \neq \delta/\delta'$. Arguments similar to the preceding ones imply that one can choose infinitely many n from the above subset such that $\delta_n = \delta_1$, $\delta_{n+1} = \delta_1'$, $d_n = d_1$, and $v_n = v_1$ are independent of n. Certainly, $\delta_{n+1}/\delta_n = \delta_1'/\delta_1 \neq \delta'/\delta$. Dividing by A_n and passing again to the limit in formula (14), we again get

$$\alpha^2 + \frac{\delta_1'}{\delta_1} = \alpha \frac{v_1}{d_1 \delta_1}.$$

However, since α is irrational, it follows that α cannot satisfy two different monic quadratic equations with rational coefficients. This argument shows that both

$$s := -\frac{\delta_{n+1}}{\delta_n} \quad \text{and} \quad r := \frac{v_n}{d_n \delta_n}$$

are constant from some n on. Since $\delta_{n+1}/\delta_n = -s$ is constant for n large enough, it follows that $\delta_n = (-s)^n c$ for some constant c and for n large enough. Since $1 \leq |\delta_n| < C$ for all $n \geq n_0$, it follows that $|s| = 1$ must hold. By formula (14), we get

$$A_{n+2} - sA_n = rA_{n+1} \qquad \text{for } n \geq n_1, \tag{15}$$

where $|s| = 1$. It remains to show that r is an integer. Assume that w is the denominator of r. By formula (15), it follows that $w \mid A_{n+1}$ for all $n \geq n_1$. From formula (15) with n replaced by $n+1$, we get

$$A_{n+3} - sA_{n+1} = rA_{n+2}.$$

Since $w \mid A_{n+1}$ and $w \mid A_{n+3}$, it follows that $w^2 \mid A_{n+2}$. By induction, one gets that $w^k \mid A_{n+k}$ for all $k \geq 1$ and $n \geq n_1$. In particular, $w^k \leq \gcd(A_{n+k}, A_{n+k+1}) < C$ for all $k \geq 1$. This shows $|w| = 1$. Hence,

$$A_{n+2} = rA_{n+1} + sA_n,$$

where r and s are integers and $|s| = 1$.

This establishes the theorem.

Proof of Corollary 4.

Let A_n be the closest integer to ab^n. Since

$$A_n = ab^n + O(b^{-n}),$$

it follows easily that $(A_n)_{n\geq 0}$ satisfies the conditions from the theorem. The remaining assertions are now obvious.

The Proof of Corollary 5.

It was shown in [1] (see also [2] for an alternative proof) that if b is an algebraic number such that the sequence $(\{b^n\})_{n\geq 1}$ tends to zero, then either $b \in (0, 1)$, or b is an integer. Consequently, Corollary 5 follows immediately from Corollary 4.

3 Acknowledgements

We thank the referee for pointing out to us Mignotte's paper [3].

Work by the second author was done while he was visiting Bielefeld. He would like to thank the Mathematics Department in Bielefeld for its hospitality during the period when this paper was written and the Alexander von Humbold Foundation for support.

We would also like to thank Friederike Jordan as well as Anton Betten and the editorial staff of these Proceedings for their technical help in preparing the final version of this manuscript.

References

1. A. Dubickas, A note on powers of Pisot numbers, *Publ. Math. (Debrecen)* **56** (2000), 141-143.
2. F. Luca, On a question of G. Kuba, *Arch. Math.* **74** (2000), 269-275.
3. M. Mignotte, A characterization of integers, *Amer. Math. Monthly* **84** (1977), 278-281.

Hereditarily Optimal Realizations: Why are they Relevant in Phylogenetic Analysis, and how does one Compute them?

Andreas Dress[1], Katharina T. Huber*[2], and Vincent Moulton**[3]

[1] FSPM-Strukturbildungsprozesse, University of Bielefeld, D-33501 Bielefeld, Germany
[2] Institute of Fundamental Sciences, Massey University, Private Bag 11 222, Palmerston North, New Zealand
[3] FMI, Mid Sweden University, Sundsvall, S 851-70, Sweden

Abstract One of the main problems in *phylogenetic analysis* (where one is concerned with elucidating evolutionary patterns between present day species) is to find good approximations of genetic distances by weighted trees. As an aid to solving this problem, it might seem tempting to consider an *optimal realization* of the metric defined by the given distances – the guiding principle being that, in case the metric is tree-like, the optimal realization obtained will necessarily be that unique weighted tree that realizes this metric. Although optimal realizations of arbitrary distances are not generally trees, but rather weighted graphs, one could still hope to obtain an informative representation of the given metric, maybe even more informative than the best approximating tree. However, optimal realizations are not only difficult to compute, they may also be non-unique. In this note we focus on one possible way out of this dilemma: *hereditarily optimal realizations*. These are essentially unique, and can also be described in an explicit way. We define hereditarily optimal realizations, discuss some of their properties, and we indicate in particular why, due to recent results on the so-called *T-construction* of a metric space, it is a straight forward task to compute these realizations for a large class of phylogentically relevant metrics.

1 Optimal Realizations

Given a metric (or, more generally, a distance function) d defined on a finite set X of taxa (i. e. a set of species) representing, say, their genetic distance, one of the main problems in phylogenetic analysis is to find a *weighted tree* $T = (V, E, w)$, with V a finite set of vertices containing X, and $w : E \to \mathbb{R}_{\geq 0}$ a weighting of the edge set E of T, so that the distance d_T on X, defined by taking the weight of the shortest path between pairs of elements in X considered as vertices of T, approximates d "optimally". Many methods have been introduced to deal with this problem (see e. g. [20, Chapter 11] for an

* The author thanks the New Zealand Marsden Fund for its support.
** The author thanks the Swedish Natural Science Research Council (NFR) for its support (grant# M12342-300).

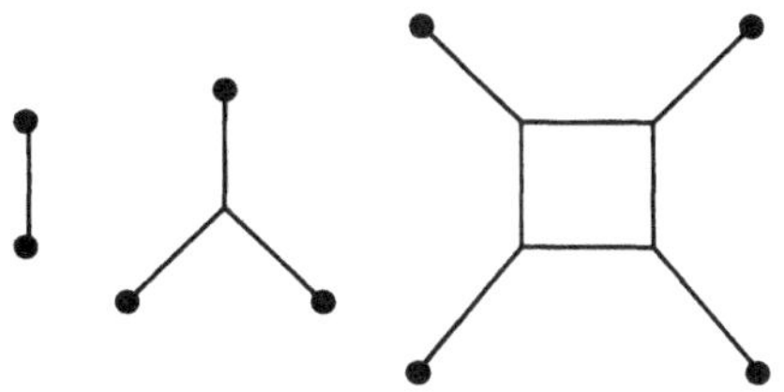

Figurel. By labeling the vertices and weighting the edges of these graphs appropriately (giving the same weight to parallel edges), one can obtain a unique optimal realization for any metric defined on two, three, or four points.

overview), and most of them satisfy the following criterion: If there is some (necessarily unique) weighted tree T with $d = d_T$, then the method returns T. As is well known [6,24], those metrics d that can actually be represented by a weighted tree – often called *tree-like* metrics – can be characterized as those satisfying the *four-point* condition:

- For all $x, y, u, v \in X$,

$$d(x, y) + d(u, v) \leq \max\{d(x, u) + d(y, v), d(x, v) + d(y, u)\}.$$

What has also been shown [22] is that the weighted tree T that represents such a metric d is actually an *optimal realization* of d, i.e. defining the *total weight* of a weighted graph $G = (V, E, w)$ to be $\|G\| = \Sigma_{e \in E} w(e)$, and saying that G *realizes* a metric d on X if $X \subseteq V$ and $d = d_G$ (the distance on X induced by using shortest paths in G) holds, the tree T is a graph that realizes d and that has minimal total weight amongst all such graphs.

Hence, since – in practice – a given metric d does not usually satisfy the (highly non-generic) four-point condition, it seems reasonable to ask for an optimal realization of d in any case. For, even though we would not necessarily obtain a tree, we would at least obtain a network realizing d that might give us a better understanding of the properties of X encoded by d. However, such a strategy has two major drawbacks: Even though, for any given metric d, an optimal realization of d is known to always exist, and even though we can obtain a unique optimal realization for d by appropriately labeling and weighting one of the graphs in Figure 1 for any X with $\#X \leq 4$, optimal realizations are hard to compute [1] and, perhaps more importantly, an optimal realization of d is not necessarily unique if $\#X \geq 5$.

In Figure 2 for example, we present two distinct optimal realizations for a metric defined on a five element set, an example that originally appeared in [9, (A 3.3)]. Indeed, it has even been shown in [1] that there exist finite metric spaces with a *continuum* of optimal realizations – see Figure 3 for an example that originally appeared in [1].

	a	b	c	d	e
a	0	7	4	5	3
b	7	0	5	6	4
c	4	5	0	7	3
d	5	6	7	0	4
e	3	4	3	4	0

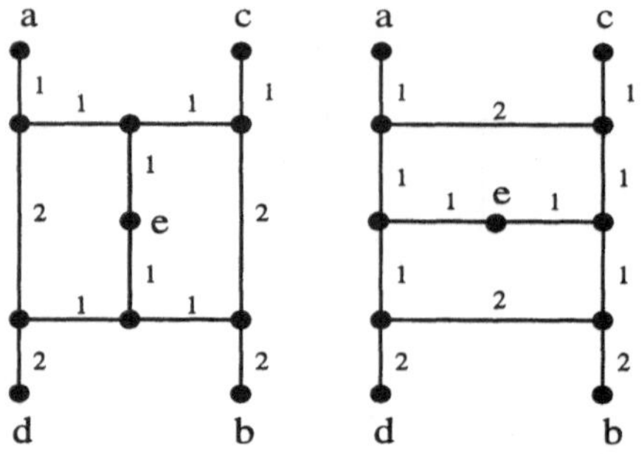

Figure2. A distance matrix with two distinct optimal realizations.

	a	b	c	d	e	f
a	0	1	2	3	3	2
b	1	0	1	3	4	3
c	2	1	0	2	3	3
d	3	3	2	0	1	2
e	3	4	3	1	0	1
f	2	3	3	2	1	0

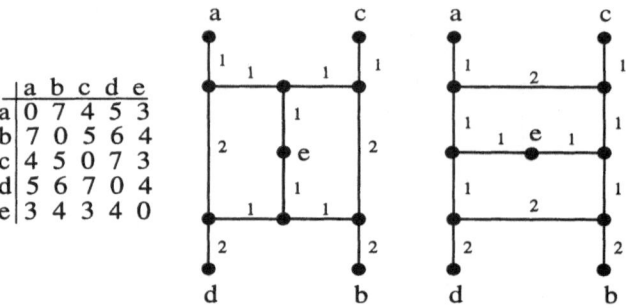

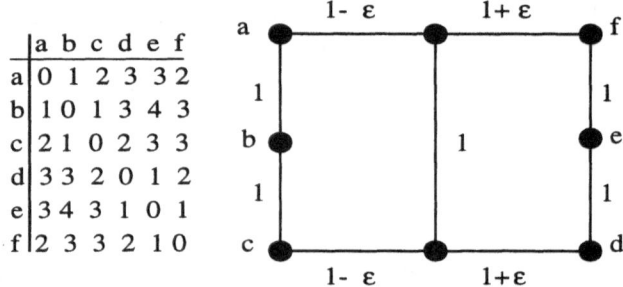

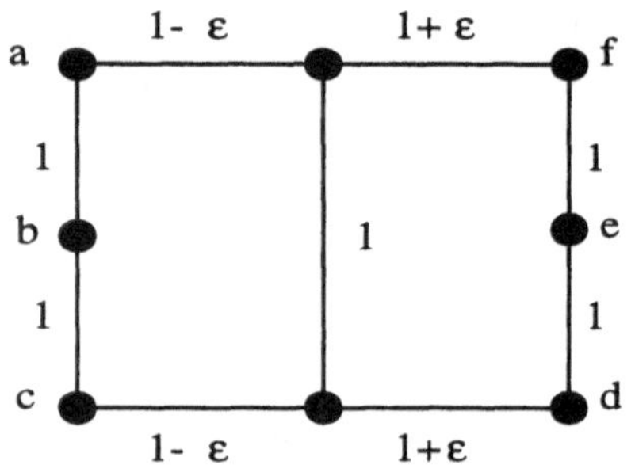

Figure3. For each ϵ with $-\frac{1}{2} \leq \epsilon \leq \frac{1}{2}$ the weighted graph on the right is an optimal realization of the metric on the left.

2 Hereditarily Optimal Realizations

Taking this into account, we now focus on one possible way out of this dilemma that also originally appeared in [9]: We define a *hereditarily optimal* – or, for short, an *h-optimal* – realization of d inductively with respect to $\#X$: If $\#X \leq 2$, then any optimal realization of d is defined to be h-optimal. If $\#X = k$ and h-optimal realizations have been defined previously for any metric defined on a set Y with $\#Y < k$, then a realization $G = (V, E, w)$ of d is defined to be h-optimal if $\|G\|$ is minimal with respect to the property that, for any $Y \subsetneq X$, there is some $E' \subseteq E$ with $Y \subseteq V' := \cup E' := \cup_{\{u,v\} \in E'} \{u, v\}$ such that $G' := (V', E', w|_{E'})$ is an h-optimal realization of $d|_Y$.

Even though, at first sight, it might appear that finding h-optimal realizations could be even harder than finding optimal ones, and that they are even less likely to be unique, the converse – actually – is true: Consider the set $P(X, d)$ consisting of those functions $f : X \to \mathbb{R}$ that satisfy the condition

$$f(x) + f(y) \geq d(x, y)$$

for all $x, y \in X$. To each $f \in P(X, d)$ associate a graph $K(f)$ that has vertex set X, and whose edge set consists of those subsets $\{x, y\} \subseteq X$ for which

$$f(x) + f(y) = d(x, y)$$

holds. Next, define a weighted graph $\Gamma_d = (V_d, E_d, w_d)$ with vertex set

$$V_d := \{f \in P(X, d) \,:\, K(f) \text{ is connected and not bipartite}\},$$

edge set

$$E_d := \{\{f, g\} \in \binom{V_d}{2} \,:\, K((f+g)/2) \text{ is connected and bipartite }\},$$

and weighting

$$w_d : E_d \to \mathbb{R}_{\geq 0}$$

defined by

$$w_d(\{f, g\}) := \sup_{x \in X} |f(x) - g(x)|.$$

Note that we can consider X as being a subset of V_d – simply associate the map $h_x : X \to \mathbb{R}$ defined by $h_x(y) := d(x, y)$ for all $y \in X$, to each element $x \in X$.

Then, it follows from [9, Theorem 7] that V_d is finite, and that the following holds:

- The graph Γ_d is an h-optimal realization of d.
- If $\Gamma = (V, E, w)$ is any other h-optimal realization of d, then Γ is essentially isomorphic to Γ_d i.e. it becomes isomorphic to Γ_d once vertices $v \in V - X$ of degree two have been deleted one by one and the corresponding edges $e_1 = \{u_1, v\}, e_2 = \{v, u_2\} \in E$ have been replaced by the single edge $\{u_1, u_2\}$ that is given weight $w(e_1) + w(e_2)$.

Thus, h-optimal realizations have the advantage that they can not only be described in an explicit way, but they are essentially unique, too. Moreover, it follows from results contained in [9] that Γ_d is isomorphic to the unique weighted tree representing d whenever d is tree-like.

3 A connection with T-theory

Although the definition of Γ_d might seem a bit strange, within *T-theory* it makes a lot of sense [19]. In fact, considering the set $P(X, d)$ as an unbounded polytope in $\mathbb{R}^X$, it is shown in [9] that Γ_d is a subgraph of the weighted graph $T_d^{(1)} = (F_0, F_1, w_P)$ that has vertex set F_0 consisting of the 0-dimensional faces (i.e. vertices) of $P(X, d)$, edge set F_1 consisting of those $\{f, g\} \in \binom{F_0}{2}$ for which f and g are the vertices of a 1-dimensional face in $P(X, d)$, and weighting w_P defined in exactly the same way as w_d was defined above. The graph

$T_d^{(1)}$ is better known as the 1-skeleton of the *tight span* $T(X,d)$ of d, a polytope consisting of the compact faces of $P(X,d)$, and the main object of study in T-theory. Thus, in theory at least, it is possible to compute $T_d^{(1)}$, and hence Γ_d, using any of the many packages that are available for computing polytopes (for example, see *http://ftp.zib.de/Packages/mathprog/polyth/index.html* for links to a selection of such packages e.g. CDD and PORTA).

We now consider a simple example of an h-optimal realization: As a consequence of [22, Theorem 3.2], the complete bipartite graph $G := K_{3,3}$ (in which every edge is assigned weight one) is an optimal realization of the metric d_G induced by G on its vertex set. Moreover, the h-optimal realization of d_G also coincides with G, but it can be seen that $\Gamma_{d_G} \neq T_{d_G}^{(1)}$ holds (see Figure 4). Hence, it may be of some interest to characterize those metrics d for which $\Gamma_d = T_d^{(1)}$ holds.

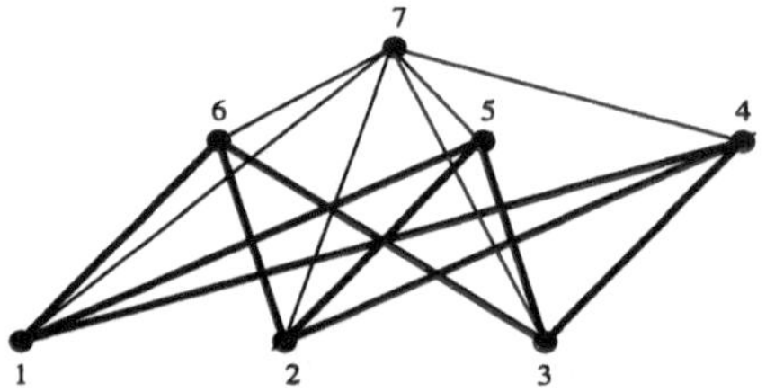

Figure4. The graph $T^{(1)}_{d_{K_{3,3}}}$ (all edges have weight one) with the subgraph $\Gamma_{d_{K_{3,3}}}$ $(= K_{3,3})$ indicated in bold.

In [14], we give an explicit answer to this question: Defining, for any four elements $u, v, x, y \in X$,

$$d(xy|uv) := \max\{d(x,u) + d(y,v), d(x,v) + d(y,u)\} - d(x,y) - d(u,v)$$

and putting $\alpha(xy|uv) := \max(d(xy|uv), 0)$, we see that those metrics d for which $\Gamma_d = T_d^{(1)}$ holds can be characterized by the following *five-point* condition:

- For all $t, x, y, u, v \in X$, the inequality

$$d(xy|uv) \leq \alpha(xt|uv) + \alpha(xy|ut)$$

holds.

Such metrics are called *totally decomposable* [2]. As a consequence of recent results appearing in [11,12,13,15], we also see in [14] that if, in addition, d satisfies the following *six-point* condition:

- For every subset Y of X of cardinality 6, there exists a pair $u, v \in Y$ of distinct elements with

$$0 \leq d(xy|uv)$$

for all $x, y \in Y - \{u, v\}$ (and hence for all $x, y \in Y$),

then the underlying graph (V_d, E_d) of $\Gamma_d = (V_d, E_d, w_d)$ is isomorphic to the well-known *Buneman graph* [10], a graph that has been used previously with quite some success in phylogenetic analysis [4,5,16]. Thus, in summary we have the following result for metrics that satisfy the above five- and six-point conditions, metrics that are called *consistent* for short [15]:

Theorem 3.1. *If d is a consistent metric, then the weighted Buneman graph Γ_d is an h-optimal realization of d.*

4 Concluding Remarks

Finally, these results give us a framework for better understanding the uses of the *split-decomposition* technique in phylogenetic analysis [2,3]: Define d to be a *Kalmanson* metric [7,8,23] on X if there exists a labeling $X = \{x_1, x_2, \ldots, x_n\}$ $(n := \#X)$ of X such that $1 \leq i < k < j < l \leq n$ implies

$$d(x_i, x_j) + d(x_k, x_l) \geq \max(d(x_i, x_k) + d(x_j, x_l), d(x_i, x_l) + d(x_j, x_k)),$$

and note that the following inclusions for the various types of metrics that we have considered above hold:

$$\text{tree-like} \subsetneq \text{Kalmanson} \subsetneq \text{consistent} \subsetneq \text{totally decomposable}.$$

Moreover, if d is a Kalmanson metric, then it can be realized by a certain *outer-planar* graph [17] that is an isometric subgraph of Γ_d and is used by the program *SplitsTree* [18,21].

Further, even though a given metric d will not in general be totally decomposable, split-decomposition theory provides us with a unique maximal totally decomposable submetric d_{split} of d that satisfies the condition $P(X, d) = P(X, d_{split}) + P(X, d - d_{split})$ [2].

In practice, d_{split} tends to be Kalmanson so that it can be represented by the outer planar graph mentioned above. If d_{split} is not Kalmanson, then some variant of the Buneman graph is currently used by SplitsTree. In light of the above results, however, the graph $\Gamma_{d_{split}}$ (or some isometric subgraph of it) might be more appropriate in this situation. Thus, in conclusion, it appears worthwhile to develop new techniques for efficiently computing the h-optimal realization of d_{split} or even that of d, a task that – as we have seen above – is closely related to computing their tight spans [13].

References

1. Althöfer, I.: On optimal realizations of finite metric spaces by graphs, Discrete Comput. Geometry **3** (1988) 103–122
2. Bandelt, H.-J., Dress, A.: A canonical decomposition theory for metrics on a finite set, Adv. in Math. **92** (1992) 47–105
3. Bandelt H.-J., Dress, A.: Split decomposition: a new and useful approach to phylogenetic analysis of distance data, Molecular Phylogenetics and Evolution **1 (3)** (1992b) 242–252
4. Bandelt, H.-J., Forster, P., Sykes, B., Richards, M.: Mitochondrial portraits of human population using median networks, Genetics **141** (October 1995) 743–753
5. Barthélemy, J., Guenoche A.: Trees and Proximity Representations, John Wiley & Sons, Chichester New York Brisbane Toronto Singapore, 1991
6. Buneman, P.: The recovery of trees from measures of dissimilarity, In F. Hodson et al., Mathematics in the Archaeological and Historical Sciences, (pp.387-395), Edinburgh University Press, 1971
7. Chepoi, V., Fichet, B.: A note on circular decomposable metrics, Geometriae Dedicata **69**(3) (March 1998) 237–240
8. Christopher, G., Farach, M., Trick, M.: The structure of circular decomposable metrics, Algorithms—ESA '96 (Barcelona), Lecture Notes in Comput. Sci., **1136**, Springer, Berlin, (1996) 486–500
9. Dress, A.: Trees, tight extensions of metric spaces, and the cohomological dimension of certain groups: A note on combinatorial properties of metric spaces, Adv. in Math. **53** (1984) 321–402
10. Dress, A., Hendy, M., Huber, K., Moulton, V.: On the number of vertices and edges in the Buneman Graph, Ann. Combin. **1** (1997) 329–337
11. Dress, A., Huber, K., Moulton, V.: Some variations on a theme by Buneman, Ann. Combin. **1** (1997) 339–352
12. Dress, A., Huber, K.T., Moulton, V.: A Comparison between two distinct continuous models in projective cluster theory: The median and the tight-span construction, Ann. Combin. **2** (1998) 299-311
13. Dress, A., Huber, K.T., Moulton, V.: An explicit computation of the injective hull of certain finite metric spaces in terms of their associated Buneman complex, Mid Sweden University Mathematics Department Report No. 6 (1999)
14. Dress, A., Huber, K.T., Moulton, V.: Hereditarily optimal realizations of consistent metrics, in preparation
15. Dress, A., Huber, K.T., Koolen, J., Moulton, V.: Six points suffice: How to check for metric consistency, Mid Sweden University Mathematics Department Report No. 8 (2000)
16. Dress, A., Huber, K.T., Lockhart, P., Moulton, V.: Lite Buneman networks: A technique for studying plant speciation, Mid Sweden University Mathematics Department Report No. 1 (1999)
17. Dress, A., Huson, D.: Computing phylogenetic networks from split systems, submitted (1999)
18. Dress, A., Huson, D., Moulton, V.: Analyzing and visualizing distance data using SplitsTree, Discrete Applied Mathematics **71** (1996) 95–110
19. Dress, A., Moulton, V., Terhalle, W.: T-theory: An Overview, Europ. J. Combinatorics **17** (1996) 161–175

20. Hillis, D., Moritz, C., Barbara, K.: Phylogenetic Inference, In Molecular Systematics, D. M. Hillis et al. , (pp.407-514), Sinauer, 1996.
21. Huson, D.: SplitsTree: a program for analyzing and visualizing evolutionary data, Bioinformatics **14 (1)** (1998) 68–73
22. Imrich, W., Simoes-Pereira, J., Zamfirescu, C.: On optimal emdeddings of metrics in graphs, Journal of Combinatorial Theory, Series B, **36, No.1**, (1984) 1–15
23. Kalmanson K.: Edgeconvex circuits and the travelling salesman problem, Canadian Jour. Math., **27** (1975) 1000–1010
24. Zaretsky, K.: Reconstruction of a tree from the distances between its pendant vertices, Uspekhi Math. Nauk (Russian Mathematical Surveys) **20** (1965) 90–92 (in Russian)

Characterization of Tensor Symmetries by Group Ring Subspaces and Computation of Normal Forms of Tensor Coordinates

Bernd Fiedler

Alfred-Rosch-Str. 13, D-04249 Leipzig, Germany
Email: Bernd.Fiedler.RoschStr.Leipzig@t-online.de

Abstract We consider the problem to determine normal forms of the coordinates of covariant tensors $T \in \mathcal{T}_r V$ of order r over a finite-dimensional $\mathbb{K}$-vector space, $\mathbb{K} = \mathbb{R}, \mathbb{C}$. A connection between such tensors and the group ring $\mathbb{K}[\mathcal{S}_r]$ can be established by assigning a group ring element $T_b := \sum_{p \in \mathcal{S}_r} T(v_{p(1)}, \ldots, v_{p(r)})\, p \in \mathbb{K}[\mathcal{S}_r]$ to every tensor $T \in \mathcal{T}_r V$ and every r-tuple $b = (v_1, \ldots, v_r) \in V^r$ of vectors. Then each symmetry class $\mathcal{T} \subseteq \mathcal{T}_r V$ of tensors can be characterized by a linear subspace $W \subseteq \mathbb{K}[\mathcal{S}_r]$ which is spanned by all T_b of the $T \in \mathcal{T}$. The elements of the orthogonal subspace $W^{\perp} \subseteq \mathbb{K}[\mathcal{S}_r]^*$ of W within the dual space $\mathbb{K}[\mathcal{S}_r]^*$ yield the set of all linear identities that are fulfilled by the coordinates of all tensors $T \in \mathcal{T}$. These identities can be used to calculate linearly independent coordinates (i.e. normal forms) of the $T \in \mathcal{T}$.

If the $T \in \mathcal{T}$ are single tensors and $\dim V \geq r$, then W is a left ideal $W = \mathbb{K}[\mathcal{S}_r] \cdot e$ generated by an idempotent e. In the case of tensor products $T_1 \otimes T_2 \otimes \ldots \otimes T_m$ or $T \otimes \ldots \otimes T$ (m-times), W is a left ideal whose structure is described by a Littlewood-Richardson product $[\alpha_1][\alpha_2] \ldots [\alpha_m]$ or a plethysm $[\alpha] \odot [m]$, respectively. We have also treated the cases in which $\dim V < r$ or the $T \in \mathcal{T}$ contain aditional contractions of index pairs. In these cases characterizing linear subspaces $W \subseteq \mathbb{K}[\mathcal{S}_r]$ with a structure $W = f \cdot \mathbb{K}[\mathcal{S}_r] \cdot e$ or $W = \sum_{i=1}^{k} a_i \cdot \mathbb{K}[\mathcal{S}_r] \cdot e$ come into play. Here $e, f \in \mathbb{K}[\mathcal{S}_r]$ are idempotents.

We have implemented a Mathematica package by which the characterizing idempotents and bases of the spaces W and the identities from $W^{\perp}$ can be determined in all above cases. This package contains an ideal decomposition algorithm and tools such as the Littlewood-Richardson rule, plethysms and discrete Fourier transforms.

1 Introduction

The investigations of this paper are motivated by normal form problems which occur within symbolic calculations with tensor expressions by means of a computer algebra system. Such calculations are very important in differential geometry, tensor analysis and general relativity theory.

Let us consider real or complex linear combinations

$$\tau = \sum_{i=1}^{n} \alpha_i T_{(i)} \quad , \quad \alpha_i \in \mathbb{R}, \mathbb{C} \tag{1}$$

of expressions $T_{(i)}$ which are formed from the coordinates of certain tensors A, B, C, ... by multiplication and, possibly, contractions of some pairs of indices. An example of such an expression is

$$A_{iabc}\, A^{a}{}_{jkd}\, B^{bd}{}_{e}\, C^{ec} \quad . \tag{2}$$

In (2) we use Einstein's summation convention. Now we aim to carry out symbolic calculations with expressions of the type (1), (2) according to the rules of the Ricci calculus.

Nowadays, there exists a whole string of computer programs[1] such as the Mathematica packages MathTensor, Ricci, CARTAN, the Maple packages GRTensor, Maple Tensor and the REDUCE package REDTEN which can be used for symbolic tensor calculations. However, the following difficulties cannot be surmounted generally by using these packages.

Assume that some of the tensors $A, B, C, ...$ possess symmetries relating to permutations of indices and/or fulfill linear identities. For instance, the Riemannian curvature tensor R of the Levi-Civita connection of a Riemannian metric g has the symmetry properties

$$R_{ijkl} \;=\; -R_{jikl} = -R_{ijlk} \;=\; R_{klij} \tag{3}$$

and fulfills the first Bianchi identity

$$R_{ijkl} + R_{iklj} + R_{iljk} = 0\,. \tag{4}$$

Another example is the identity[2]

$$C_{iabc}\, C_{j}{}^{abc} - \tfrac{1}{4}\, g_{ij}\, C_{abcd}\, C^{abcd} = 0 \tag{5}$$

for the Weyl tensor C of a 4-dimensional pseudo-Riemannian manifold. Relation (5) can be interpreted as a linear identity for a tensor with coordinates $g_{i_1 i_2} C_{i_3 i_4 i_5 i_6} C_{i_7 i_8 i_9 i_{10}}$.

If relations of the type (3) – (5) exist for the tensors $A, B, C, \ldots$ in (1), then there is a possibility to express some of the $T_{(i)}$ in (1) by the others. This is a hard problem[3] within symbolic computer calculations with tensor expressions. We need efficient algorithms to detect dependences between the

[1] We refer to [6, Appendix A.6] for references of these program packages.

[2] See V. Wünsch [21]. A large collection of identities for the curvature tensor, Ricci tensor, Weyl tensor, etc. and covariant derivatives of these tensors is contained in the appendix of the book [8] of P. Günther.

[3] This problem is caused not only by symmetries and identities but also by simple permutations of names of indices and permutations of factors within a tensor expression. For instance, the two expressions $T_{abc}\, T^{c}{}_{de}\, T^{e}{}_{f}{}^{a}\, T^{bdf}$ and $T_{abc}\, T_{de}{}^{a}\, T_{f}{}^{eb}\, T^{cfd}$ are equal. This becomes visible, if we rename the indices of the first expression according to the rule $a \to c\,,\ b \to f\,,\ c \to d\,,\ d \to e\,,\ e \to a\,,\ f \to b$ and raise or lower suitable indices. The determination of such transformations is non-trivial.

$T_{(i)}$ and to reduce sums (1) to linear combinations of linearly independent $T_{(i)}$. This problem is connected with the problem of the determination of normal forms of tensor expressions of type (2).

The connection between tensors and the representation theory of symmetric groups has been considered already in books [18] of J.A. Schouten (1924), [20] of H. Weyl (1939) and [1] of H. Boerner (1955). In the 1940s Littlewood has developed and used tools such as the Littlewood-Richardson rule and plethysms for the investigation of tensors (see references in [7] and [13, Appendix]). Applying the same methods, Fulling, King, Wybourne and Cummins [7] have calculated lists of normal form terms of polynomials of the Riemann curvature tensor and its derivatives.

In [7] they have formulated the following steps to solve the above term combination problem for tensors:

(a) Generate the space spanned by the set of homogeneous monomials of a definite 'order' or 'degree of homogeneity' formed from the coordinates of tensors of relevance by multiplication and index-pair contraction.
(b) Construct a basis of this space.
(c) Present an algorithm for expressing an arbitrary element of the space in terms of the basis.

The results of our present paper yield a solution of (b) and (c) for arbitrary tensors. A short description of details is given in the summary.

2 Tensors and the Group Ring of a Symmetric Group

Let $\mathbb{K}$ be one of the fields $\mathbb{R}$ or $\mathbb{C}$, which are usually used in differential geometry.

Definition 2.1. Let V be a finite-dimensional vector space over $\mathbb{K}$, V^* be the dual space of V and let $\kappa := V_1 \times \ldots \times V_r$ be a cartesian product in which p factors conincide with V^* $(1 \leq p \leq r)$ and $q := r - p$ factors coincide with V. A multilinear mapping $T : \kappa = V_1 \times \ldots \times V_r \to \mathbb{K}$ is called a *tensor of type* κ over V.

The set $\mathcal{T}_\kappa V$ of all tensors of type κ over V is a $\mathbb{K}$-vector space. The elements of $\mathcal{T}_\kappa V$ are also named *p-times contravariant* and *q-times covariant* tensors. We denote by $\mathcal{T}_r V$ the space of all covariant tensors of order r that belongs to the type $\kappa = V \times \ldots \times V =: V^r$.

Let $\{n_1, \ldots, n_d\}$ be a basis of V and $\{n^1, \ldots, n^d\}$ be the corresponding *dual basis* of V^*, which fulfills $n^i(n_j) = \delta^i_j$. Then the *coordinates* of a tensor $T \in \mathcal{T}_\kappa V$ are obtained by putting these basis vectors into the multilinear mapping T. For instance a tensor T of type $\kappa = (V^*)^p \times V^q$ has the coordinates

$$T^{i_1 \ldots i_p}{}_{j_1 \ldots j_q} := T(n^{i_1}, \ldots, n^{i_p}, n_{j_1}, \ldots, n_{j_q}) \in \mathbb{K}\,. \tag{6}$$

Definition 2.2. A *fundamental tensor* or *metric tensor* is a covariant tensor $g \in \mathcal{T}_2 V$ that is symmetric and non-singular, i.e., its coordinates fulfill

$$g_{ij} = g_{ji} \quad \text{and} \quad \det[g_{ij}] \neq 0\,.$$

The elements g^{ij} of the inverse matrix $[g_{ij}]^{-1}$ are the coordinates of a symmetric, non-singular, contravariant tensor of order 2 over V. If a fundamental tensor is given on V, then we can *raise* or *lower* indices of an arbitrary tensor $T \in \mathcal{T}_\kappa V$ by means of g^{ij} and g_{ij} according to the rules

$$T_{*\ldots*}{}^{i}{}_{*\ldots*} := g^{ij}\, T_{*\ldots*\,j\,*\ldots*} \quad , \quad T_{*\ldots*\,i\,*\ldots*} := g_{ij}\, T_{*\ldots*}{}^{j}{}_{*\ldots*}\,. \tag{7}$$

The stars "$*$" represent upper or lower indices of T that remain fix. The transformations (7) are vector space isomorphisms $\mathcal{T}_\kappa V \to \mathcal{T}_{\kappa'} V$ of those spaces of tensors which contain the tensors involved. A related operation is the *contraction* of a tensor on an upper and a lower index

$$T_{*\ldots*\,i\,*\ldots*}{}^{i}{}_{*\ldots*} \text{ or } T_{*\ldots*}{}^{i}{}_{*\ldots*\,i\,*\ldots*} \mapsto S_{*\ldots**\ldots**\ldots*}\,, \tag{8}$$

which transforms a p-times contravariant and q-times covariant tensor T into a $(p-1)$-times contravariant and $(q-1)$-times covariant tensor S.

Since every finite-dimensional vector space can be provided with a metric tensor g, we can reduce the investigation of tensors T of an arbitrary type to the investigation of covariant tensors by lowering of all upper indices of T.

Now we make use of the following connection between tensors from $\mathcal{T}_r V$ and elements of the *group ring* $\mathbb{K}[\mathcal{S}_r]$ of a symmetric group $\mathcal{S}_r$ over $\mathbb{K}$.

Definition 2.3. Any tensor $T \in \mathcal{T}_r V$ and any r-tuple $b := (v_1, \ldots, v_r) \in V^r$ of r vectors from V induce a function $T_b : \mathcal{S}_r \to \mathbb{K}$ according to the rule

$$T_b(p) := T(v_{p(1)}, \ldots, v_{p(r)}) \quad , \quad p \in \mathcal{S}_r\,. \tag{9}$$

We identify this function with the group ring element $\sum_{p\in\mathcal{S}_r} T_b(p)\, p \in \mathbb{K}[\mathcal{S}_r]$, which we denote by T_b, too.

We allow the linear dependence of the v_i and repetitions of vectors in the above r-tuple b. Obviously, two tensors $S, T \in \mathcal{T}_r V$ fulfill $S = T$ iff $S_b = T_b$ for all r-tuples $b \in V^r$.

Definition 2.4. If $T \in \mathcal{T}_r V$ and $a = \sum_{p\in\mathcal{S}_r} a(p)\, p \in \mathbb{K}[\mathcal{S}_r]$, then we denote by aT that tensor from $\mathcal{T}_r V$ which is defined by

$$(aT)(v_1, \ldots, v_r) := \sum_{p\in\mathcal{S}_r} a(p)\, T(v_{p(1)}, \ldots, v_{p(r)}) \quad , \quad v_i \in V. \tag{10}$$

By rule (10), every group ring element $a \in \mathbb{K}[\mathcal{S}_r]$ acts as so-called *symmetry operator* $a : T \mapsto aT$ on $\mathcal{T}_r V$. Equation (10) is equivalent to

$$(aT)_{i_1\ldots i_r} = \sum_{p\in\mathcal{S}_r} a(p)\, T_{i_{p(1)}\ldots i_{p(r)}}\,. \tag{11}$$

Definition 2.5. We denote by $* : \mathbb{K}[\mathcal{S}_r] \to \mathbb{K}[\mathcal{S}_r]$ the mapping $a = \sum_{p \in \mathcal{S}_r} a(p)\, p \;\mapsto\; a^* := \sum_{p \in \mathcal{S}_r} a(p)\, p^{-1}$. Furthermore, if $p \in \mathcal{S}_r$ and $b = (v_1, \ldots, v_r) \in V^r$, then we denote by pb the r-tuple $pb := (v_{p(1)}, \ldots, v_{p(r)})$.

Many of our calculations are based on

Lemma 2.6. *If $a = \sum_{q \in \mathcal{S}_r} a(q)\, q \in \mathbb{K}[\mathcal{S}_r]$, $T \in \mathcal{T}_r V$, $p, q \in \mathcal{S}_r$ and $b = (v_1, \ldots, v_r) \in V^r$, then we have*

$$T_b(p \circ q) = (qT)_b(p) \;=\; T_{pb}(q) \qquad (12) \qquad\qquad (aT)_b = T_b \cdot a^* \qquad (14)$$

$$q(pb) = (p \circ q)b \qquad (13) \qquad\qquad T_b = p \cdot T_{pb}\,. \qquad (15)$$

Proof. Let us use the notation $pb = (w_1, \ldots, w_r)$ with $w_i = v_{p(i)}$. Then equation (13) follows from

$$q(pb) \;=\; (w_{q(1)}, \ldots, w_{q(r)}) \;=\; (v_{p(q(1))}, \ldots, v_{p(q(r))}) \;=\; (p \circ q)b\,.$$

Equation (12) is a consequence of

$$T_b(p \circ q) \;=\; T(v_{p\circ q(1)}, \ldots, v_{p\circ q(r)}) \;=\; T(w_{q(1)}, \ldots, w_{q(r)}) \;=\; T_{pb}(q)$$

and

$$T(w_{q(1)}, \ldots, w_{q(r)}) \;=\; (qT)(w_1, \ldots, w_r) \;=\; (qT)(v_{p(1)}, \ldots, v_{p(r)}) \;=\; (qT)_b(p)\,.$$

Now equation (14) can be proved by the calculation

$$\begin{aligned}(aT)_b = \sum_{q \in \mathcal{S}_r} a(q)\,(qT)_b \;&=\; \sum_{p,q \in \mathcal{S}_r} a(q)\,(qT)_b(p)\, p \;=\; \sum_{p,q \in \mathcal{S}_r} a(q)\, T_b(p \circ q)\, p \\ &=\; \sum_{q,s \in \mathcal{S}_r} a(q)\, T_b(s)\, s \circ q^{-1} \;=\; T_b \cdot a^*\,.\end{aligned}$$

Finally, we obtain equation (15) from

$$p \cdot T_{pb} \;=\; \sum_{q \in \mathcal{S}_r} T_{pb}(q)\, p \circ q \;=\; \sum_{q \in \mathcal{S}_r} T_b(p \circ q)\, p \circ q \;=\; T_b\,. \;\square$$

Clearly, the group ring elements of the type T_b generate a linear subspace $\mathcal{L}_{\mathbb{K}}\{T_b \mid T \in \mathcal{T}_r V\,, b \in V^r\}$ of $\mathbb{K}[\mathcal{S}_r]$. (Here $\mathcal{L}_{\mathbb{K}}$ denotes the forming of the linear closure with respect to $\mathbb{K}$.) Moreover, we have

Proposition 2.7. *If $\dim V \geq r$, then $\mathcal{S}_r \subset \{T_b \mid T \in \mathcal{T}_r V\,, b \in V^r\}$. This means that both $\mathbb{R}[\mathcal{S}_r]$ and $\mathbb{C}[\mathcal{S}_r]$ are spanned by the set of all T_b for arbitrary choice of the field $\mathbb{K} \in \{\mathbb{R}\,, \mathbb{C}\}$ over which the tensors are defined.*

Proof. Let $n_1, \ldots, n_d$ be a basis of V. For $1 \leq j_1, \ldots, j_r \leq d$ we assign to every fixed r-tuple $b := (n_{j_1}, \ldots, n_{j_r}) \in V^r$ a tensor $T^{(b)} \in \mathcal{T}_r V$ which fulfills $T^{(b)}(n_{i_1}, \ldots, n_{i_r}) = 1$ if $(n_{i_1}, \ldots, n_{i_r}) = b$ and $T^{(b)}(n_{i_1}, \ldots, n_{i_r}) = 0$ else. If $\dim V \geq r$, we use the r-tuple $b := (n_1, \ldots, n_r)$ of the first r basis vectors. Then $T_b^{(pb)} = p$ for all $p \in \mathcal{S}_r$. $\square$

3 Symmetries of Tensors

The following symmetry concepts for tensors can be found in the literature.

Definition 3.1. (i) Let $C \subseteq \mathcal{S}_r$ be a subgroup of $\mathcal{S}_r$ and $\epsilon : C \to \mathbb{K}^\times$ be a homomorphism of C onto a finite subgroup of the multiplicative subgroup $\mathbb{K}^\times := \mathbb{K} \setminus \{0\}$ of $\mathbb{K}$. We say that a tensor $T \in \mathcal{T}_r V$ possesses the *commutation symmetry* (C, ϵ) if

$$\forall c \in C : \quad cT = \epsilon(c)\, T\,. \tag{16}$$

(ii) Let $G \subseteq \mathcal{S}_r$ be a subgroup of $\mathcal{S}_r$ and χ be a $\mathbb{K}$-valued irreducible character of G. We form the group ring element[4] $\tilde{\chi} := \frac{\chi(\mathrm{id})}{|G|} \sum_{g \in G} \chi(g)\, g$ from χ. A tensor $T \in \mathcal{T}_r V$ is called *symmetric*[5] with respect to G and χ if

$$\tilde{\chi}^* T = T\,. \tag{17}$$

(iii) Let $\mathfrak{r} \subseteq \mathbb{K}[\mathcal{S}_r]$ be a right ideal of $\mathbb{K}[\mathcal{S}_r]$ for which an $a \in \mathfrak{r}$ and a $T \in \mathcal{T}_r V$ exist such that $aT \neq 0$. Then the tensor set $\mathcal{T}_\mathfrak{r} := \{aT \mid a \in \mathfrak{r}\,,\, T \in \mathcal{T}_r V\}$ is called the *symmetry class*[6] of tensors defined by $\mathfrak{r}$.

(iv) Let $a_1, \ldots, a_n \in \mathbb{K}[\mathcal{S}_r]$ be a finite set of group ring elements. We say that a tensor $T \in \mathcal{T}_r V$ possesses a *symmetry*[7] defined by $a_1, \ldots, a_n$ if T satisfies the linear equation system

$$a_i T = 0 \qquad (i = 1, \ldots, n)\,. \tag{18}$$

It is easy to see that every symmetry concept (k) in this definition is a special case of the next symmetry concept (k + 1) (see [6, Sec. III.2.1]).

Lemma 3.2. *Let e be a generating idempotent of the right ideal $\mathfrak{r} \subseteq \mathbb{K}[\mathcal{S}_r]$. Then $\mathcal{T}_\mathfrak{r}$ fulfills $\mathcal{T}_\mathfrak{r} = \{eT \mid T \in \mathcal{T}_r V\}$ and a tensor $T \in \mathcal{T}_r V$ belongs to $\mathcal{T}_\mathfrak{r}$ iff*

$$eT = T\,. \tag{19}$$

Furthermore, if (C, ϵ) is a commutation symmetry, then it can be shown that $\epsilon(C) \subseteq \mathbb{C}^\times$ is a finite cyclic subgroup of complex units and ϵ is a linear (1-dimensional) character of C. Moreover, $\tilde{\epsilon} := \frac{1}{|C|} \sum_{c \in C} \epsilon(c)\, c$ is an idempotent of $\mathbb{C}[C] \subseteq \mathbb{C}[\mathcal{S}_r]$. If $\mathbb{K} = \mathbb{R}$, then $\epsilon(C) \subseteq \{1, -1\}$ is a subgroup of the multiplicative group $\{1, -1\}$ and $\tilde{\epsilon}$ possesses only rational coefficients.

Lemma 3.3. *Let (C, ϵ) be a commutation symmetry. Then a tensor $T \in \mathcal{T}_r V$ possesses the symmetry (C, ϵ) iff T satisfies $\tilde{\epsilon}^* T = T$.*

[4] Every such $\tilde{\chi}$ is a centrally primitive idempotent in $\mathbb{K}[G]$.

[5] See [15, pp.151,153,157]. In [15] this symmetry is defined for more general multilinear mappings $\Phi : V^r \to W$ which take their values in a vector space W.

[6] See [1, p.127]. In [1], the case $\mathbb{K} = \mathbb{C}$ is mainly considered.

[7] See [3, p.601].

We refer to [6, Sec. III.2.1] for details of the easy proofs of all statements made in this Section. From the above Lemmas we see the hierarchy (i) $\subset$ (ii) $\subset$ (iii) of the symmetry concepts in Definition 3.1. Moreover, (i), (ii) and (iii) are special cases of (iv) since (16), (17) and (19) can be written as homogeneous linear equations $(c - \epsilon(c)\,\mathrm{id})T = 0$, $(\tilde{\chi}^* - \mathrm{id})T = 0$ and $(e - \mathrm{id})T = 0$.

In [6, Sec. III.2.2] we have determined a complete list of all commutation symmetries (i) which are possible for tensors $T \in \mathcal{T}_r V$ with $r \leq 6$.

4 Ideals or Subspaces Characterizing Tensors

In this Section we show that the T_b of tensors $T \in \mathcal{T}_r V$ which possesses certain symmetry properties lie in certain linear subspaces W of $\mathbb{K}[\mathcal{S}_r]$. We construct such subspaces for all cases that are important for symbolic tensor calculations. For the sake of shortness, we give only essential parts of the proofs and refer for more details to [6, Sec. III.3].

4.1 Characterizing Left Ideals of "Simple" Tensors

In this Section we consider single covariant tensors $T \in \mathcal{T}_r V$.

Proposition 4.1. *Let $e \in \mathbb{K}[\mathcal{S}_r]$ be an idempotent. Then a $T \in \mathcal{T}_r V$ fulfills the condition[8] $eT = T$ iff all T_b of T lie in the left ideal $\mathfrak{l}$ generated by e^*:*

$$\forall\, b = (v_1, \ldots, v_r) \in V^r : \quad T_b \;\in\; \mathfrak{l} := \mathbb{K}[\mathcal{S}_r] \cdot e^* . \tag{20}$$

Proof. An application of (14) shows that $T = eT$ iff $T_b = T_b \cdot e^*$ for all $b \in V^r$. □

Proposition 4.2. *Let $a_1, \ldots, a_m \in \mathbb{K}[\mathcal{S}_r]$ be given group ring elements. A tensor $T \in \mathcal{T}_r V$ satisfies a system of linear identities[9]*

$$a_i T = 0 \quad , \quad a_i \in \mathbb{K}[\mathcal{S}_r]\,,\; i = 1, \ldots, m \tag{21}$$

iff all $T_b \in \mathbb{K}[\mathcal{S}_r]$ of T lie in the left annihilator ideal $\mathfrak{l}$ of the set $\{a_1^*, \ldots, a_m^*\}$:

$$\forall\, b \in V^r : \;\; T_b \;\in\; \mathfrak{l} := \{x \in \mathbb{K}[\mathcal{S}_r] \mid x \cdot a_i^* = 0\,,\, i = 1, \ldots, m\} . \tag{22}$$

Proof. Proposition 4.2 results from an application of (14) to (21). □

Proposition 4.3. *Let $\mathfrak{l} \subseteq \mathbb{K}[\mathcal{S}_r]$ be a left ideal which characterizes the symmetry class $\mathcal{T}_{\mathfrak{l}^*}$ of tensors (see (iii) in Definition 3.1). Then we have*

$$T \in \mathcal{T}_{\mathfrak{l}^*} \iff \forall\, b \in V^r : \;\; T_b \in \mathfrak{l} . \tag{23}$$

[8] This condition means that T has a symmetry (i), (ii) or (iii) from Definition 3.1.
[9] This means that T possesses a symmetry of type (iv) from Definition 3.1.

Proof. If e^* is a generating idempotent of $\mathfrak{l}$, then Proposition 4.3 follows from the conclusions

$$\forall b \in V^r : T_b \in \mathfrak{l} \quad \Leftrightarrow \quad \forall b \in V^r : T_b = T_b \cdot e^* = (eT)_b \quad \Leftrightarrow \quad T = eT \,. \square$$

Proposition 4.4. *If* $\dim V \geq r$, *then every left ideal* $\mathfrak{l} \subseteq \mathbb{K}[\mathcal{S}_r]$ *fulfills* $\mathfrak{l} = \mathcal{L}_{\mathbb{K}}\{T_b \mid T \in \mathcal{T}_{\mathfrak{l}^*}\,,\, b \in V^r\}$.

Proof. Proposition 2.7, Lemma 3.2 and (14) yield $\{T_b \mid T \in \mathcal{T}_{\mathfrak{l}^*}\,,\, b \in V^r\} \supset \mathcal{S}_r \cdot e^*$ where e^* is a generating idempotent of $\mathfrak{l}$. From this the statement follows. □

4.2 Characterizing Left Ideals of Tensor Products

Now we consider left ideals $\mathfrak{l} \subseteq \mathbb{K}[\mathcal{S}_r]$ which characterize symmetries of tensor products of tensors.

Proposition 4.5. *Let* $\mathfrak{l}_i \subseteq \mathbb{K}[\mathcal{S}_{r_i}]$ $(i = 1, \ldots, m)$ *be left ideals and* $T^{(i)} \in \mathcal{T}_{\mathfrak{l}_i^*} \subseteq \mathcal{T}_{r_i}V$ *be* r_i*-times covariant tensors from the symmetry classes characterized by the* $\mathfrak{l}_i$. *Consider the product*

$$T := T^{(1)} \otimes \ldots \otimes T^{(m)} \in \mathcal{T}_r V \quad , \quad r := r_1 + \ldots + r_m \,. \tag{24}$$

For every i *we define an embedding*

$$\iota_i : \mathcal{S}_{r_i} \to \mathcal{S}_r \quad , \quad (\iota_i s)(k) := \begin{cases} \Delta_i + s(k - \Delta_i) & \text{if } r_{i-1} < k \leq r_i \\ k & \text{else} \end{cases} \tag{25}$$

where $\Delta_i := r_0 + \ldots + r_{i-1}$ *and* $r_0 := 0$. *Then the* T_b *of the tensor (24) fulfill*

$$\forall b \in V^r : T_b \in \mathfrak{l} := \mathbb{K}[\mathcal{S}_r] \cdot \mathcal{L}\{\tilde{\mathfrak{l}}_1 \cdot \ldots \cdot \tilde{\mathfrak{l}}_m\} = \mathbb{K}[\mathcal{S}_r] \cdot (\tilde{\mathfrak{l}}_1 \otimes \ldots \otimes \tilde{\mathfrak{l}}_m) \tag{26}$$

where $\tilde{\mathfrak{l}}_i := \iota_i(\mathfrak{l}_i)$ *are the embeddings of the* $\mathfrak{l}_i$ *into* $\mathbb{K}[\mathcal{S}_r]$ *induced by the* ι_i. *If* $\dim V \geq r$, *then the above left ideal* $\mathfrak{l}$ *is generated by all* $T_b \in \mathbb{K}[\mathcal{S}_r]$ *which are formed from tensor products (24) of arbitrary tensors* $T^{(i)} \in \mathcal{T}_{\mathfrak{l}_i^*}$.

Proof. We form the Young subgroup $G := G_1 \cdot \ldots \cdot G_m \subseteq \mathcal{S}_r$ with $G_i := \iota_i(\mathcal{S}_{r_i}) \subseteq \mathcal{S}_r$ and a system $\mathcal{R}$ of representatives of the left cosets $p \circ G$ of $\mathcal{S}_r$ relative to G. Then we can carry out the following calculation for every r-tuple $b \in V^r$:

$$\begin{aligned}
T_b &= \sum_{p \in \mathcal{R}} \sum_{g \in G} T_b(p \circ g)\, p \circ g = \sum_{p \in \mathcal{R}} p \cdot \Big(\sum_{g \in G} T_{pb}(g)\, g \Big) \\
&= \sum_{p \in \mathcal{R}} p \cdot \Big(\sum_{i=1}^{m} \sum_{g_i \in G_i} T_{pb}(g_1 \cdot \ldots \cdot g_m)\, g_1 \cdot \ldots \cdot g_m \Big) \\
&= \sum_{p \in \mathcal{R}} p \cdot \Big(\sum_{\substack{s_i \in \mathcal{S}_{r_i} \\ i = 1, \ldots, m}} T^{(1)}_{(pb)_1}(s_1) \cdots T^{(m)}_{(pb)_m}(s_m)\, \iota_1(s_1) \cdot \ldots \cdot \iota_m(s_m) \Big) \\
&= \sum_{p \in \mathcal{R}} p \cdot \iota_1 \Big(\sum_{s_1 \in \mathcal{S}_{r_1}} T^{(1)}_{(pb)_1}(s_1)\, s_1 \Big) \cdots \iota_m \Big(\sum_{s_m \in \mathcal{S}_{r_m}} T^{(m)}_{(pb)_m}(s_m)\, s_m \Big) \\
&= \sum_{p \in \mathcal{R}} p \cdot \iota_1 \big(T^{(1)}_{(pb)_1} \big) \cdot \ldots \cdot \iota_m \big(T^{(m)}_{(pb)_m} \big) \,.
\end{aligned}$$

In the first line we have used (12). The third line results after substitutions $g_i = \iota_i(s_i)$, $s_i \in \mathcal{S}_{r_i}$. By $(pb)_i$ we denote the subsequence of vectors from pb which are put into the tensor $T^{(i)}$. The elements of every such $(pb)_i$ are numbered with $1, \ldots, r_i$.

Since $T^{(i)}_{(pb)_i} \in \mathfrak{l}_i$ and $\iota_i(T^{(i)}_{(pb)_i}) \in \tilde{\mathfrak{l}}_i$ for every i, statement (26) follows from the last line of the calculation. The rest of the proof can be found in [6, Sec. III.3.2]. □

The proof of the next Proposition is also contained in [6, Sec. III.3.2].

Proposition 4.6. *Let $\mathfrak{l}_0 \subseteq \mathbb{K}[\mathcal{S}_m]$ be a left ideal and $T \in \mathcal{T}_{\mathfrak{l}_0^*} \subseteq \mathcal{T}_m V$ be a tensor of order m from the symmetry class $\mathcal{T}_{\mathfrak{l}_0^*}$. Consider the product*

$$\hat{T} := \underbrace{T \otimes \ldots \otimes T}_{n} \in \mathcal{T}_{mn} V . \tag{27}$$

Then all $\hat{T}_b$, $b \in V^{mn}$, lie in the left ideal

$$\mathfrak{l} := \mathbb{K}[\mathcal{S}_{mn}] \cdot \mathcal{L}\{\mathfrak{l}_1 \cdot \ldots \cdot \mathfrak{l}_n \cdot \mathfrak{l}'\} = \mathbb{K}[\mathcal{S}_{mn}] \cdot (\mathfrak{l}_1 \otimes \ldots \otimes \mathfrak{l}_n \otimes \mathfrak{l}') \tag{28}$$

where $\mathfrak{l}_i := \iota_i(\mathfrak{l}_0)$ are embeddings of $\mathfrak{l}_0$ into $\mathbb{K}[\mathcal{S}_{mn}]$ which are formed by means of mappings (25) with $r_1 = \ldots = r_n = m$ and $r = mn$. Further $\mathfrak{l}'$ denotes the 1-dimensional ideal $\mathfrak{l}' := \mathcal{L}\{\sum_{q \in Q} q\}$ of $\mathbb{K}[Q]$ where $Q \subset \mathcal{S}_{mn}$ is the subgroup

$$Q := \left\{ q = \begin{pmatrix} k \cdot m - l \\ s(k) \cdot m - l \end{pmatrix}_{\substack{1 \le k \le n \\ 0 \le l \le m-1}} \in \mathcal{S}_{mn} \;\middle|\; s \in \mathcal{S}_n \right\} \cong \mathcal{S}_n . \tag{29}$$

If $\dim V \geq m \cdot n$, then the above left ideal $\mathfrak{l}$ is generated by all $\hat{T}_b \in \mathbb{K}[\mathcal{S}_{mn}]$ which are formed from tensor products (27) of arbitrary tensors $T \in \mathcal{T}_{\mathfrak{l}_0^*}$.

Let $\breve{\omega}_G : G \to GL(\mathbb{K}[G])$ denote the *regular representation* of a finite group G defined by $\breve{\omega}_g(f) := g \cdot f$, $g \in G$, $f \in \mathbb{K}[G]$. If we use the above left ideals $\mathfrak{l}_i$, $\mathfrak{l}_0$, $\mathfrak{l}$ to define subrepresentations $\alpha_i := \breve{\omega}_{\mathcal{S}_{r_i}}|_{\mathfrak{l}_i}$, $\alpha := \breve{\omega}_{\mathcal{S}_m}|_{\mathfrak{l}_0}$, $\beta := \breve{\omega}_{\mathcal{S}_r}|_{\mathfrak{l}}$, then the representation β is equivalent to a Littlewood-Richardson product or a plethysm[10], respectively (see [6, Sec. III.3.2]):

$$\mathfrak{l} \text{ according to (26)} \implies \beta \sim \alpha_1 \# \ldots \# \alpha_m \uparrow \mathcal{S}_r \tag{30}$$

$$\mathfrak{l} \text{ according to (28)} \longrightarrow \beta \sim \alpha \odot [n] . \tag{31}$$

These results correspond to statements in [7]. (30) and (31) yield valuable information about multiplicities if one has to determine a generating idempotent of $\mathfrak{l}$ and its decomposition into pairwise orthogonal, primitive idempotents by means of our decomposition algorithm described in [4,5,6].

4.3 Subspaces Characterizing Tensors on Low Dimensional Vector Spaces

Now we consider tensors $T \in \mathcal{T}_{\mathfrak{l}^*} \subseteq \mathcal{T}_r V$ over a vector space V with $\dim V < r$. In this case the T_b of tensors $T \in \mathcal{T}_{\mathfrak{l}^*}$ do not span the full left ideal $\mathfrak{l}$ that determines the symmetry class, but only a linear subspace W of $\mathfrak{l}$.

10 See [10,11,9,16], [13,14] and [7] for the Littlewood-Richardson rule and plethysms.

Definition 4.7. If $\lambda = (\lambda_1, \ldots, \lambda_k) \vdash r$ is a partition with length $|\lambda| = k$ and $(v_1, \ldots, v_k) \in V^k$ is a k-tuple of vectors, then we denote by $\langle \lambda; v_1, \ldots, v_k \rangle$ or short $\langle \lambda; v_i \rangle$ that r-tuple from V^r which has the structure

$$\langle \lambda; v_1, \ldots, v_k \rangle := (\underbrace{v_1, \ldots, v_1}_{\lambda_1}, \underbrace{v_2, \ldots, v_2}_{\lambda_2}, \ldots, \underbrace{v_k, \ldots, v_k}_{\lambda_k}) \in V^r . \tag{32}$$

For every r-tuple $b = (v_1, \ldots, v_r) \in V^r$, there exists a unique partition $\lambda \vdash r$ and a permutation $q \in \mathcal{S}_r$ such that b can be written as $b = q\langle \lambda; w_1, \ldots, w_{|\lambda|} \rangle$ where $w_1, \ldots, w_{|\lambda|}$ are the pairwise different, suitably renumbered vectors from b. We call $\langle \lambda; w_1, \ldots, w_{|\lambda|} \rangle$ a *grouping* of b and λ the *grouping partition* of b, which we also denote by $\lambda = b^{\vdash}$.

Proposition 4.8. *Let $\mathfrak{l} \subseteq \mathbb{K}[\mathcal{S}_r]$ be a left ideal with generating idempotent $e \in \mathbb{K}[\mathcal{S}_r]$, $T \in \mathcal{T}_{\mathfrak{l}^*} \subseteq \mathcal{T}_r V$ be a tensor from the symmetry class $\mathcal{T}_{\mathfrak{l}^*}$ and $b \in V^r$ be an r-tuple of vectors. We determine a grouping $\langle \lambda; w_i \rangle$ of b and a permutation $q \in \mathcal{S}_r$ which yields $b = q\langle \lambda; w_i \rangle$. Then we have*

$$T_b \in q^{-1} \cdot 1_{\mathcal{H}_t} \cdot \mathbb{K}[\mathcal{S}_r] \cdot e \tag{33}$$

where $\mathcal{H}_t$ is the group of the horizontal permutations of the lexicographically smallest standard tableau t of $\lambda = b^{\vdash}$ and $1_{\mathcal{H}_t} := \sum_{s \in \mathcal{H}_t} s$.

Proof. Let $\mathcal{R}$ be a system of representatives of the right cosets of $\mathcal{S}_r$ relative to $\mathcal{H}_t$. Taking into account $b = q\langle \lambda; w_i \rangle$, (12), (15) and $T_b = T_b \cdot e$ (which follows from $T_b \in \mathfrak{l}$), we obtain

$$\begin{aligned} T_b = T_b \cdot e \; &= \; T_{q\langle \lambda; w_i \rangle} \cdot e \; = \; q^{-1} \cdot T_{\langle \lambda; w_i \rangle} \cdot e \\ &= q^{-1} \cdot \sum_{s \in \mathcal{H}_t} \sum_{p \in \mathcal{R}} T_{\langle \lambda; w_i \rangle}(s \circ p)\, s \cdot p \cdot e \; = \; q^{-1} \cdot \sum_{s \in \mathcal{H}_t} \sum_{p \in \mathcal{R}} T_{s\langle \lambda; w_i \rangle}(p)\, s \cdot p \cdot e \\ &= q^{-1} \cdot 1_{\mathcal{H}_t} \cdot \Big(\sum_{p \in \mathcal{R}} T_{\langle \lambda; w_i \rangle}(p)\, p \Big) \cdot e . \end{aligned}$$

The last step is correct since every $s \in \mathcal{H}_t$ permutes only such vectors from $\langle \lambda; w_i \rangle$ which are equal to each other, i.e. $s\langle \lambda; w_i \rangle = \langle \lambda; w_i \rangle$ for all $s \in \mathcal{H}_t$. □

Proposition 4.9. *Consider the quantities from Proposition 4.8. Let $\lambda \vdash r$ be a partition with $|\lambda| \leq \dim V$ and $b = \langle \lambda; n_1, \ldots, n_{|\lambda|} \rangle \in V^r$ be a grouping of basis vectors of V. Then the linear subspace $1_{\mathcal{H}_t} \cdot \mathbb{K}[\mathcal{S}_r] \cdot e$ in (33) is generated by all group ring elements $T_{\langle \lambda; n_1, \ldots, n_{|\lambda|} \rangle}$ of tensors $T \in \mathcal{T}_{\mathfrak{l}^*}$, i.e.*

$$1_{\mathcal{H}_t} \cdot \mathbb{K}[\mathcal{S}_r] \cdot e = \mathcal{L}_{\mathbb{K}} \big\{ T_{\langle \lambda; n_1, \ldots, n_{|\lambda|} \rangle} \;\big|\; T \in \mathcal{T}_{\mathfrak{l}^*} \big\} . \tag{34}$$

The proof is given in [6, Sec. III.3.3].

4.4 Subspaces Characterizing Tensors with Index Contractions

First we recall *Silvester's law of inertia* for metric tensors (see [19, S90]).

Lemma 4.10. *Let $g \in \mathcal{T}_2 V$ be a fundamental tensor on V. Then we can choose such a basis*[11] *$\mathcal{B} = \{n_1, \ldots, n_d\}$ of V that the $(d \times d)$-matrix of the*

[11] $\mathcal{B}$ is then called an *orthonormal basis* with respect to g.

coordinates $g_{ij} = g(n_i, n_j)$ *of* g *becomes a diagonal matrix*

$$[g(n_i, n_j)]_{d,d} = \mathrm{diag}(\underbrace{1, \ldots, 1}_{p}, \underbrace{-1, \ldots, -1}_{q}). \qquad (35)$$

The non-negative integers p *and* q *are independent of the choice of* $\mathcal{B}$. *In the case* $\mathbb{K} = \mathbb{C}$, *we have* $q = 0$ *and* $p = d = \dim V$.

Definition 4.11. Let $\mathcal{B} = \{n_1, \ldots, n_d\}$ be an orthonormal basis of V with respect to a fundamental tensor $g \in \mathcal{T}_2 V$. Further let r, l be integers with $2 \le 2l < r$ and $b_0 = (v_{2l+1}, \ldots, v_r) \in \mathcal{B}^{r-2l}$ be a fixed $(r-2l)$-tuple of vectors from $\mathcal{B}$. Then we denote by $\mathfrak{B}_{b_0}$ the set[12] of r-tuples of basis vectors

$$\mathfrak{B}_{b_0} := \{(w_1, w_1, w_2, w_2, \ldots, w_l, w_l, v_{2l+1}, \ldots, v_r) \in \mathcal{B}^r \mid (w_1, \ldots, w_l) \in \mathcal{B}^l\}.$$

Moreover, we set $\gamma_b := \prod_{i=1}^{l} g(w_i, w_i) \in \{1, -1\}$ for every $b \in \mathfrak{B}_{b_0}$.

In the case of tensor expressions with index contractions the role of the T_b is played by sums $\sum_{b \in \mathfrak{B}_{b_0}} \gamma_b T_b$.

Proposition 4.12. *Let* $T \in \mathcal{T}_r V$ *be a tensor and* $g \in \mathcal{T}_2 V$ *be a fundamental tensor. We determine all tensor coordinates with respect to an orthonormal basis* $\mathcal{B} = \{n_1, \ldots, n_d\}$ *of* V. *Let* $b_0 = (n_{i_{2l+1}}, \ldots, n_{i_r}) \in \mathcal{B}^{r-2l}$ *be a fixed* $(r-2l)$*-tuple of basis vectors. Then*

$$\sum_{p \in \mathcal{S}_r} (pT)_{j_1}{}^{j_1}{}_{j_2}{}^{j_2}{}_{\ldots}{}^{\ldots}{}_{j_l}{}^{j_l}{}_{i_{2l+1} \ldots i_r}\, p = \sum_{b \in \mathfrak{B}_{b_0}} \gamma_b T_b . \qquad (36)$$

Proof. Taking into account $g_{ij} = \gamma_{(i)} \delta_{ij}$ and $g^{ij} = \gamma_{(i)} \delta^{ij}$ with $\gamma_{(i)} := g(n_i, n_i) = \pm 1$, we can carry out the following calculation for every $p \in \mathcal{S}_r$:

$$\begin{aligned}
(pT)_{j_1}{}^{j_1}{}_{j_2}{}^{j_2}{}_{\ldots}{}^{\ldots}{}_{j_l}{}^{j_l}{}_{i_{2l+1} \ldots i_r} &= g^{j_1 k_1} \cdots g^{j_l k_l} (pT)_{j_1 k_1 \ldots j_l k_l i_{2l+1} \ldots i_r} \\
&= \sum_{j_1, \ldots, j_l} (\gamma_{(j_1)} \cdots \gamma_{(j_l)}) \times \\
&\qquad (pT)(n_{j_1}, n_{j_1}, \ldots, n_{j_l}, n_{j_l}, n_{i_{2l+1}}, \ldots, n_{i_r}) \\
&= \sum_{b \in \mathfrak{B}_{b_0}} \gamma_b (pT)_b(\mathrm{id}) = \sum_{b \in \mathfrak{B}_{b_0}} \gamma_b T_b(p) .
\end{aligned}$$

In the last line we have used the relations $b = (n_{j_1}, n_{j_1}, .., n_{j_l}, n_{j_l}, n_{i_{2l+1}}, .., n_{i_r})$, $\gamma_b = \gamma_{(j_1)} \cdots \gamma_{(j_l)}$ and (12). If we now multiply both sides of the calculation by p and sum over all $p \in \mathcal{S}_r$, we gain (36). □

Now we give a universal linear subspace which contains the group ring elements $\sum_{b \in \mathfrak{B}_{b_0}} \gamma_b T_b$ of a tensor T with l index contractions for every value of $\dim V$.

[12] We set $b_0 := \emptyset$ and $\mathfrak{B}_\emptyset := \{(w_1, w_1, w_2, w_2, \ldots, w_l, w_l) \in \mathcal{B}^r \mid (w_1, \ldots, w_l) \in \mathcal{B}^l\}$ in the case $r = 2l > 0$.

Theorem 4.13. *Let $V, \mathcal{B}, r, l, g, b_0$ have the meaning given in Proposition 4.12. Consider the partition $\lambda_0 := (2^l, 1^{r-2l}) \vdash r$ and the lexicographically smallest standard tableau t of λ_0. Form the group*[13] $G := \mathcal{H}_t \cdot Q$ *where $\mathcal{H}_t$ is the group of all horizontal permutations of t and $Q \subset \mathcal{V}_t$ is the subgroup of all such vertical permutations of t which only permute full rows of t with length 2. Then every tensor $T \in \mathcal{T}_{\mathfrak{l}^*} \subseteq \mathcal{T}_r V$ ($\mathfrak{l} = \mathbb{K}[\mathcal{S}_r] \cdot e$, e idempotent) fulfills*

$$\sum_{b \in \mathfrak{B}_{b_0}} \gamma_b T_b \in 1_G \cdot \mathbb{K}[\mathcal{S}_r] \cdot e \quad , \quad 1_G := \sum_{g \in G} g \, . \tag{37}$$

Furthermore, if $\dim V \geq r - l$, *then there is such a $b_0 \in \mathcal{B}^{r-2l}$ that*[14]

$$1_G \cdot \mathbb{K}[\mathcal{S}_r] \cdot e = \mathcal{L}_{\mathbb{K}}\Big\{ \sum_{b \in \mathfrak{B}_{b_0}} \gamma_b T_b \,\Big|\, T \in \mathcal{T}_{\mathfrak{l}^*} \Big\} . \tag{38}$$

The proof can be found in [6, Sec. III.3.4]. If $\dim V < r - l$, then the $\sum_{b \in \mathfrak{B}_{b_0}} \gamma_b T_b$ will span only a linear subspace of $1_G \cdot \mathbb{K}[\mathcal{S}_r] \cdot e$ in general. To describe this subspace, we need the following concepts:

Definition 4.14. Let $\mathcal{B}$ be an orthonormal basis with respect to a fundamental tensor $g \in \mathcal{T}_2 V$. We call $(n_{i_1}, \ldots, n_{i_{r'}}) \in \mathcal{B}^{r'}$ smaller than $(n_{j_1}, \ldots, n_{j_{r'}}) \in \mathcal{B}^{r'}$ if the first non-vanishing difference $j_k - i_k$ fulfills $j_k - i_k > 0$. If $\langle \lambda; w_1, \ldots, w_{|\lambda|} \rangle$ and $\langle \lambda; w'_1, \ldots, w'_{|\lambda|} \rangle$ are two groupings of a fixed r-tuple $b \in \mathcal{B}^r$ of basis vectors, then we call $\langle \lambda; w_1, \ldots, w_{|\lambda|} \rangle$ smaller than $\langle \lambda; w'_1, \ldots, w'_{|\lambda|} \rangle$ if the $|\lambda|$-tuple $(n_{i_1}, \ldots, n_{i_{|\lambda|}}) := (w_1, \ldots, w_{|\lambda|})$ is smaller than the $|\lambda|$-tuple $(n_{j_1}, \ldots, n_{j_{|\lambda|}}) := (w'_1, \ldots, w'_{|\lambda|})$.

For every r-tuple $b \in \mathcal{B}^r$ there exists a permutation $p_b \in \mathcal{S}_r$ such that b has a representation $b = p_b \langle \lambda; w_1, \ldots, w_{|\lambda|} \rangle$ where $\langle \lambda; w_1, \ldots, w_{|\lambda|} \rangle$ is the smallest grouping of b and $\lambda = b^{\vdash}$. We denote by $\mathfrak{p}$ a single-valued mapping $\mathfrak{p} : \mathcal{B}^r \to \mathcal{S}_r, b \mapsto \mathfrak{p}(b) := p_b$ which assigns exactly one of such permutations p_b to b.

Let $b_0 \in \mathcal{B}^{r-2l}$ be an $(r-2l)$-tuple of vectors from the basis $\mathcal{B}$. We denote by Λ_{b_0} the set $\Lambda_{b_0} := \{\lambda \vdash r \mid \exists b \in \mathfrak{B}_{b_0} : \lambda = b^{\vdash}\}$. Furthermore, we assign to every partition $\lambda \in \Lambda_{b_0}$ the lexicographically smallest standard tableau t_λ of λ and the set $\mathcal{M}_{b_0,\lambda} := \{\mathfrak{p}(b)^{-1} b \in \mathcal{B}^r \mid b \in \mathfrak{B}_{b_0} \text{ with } b^{\vdash} = \lambda\}$ of such r-tuples which are the smallest groupings of the $b \in \mathfrak{B}_{b_0}$ with given grouping partition λ.

---

[13] G is a semidirect product $\mathcal{H}_t \rtimes Q$ and isomorphic to the wreath product $\mathcal{S}_2 \wr \mathcal{S}_l$.

[14] Corollary: If $r - 2l = 0$, then the decomposition of $1_G \cdot \mathbb{K}[\mathcal{S}_r]$ into minimal right ideals is characterized by a plethysm $[2] \odot [l] \sim \sum_{\mu \vdash l} [2\mu]$. Thus the number I of linearly independent *invariants* of T is bounded by the sum M of the multiplicities of minimal left ideals $\mathfrak{l}_i$ belonging to partitions 2μ, $\mu \vdash l$, in $\mathbb{K}[\mathcal{S}_r] \cdot e = \bigoplus \mathfrak{l}_i$. If $\dim V \geq l$, then $I = M$. (See [6, Sec.III.4.2]. Compare [7].)

Theorem 4.15. *Let $V, \mathcal{B}, r, l, g, b_0, \mathcal{T}_{\mathfrak{l}^*}$ and e have the meaning given in Theorem 4.13 and $\mathfrak{p} : \mathcal{B}^r \to \mathcal{S}_r$ be a mapping of the type described in Definition 4.14. Then we have*

$$\mathcal{L}_{\mathbb{K}}\Big\{ \sum_{b \in \mathfrak{B}_{b_0}} \gamma_b T_b \,\Big|\, T \in \mathcal{T}_{\mathfrak{l}^*} \Big\} = \sum_{\lambda \in \Lambda_{b_0}} \; \sum_{\langle \lambda; w_i \rangle \in \mathcal{M}_{b_0;\lambda}} a_{\langle \lambda; w_i \rangle} \cdot 1_{\mathcal{H}_{t_\lambda}} \cdot \mathbb{K}[\mathcal{S}_r] \cdot e$$

where

$$a_{\langle \lambda; w_i \rangle} := \sum_{\substack{b \in \mathfrak{B}_{b_0} \\ \mathfrak{p}(b)^{-1} b = \langle \lambda; w_i \rangle}} \gamma_b \, \mathfrak{p}(b)^{-1} .$$

The proof is given in [6, Sec. III.3.4].

5 The Equation System for Linear Identities

Now we explain how the above results can be used for doing symbolic tensor calculations and investigations of tensor symmetries.

The term combination problem from the Introduction can be reformulated as the problem to find all linear identities between the summands of given tensor expressions

$$\mathcal{T}_{i_1 \ldots i_r} := \sum_{p \in \mathcal{P}} c_p \, T_{i_{p(1)} \cdots i_{p(r)}} \quad , \quad c_p \in \mathbb{K} \, , \, \mathcal{P} \subseteq \mathcal{S}_r \quad \text{or} \tag{39}$$

$$\mathcal{T}_{i_{2l+1} \ldots i_r} := \sum_{p \in \mathcal{P}} c_p \, (pT)_{j_1 \; j_2 \; \ldots \; j_l \; i_{2l+1} \ldots i_r}^{\;\; j_1 \; j_2 \; \ldots \; j_l} \quad , \quad c_p \in \mathbb{K} \, , \, \mathcal{P} \subseteq \mathcal{S}_r \tag{40}$$

where $T \in \mathcal{T}_{\mathfrak{l}^*}$ is a tensor from a given symmetry class. We assume that (39) and (40) are results of symbolic computer calculations. $\mathcal{P}$ is a subset of permutations, which is determined by the concrete form of the given expressions (39) or (40).

In Propositions 4.1, 4.2, 4.5, 4.6, 4.9 and Theorems 4.13, 4.15 we have presented linear subspaces $W \subseteq \mathbb{K}[\mathcal{S}_r]$ which contain all T_b or $\sum_{b \in \mathfrak{B}_{b_0}} \gamma_b T_b$ of T. If we consider the *orthogonal subspace* $W^\perp := \{ x \in \mathbb{K}[\mathcal{S}_r]^* \mid \forall w \in W : \langle x, w \rangle = 0 \}$ of such a space W, then every $x \in W^\perp$ yields a linear identity for the coordinates of T since

$$\begin{aligned} 0 = \langle x, T_b \rangle &= \sum_{p \in \mathcal{S}_r} x_p \, T_b(p) = \sum_{p \in \mathcal{S}_r} x_p \, T_{i_{p(1)} \cdots i_{p(r)}} \quad \text{or} \\ 0 = \langle x, \sum_{b \in \mathfrak{B}_{b_0}} \gamma_b T_b \rangle &= \sum_{\substack{b \in \mathfrak{B}_{b_0} \\ p \in \mathcal{S}_r}} \gamma_b \, T_b(p) \, x_p = \sum_{p \in \mathcal{S}_r} x_p \, (pT)_{j_1 \; \ldots \; j_l \; i_{2l+1} \ldots i_r}^{\;\; j_1 \; \ldots \; j_l} \end{aligned} \tag{41}$$

where $x_p := \langle x, p \rangle$, $p \in \mathcal{S}_r$. (The last steps are correct if all b occurring in (41) are r-tuples of basis vectors of V.) Every identity (41) can be used to eliminate certain summands in (39), (40). If W is spanned by all T_b or $\sum_{b \in \mathfrak{B}_{b_0}} \gamma_b T_b$

of the tensors considered, then $W^{\perp}$ contains all linear identities which are possible between summands of expressions (39), (40).

If a basis $\{h_1, \ldots, h_k\}$ of W is known, then the coefficients x_p of the $x \in W^{\perp}$ can be obtained from the linear equation system

$$\langle x, h_i \rangle = \sum_{p \in \mathcal{S}_r} h_i(p)\, x_p = 0 \qquad (i = 1, \ldots, k). \tag{42}$$

Thus an important goal is to find such a basis $\{h_1, \ldots, h_k\}$ of W.

Note that (42) is a very large system with a $(k \times r!)$-coefficient matrix, $k = \dim W$. However, since we only need identities to reduce sums (39), (40), we can restrict us to solutions of (42) which fulfill $x_p = 0$ for $p \in \mathcal{S}_r \setminus \mathcal{P}$. This reduces the number of columns to $|\mathcal{P}|$. Furthermore, every of our spaces W is a linear subspace of a left ideal $\mathfrak{l} = \mathbb{K}[\mathcal{S}_r] \cdot e$. A decomposition $e = e_1 + \ldots + e_m$ of the generating idempotent into pairwise orthogonal, primitive idempotents e_i induces a decomposition $W = W_1 \oplus \ldots \oplus W_m$ with $W_i \subseteq \mathbb{K}[\mathcal{S}_r] \cdot e_i$ and a decomposition of the tensors $T \in \mathcal{T}_{\mathfrak{l}^*}$: $T = e_1^* T + \ldots + e_m^* T$. Then we can transform (39), (40) into expressions formed from the $e_i^* T$, for instance

$$(40) \quad \Rightarrow \quad \tau_{i_{2l+1} \ldots i_r} := \sum_{i=1}^{m} \sum_{p \in \mathcal{P}} c_p \, (p(e_i^* T))_{j_1 \; j_2 \; \ldots \; j_l \; i_{2l+1} \ldots i_r}^{\; j_1 \; j_2 \; \ldots \; j_l} ,$$

and use the smaller equation systems of the smaller spaces W_i to determine linear identities for the coordinates of the $e_i^* T$.

An above decomposition $e = e_1 + \ldots + e_m$ can be constructed by means of a decomposition algorithm which we have developed in [4,5,6]. Moreover, this algorithm can be used to determine idempotents e or f such that the spaces from Proposition 4.2 or Theorem 4.15 can be written as $\mathbb{K}[\mathcal{S}_r] \cdot e$ or $f \cdot \mathbb{K}[\mathcal{S}_r] \cdot e$, respectively. In the case of a large r calculations in $\mathbb{K}[\mathcal{S}_r]$ are very expensive. Then *fast discrete Fourier transforms* $D : \mathbb{K}[\mathcal{S}_r] \to \bigoplus_{\lambda \vdash r} \mathbb{K}^{n_\lambda \times n_\lambda}$ are important tools, which further make calculations possible under such conditions (see [2] and [6]). Moreover, the determination of a basis of $D(W)$ instead of a basis of W can be carried out very efficient in $\bigoplus_{\lambda \vdash r} \mathbb{K}^{n_\lambda \times n_\lambda}$ (see [6]). We have tested our methods in computer calculations for which we have used SYMMETRICA [12], GAP [17] and an own Mathematica package PERMS [6] mentioned in the Summary.

References

1. Boerner, H.: *Darstellungen von Gruppen* (Die Grundlehren der mathematischen Wissenschaften in Einzeldarstellungen: Vol. 74). Berlin, Göttingen, Heidelberg: Springer-Verlag, 1955.
2. Clausen, M. and U. Baum: *Fast Fourier Transforms.* Mannheim, Leipzig, Wien, Zürich: BI Wissenschaftsverlag, 1993.

3. Eisenreich, G.: *Lexikon der Algebra.* Berlin: Akademie-Verlag, 1989.
4. Fiedler, B.: *A use of ideal decomposition in the computer algebra of tensor expressions.* Z. Anal. Anw. 16, 1 (1997), 145 – 164.
5. Fiedler, B.: *An algorithm for the decomposition of ideals of the group ring of a symmetric group.* In: Actes 39e Séminaire Lotharingien de Combinatoire, Thurnau, 1997. Ed.: Kerber, A. (Publ. I.R.M.A. Strasbourg). Institut de Recherche Mathématique Avancée, Université Louis Pasteur et C.N.R.S. (URA 01), 1998. Electronically published: `http://cartan.u-strasbg.fr:80/~slc`. B39e, 26 pp.
6. Fiedler, B.: *An algorithm for the decomposition of ideals of semi-simple rings and its application to symbolic tensor calculations by computer.* Habilitationsschrift. Leipzig, Germany: Universität Leipzig, November 1999. Fakultät für Mathematik und Informatik.
7. Fulling, S., King, R., Wybourne, B. and C. Cummins: *Normal forms for tensor polynomials: I. The Riemann tensor.* Class. Quantum Grav. 9 (1992), 1151 – 1197.
8. Günther, P.: *Huygens' Principle and Hyperbolic Equations* (Perspectives in Mathematics: Vol. 5). Boston, San Diego, New York, Berkeley, London, Sydney, Tokyo, Toronto: Academic Press, Inc., 1988.
9. James, G. D. and A. Kerber: *The Representation Theory of the Symmetric Group* (Encyclopedia of Mathematics and its Applications: Vol. 16). Reading, Mass., London, Amsterdam, Don Mills, Ont., Sidney, Tokyo: Addison-Wesley Publishing Company, 1981.
10. Kerber, A.: *Representations of Permutation Groups* (Lecture Notes in Mathematics: Vol. 240, 495). Berlin, Heidelberg, New York: Springer-Verlag, 1971, 1975.
11. Kerber, A.: *Algebraic combinatorics via finite group actions.* Mannheim, Wien, Zürich: BI-Wiss.-Verl., 1991.
12. Kerber, A., Kohnert, A. and A. Lascoux: *SYMMETRICA, an object oriented computer-algebra system for the symmetric group.* J. Symbolic Computation 14 (1992), 195 – 203.
13. Littlewood, D.: *The Theory of Group Characters and Matrix Representations of Groups* (2. Ed.). Oxford: Clarendon Press, 1950.
14. Macdonald, I.: *Symmetric Functions and Hall Polynomials.* Oxford: Clarendon Press, 1979.
15. Merris, R.: *Multilinear Algebra* (Algebra, Logic and Applications Series: Vol. 8). Amsterdam: Gordon and Breach Science Publishers, 1997.
16. Sänger, F.: *Plethysmen von irreduziblen Darstellungen symmetrischer Gruppen.* Dissertation. Aachen: Rheinisch-Westfälische Technische Hochschule, Mathematisch-Naturwissenschaftliche Fakultät, 1980.
17. Schönert, M. et al.: *GAP – Groups, Algorithms, and Programming* (fifth Ed.). Aachen, Germany: Lehrstuhl D für Mathematik, Rheinisch Westfälische Technische Hochschule, 1995.
18. Schouten, J. A.: *Der Ricci-Kalkül* (Die Grundlehren der Mathematischen Wissenschaften: Vol. 10). Berlin: Verlag von Julius Springer, 1924.

19. van der Waerden, B.: *Algebra.* 9., 6. Ed., Vol. I, II. Berlin, Heidelberg, New York, London, Paris, Tokyo, Hong Kong, Barcelona, Budapest: Springer-Verlag, 1993.
20. Weyl, H.: *The Classical Groups, their Invariants and Representations.* Princeton, New Jersey: Princeton University Press, 1939.
21. Wünsch, V.: *Über selbstadjungierte Huygenssche Differentialgleichungen mit vier unabhängigen Veränderlichen.* Math. Nachr. 47 (1970), 131 - 154.

Facts and Conjectures about Fullerene Graphs: Leapfrog, Cylinder and Ramanujan Fullerenes

Patrick W. Fowler[1], Kevin M. Rogers[1], Siemion Fajtlowicz[2], Pierre Hansen[3], and Gilles Caporossi[3]

[1] School of Chemistry, University of Exeter, Stocker Road, Exeter EX4 4QD, UK
[2] Department of Mathematics, University of Houston, Houston, TX 77204, USA
[3] GERAD and École des Hautes Études Commerciales, 3000 Chemin de la Côte-Sainte-Catherine, Montréal H3T 2A7, Québec, Canada

Abstract The definition of a fullerene as a cubic polyhedron made up entirely of pentagons and hexagons is compatible with a huge variety of isomeric forms for structures of chemically achievable size ($n \sim 100$ or fewer vertices $\equiv$ carbon atoms). Generation of complete sets of structures in this size range allows evaluation of conjectures, both chemical and mathematical, on energetics and graph-theoretical properties of this class of molecular graphs. Counterexamples to conjectures of GRAFFITI, including some on fullerenes that are Ramanujan graphs (ramafullerenes) are provided. Graph-theoretical indicators for closure of π shells and low overall energy of fullerenes are also briefly discussed.

1 Introduction

With the discovery of the fullerenes and related novel forms of carbon, the chemistry, physics and materials science of this central element have undergone a revolution in the past 15 years [1]. The new carbons give opportunities for collaboration between mathematicians and chemists, as they are examples of discrete mathematical structures where graph theory, combinatorics and symmetry may generate qualitative chemical understanding [2]. The fullerenes themselves, the subjects of the present article, are mathematically well defined objects, being pseudospherical polyhedral shells of carbon atoms in which each atom (vertex) is linked by a bond (edge) to three nearest neighbours, and all rings (faces) are either pentagonal or hexagonal. Simple reasoning based on the Euler equation shows that such a cage with n vertices, C_n, will have twelve pentagonal and $h = (n/2 - 10)$ hexagonal faces, and it is known that at least one fullerene can be realised for any even value of $n \geq 20$, with the exception of $n = 22$ for which the number of hexagons would have to be one [3]. Once past the first few small cases, the series C_n with $n = (20 + 2h)$ displays the usual combinatorial explosion, and $\#(n)$, the number of conceivable structural isomers grows as a high power of n (e.g. $\#(20) = 1$, $\#(40) = 40$, $\#(60) = 1,812$, $\#(80) = 31,924$, $\#(100) = 285,913, \ldots$) and some discussion of the asymptotics is given in the fullerene literature [2,4,5]. However, very few of the mathematically possible isomers are found in experiment. Confirmed characterisations of about a dozen isomers of the higher fullerenes, with $n \geq 60$ and all twelve pentagons disjoint, have been reported in the chemical literature, and one pressing task of theory is to identify the factors that determine this selection of a small subset from the millions of possibilities.

Construction of isomer sets and elucidation of structure–stability rules of thumb are central to this task and these topics will be treated briefly here. Two methods for fullerene isomer construction have been used in the Exeter group. One is based on the early work of Goldberg on medial and multi-symmetric polyhedra [6,7] and uses the equivalence between fullerene tessellations of the sphere and construction of nets from their dual triangulations on the equilateral triangular planar lattice. Within any accessible symmetry group, isomers are found by solving bilinear equations for integer parameters that describe the net; this is an easy task for high symmetries but impractical for low symmetries and especially for the C_1 group to which the majority of fullerene isomers belong [8]. A second method, which is independent of molecular symmetry, is the spiral construction first proposed by Manolopoulos and co-workers [9]. Here a fullerene isomer is 'peeled' to yield a continuous spiral strip of faces; reversing the procedure, such sequences of faces yield fullerene isomers when wrapped back onto the sphere. Construction and testing of spirals is easily automated. The method has the disadvantage that it misses some isomers at large n ($n \geq 380$) [2,10,11], but as these appear to be a tiny proportion of the total set, beyond the present and foreseeable size range of chemical synthesis, and energetically disfavoured by their crowded arrangement of pentagons, this is more a technicality than a significant problem. The spiral construction gives a compact code for each fullerene that it can represent, as an $(n/2+2)$-digit number, twelve of whose digits are 5 and $n/2 - 10$ are 6, or simply as the set of twelve pentagon positions in this sequence. Lexicographic spiral codes have been adopted in IUPAC recommendations for the nomenclature of fullerenes [12], and can be supplemented to deal with the exceptional 'unspirallable' fullerenes [13].

A mathematically more satisfactory solution of the isomer problem, which is both efficient and complete, is the PentHex puzzle algorithm of Brinkmann and Dress [14]. This represents the state-of-the-art in fullerene enumeration and it is to this method that we owe the certainty that the spiral algorithm is 'safe' up to 176 vertices. In view of the exact equivalence in the range, and the convenience of the spiral codes for generating adjacency matrices, the spiral algorithm will be used here for our search for isomers C_n ($n \leq 100$).

Quite apart from their chemical utility, lists of structures allow testing of the many mathematical conjectures that have been put forward for fullerenes. The GRAFFITI program [15] is a fertile source of these conjectures, and one such, that all fullerene graphs have at least $n/2$ non-negative eigenvalues, has already been disproved by a counterexample from the list of tetrahedrally symmetric fullerenes generated by the extended Goldberg method [16]. The present article will include further 'experimental mathematics' investigations of GRAFFITI conjectures, but first reviews some known facts and chemical rules of thumb for fullerenes.

2 Fullerene Energetics

The molecular graph of a fullerene is that of a cubic polyhedron, which in chemical terms implies an unsaturated molecule with a system of π electrons delocalised (at least to some extent) over the surface of the spherical framework of σ bonds defining the edges of the polyhedron. Approximations to the energy levels of the π system are to be found by diagonalisation of the adjacency matrix of the polyhedron according to the usual prescription of Hückel theory [17]. Once the eigenvalues $\{\lambda_i\}$ are found and ordered as $+3 = \lambda_1 > \lambda_2 \geq \lambda_3 \geq \ldots \geq \lambda_n > -3$, the π-electronic configuration is determined by filling the orbitals with the n π electrons according to the three chemical rules: the Aufbau principle (fill the stack of eigenvalues in decreasing order of λ_i), the Pauli principle (put no more than two electrons in any level λ_i) and Hund's rule of maximum multiplicity (when a multiple eigenvalue is reached, place up to one electron in each before adding a second to any member of the set).

The possible configurations for neutral C_n (n π electrons) are: properly closed ($\lambda_{n/2} \neq \lambda_{n/2+1}$, $\lambda_{n/2} > 0$, $\lambda_{n/2+1} \leq 0$); pseudo-closed ($\lambda_{n/2} \neq \lambda_{n/2+1}$, $\lambda_{n/2+1} > 0$); meta-closed ($\lambda_{n/2} \neq \lambda_{n/2+1}$, $\lambda_{n/2} \leq 0$); open ($\lambda_{n/2} = \lambda_{n/2+1}$). The properly closed π shell is in a sense an ideal one for the π electrons: all electrons are in bonding levels, and no bonding capacity is 'wasted' in unfilled but potentially bonding levels. Open shells, on the other hand, are particularly undesirable as they will correspond to species that are highly reactive (and probably of high energy). Chemical intuition suggests that pseudo-closed shells will be common, as pentagons are electron-deficient (mathematically, a C_5 ring has three positive eigenvalues and hence the capacity for six electrons). This expectation turns out to be well founded. Overwhelmingly most fullerenes have the pseudo-closed π configuration. The three cases where ideal properly closed π shells are found are reviewed below.

(i) Leapfrogs. All *leapfrog* fullerenes have closed π shells [18]. A leapfrog is obtained by omnicapping and then dualising a parent fullerene, giving a new fullerene of the same symmetry, with a larger separation of its pentagons (in particular, they are all disjoint) and with $3n$ atoms [18]. Leapfrog fullerenes therefore occur at $n = 60 + 6k$ ($k \neq 1$) with the number of leapfrogs at n equal to the total number of fullerene isomers at $n/3$. The split nature of the eigenvalue spectrum of leapfrog fullerenes can be proved by detailed arguments [19] but is readily rationalised once it is realised that each edge of the parent has a descendant that carries a double bond in a fully symmetric Kekulé structure in the leapfrog [20]. The result for leapfrog fullerenes is part of a full classification for leapfrogs of cubic maps. Analogues of leapfrog fullerenes on the torus, Klein-bottle and elliptic plane, for example, have four, two and zero non-bonding (zero) eigenvalues, respectively, and leapfrog toroidal and Klein-bottle fullerenes have open shells if considered as neutral chemical species [21].

(ii) Carbon Cylinders. Carbon cylinder fullerenes have closed π shells with a positive eigenvalue $\lambda_{n/2}$ and $\lambda_{n/2+1} = 0$ [22]. The carbon cylinders are formed by tubular extension along the high-symmetry axis of the truncated icosahedron, or of the unique C_{72} leapfrog structure. Analysis in terms of standing waves on a cylinder shows that closed shells of this type occur for every third member of the series, at the vertex numbers $n = 70 + 30k$ ($k = 0, 1, 2, \ldots$) and $n = 84 + 36k$ ($k = 0, 1, 2, \ldots$).

(iii) 'Sporadic' Closed Shells. These extra closed-shell isomers occur outside series (i) and (ii) for large enough values of n [2]. If their number is $n(\#s)$ at n vertices, the pairs for $n \leq 140$ are 112(1), 116(1), 120(1), 122(1), 124(3), 128(3), 130(3), 132(4), 134(7), 136(9), 138(4), 140(12). All members of this set have very small first negative eigenvalues (typically $\sim -10^{-3}$) and in the absence of apparent distinguishing features it is thought that they are 'numerical accidents', unlikely to be different in chemical properties from the pseudo-closed fullerenes with similar gaps between occupied (HOMO) and unoccupied (LUMO) levels.

Series (i) and (ii) are chemically satisfactory in that both of the two most easily produced fullerenes, C_{60} and C_{70}, appear as parents of infinite families of isolated-pentagon isomers. Leapfrog fullerenes have non-zero HOMO–LUMO gaps, $\lambda_{n/2} - \lambda_{n/2+1}$. As the size of the fullerene increases, the typical gap is expected on physical grounds to decrease, as graphite itself has a zero gap; numerical evidence confirms this general expectation but gives no detailed picture of the asymptotics. One chemically plausible conjecture is:

Conjecture 2.1. For numbers n at which a leapfrog fullerene is possible, i.e. $n = 60 + 6k$ ($k \neq 1$), the greatest HOMO–LUMO gap $\lambda_{n/2} - \lambda_{n/2+1}$ will belong to a leapfrog isomer.

For C_n ($n < 100$) the maximum gaps (Δ) over all fullerene isomers are indeed found for leapfrogs: (n:m:m', Δ) = (60:1812:1, 0.75660), (72:11190:1, 0.70229), (78:24108:4, 0.63331), (84:51588:20, 0.69620), (90:99888:16, 0.64994), (96:191788:136, 0.64184), where the isomer on n vertices is mth in the spiral sequence for general fullerenes and m'th in the sequence of isolated-pentagon fullerenes. Beyond $n = 100$ a leapfrog is known to have the largest gap amongst isolated-pentagon fullerenes in all cases tested, for example, C_{120} where the winning candidate is 120:10666, the leapfrog of 40:39 [2].

Thus the systematics of the π energies of fullerenes is broadly understood. Unfortunately, the π system accounts for only a quarter of the valence electrons of a fullerene, and for only part of the total energy. In addition, direct calculation of the relative energies of fullerene isomers, for which many quantum-chemical methods give very similar orderings, shows a poor correlation with π energy. The experimental structures of, for example, C_{84} are not the π-ideal leapfrog or cylinder isomers, but are those with lowest calculated energies. What determines this energy ordering?

Graph-theoretical arguments, though no longer those of Hückel theory, are useful here too. The key observation is that pentagon crowding, presumed to be energetically unfavourable, is measured by N_p, the number of pentagon fusions. The quantum-mechanically calculated energies show a tendency to rise with N_p: if ΔE (energy relative to the best isomer at the same vertex count) is plotted against ΔN_p (difference of N_p from the minimal achievable value at the same vertex count) for several thousand isomers (C_n, $n \leq 50$ and all $1,812$ for $n = 60$), a convincing universal linear correlation emerges [23]. Physically, this is equivalent to determining an effective energy penalty per pentagon adjacency ($\sim$ 0.7 to 1.5 eV depending on method [23,24]). The dependence is not expected to be strictly linear, as N_p may still conceal significant variation in geometry, but it implies that a major determinant of energy has been found. Minimisation of N_p is therefore a driving force to low energy for lower fullerenes ($n < 60$). A direct consequence for higher fullerenes of this approximate relation is the isolated-pentagon rule (IPR):

Conjecture 2.2. Where fullerene isomers with disjoint pentagons are possible ($n = 60$, $n \geq 70$), one of them will be the isomer of lowest total energy.

Distinction *between* IPR isomers can also be made on graph-theoretical grounds. A rule in which the topological invariant to be minimised is the second moment of the hexagon-neighbour signature, $H = \sum_k k^2 h_k$, where h_k is the number of hexagons with exactly k hexagonal neighbours, appears to be highly selective in the range $60 \leq n \leq 140$. Minimisation of H is a way of satisfying Raghavachari's criterion of maximum similarity between environments of hexagonal rings [2,25,26]. Again, strong correlations between H and energy are found in calculations suggesting H as a useful invariant for distinguishing IPR structures within the range $60 \leq n \leq 140$. Beyond $n = 140$, nearly all IPR structures have minimal H and invariants sensitive to the longer-range ordering of the pentagons will be required.

3 Some Conjectures on Fullerenes

The GRAFFITI and MINUTEMAN programs [15] have produced several conjectures about relationships between mathematical properties of fullerene graphs (such as their second eigenvalues), and the stability (in a chemical sense) of the associated carbon frameworks. The present section reports some new calculations of fullerene eigenvalue spectra and other invariants which furnish counterexamples to some of the conjectures and in turn suggest new generalisations.

Some terms and definitions are required (see [15,27]). As before, the n adjacency eigenvalues of the fullerene graph, arranged in non-increasing order are $+3 = \lambda_1 > \lambda_2 \geq \lambda_3 \geq \ldots \geq \lambda_n > -3$. A *Ramanujan* graph is a finite regular graph of degree k for which all eigenvalues (other than $\pm k$) have modulus at most $2\sqrt{k-1}$. Ramanujan fullerenes (*ramafullerenes*, for short) are therefore those with one eigenvalue $\lambda_1 = +3$, and for all others $|\lambda_i| \leq 2\sqrt{2}$. We call here *positive* ramafullerenes those that fail to meet the Ramanujan criterion only at the end of the ordered eigenvalue spectrum, i.e. those for which $\lambda_2 \leq 2\sqrt{2}$ but $\lambda_n < -2\sqrt{2}$. Likewise, *negative* ramafullerenes fail at the beginning of the spectrum, with $\lambda_n \geq -2\sqrt{2}$ but $\lambda_2 > 2\sqrt{2}$.

The *separator*, $s(G)$, of the adjacency matrix of a graph G is the difference between its largest and second largest eigenvalues, i.e. $s(G) = \lambda_1 - \lambda_2$. The *radius*, $r(G)$, of a graph G is the minimum *eccentricity*, where the eccentricity of a vertex is the maximum number of edges required to reach *any* other vertex by the shortest path from the chosen vertex. The *diameter*, $d(G)$, of a graph is the maximum eccentricity taken over all vertices. $\overline{l(G)}$ is the *average distance* of the graph, taken over all pairs of vertices; this is related to the *Wiener index*, $W(G)$, by $\overline{l(G)} = 2W(G)/(n(n-1))$. The set of vertices of minimum eccentricity is called the *centre* of the graph; any vertex can be reached in $r(G)$ or fewer steps from some central vertex. Another subset of vertices, which may be called the *perimeter*, consists of those vertices with the largest eccentricity; vertices in this set are called *extreme*. $E(G)$ will denote the number of extreme vertices of a graph G. The *independence number*, $I(G)$, of a graph G is defined by a colouring of the vertices: colour vertices black or white, such that no two black vertices are adjacent; $I(G)$ is then the size of the largest possible set of black vertices.

The relevant conjectures made by the GRAFFITI program are:

895: If G is a fullerene, then $s(G) \leq 1$;

896: The separator of a fullerene with n atoms is at most $1 - 3/n$;

898: If the second largest eigenvalue of a fullerene G is $\lambda_2 \leq 2\sqrt{2}$, then G is a ramafullerene, i.e. in our terms, no positive ramafullerenes exist;

902: If G is a ramafullerene, then $r(G)/s(G) \leq 2I(G)$ [this conjecture also appears as **910**: For every ramafullerene, $I(G)/r(G) \geq (2s(G))^{-1}$];

903: If G is a ramafullerene, and $S_+(G)$ is the sum of the positive eigenvalues, then $r(G)/s(G) \leq 2S_+(G)$;

905: If G is a fullerene, $d(G)s(G) \leq E(G) + 1$;

907: If G is a fullerene, $I(G)/(d(G)s(G)) \leq 1 + \lambda_n + \overline{l(G)}/s(G)$;

908: If G is a fullerene, $I(G)/r(G) \leq r(G)$;

909: If G is a fullerene, $I(G)/r(G) \leq \overline{l(G)}$;

911: For every fullerene, $I(G)/r(G) \geq \log(n/2-1)$;

912: For every fullerene, $I(G) \geq 2(d(G)-1))$.

A hypothesis in Written on the Wall [15], which is not given as a conjecture in the formal sense, is that the (physical) stability of fullerenes (for a given number of atoms) may be an increasing function of their separators. This hypothesis and the conjectures were tested by explicit computation of the invariants for the full set of fullerene graphs on up to 100 vertices, and our observations are now reported. Tables 1 and 2 show the distribution of Ramanujan graphs within the test set.

A trivial first observation is that the number of ramafullerenes is finite. From Table 1, all fullerenes on $n \leq 46$ vertices are ramafullerenes, but from then on the proportion as a percentage of all fullerenes on n vertices starts to fall, reaching zero at $n = 86$. The fact that the class of ramafullerenes is finite follows from the result [28] that the class of planar, cubic Ramanujan graphs is finite (SF thanks Noga Alon for discussion on this point). The tapering to zero of the counts for $n \geq 86$ suggests to us that we have found all of the ramafullerenes.

Figure 1 identifies the first few non-ramafullerenes, i.e. the exceptional cases at 48 and 50 vertices, and Fig. 2 gives what we conjecture to be the complete list of 14 IPR ramafullerenes.

Conjecture 3.1. All ramafullerenes have 84 or fewer vertices. There are 20,175 general ramafullerenes, of which all but 14 have pentagon adjacencies.

The statistics in Table 1 allow an immediate comment on Conjectures **895** and **896**. All fullerenes within the test set easily obey the relation $s(G) \leq 1-3/n$. In fact, all would obey a stronger rule such as $s(G) \leq 3-\sqrt{5}$, as the dodecahedron appears to have the largest separator of any fullerene. A plot of extrema of $s(G)$ against n (Fig. 3) settles down, after initial undulations, to a set of more-or-less monotonically decreasing curves. At $n = 68$ for general fullerenes, there is a small rise in maximum $s(G)$ (of only 5×10^{-5}). Some bumps in the curves are due to cases with high symmetries where $\lambda_2 =$

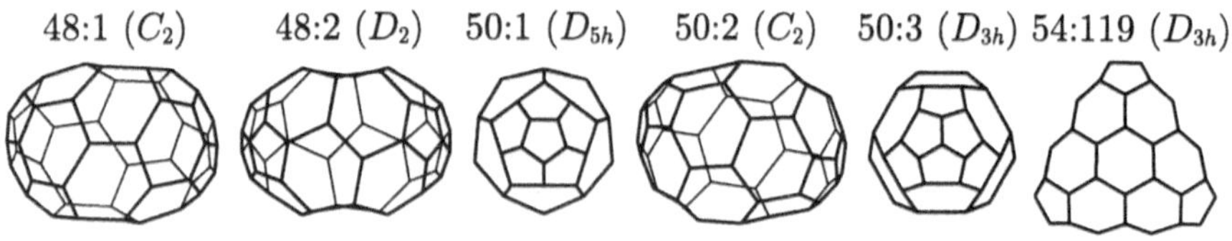

Figurel. The five smallest non-ramafullerenes and the smallest positive ramafullerene (54:119) labelled n:m (G), where n is the number of vertices, m the isomer position in the list of lexicographically ordered spirals and G the maximal point group [2]

Table1. Ramafullerene statistics for general fullerenes on $20 \leq n \leq 100$ vertices. N is the total number of distinct fullerene isomers (counting each enantiomer pair as a single isomer), N_r is the number of ramafullerenes, N_+ and N_- are the numbers of positive and negative ramafullerenes. s is the value of the largest separator at the given n, marked with an asterisk when achieved by an isolated-pentagon fullerene. At 74 and 84 vertices, the maximum separator is shared by two isomers; both are IPR for $n = 84$ but for $n = 74$, one has pentagon adjacencies.

n	N	N_r	N_+	N_-	s	
20	1	1	0	0	0.7639	
24	1	1	0	0	0.5691	
26	1	1	0	0	0.4859	
28	2	2	0	0	0.5272	
30	3	3	0	0	0.4558	
32	6	6	0	0	0.4012	
34	6	6	0	0	0.3870	
36	15	15	0	0	0.3916	
38	17	17	0	0	0.3768	
40	40	40	0	0	0.3614	
42	45	45	0	0	0.3289	
44	89	89	0	0	0.3261	
46	116	116	0	0	0.3014	
48	199	197	0	1	0.2934	
50	271	268	0	1	0.2845	
52	437	424	0	11	0.2799	
54	580	554	1	22	0.2579	
56	924	853	1	60	0.2527	
58	1,205	1,076	0	113	0.2451	
60	1,812	1,456	3	313	0.2434	*
n	N	N_r	N_+	N_-	s	
62	2,385	1,772	5	549	0.2263	
64	3,465	2,180	25	1,098	0.2219	
66	4,478	2,276	8	1,863	0.2128	
68	6,332	2,527	27	3,018	0.2128	
70	8,149	2,292	25	4,592	0.2031	
72	11,190	1,723	28	6,958	0.1991	
74	14,246	1,300	11	8,823	0.1923	*
76	19,151	745	19	11,202	0.1913	*
78	24,109	156	4	13,041	0.1845	
80	31,924	25	3	14,265	0.1819	*
82	39,718	7	1	14,508	0.1743	
84	51,592	2	0	14,577	0.1721	*
86	63,761	0	0	14,229	0.1672	
88	81,738	0	0	13,657	0.1643	*
90	99,918	0	0	12,433	0.1593	
92	126,409	0	0	11,100	0.1576	*
94	153,493	0	0	9,688	0.1519	
96	191,839	0	0	8,349	0.1470	
98	231,017	0	0	6,721	0.1477	*
100	285,913	0	0	5,615	0.1446	

Table2. Ramafullerene statistics for isolated-pentagon fullerenes on $60 \le n \le 100$ vertices. Symbols are as in Table 1, but apply to the IPR subset of fullerenes.

n	N	N_r	N_+	N_-	s	n	N	N_r	N_+	N_-	s
60	1	1	0	0	0.2434	86	19	0	0	19	0.1651
70	1	1	0	0	0.1864	88	35	0	0	35	0.1643
72	1	1	0	0	0.1852	90	46	0	0	45	0.1546
74	1	1	0	0	0.1923	92	86	0	0	85	0.1576
76	2	1	0	1	0.1913	94	134	0	0	134	0.1483
78	5	2	0	3	0.1794	96	187	0	0	179	0.1462
80	7	3	0	4	0.1819	98	259	0	0	252	0.1477
82	9	2	0	7	0.1741	100	450	0	0	419	0.1442
84	24	2	0	22	0.1721						

$\lambda_3 = \lambda_4$ (tetrahedral or icosahedral) or $\lambda_2 = \lambda_3$ (dihedral groups) where the separator is larger than when all three eigenvalues are distinct.

The data in Table 1 provide an immediate refutation of Conjecture **898**: positive ramafullerenes exist and each is a counterexample to the conjecture. Positive ramafullerenes are comparatively rare, appearing first at $n = 54$ with a D_{3h} isomer containing three fused quadruples of pentagons (Fig. 1), and apparently disappearing for $n > 82$. It seems likely that the 161 cases found are the only positive ramafullerenes. All the positive ramafullerenes found in the search have pentagon adjacencies, and so a weaker version of Conjecture **898** is consistent with the facts:

898′: If the second largest eigenvalue of an isolated-pentagon fullerene G is $\lambda_2 \le 2\sqrt{2}$, then G is a ramafullerene.

Negative ramafullerenes are encountered much more frequently, rising to 62% of all fullerenes on 72 vertices, but falling as a proportion, and eventually in absolute terms, as n continues to increase. It seems reasonable to suppose that this class too is finite. The negative ramafullerenes include many isolated-pentagon cases (Table 2). The fact that negative ramafullerenes are more common than positive is consistent with the chemical observation that fullerenes tend to be electron-deficient, i.e. having a spectrum with (usually) more positive than negative eigenvalues [29].

It is perhaps worth noting that no case is found of an 'exact' ramafullerene, i.e. a ramafullerene with $|\lambda_2| = 2\sqrt{2}$ or $|\lambda_n| = -2\sqrt{2}$. Dias [30] has remarked that examples of chemical graphs of maximum degree three having an eigenvalue $|\lambda| = \sqrt{l}$ are known for all integer values $1 \le l \le 9$ except for $l = 8$. The present search has not changed that position.

The computations of separator values also show that the correlation suggested in [15] between maximum separator and maximum stability at a given n is not generally true. It is known that the fullerene isomers of lowest total energy at a given value of n have the minimal number of pentagon adjacencies, and that this minimum-adjacency rule lies behind the isolated-pentagon

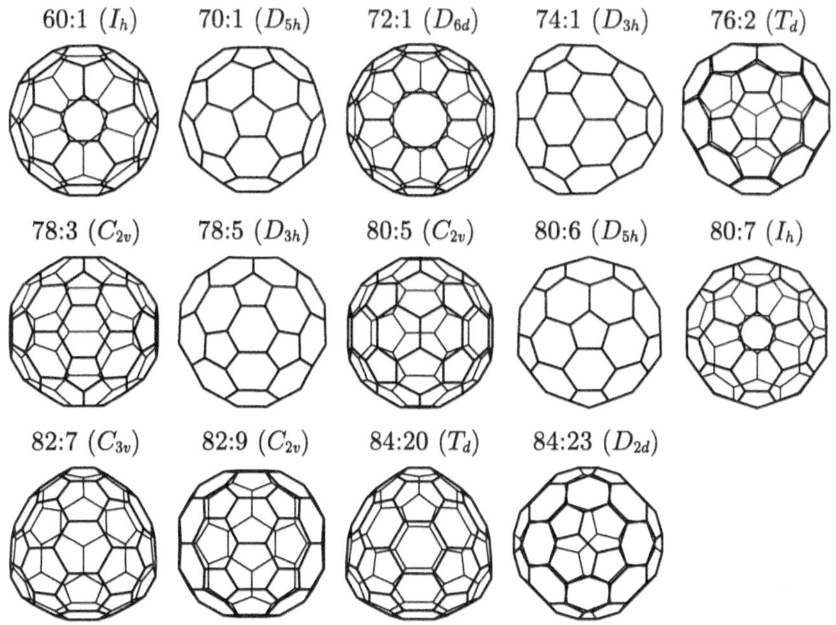

Figure2. The complete list of 14 isolated-pentagon ramafullerenes

rule (IPR) for fullerenes (Conjecture 2). Thus, of 1,812 isomers of C_{60}, the I_h IPR isomer 60:1812 is significantly lower in energy than its nearest (and as yet unobserved) rival with two pentagon adjacencies. For C_{70}, the low-energy isomer is the experimentally characterised D_{5h} IPR isomer 70:8149. However, whilst 60:1812 has the largest separator of all 60-vertex fullerenes, 70:8149 has only the 269th largest separator of all fullerenes on 70 vertices. Again at $n = 84$, the two fullerenes with largest separator are 84:20 and 84:23, of which 84:23 is in fact a low-energy cage, but is isoenergetic with 84:22 which has only 5th largest separator of the 24 IPR isomers. Table 1 shows that the fullerene of largest separator is often one with adjacent pentagons, even in the IPR range. Amongst the lower fullerenes there are also many cases where $s(G)$ fails to predict the low-energy isomer; for example, at C_{40} the T_d isomer 40:40 has the largest separator $s = 0.3614$ and 12 pentagon adjacencies, whereas isomer 40:38, favoured by all quantum-mechanical methods [24], has only 10 pentagon adjacencies and a smaller separator of $s = 0.3362$.

From the chemical point of view, there is no particular reason to expect a correlation of s with either total energy or reactivity, as the reactivity is determined by the frontier π orbitals (i.e. those around the middle of the eigenvalue spectrum) and the total energy is heavily influenced by strain in the σ framework. A chemically more plausible correlation would be between stability and the range of the positive part of the eigenvalue spectrum; in

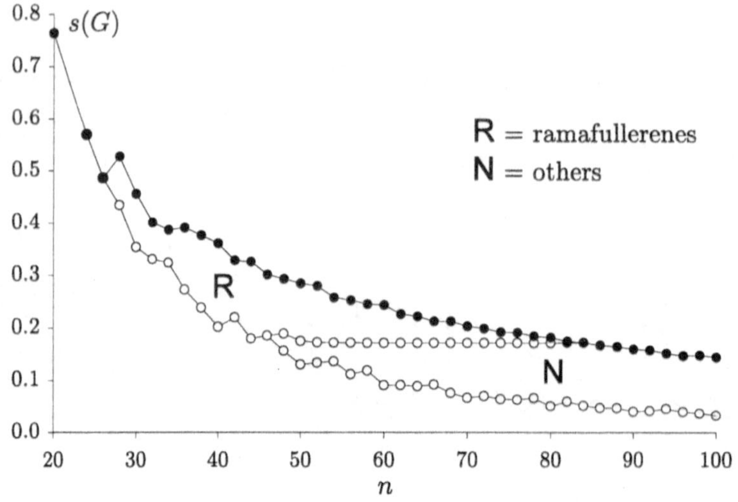

Figure3. Maximum (•) and minimum (∘) values of the separator $s(G)$ for fullerenes

this respect, C_{70} is an interesting case where the most stable isomer has the smallest range, although this correlation does not hold generally.

Conjectures **902** and **903** refer to ramafullerenes only. Direct calculation of the 'balance functions' $b_{902}(G) = 2I(G) - r(G)/s(G)$ and $b_{903}(G) = 2S_+(G) - r(G)/s(G)$ shows that both are positive for all ramafullerenes. The first function is always at least ~ 3.690 (isomer 52:6), and $b_{903}(G)$ at least $(49+19\sqrt{5})/4 \approx 22.871$ [$= b_{903}(C_{20})$]. If Conjecture 3, that all ramafullerenes have $n \leq 84$ vertices is correct, then both **902** and **903** have been verified by exhaustion, and **903** could be trivially strengthened by subtraction of the dodecahedral balance term. Conjectures **902** and **903** are not fully independent, however, as a simple argument based on uniform distribution of eigenvalues would predict $S_+(G)/I(G) \geq 1.5$ for fullerenes; in fact this ratio lies between ~ 2.001 and ~ 1.625 for all fullerenes with $n \leq 100$ vertices.

Neither conjecture extends to general fullerenes, as the term with $s(G)$ in the denominator can take very large values. Amongst the general fullerenes, $b_{902}(G)$ first becomes negative at $n = 50$ and $b_{903}(G)$ at $n = 70$. Only a small number of isomers would fail the general-fullerene version of **903**: there are $1, 1, 1, 1, 0, 4, 1, 3, 3, 8, 11, 19, 19, 30, 48$ and 52 failures for $n = 70, 72, \ldots, 100$. Fullerenes with $n \leq 100$ would obey a version of **903** where 'ramafullerene' is replaced by 'isolated-pentagon fullerene', but the falling balance in the range indicates a probable breakdown at large n for IPR fullerenes.

The remaining six conjectures are not restricted to ramafullerenes. No counterexamples to any of them were found in the range. Conjecture **905** comes close to failure for isomer 26:1, where the balance $b_{905}(G) = E(G) + 1 - d(G)s(G)$ is only 0.085. It is interesting to note that the simpler inequality

$d(G)s(G) \leq 3$, the truth of which would imply Conjecture **905**, fails for just two fullerenes with $n \leq 100$ – dodecahedral 20:1 and tetrahedral 28:2. The inequality in Conjecture **912** is sharp for 20:1.

In summary, this search has disproved one conjecture of GRAFFITI, given a number of directions for future exploration of others and provided what we believe is the complete list of the family of ramafullerenes, showing that fullerene enumeration can have mathematical as well as chemical interest.

References

1. Kroto, H. W.: The celestial sphere that fell to earth. Angew. Chem. Int. Ed. Engl. **31** (1992) 111–129
2. Fowler, P. W., Manolopoulos, D. E.: An atlas of fullerenes. Oxford University Press, Oxford (1995)
3. Grünbaum, B., Motzkin, T. G.: The number of hexagons in and the simplicity of geodesics on certain polyhedra. Can. J. Math. **15** (1963) 744–751
4. Sah, C. H.: A generalised leapfrog for fullerene structures. Fullerene Sci. Technol. **2** (1994) 445–458
5. Rutherford, J. S.: Generating functions for the cage isomers of the C_{20n} icosahedral fullerenes. J. Math. Chem. **14** (1993) 385–390
6. Goldberg, M.: The isoperimetric problem for polyhedra. Tôhoku Math. J. **40** (1934) 226–231
7. Goldberg, M.: A class of multi-symmetric polyhedra. Tôhoku Math. J. **43** (1937) 104–108
8. Fowler, P. W., Cremona, J. E., Steer, J. I.: Systematics of bonding in non-icosahedral carbon clusters. Theor. Chim. Acta **73** (1988) 1–26
9. Manolopoulos, D. E., May, J. C., Down, S. E.: Theoretical studies of the fullerenes – C_{34} to C_{70}. Chem. Phys. Lett. **181** (1991) 105–111
10. Manolopoulos, D. E., Fowler, P. W.: A fullerene without a spiral. Chem. Phys. Lett. **204** (1993) 1–7
11. Yoshida, M., Fowler, P. W.: Dihedral fullerenes of threefold symmetry with and without face spirals. J. Chem. Soc. Faraday Trans. **93** (1997) 3289–3294
12. Godly, E. W., Taylor, R.: Nomenclature and terminology of fullerenes: a preliminary survey. Pure Appl. Chem. **69** (1997) 1411–1434
13. Fowler, P. W., Pisanski, T., Graovac, A., Žerovnik, J.: A generalized ring spiral algorithm for coding fullerenes and other cubic polyhedra. DIMACS series on Discrete Mathematics and Theoretical Computer Science (to appear)
14. Brinkmann, G., Dress, A. W. M.: A constructive enumeration of fullerenes. J. Algorithms **23** (1997) 345–358
15. Fajtlowicz, S.: Written on the wall: a list of conjectures made by the Graffiti program. The list is available from the author at siemion@@math.uh.edu or Dept. of Mathematics, University of Houston, Houston, TX 77204, USA
16. Fowler, P. W.: Fullerene graphs with more negative than positive eigenvalues: the exceptions that prove the rule of electron deficiency? J. Chem. Soc. Faraday Trans. **93** (1997) 1–3
17. Streitwieser, A.: Molecular orbital theory for organic chemists. Wiley, New York (1961)

18. Fowler, P. W., Steer, J. I.: The leapfrog principle – a rule for electron counts of carbon clusters. J. Chem. Soc. Chem. Comm. (1987) 1403–1405
19. Manolopoulos, D. E., Woodall, D. R., Fowler, P. W.: Electronic stability of fullerenes – eigenvalue theorems for leapfrog carbon clusters. J. Chem. Soc. Faraday Trans. **88** (1992) 2427–2435
20. Fowler, P. W.: Localised models and leapfrog structures of fullerenes. J. Chem. Soc. Perkin Trans. 2 (1992) 145–146
21. Deza, M., Fowler, P. W., Rassat, A., Rogers, K. M.: Fullerenes as tilings of surfaces. J. Chem. Inf. Comp. Sci. (to appear)
22. Fowler, P. W.: Carbon cylinders – a class of closed-shell clusters. J. Chem. Soc. Faraday Trans. **86** (1990) 2073–2077
23. Campbell, E. E. B., Fowler, P. W., Mitchell, D., Zerbetto, F.: Increasing cost of pentagon adjacency for larger fullerenes. Chem. Phys. Lett. **250** (1996) 544–548
24. Albertazzi, E., Domene, C., Fowler, P. W., Heine, T., Seifert, G., Van Alsenoy, C., Zerbetto, F.: Pentagon adjacency as a determinant of fullerene stability. Phys. Chem. Chem. Phys. **1** (1999) 2913–2918
25. Raghavachari, K.: Ground state of C_{84} – 2 almost isoenergetic isomers. Chem. Phys. Lett. **190** (1992) 397–400
26. Achiba, Y., Fowler, P. W., Mitchell, D., Zerbetto, F.: Structural predictions for the C-116 molecule. J. Phys. Chem. A **102** (1998) 6835–6841
27. Cvetković, D. M., Doob, M., Sachs, H.: Spectra of graphs: theory and application. Academic Press, New York (1979)
28. Alon, N., Milman, V. D.: λ_1, isoperimetric inequalities for graphs and superconcentrators. J. Comb. Theory B **38** (1985) 73–88
29. Fowler, P. W., Ceulemans, A.: Electron deficiency of the fullerenes. J. Phys. Chem. **99** (1995) 508–510
30. Dias, J. R.: Techniques in facile calculation of molecular orbital parameters and related conceptualizations – molecular orbital functional groups. J. Mol. Struct. (Theochem) **417** (1997) 49–67

Magnetic Configurations, Riggings, Bethe Ansatz and Robinson-Schensted-Knuth Algorithm

D. Golojuch, R. Bal and P. Jakubczyk

Institute of Physics, The Pedagogical University
ul. Rejtana 16A, 35-359 Rzeszów, Poland

Abstract The one - dimensional Heinsenberg model of a magnet is considered within the Bethe Ansatz approach. We provide here a combinatorial description of the following sets of interest: (i) the set of all magnetic configurations for the spin 1/2 and $N = 10$ crystal nodes, (ii) the set of appropriate standard bitableaux, (iii) the set of all configurations of strings (i.e. systems of coupled spin deviations, defined in terms of a special collection of spectral parameters) and their riggings. An application of the Robinson - Schensted - Knuth algorithm is demonstrated.

1 Introduction

The Heinsenberg model of magnetism is a good example of an applications of combinatoric notions in physics of condensed matter.The solutions of the eigenproblem of the isotropic Heinsenberg Hamiltonian in one dimension , i.e. for linear magnetic ring consisting of N nodes, each with the spin 1/2,was given by Bethe [1] in 1931. Nowadays it is known as the Bethe Ansatz BA. An important ingredient of BA is the hypothesis of strings, which yields a classification of BA eigenfunctions in terms of some combinatorial objects, deseribed by Kerov, Kirillov and Reshetikhin [2] as rigged string configurations. The aim of this paper is a presentation of appropriate combinatorial notions related to three sets of interest: (i) the set of all magnetic configurations, which provides the initial basis of the space of all quantum states of the magnet, (ii) the set of all standard bitableaux, which is obtained from the duality of Weyl [3] in terms of the Robinson-Schensted-Knuth algorithm [4,5,6,7] and thus provides an intermediate basis, adapted to the spherical symmetry of the model, (iii) the set of all configurations of strings and their riggings [2,8,9,10,11,12]. We present all these notions on an example of a magnet with $N = 10$ nodes, which proves to be sufficiently large for a transparent and non-trivial exposition, and does not yet reach any combinatoric explosion. We also restrict ourselves to the case of the single-node spin 1/2. It corresponds to considering only "monochromatic", i.e. single colour strings.

2 Magnetic configurations

A magnetic configuration is defined if each node of the magnet has a prescribed value of z-projection of the single-node spin, which might be $\pm 1/2$. Formally, we introduce the set

$$\tilde{10} = \{1, 2, ..., 10\}$$

of all nodes of our crystal, together with the set

$$\tilde{2} = \{+, -\}$$

of the single-node spin projections, and define a magnetic configuration as any mapping $f : \tilde{10} \to \tilde{2}$. The set

$$\tilde{2}^{\tilde{10}} = \{f : \tilde{10} \to \tilde{2}\}$$

of all such mappings has $2^{10} = 1024$ elements, and spans the linear unitary space $\mathcal{H}$ of all quantum states of the magnet. Let Σ_{10} be the symmetric groups on the set $\tilde{10}$. The natural action $P : \Sigma_{10} \times \tilde{2}^{\tilde{10}} \to \tilde{2}^{\tilde{10}}$ decomposes the set 2^{10} into orbits, labelled by the bipartition

$$\mu = \{N - r, r\}, r \in \{0, 1, 2, ..., 10\} .$$

The bipartition μ corresponds to the Young subgroup

$$\Sigma^{\mu} = \Sigma_{N-r} \times \Sigma_r ,$$

which is a stabiliser on the corresponding orbit, so that the restriction of the action P to this orbit is a transitive representation of the group Σ_{10}, denoted by

$$P\mid_r = R^{\Sigma_{10}:\Sigma^{\mu}} ,$$

or, more shortly, by R^{μ}. Thus we have

$$P = 2 \cdot R^{\{10\}} + 2 \cdot R^{\{91\}} + 2 \cdot R^{\{82\}} + 2 \cdot R^{\{73\}} + 2 \cdot R^{\{64\}} + R^{\{5^2\}}$$

and the dimensional check reads

$$2^{10} = 1024 = 2 \cdot 1 + 2 \cdot 10 + 2 \cdot 45 + 2 \cdot 120 + 2 \cdot 210 + 252 .$$

Clearly, the total magnetization M, i.e. the sum of all z-projections of spins,

$$M = \frac{N}{2} - r$$

is constant on each orbit r of the action P of Σ_{10}.

3 The total spin and Robinson-Schensted-Knuth algorithm

We consider the XXX model, in which the total spin S is an exact quantum number. Physically conservation of S corresponds to the spherical symmetry of the Heisenberg Hamiltonian: the interaction between spins is invariant under the group $SO(3,R)$ of all three-dimensional relations. The corresponding action $Q : SU(2) \times \mathcal{H} \to \mathcal{H}$ of the group $SU(2)$, the universal covering group for $SO(3,R)$ in the space $\mathcal{H}$ yields the decomposition

$$Q = (D^{1/2})^{\otimes 10} = 42D^0 + 90D^1 + 75D^2 + 35D^3 + 9D^4 + D^5$$

with the dimensional check

$$2^{10} = 42 \cdot 1 + 90 \cdot 3 + 75 \cdot 5 + 35 \cdot 7 + 9 \cdot 9 + 1 \cdot 11$$

Each term D^S in Eq.(3.1) corresponds to the total spin S, and can be labelled by the partition

$$\lambda = \{N - r^{\text{‘}}, r^{\text{‘}}\}$$

such that

$$S = \frac{N}{2} - r^{\text{‘}} .$$

The intersection of the space of all states with a given total spin S with the carrier space of R^{μ}, correspond to all states with $r = \frac{N}{2} - M$ and $r^{\text{‘}} = \frac{N}{2} - S$, $0 \le r^{\text{‘}} \le r$, and thus, by the duality of Weyl, to the carier space of the irreducible representation Δ^{λ} of the symetric group Σ_{10}. We have therefore

$$R^{\{N-r,r^{\text{‘}}\}} = \sum_{r^{\text{‘}}=0}^{r} \oplus \Delta^{\{N-r^{\text{‘}},r^{\text{‘}}\}}$$

This formula corresponds to decomposition of the space of all states with r spin deviations into subspaces with different total spin S. The multiplicities in Eq.(3.5) are Kostka numbers $\kappa_{\mu\lambda}$, which take here only values 1 or 0. In particular, the last term in Eq.(3.5), i.e. that with $\lambda = \mu$, or $r = r^{\text{‘}}$, or equivalently, $M = S$, corresponds to the highest weight vectors, referred hereafter to as Bethe vectors. For example, for $N = 10$ and $r = 5$ we have

$$R^{\{5^2\}} = \Delta^{\{10\}} + \Delta^{\{91\}} + \Delta^{\{82\}} + \Delta^{\{73\}} + \Delta^{\{64\}} + \Delta^{\{5^2\}} ,$$

with the dimensional check

$$\binom{10}{5} = 1 + 9 + 35 + 75 + 90 + 42 = 252 .$$

The space of Bethe vectors for this case has the dimension 42. The decomposition (3.5) can be applied to each transitive representation R^{μ} for $N = 10$,

$$R^{\{10\}} = \Delta^{\{10\}}$$

$$R^{\{91\}} = \Delta^{\{10\}} + \Delta^{\{91\}}$$

$$R^{\{82\}} = \Delta^{\{10\}} + \Delta^{\{91\}} + \Delta^{\{82\}}$$

$$R^{\{73\}} = \Delta^{\{10\}} + \Delta^{\{91\}} + \Delta^{\{82\}} + \Delta^{\{73\}}$$

$$R^{\{64\}} = \Delta^{\{10\}} + \Delta^{\{91\}} + \Delta^{\{82\}} + \Delta^{\{73\}} + \Delta^{\{64\}}$$

$$R^{\{5^2\}} = \Delta^{\{10\}} + \Delta^{\{91\}} + \Delta^{\{82\}} + \Delta^{\{73\}} + \Delta^{\{64\}} + \Delta^{\{5^2\}}$$

$$R^{\{64\}} = \Delta^{\{10\}} + \Delta^{\{91\}} + \Delta^{\{82\}} + \Delta^{\{73\}} + \Delta^{\{10\}}$$

$$R^{\{73\}} = \Delta^{\{10\}} + \Delta^{\{91\}} + \Delta^{\{82\}} + \Delta^{\{73\}}$$

$$R^{\{82\}} = \Delta^{\{10\}} + \Delta^{\{91\}} + \Delta^{\{82\}}$$

$$R^{\{91\}} = \Delta^{\{10\}} + \Delta^{\{91\}}$$

$$R^{\{10\}} = \Delta^{\{10\}}$$

Figure1. The scheme of decomposition of transitive representations R^{μ} into irreducible representations Δ^{λ} for $N = 10$

r = 10	1	=1
r = 9	1 + 9	=10
r = 8	1 + 9 + 35	=45
r = 7	1 + 9 + 35 + 75	=120
r = 6	1 + 9 + 35 + 75 + 90	=210
r = 5	1 + 9 + 35 + 75 + 90 + 42	=252
r = 4	1 + 9 + 35 + 75 + 90	=210
r = 3	1 + 9 + 35 + 75	=120
r = 2	1 + 9 + 35	=45
r = 1	1 + 9	=10
r = 0	1	=1
		2^{10}

Figure2. The sum rules corresponding to Fig.3.1

and the result is presented in Fig.3.1. Each row correspond to a particular decomposition of R^{μ} (the first colum in Fig.3.1), and subspaces with a given spin S are arranged into columns. The rightmost term in each row corresponds to Bethe vectors. The corresponding sum rules are given in Fig.3.2.

The duality of Weyl implies a new basis in the space $\mathcal{H}$, i.e. the basis which is irreducible under both groups, Σ_{10} and $SU(2)$. It is given by the set

$$b_{irr} = \{|\lambda y_W y_Y > \mid \lambda = \{N - r', r'\}, r' \in \{0, 1, 2, ..., 10\} ,$$

where y_W runs over all standards Weyl tableaux of the shape λ, i.e. all semistandard fillings of the Young diagram λ by elements the set $\tilde{2}$, y_Y runs over all standard Young tableaux, i.e. all standard fillings of λ by elements of the dual set $\tilde{10}$, and λ runs over all bipartitions of 10. The canonical bijection between the sets $\tilde{2}^{\tilde{10}}$ and b_{irr} is given by the Robinson-Schensted-Knuth (RSK) algorithm. We code the magnetic configuration

$$|f >= |f(1), f(2), ..., f(10) >$$

in terms of a 2×10 matrix m whose j-th column is

$$m_{\bullet j} = \begin{cases} \binom{1}{0} \text{ if } f(j) = + , \\ \binom{0}{1} \text{ if } f(j) = - . \end{cases}$$

For example, the configuration $| + + - + - + - +- >$ is represented by the matrix $m = \binom{1101010101}{0010101010}$. It follows from this construction that the set $M(\mu, 1^{10})$ of all integer non-negative matrices with the row sums μ and column sums 1^{10} coincides with the orbit of the action P, corresponding to the partition μ, and thus to the transitive representation R^{μ}. The RSK bijection can be presented in the form

$$RSK : \tilde{2}^{\tilde{10}} \to \bigcup_{\mu} \bigcup_{\lambda \geq \mu} SBT^{\lambda}(\mu, 1^{10}) ,$$

where the first sum runs over all rows of Fig.1, and the second over each term λ in the row μ (and thus over all partitions λ greater than μ in the partial order of dominance),

$$SBT^{\lambda}(\mu, 1^{10}) = ST^{\lambda}(\mu) \times ST^{\lambda}(1^N) .$$

is the set of all standard bitableaux of the shape λ and the two contents, μ and 1^N corresponding respectively to standard Weyl and Young tableaux. This algorithm has ben carried our for all magnetic configurations in the set $\tilde{2}^{\tilde{10}}$ along the description given in the monograph of Kerber [7]. An example for an orbit on the cyclic groups $C_{10} \subset \Sigma_{10}$ (the translation symmetry group of the crystal $\tilde{10}$) is given in Table 1.

Clearly, the RSK algorithm provides only a bijection between sets, as indicated in Eq.(3.11). Moreover, this bijection is compatible with the classification scheme shown in Fig.1, and resulting from the duality of Weyl.

Table1. The RSK algorithm for the orbit of the cyclic group $C_{10} \subset \Sigma_{10}$, generated from the configuration $|----+-++++>(\mu = \{5^2\})$. The first column gives the magnetic configuration f with the signs " + " omitted, the second - the corresponding matrix $m \in M(\{5^2\},\{1^{10}\})$, the thrid - the double sequence z (cf. [7]), the fourth - the Weyl tableau y_W, the fifth - the Young tableau y_Y, and the sixth - the shape λ

f	m	z	y_w	y_y	λ
---- -	0000101111	11111 22222	11111	12346	{55}
	1111010000	57891012346	22222	578910	
---- -	1000010111	11111 22222	111112	1234 57	{64}
	0111101000	15891023457	2222	68910	
---- -	1100001011	11111 22222	1111122	123 4568	{73}
	0011110100	12791034568	222	7910	
---- -	1110000101	11111 22222	11111222	12 345679	{82}
	0001111010	12381045679	22	810	
---- -	1111000010	1111122222	111112222	1234567810	{91}
	0000111101	12349567810	2	9	
- ----	0111100001	11111 22222	1111122222	13 456789	{82}
	1000011110	23451016789		210	
- ----	1011110000	1111122222	111112222	1245678910	{91}
	0100001111	13456278910	2	3	
- - ---	0101111000	1111122222	11111222	135678910	{82}
	1010000111	24567138910	22	24	
-- - --	0010111100	1111122222	1111122	12478910	{73}
	1101000011	35678124910	222	356	
---- -	0001011110	1111122222	111112	1235910	{64}
	1110100001	46789123510	2222	4678	

However, for purposes of diagonalization of the Heinsenberg Hamiltonian, we need a definite linear mapping, along the formula

$$|\lambda y_W y_Y> = \sum_{1 \leq j_1 < j_2 < \ldots < j_r \leq 10} U^{RSK}_{j_1 \ldots j_r, y_w y_y} |j_1 \ldots j_r> ,$$

where $|j_1 \ldots j_r>$ is the configuration $|f>$, written in terms of positions of spin deviations (e.g. $|----+-++++>$ is denoted by $|12346>$). The transformation matrix U^{RSK} should be given by the representation theory of the symetric group Σ_{10}. Essentially, the basis(3.13) corresponds to eigenvectors of the so called "corner transfer matrix"within the context of exactly solvable models of statistical physics in two dimensions (cf.terminology of Refs. [10,11]).

4 Bethe Ansatz, string configurations, and riggings

From now on, we consider only these subspaces in the space $\mathcal{H}$ of all quantum states of the magnet, which correspond to the highest weight vectors $S = M$, that is, to irreducible representations Δ^{λ} with $\lambda = \mu = \{N-r, r\}$ (cf. Fig.1). Having known all eigenfunctions of the Heinsenberg Hamiltonian in these subspaces, one easily derives all the remaining eigenfunctions by use of appropriate step operators of the angular momentum theory. Thus our spaces of interest have dimensions $1, 9, 35, 75, 90$, and 42 for $r = 0, 1, 2, 3, 4$, and 5, respectively. For each r, BA eigenfunctions are labelled by the set $\{\lambda_1, ..., \lambda_r\}$ of r numbers, possibly complex, referred to as spectral parameters, and obtained as non-zero, finite, and pairwise distinct ($\lambda_\alpha \neq \lambda_{\alpha'}$) solutions of the so called Bethe system of equations

$$\left(\frac{\lambda_\alpha - \frac{i}{2}}{\lambda_\alpha + \frac{i}{2}}\right)^{10} = \prod_{\substack{\alpha' \in \tilde{r} \\ \alpha' \neq \alpha}} \frac{\lambda_\alpha - \lambda_{\alpha'} - i}{\lambda_\alpha - \lambda_{\alpha'} + i}, \quad \alpha \in \tilde{r}\,.$$

where

$$\tilde{r} = \{\alpha = 1, 2, ..., r\}$$

is the set of labels of spin deviations (Bethe pseudoparticles). According to the hypothesis of Bethe [1,2,13,14], in the asymptotic limit $N \to \infty$, elements λ_α of the set $\{\lambda_1, ..., \lambda_r\}$ are organized into *strings*. Normally, each spectral parameter achieves the form

$$\lambda_m^{lv} = \lambda^{lv} + im + O(e^{-\delta N})\,,$$

where λ^{lv} is a real number, the imaginary part is a multiple of $1/2$, specifed below, and $O(e^{-\delta N})$ is the reminder which vanishes exponentially ($\delta > 0$). Moreover, the real part λ^{lv} is constant for $l \leq r$ spectral parameters, and the integer l is called the lenght of a string. A string itself is an l-element subset of the set $\{\lambda_1, ..., \lambda_r\}$, with the constituent elements labelled by $m \in \{-s, -s+1, ..., s\}$, where

$$s = \frac{l-1}{2}$$

(so that $l = 2s+1$) is referred to as the spin of the string. Thus m takes on the integer and half-integer values for l odd and even, respectively. Clearly, the Bethe hypothesis of strings defines a partition ν of the set $\{\lambda_1, ..., \lambda_r\}$ of spectral parameters which classify BA eigenfunctions. This partition is presented in a form

$$r = \sum_{l=1}^{\infty} l m_l\,,$$

where m_l is the number of strings of the lenght l, and referred to as a *string configuration*. The label v in Eq.(4.3) ranges over the set

$$\tilde{m_l} = \{v = 1, 2, ..., m_l\}\,,$$

which completes the classification of spectral parameters within the picture of strings. Each string can be looked at as a composite system, consisting of l

Bethe pseudoparticles. For a given string configurations ν, the total number of strings is

$$q = \sum_{l=1}^{\infty} m_l \, .$$

Clearely, $q \leq r$. Each string $i = 1, 2, ..., q$, can exist, in general, in different states, labelled by a quantum number L_i. The set

$$L = \{L_i | i = 1, 2, ..., q\}$$

of all such quantum numbers is called a *rigging* of the configuration ν. BA equations imply the following selection rules for admissible riggings: (i) For a given string configuration ν, riggings for each block of all strings of a given lenght l should be done seperately. (ii) Within each block of m_l strings of the lenght l, the corresponding riggings should be integers which satisfy

$$0 \leq L_v \leq L_{v+1} \leq P_l \, ,$$

where the range P_l is obtained from the combinatorial condition

$$P_l = N - 2Q_l \, ,$$

with Q_l being the number of boxes in the first l columns of the Young diagram ν. The total number of riggins of a given string configuration ν is readily obtained as

$$z(\nu) = \prod_{l=1}^{\infty} \binom{P_l + m_l}{m_l} \, .$$

Moreover, the Weyl duality, taken together with the hypothesis of strings, imply

$$dim \Delta^{\{N-r,r\}} = \sum_{\nu \vdash r} z(\nu) \, .$$

We consider as an example the case of the irreducible representation $\Delta^{\{5^2\}}$, corresponding to the case $S = 0, r = 5$. Now the string configurations ν are possible partitions of $r = 5$. For $\nu = \{1^5\}$ we get $l = 1, m_1 = 5, P_1 = 0$. Thus we have here 5 strings, each of the length 1. Each string has a rigging $L_i, i = 1, ..., 5$, but the condition (4.9) implies $L_1 = L_2 = ... = L_5 = 0$, Thus $z(1^5) = 1$, i.e. we have only one admissible rigging in this case. It can be presented in a form

$$\nu = \{1^5\} = \begin{array}{|c|l} \hline \; & L_1 0 \\ \hline \; & L_2 0 \\ \hline \; & L_3 0 \\ \hline \; & L_4 0 \\ \hline \; & L_5 0 \\ \hline \end{array}$$

Similary, for $\nu = \{21^3\}$, we have one string of the length 2, and three strings of the length 1, so that $P_1 = 2$, $P_2 = 0$, which yields the following riggings

$\nu = \{21^3\} =$ 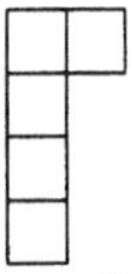
$L_1 0000000000$
$L_2 0000001112$
$L_3 0001121122$
$L_4 0121221222$

The other riggings for this case are the following

$\nu = \{2^2 1\} =$ 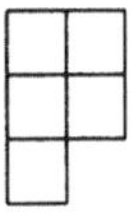
$L_1 00000$
$L_2 00000$
$L_3 01234$

$\nu = \{31^2\} =$ 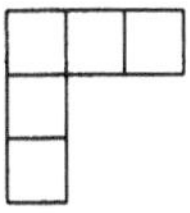
$L_1 000000000000000$
$L_2 000001111222334$
$L_3 012341234234344$

$\nu = \{32\} =$ 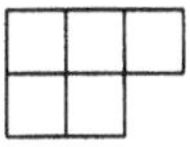
$L_1 000$
$L_2 012$

$\nu = \{41\} =$ 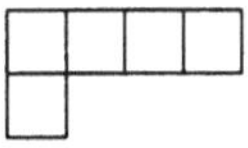
$L_1 0000000$
$L_2 0123456$

$\nu = \{5\} =$ 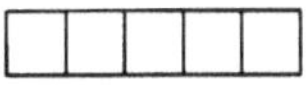 $L0$

The dimensional check for this case reads

$$z(1^5) + z(21^3) + z(2^2 1) + z(31^2) + z(32) + z(41) + z(5) = \dim \Delta^{\{5^2\}} ,$$

that is

$$1 + 10 + 5 + 15 + 3 + 7 + 1 = 42$$

In this way, BA offers the basis in the space $\mathcal{H}$, consisting of eigenstates of the Heinsenberg Hamiltonian. The highest weight eigenvectors are labelled by $|r\nu L>$, where r is the carrier space of the irreducible representation $\Delta^{\{N-r,r\}}$, corresponding to Bethe vectors, ν ranges over all partitions of r, and L-over all admissible riggings of the string configuration ν. Let b_r be the set of all vectors $|rvL>$ with r fixed. Kerov, Kirillov and Reshetikin [2] have constructed a bijection

$$KKR : ST^{\{N-r,r\}}(1^N) \to b_r ,$$

which is an extension of the RSK algorithm to rigged string configurations. The transformation matrix U^{KKR}, defined as the rigged counterpart of Eq. (3.13), can be, in principle, determined by an explicit use of solutions of the BA equations (4.1). The basis b_r corresponds to eigenvectors of the "row transfer matrix"of the paper of Dasmahapatra and Foda [11] , whereas the bijection (4.15) points out the mediating role of standard Young tableaux, posed in this paper.

5 Conclusion

We have presented combinatoric constructions, relevant in application of the Bethe Ansatz to the particular case of a linear magnetic ring with

$N = 10$ nodes and the spin 1/2. . The construction are related to different orthonormal bases in the spaces $\mathcal{H}$ of all quantum states of the magnet: (i) the set $\tilde{2}^{\tilde{10}}$ of all magnetic configurations, (ii) the sets $SBT^{\lambda}(\mu, 1^N)$ of appropriate standard bitableaux, emerging from the duality of Weyl, and (iii) the sets b_r of rigged string configurations which parametrize the BA solutions. The RSK algorithm provides a bijection between sets (i) and (ii), whereas Kerov, Kirillov and Reshetikhin propose a bijection between sets (ii) and (iii). Both bijections can be accompanied by appropriate transformation matrices.

We would like to thank to Prof. T. Lulek for important discussions and remarks, which help us to write this paper and for his patience during our questions. It was a pleasure for us to work together.

References

1. H. Bethe, Z. Phys. **71**, 205 (1931) (English translation in: D.C. Mattis, An Encyclopedia of Exactly Solved Models in One Dimension, World Sci., Singapore 1993, pp. 689-716)
2. S.V. Kerov, A.N. Kirillov, and N. Yu. Reshetikhin, J. Sov. Math. **41**, 916 (1988)
3. H. Weyl, The Theory of Groups and Quantum Mechanics, Princeton 1931.
4. G. de B. Robinson, Representation Theory of The Symetric Group, Edinburgh Univ. Pres, UK 1961
5. C.Schensted, Canad. J.Math. **13**, 179(1961)
6. D.E.Knuth, Pacific J. Math. **34**, 709(1970)
7. A. Kerber, Algebraic Combinatorics via Finite Group Actions, BI Wissenschaftverlag, Berlin 1991.
8. A. N. Kirillov, Zap. Nauch. Semin. LOMI **131**, 88(1983)(in Russian; English transl: J. Soviet Math. 30, 2298(1985)).
9. A. N. Kirillov, Zap. Nauch. Semin. LOMI **134**, 169(1984)(in Russian; English transl: J. Soviet Math. 36, 115(1987)).
10. A.Lascoux, B.Laclerc and J.-Y. Thibon, J. Math. Phys.**38**, 1041-68(1997).
11. S.Dasmahapatra, O.Foda, Strings, paths,and standard tableaux, Int. J. Mod. Phys. A 13, 1998.
12. A.Schilling and S. O. Warnaar, Comm. Math. Phys. **202** 359-401(1999).
13. M. Takahaski, Progr. Theor. Phys. **46**, 401-15(1971)
14. L. D. Faddeev and L. A. Takhtadzhyan, zap. Nauch. Semin. LOMI **109**, 134-73 (1981) (in Russian; English translation: J. Soviet Math. **24**, 241-67(1984)).

Computing Resolutions Over Finite p-Groups

Johannes Grabmeier[1] and Larry A. Lambe[2]

[1] IBM Deutschland Informationssysteme GmbH, Postfach 103068, D-69020 Heidelberg, Germany, grabm@de.ibm.com
[2] Centre for Innovative Computation University of Wales, Bangor, Gwynedd, LL57 1UT U.K. l.lambe@bangor.ac.uk

Dedicated to Professor Adalbert Kerber at the occasion of his 60th birthday.

Abstract A uniform and constructive approach for the computation of resolutions and for (co)homology computations for any finite p-group is detailed. The resolutions we construct ([32]) are, as vector spaces, as small as the minimal resolution of $\mathbb{F}_p$ over the elementary abelian p-group of the same order as the group under study. Our implementations are based on the development of sophisticated algebraic data structures. Applications to calculating functional cocycles are given and the possibility of constructing interesting codes using such methods is presented.

1 Introduction

In this paper, we present a uniform constructive approach to calculating relatively small resolutions over finite p-groups. The algorithm we use comes from [32, 8.1.8 and the penultimate paragraph of 9.4]. There has been a massive amount of work done on the structure of p-groups since the beginning of group theory. A good introduction is [22].

We combine mathematical and computer methods to construct the uniform resolutions in this paper. These resolutions are much smaller than the bar construction [34, Chapter IV, S5] (or Sect. 3.4), but in fact, actually use the bar construction in an essential way. As vector spaces over $\mathbb{F}_p$ (the field with p elements), the resolutions are as small as the minimal resolution of $\mathbb{F}_p$ over the elementary abelian p-group G_+ of the same order as the group G under study.

In low degrees, (e.g. less than or equal to 7 depending on the size of the group), these resolutions can be used to explicitly calculate not only homology and cohomology, but explicit cycle and (functional) cocycle representives of classes. This takes things a step further than one can go with, e.g. GAP or MAGMA where one can only get at a basis for first and second (co)homology. Having functional cocycles in hand allows one to examine interesting combinatorial properties. In this way we mention briefly how certain codes arise from some explicit cocycles. Also see [9] and [10].

We note that, in general, in order to increase practicality further, one needs to devise "reduction strategies" along the lines of [27] to reduce the size of the resolution for a general group. Our moderate sized resolutions are a good starting point for these methods, but such reductions are not within the scope of the current paper and will be discussed elsewhere.

A wide variety of algebraic data structures were required for our implementations. We briefly recall the mathematical setting for all these algebraic data structures and we also discuss them from the viewpoint of implementation. In so doing, we realized that it is well worth and by no means trivial to design a GENERIC LANGUAGE (GL) for the description and natural implementations of sophisticated algebraic objects, algorithms and data structures. However, these fundamental considerations combining mathematical and computer science methods and techniques would be far beyond the scope of the current paper and hence will be developed and discussed elsewhere [5]. For now, we use the computer algebra system AXIOM [24] which consists of a language, compiler, interpreter and a user interface to accomplish our goals.

The algorithm we present in Sect. 5 is of a recursive nature and can be applied naturally to p-groups G given as a polynomial perturbation of the elementary abelian group G_+ (Sect. 2.4). If the group is not given this way, we use the theorems of Jennings and Birkhoff-Poincaré-Witt to construct an appropriate isomorphism $\Xi : \mathbb{F}_p G_+ \longrightarrow \mathbb{F}_p G$ as vector spaces. This is related to the mod-p lower central series of G.

2 Finite p-groups

In this section, we give an exposition of some well-known properties of finite p-groups needed to understand the algorithms presented for practical applications of the main theorems in Sect. 4.

2.1 The mod-p Lower Central Series

For each finite p-group G, an n-dimensional *mod-p restricted* Lie algebra $\mathrm{gr}_p G$ can be defined ([25]) using the *mod-p lower central* series $G = Z_1 \geq Z_2 \geq \ldots$. Here Z_i is defined by

$$Z_i = \langle (x_1, (x_2, (\ldots (x_{j-1}, x_j) \ldots) y^{p^k} | j p^k \geq i \rangle$$

or equivalently, for $i > 1$, $Z_i = (Z_{i-1}, Z_1) Z_j{}^p$ where j is the smallest integer greater than or equal to $\frac{i}{p}$.

As $(Z_i)_{i \geq 1}$ in (2.1) is a *p-filtration* of the group G, i.e. the commutator group (Z_i, Z_j) is contained in Z_{i+j} and $Z_i^p \subseteq Z_{pi}$, the groups Z_i/Z_{i+1} (of order say p^{d_i}) are abelian and elementary, and hence we are able to define the $\mathbb{F}_p$-vector space $\mathrm{gr}_p G = \oplus_{i \geq 1} Z_i / Z_{i+1}$ and recursively can choose elements $\{x_{i,k} | 1 \leq k \leq d_i\}$ in the group, use their remainder classes as an $\mathbb{F}_p$-basis and define the Lie algebra multiplication

$$[x_{i,k}, x_{j,l}] = \overline{(x_{i,k}, x_{j,l})}$$

by using the commutator $(x_{i,k}, x_{j,l})$ in the group and reducing modulo the next subgroup Z_{i+j+1} in the filtration. More precisely, let ρ_i denote the natural surjection $\rho_i : Z_i \longrightarrow Z_i/Z_{i+1}$. Then the Lie bracket for elements $\rho_i(x) \in Z_i/Z_{i+1}$ and $\rho_j(y) \in Z_j/Z_{j+1}$ is defined by

$$[\rho_i(x), \rho_j(y)] = \rho_{i+j}((x, y))$$

while the definition

$$\rho_i(x)^p = \rho_{ip}(x^p)$$

satisfies the identities given on pages 91 – 93 in [25] (also see [23]), hence it is a *p-restriction* and this indeed yields a *p-restricted* Lie algebra. Let $\epsilon : \mathbb{F}_p G \longrightarrow \mathbb{F}_p, \sum_g a_g g \mapsto \sum_g a_g$ (cf. 3.3) be the augmentation of the group algebra $\mathbb{F}_p G$ and $I = \ker(\epsilon)$ the augmentation ideal. Then note also that the important relation used below (2.3):

$$Z_i = \{g \in G | (g-1) \in I^i\}$$

holds ([25]).

2.2 The Universal Enveloping Algebra of a p-Restricted Lie Algebra and the Theorem of Birkhoff-Poincaré-Witt

Let L be an ordinary Lie algebra over $\mathbb{F}_p$. One has the universal enveloping algebra $T(L)/J$, where $T(L)$ is the tensor algebra of the underlying vector space structure of L and J is the ideal generated by $\{x \otimes y - y \otimes x - [x, y] \mid x, y \in L\}$ (see e.g. [37]) – there is a universal enveloping algebra for p-restricted Lie algebras. One adds the additional relations $\underbrace{x \otimes \ldots \otimes x}_{p} - x^p$ for $x \in L$ to the ideal J. We denote this algebra by $\mathcal{V}(L)$ for a p-restricted Lie algebra L. Similar to the ordinary case, if A is any associative algebra over $\mathbb{F}_p$, there is a functor $\mathcal{L}$ such that $\mathcal{L}(A)$ is a p-restricted Lie algebra. The underlying vector space structure is that of A, the bracket is $[x, y] = xy - yx$, and the restriction is given by the p-th power in A. Furthermore, the universal property for $\mathcal{V}(L)$ holds for any p-restricted Lie algebra, viz. any map $f : L \longrightarrow \mathcal{L}(A)$ extends to a unique algebra map $\mathcal{V}(f) : \mathcal{V}(L) \longrightarrow A$ (see [23]).

$\mathcal{V}(L)$ has a natural grading induced by the filtering by length, and hence we can form the associated graded algebra, i.e. $E_0(\mathcal{V}(L)) = \sum_{i \geq 0} V_i/V_{i-1}$, where V_i for $i \geq 0$ is the submodule consisting of all elements, which are images of elements in the tensor algebra of (total) length less or equal to i and $V_{-1} = 0$. This construction is important for computation in the universal enveloping algebra $\mathcal{V}(\mathrm{gr}_p G)$ of the p-restricted Lie algebra $\mathrm{gr}_p G$ as one can make use of the p-modular version of the Birkhoff-Poincaré-Witt theorem.

Theorem 2.1. *Let L be a p-restricted Lie algebra with basis $\{e_1, \ldots, e_n\}$. The associated graded algebra $E_0(\mathcal{V}(L))$ of the universal enveloping algebra $\mathcal{V}(L)$ given by the length filtration is isomorphic as a graded $\mathbb{F}_p$-algebra to the algebra $\mathbb{F}_p[e_1, \ldots, e_n]/(e_1{}^p, \ldots, e_n{}^p)$ of truncated polynomials.*

The proof is given in [23].

2.3 The Theorem of Jennings

The theorem of Jennings, [25], is the final link between these constructions.

Theorem 2.2. *Let $(Z_i)_{i>0}$ be the* mod $-$ *p lower central series of a finite p-group G and let $I = \ker(\epsilon)$ be the augmentation ideal in the group ring $\mathbb{F}_p G$. Let $E_0(\mathbb{F}_p G) = \sum_{i\geq 0} I^i/I^{i+1}$ be the* associated graded algebra *with respect to the filtration given by powers of the augmentation ideal I. Let ρ_i denote both the canonical surjections $Z_i \longrightarrow Z_i/Z_{i+1}$ and $I^i \longrightarrow I^i/I^{i+1}$. Then $\rho_i(x) \mapsto \rho_i(x-1)$ induces an homomorphism of p-restricted Lie algebras $\mathrm{gr}_p G \longrightarrow \mathcal{L}(E_0(\mathbb{F}_p G))$ and its extension to a map*

$$\mathcal{V}(\mathrm{gr}_p G) \longrightarrow E_0(\mathbb{F}_p G)$$

is a graded algebra isomorphism.

Note that from the construction of $x_{i,k}$ in 2.1 it is clear that for each element g of the group there is a unique sequence of exponents $0 \leq \epsilon_{i,j} < p$ such that $g = \prod_{i\geq 1} \prod_{k=1}^{d_i} x_{i,k}^{\epsilon_{i,k}}$, where the order of the multiplied elements is lexicographic w.r.to (i,k). The crucial step in Jennings' work and particularly important for implementations is his proof that

$$x^{\epsilon} = \prod_{i\geq 1} \prod_{k=1}^{d_i} (x_{i,k} - 1)^{\epsilon_{i,k}}$$

in $E_0(\mathbb{F}_p G)$ for $0 \leq \epsilon_{i,k} < p$ (same ordering) is an $\mathbb{F}_p$-Basis. Note that all these basis elements are homogeneous, where the degree is computed by $\deg(x^{\epsilon}) = \sum_{i\geq 1} \sum_{k=1}^{d_i} i\epsilon_{i,k}$, and all the basis elements of degree i form a basis of I^i/I^{i+1} ([25]).

This theorem gives a close connection between restricted Lie algebras and finite p-groups. It is an important relation. It was generalized by Quillen to all groups – not just p-groups, see [36].

2.4 Polynomial Group Laws

It is not hard to see that any group G of order p^n is isomorphic to a group of the form $(\mathbb{F}_p^n, \rho)$, where the group law $\rho : \mathbb{F}_p^n \times \mathbb{F}_p^n \longrightarrow \mathbb{F}_p^n$ is a polynomial function. That is, the i-th component function $\rho_i(a,b)$ is given by a polynomial in the coordinates $(a_1, \ldots, a_n, b_1, \ldots, b_n)$ with coefficients in $\mathbb{F}_p$. Furthermore, the ρ_i may be chosen to satisfy

$$\rho_i(a,b) = a_i + b_i + \mu_i(a_1, \ldots, a_{i-1}, b_1, \ldots, b_{i-1}) \tag{1}$$

where μ_i is zero if any argument is zero. It is clear that the identity element is the zero vector and that μ_i cannot have a constant term.

Letting $e_n = (0, \dots, 0, 1)$, it is clear that e_n has order p, $e_n^j = (0, \dots, 0, j)$ and all these elements are in the center of the group.

The proof is by induction. Clearly the result is true for an elementary abelian p-group since its operation is just $+$. Assume inductively that the result is true for all finite p-groups G of order p^n. Given a finite p-group $\widetilde{G}$ of order p^{n+1}, as is well-known, there is a non-trivial element in the center and it can be chosen to have order p. We therefore have a central extension

$$0 \longrightarrow (\mathbb{F}_p, +) \xrightarrow{\alpha} \widetilde{G} \xrightarrow{\beta} G \longrightarrow 1.$$

As is also well-known ([34]), a 2-cocycle μ arises by choosing a right inverse $u : G \longrightarrow \widetilde{G}$ for β with $u(1) = 1$ and taking $G \times G \xrightarrow{\mu} \mathbb{F}_p$ to be $\mu(a, b) = u(a)u(b)u(ab)^{-1}$.

Assuming inductively that G and the group law ρ of G are of the desired form and noting that the group $\widetilde{G}$ is isomorphic to the group given by $G \times \mathbb{F}_p$ with group law

$$\widetilde{\rho}((a, a_{n+1}), (b, b_{n+1})) = (\rho(a, b), a_{n+1} + b_{n+1} + \mu_{n+1}(a, b)),$$

the result follows.

A natural class of examples is given by the upper triangular $n \times n$ matrix groups over $\mathbb{F}_p$, where n is any positive integer.

$$UT_n(p) = \left\{ \begin{pmatrix} 1 & a_1 & a_n & \cdots & a_m \\ 0 & 1 & a_2 & \ddots & \vdots \\ \vdots & \ddots & \ddots & \ddots & a_{2n-3} \\ 0 & 0 & \ddots & \ddots & a_{n-1} \\ 0 & 0 & 0 & \ddots & 1 \end{pmatrix} \mid a_{i,j} \in \mathbb{F}_p \right\}.$$

Clearly, $UT_n(p) \cong (\mathbb{F}_p^m, \rho)$ where $m = \binom{n}{2}$ and the group law (matrix multiplication) is a polynomial function of the required form. Note also that any finite p-group can be embedded in $UT_n(p)$ for some n (see [22]).

For the class of cyclic groups C_{2^n} of order 2^n it is also easy to write down a polynomial group law. Define $c_1 = 0$ and for any positive integer i, let $\rho_i = a_i + b_i + c_i$ where

$$c_i(a_1, \dots, a_{i-1}, b_1, \dots, b_{i-1}) = \sum_{\gamma=1}^{i-1} a_\gamma b_\gamma \prod_{\kappa=\gamma+1}^{i-1} (a_\kappa + b_\kappa). \tag{2}$$

Then for all positive integers n, $\rho = (\rho_1, \dots, \rho_n)$ is a polynomial group law and in fact, the group $(\mathbb{F}_2^n, \rho)$ determined by ρ is a cyclic group of order 2^n. Note that this group is generated by $(1, 0, \dots, 0)$. More generally, if e_i is the ith standard basis element vector, then $e_i e_i = e_{i+1}$ and e_i generates a subgroup of order 2^{n+1-i}. Moreover,

$$(e_1)^j = (j_0, j_1, \ldots, j_{n-2}, j_{n-1})$$

for $0 \leq j < 2^n$ and $j = \sum_{\nu=0}^{n-1} j_\nu 2^\nu$ is the representation of j as a binary number.

Note also that $c_0 = 0$ and $c_{i+1} = c_i(a_i + b_i) + a_i b_i$. It is easily seen, that the polynomial group law precisely describes the addition of binary numbers. Contrary to the usual we have to reverse the order of the digits in this situation. A proof consists in verifying that adding three natural numbers a_i, b_i and a carry c_i from the position before from $\{0,1\}$ to get $(c_{i+1}, \rho_i)_2$ can be realized recursively by $\rho_i = a_i + b_i + c_i$ (sum bits of two half adders) and the carry bit $c_{i+1} = a_i b_i + (a_i + b_i)c_i = a_i b_i \vee (a_i + b_i)c_i$ as $a_i b_i$ and $(a_i + b_i)c_i$ are never both equal to 1. This shows that we indeed have the usual implementation of a full adder.

For another natural example, let i be a positive integer and define $\rho_i = a_i + b_i + c_{i-1,2} + \delta_i a_1$, where $c_{i-1,2}$ is the cyclic cocycle for the positions $2, \ldots, i$ and $\delta_i = \sum_{\kappa=1}^{i} p_\kappa$, where p_κ is recursively defined by

$$p_1 = 0, \; p_2 = 0, \; p_{\kappa+1} = (b_\kappa + a_\kappa)\delta_\kappa + c_{\kappa-1,2} + b_\kappa.$$

Then for all positive integers n, $\rho = (\rho_1, \ldots, \rho_n)$ is a polynomial group law and the group $(\mathbb{F}_2^n, \rho)$ is a dihedral group D_{2^n} of order 2^n. Its cyclic subgroup of order 2^{n-1} can be generated by $c = (0, 1, 0, \ldots, 0)$, while $x = (1, 0, 0, \ldots, 0)$ is a reflection.

This can be verified by direct computation for the products $xa = (1 + a_1, \overline{a_2}, \ldots, \overline{a_{n-2}})$, $ax = e_1 + a$ for $a \in D_{2^n}$ where, as above, $e_1 = (1, 0, \ldots, 0)$ and $(x(c^i))(x(c^i)) = 1$ as well relating the recursion of p and ρ to the computation of a^{-1}. Note that for $i = i_2 + i_3 2 + \ldots + i_n 2^{n-2}$ we denoted by $\overline{i_2} + \overline{i_3} 2 + \ldots + \overline{i_n} 2^{n-2}$ the element $-i$ modulo 2^{n-1}.

3 Some Homological Algebra

We recall some basic facts from homological algebra needed to understand the purpose of the algorithms presented in Sect. 5. Throughout this section, R will denote a commutative ring with unit.

3.1 Chain/Cochain Complexes

A *chain complex* over R is a sequence of R-modules and R-linear maps

$$\cdots \xrightarrow{d_{n+1}} X_n \xrightarrow{d_n} X_{n-1} \longrightarrow \cdots$$

such that for all n, $d_n d_{n+1} = 0$. Following the usual conventions, such a chain complex will be denoted by (X, d) or simply X when the context is clear. The map d is called the *differential*. If it needs to be stressed, we will write the differential in X as d_X. Elements of X_n are said to have degree n and if $x \in X_n$, we write $|x|$ for its degree. Since d lowers the degree by one, we say it has degree -1 and we write $|d| = -1$.

The nth homology module of X, denoted by $H_n(X)$ is, by definition, the quotient module $\ker(d_n)/\mathrm{im}(d_{n+1})$, the homology of X is $H_*(X) = \oplus_n H_n(X)$.

A *cochain complex* over R is a sequence of R-modules and R-linear maps

$$\cdots \xleftarrow{d_n} X_n \xleftarrow{d_{n-1}} X_{n-1} \longleftarrow \cdots$$

such that for all n, $d_n d_{n-1} = 0$. The nth cohomology module of X, denoted by $H^n(X)$ is, by definition, the quotient module $\ker(d_n)/\mathrm{im}(d_{n-1})$. The cohomology of X is defined to be $H^*(X) = \oplus_n H^n(X)$.

Note that if X is chain complex, then the linear dual $X^* = \mathrm{Hom}_R(X, R)$ is a cochain complex in the obvious way.

3.1.1 Chain Maps and Homotopies

A *chain map* $f : X \longrightarrow Y$ is a sequence of R-linear maps making the diagram

$$\begin{array}{ccc} X_n & \xrightarrow{f_n} & Y_n \\ {\scriptstyle d_n}\downarrow & & \downarrow{\scriptstyle d_n} \\ X_{n-1} & \xrightarrow[f_{n-1}]{} & Y_{n-1} \end{array}$$

commute. Its easy to see that this condition causes any chain map to induce an R-linear map on homology $H_*(f) : H_*(X) \longrightarrow H_*(Y)$ in the obvious way.

Note that the identity map on X which we will denote by 1_X is a chain map.

Two chain maps $f, g : X \longrightarrow Y$ are said to be *chain homotopic* (by ϕ) if there is an R-linear map $\phi_n : X_n \longrightarrow Y_{n+1}$ such that $\phi_{n-1} d_n + d_{n+1}\phi_n = f_n - g_n$ for all n. Following conventions, this condition is simply written $d\phi + \phi d = f - g$. The (degree $+1$) map ϕ is called a *chain homotopy* between f and g. Its easy to see that if f and g are chain homotopic, then they induce the same map in homology.

Note that these notions clearly have analogues for cochain complexes.

3.1.2 Strong Deformation Retracts

Let X and Y be chain complexes, $\nabla : X \longrightarrow Y$, $f : Y \longrightarrow X$ be chain maps and let $\phi : Y \longrightarrow Y$ be a degree $+1$ R-linear map such that $f\nabla = 1_X$ and $d\phi + \phi d = 1_Y - \nabla f$. Thus, f and ∇ compose to the identity, but the composition the other way around is only chain homotopic to the identity. When these conditions hold, we say that this collection of data forms a *strong deformation retraction* (SDR) and we write

$$X \underset{f}{\overset{\nabla}{\rightleftarrows}} (Y, \phi). \tag{3}$$

Crucial to computations are the *side conditions* ([30])

$$\phi^2 = 0, \quad \phi\nabla = 0, \quad \text{and,} \quad f\phi = 0.$$

In fact, it can be shown with a bit of computation that these may always be assumed to hold: if the last two do not hold, replace ϕ by $\phi' = D(\phi)\phi D(\phi)$ where $D(\phi) = \phi d + d\phi$ and the last two conditions will now hold with respect to ϕ'. If the first condition does not hold for ϕ', replace it by $\phi'' = \phi' d\phi'$ and all three conditions will hold for the chain homotopy ϕ''.

3.2 The Perturbation Lemma

Given the SDR (3) and in addition a second differential d'_Y on Y, let $t = d'_Y - d_Y$. The map t is called the *initiator* ([1]). The *perturbation lemma*, [2], [12], [1] states that if we set $t_n = (t\phi)^{n-1}t, \quad n \geq 1$ and, for each n, define new maps on X:

$$\partial_n = d + f(t_1 + t_2 + \cdots + t_{n-1})\nabla \tag{4}$$
$$\nabla_n = \nabla + \phi(t_1 + t_2 + \cdots + t_{n-1})\nabla. \tag{5}$$

On Y:

$$f_n = f + f(t_1 + t_2 + \cdots + t_{n-1})\phi \tag{6}$$
$$\phi_n = \phi + \phi(t_1 + t_2 + \cdots + t_{n-1})\phi. \tag{7}$$

then in the limits (provided they exist), we have new SDR data

$$(X, \partial_\infty) \underset{f_\infty}{\overset{\nabla_\infty}{\rightleftarrows}} ((Y, d'_Y), \phi_\infty). \tag{8}$$

Note that the limits will certainly exist if $t\phi$ is nilpotent in each degree.

Examples are given in [2], [12], [28], [33], [31], [29], [30], [13], [14], [21], [19], [20], for example. In particular, this paper discusses an implementation of the algorithm given at the end of section (9.4) in [32].

3.3 Homology and Cohomology of Algebras

Let A be an algebra over R. For a left A-module M, a projective resolution of M over A is an exact sequence of projective A-modules

$$\cdots \xrightarrow{d_2} X_1 \xrightarrow{d_1} X_0 \xrightarrow{\epsilon} M \longrightarrow 0. \tag{9}$$

In particular, there is an associated chain complex X:

$$\cdots \xrightarrow{d_2} X_1 \xrightarrow{d_1} X_0 \longrightarrow 0$$

such that $H_i(X) = 0$ if $i > 0$ and $H_0(X) = M$. Note that we will often write (9) as $X \xrightarrow{\epsilon} M \longrightarrow 0$

If N is another A-module and X is a projective resolution, $X \otimes_A N$ is a chain complex whose homology is $\mathrm{Tor}^A(M, N)$ and the (co)homology of the chain complex $\mathrm{Hom}_A(M, N)$ is $\mathrm{Ext}_A(M, N)$. In the special case that A is *augmented* over R, i.e. comes equipped with a ring homomorphism $A \xrightarrow{\epsilon} R \longrightarrow 0$, we can make R a left A-module via the action $ar = \epsilon(a)r$, for $a \in A$, $r \in R$. In this special case, it is conventional to denote $\mathrm{Tor}^A(A, R)$ by $H_*(A)$ and call it the homology of A and in addition, denote $\mathrm{Ext}_A(A, R)$ by $H^*(A)$ and call it the cohomology of A.

3.3.1 Homology and Cohomology of Groups If G is any group, let RG be the group ring over R, i.e. the free R-module over R with basis G and multiplication given by the bilinear extension of the product in G to all of RG. This algebra is augmented over R by $\epsilon(\sum r_g g) = \sum r_g$. The homology and cohomology of RG (in the sense above) are simply denoted by $H_*(G)$ and $H^*(G)$ in this case. In this paper, we are concerned with the case that G is a finite p-group and $R = \mathbb{F}_p$.

3.4 The Standard Resolution (Bar Construction)

Let A be an R-Algebra with augmentation ϵ as above. The *bar construction* $B(A)$ ([34], [3]) is a particular A-free resolution of R. It is of the form $B(A) = A \otimes_R \bar{B}(A)$ where

$$\bar{B}(A) = \sum_{n \geq 0} \otimes_R^n \bar{A},$$

and $\bar{A} = A/R$ (thinking of R as a submodule of A via the unit). Following convention, we write $a[a_1|\ldots|a_n] = a \otimes a_1 \otimes \ldots \otimes a_n$ for an element of $A \otimes_R \bar{B}(A)$ and we think of the a_i as coming from A with the convention that $[a_1|\ldots|a_n] = 0$ if one of the a_i is in R. Also, the class of the identity element of A in $\bar{B}_0(A) = \bar{A}$ is denoted by $[]$. Elements of $\bar{B}_n(A) = \otimes_R^n \bar{A}$ are called *reduced elements*.

Define an R-linear map $B(A) \xrightarrow{s} B(A)$ by

$$s(a[a_1|\ldots|a_n]) = [a|a_1|\ldots|a_n]$$

and extend the augmentation map to all of $B(A)$ by taking $\epsilon(a[]) = \epsilon(a)$ and $\epsilon(a[a_1|\ldots|a_n]) = 0$ if $n \geq 1$. Let $R \xrightarrow{\sigma} A$ denote the unit map. Now consider the equation

$$\partial s + s\partial = 1_{B(A)} \tag{10}$$

in ∂ for chain maps. Let $\partial_0(a[]) = []\epsilon(a)$, and $s_{-1} : R \longrightarrow B_0(A) = A$ be given by $s_{-1}(r) = []\sigma(r)$. The formula (10) then inductively determines ∂_n for all $n \geq 1$. It is straightforward to derive the formula

$$\partial([a_1|\ldots|a_n]) = a_1[a_2|\ldots|a_n] + \sum_{j=1}^{n-1}(-1)^j[a_1|\ldots|a_j a_{j+1}|\ldots|a_n] \\ + (-1)^n[a_1|\ldots|a_{n-1}]\epsilon(a_n) \qquad (11)$$
$$(12)$$

(extend A-linearly over all of $B(A)$). In fact, it is not hard to prove that $\partial^2 = 0$ so that we have a free resolution of the A-module R (via augmentation). In fact, it is not hard to see from the definitions that we actually have an explicit SDR

$$R \underset{\epsilon}{\overset{s_{-1}}{\rightleftarrows}} (B(A), s). \qquad (13)$$

3.4.1 Functional Cocycles Let G be a group and $A = RG$ be the group ring. The dual of the bar construction

$$C(G) = \mathrm{Hom}_A(B(A), A) \qquad (14)$$

is a complex whose cohomology is the cohomology of G (as defined in section (3.3). It is not hard to see that

$$C^n(G) = \mathrm{Hom}_R(\bar{B}_n(A), R) \cong F^n \qquad (15)$$

where $F^n = \{G^n \xrightarrow{f} R \mid f(g_1, \ldots, g_n) = 0, \text{ if } g_\iota \in R \text{ for some } \iota\}$. See [34] for details. We shall identify $C^n(G)$ with such functions from G^n to R. Note that the differential in this context is given as follows. If $f : G^n \longrightarrow R$, then $\delta(f) : G^{n+1} \longrightarrow R$ is the function

$$\delta(f)(g_1, \ldots, g_{n+1}) = \\ f(g_2, \ldots, g_{n+1}) + \sum(-1)^k f(g_1, \ldots, g_k g_{k+1}, \ldots, g_{n+1}) \\ + (-1)^{n+1} f(g_1, \ldots, g_n) \qquad (16)$$

The algorithms given in the next sections can be used to explicitly compute such functional cochain representatives for the cohomology of any finite p-group. In fact, as we will see, the algorithm actually produces polynomials that represent cocycles.

3.5 The Comparison Theorem

The comparison theorem in homological algebra [34], [3] states that if X and Y are and two projective resolutions of M over A, they are chain homotopy equivalent, i.e. there are chain maps $f : X \longrightarrow Y$ and $g : Y \longrightarrow X$ such that fg and gf are both chain homotopy equivalent to the identity map. We will need a constructive version of this for free resolutions that essentially goes back to [4]. In fact, we are interested only in the case when $Y = B(A)$ and $fg = 1_X$, so that we actually obtain an SDR. The explicit formulae and discussion were given in [33]. We will simply repeat the formulas here for the reader's convenience.

Given a resolution of the form $X = A \otimes_R \bar{X} \longrightarrow R \longrightarrow 0$ with an explicit contracting homotopy $\psi : X \longrightarrow X$, construct maps $\nabla : X \longrightarrow B(A)$, $f : B(A) \longrightarrow X$, and $\phi : B(A) \longrightarrow B(A)$ inductively as follows:

$$\nabla_0 = 1_A, \quad f_0 = 1_A, \quad \phi_0 = 0, \tag{17}$$

and for $n > 0$, extend A-linearly the map defined on $\bar{X}$ given by

$$\nabla_n = s_{n-1} \nabla_{n-1} d_n \tag{18}$$

and extend A-linearly the maps defined on $\bar{B}(A)$ given by

$$f_n = \psi_{n-1} f_{n-1} \partial_n, \tag{19}$$

$$\phi_n = s_n (1_{\bar{B}_n(A)} - \nabla_n f_n - \phi_{n-1} \partial_n). \tag{20}$$

Conditions under which ∇ constructed this way is one-one are given in [35]. When they are satisfied, the maps above produce an SDR as is discussed in [33]. In this paper, it will be clear that the maps ∇ we define below are one-one by construction (and hence SDR data results).

The formulae for f and ϕ can easily be worked out ([33]) as follows. The A-linear map f is given recursively by

$$f([\,]) = 1, \; f([b_1| \dots |b_n]) = \psi(b_1 f([b_2| \dots |b_n])). \tag{21}$$

The A-linear map ϕ is given recursively by

$$\phi([\,]) = 0, \; \phi([b_1| \dots |b_n]) = s \nabla f([b_1| \dots |b_n]) + s(b_1 \phi([b_2| \dots |b_n])). \tag{22}$$

3.6 The Minimal Resolution of $\mathbb{F}_p^n$ Over $\mathbb{F}_p$

Let G_+ be the underlying abelian group of $\mathbb{F}_p$. H. Cartan [4] gave a resolution of $\mathbb{F}_p$ over $\mathbb{F}_p G_+^n$ and an explicit contracting homotopy which we will recall in this subsection. First, we will need to recall some standard algebras. The ordinary polynomial algebra is denoted by $\mathbb{F}_p[t_1, \dots, t_n]$.

3.6.1 The Divided Power Algebra The divided power algebra $\Gamma_p[y]$, for an even degree generator y has $\mathbb{F}_p$-basis $\{\gamma_i(y) | i = 0, \dots\}$. The multiplication is determined by extending

$$\gamma_i(y)\gamma_j(y) = \binom{i+j}{j}_p \gamma_{i+j}(y)$$

bilinearly over all of $\Gamma_p[y]$ where $\binom{i+j}{j}_p$ is the binomial coefficient mod p. Note that $\gamma_0(y) = 1$ and, by convention, we write $y = \gamma_1(y)$. For even degree generators, we define

$$\Gamma_p[y_1, \ldots, y_n] = \Gamma_p[y_1] \otimes \cdots \otimes \Gamma_p[y_n]$$

(tensor product algebra). We omit writing the tensor sign for elements when convenient. The degree is given by $|\gamma_{i_1}(y_1) \ldots \gamma_{i_n}(y_n)| = \sum i_\nu |y_\nu|$.

3.6.2 The Exterior Algebra The exterior algebra $\Lambda_p[x]$, for an odd degree generator x, is the quotient $\mathbb{F}_p[x]/(x^2)$. For odd degree generators, $x_1, \ldots, x_n$, we take

$$\Lambda_p[x_1, \ldots, x_n] = \Lambda_p[x_1] \otimes \cdots \otimes \Lambda_p[x_n].$$

We also omit writing tensor signs for elements when convenient. Note that every element of $\Lambda_p[x_1, \ldots, x_n]$ can be written uniquely as an $\mathbb{F}_p$-linear combination of elements of the form $x_1^{i_1} \ldots x_n^{i_n}$ where $i_\nu \in \{0, 1\}$ and $|x_1^{i_1} \ldots x_n^{i_n}| = \sum i_\nu |x_\nu|$.

3.6.3 Cartan's "Little Resolution" in Any Characteristic Note first of all that as an algebra, $\mathbb{F}_p G_+ \cong \mathbb{F}_p[t]/(t^p - 1)$. We assign all elements the degree zero.

Let $C = \mathbb{F}_p G_+ \otimes \Gamma_p[y] \otimes \Lambda_p[x]$ (as an algebra) where $|y| = 2$ and $|x| = 1$. Again we omit tensor signs in writing elements when convenient. Extend the grading to all of C in the usual way, i.e. $|cc'| = |c| + |c'|$.

C is augmented by the map ϵ given by extending the assignments

$$\epsilon(t) = 1, \ \epsilon(x) = 0, \ \text{and} \ \epsilon(\gamma_i(y)) = 0$$

to an algebra map to $\mathbb{F}_p$. Note that the unit map is $\sigma(r) = r \otimes 1 \otimes 1$.

The differential d on C is the unique $\mathbb{F}_p G_+$-linear graded derivation determined by

$$dt = 0, \quad dx = t - 1, \quad d\gamma_i(y_1) = t^{[p]} \otimes \gamma_{i-1}(y) \otimes x \tag{23}$$

where we use the notation $t^{[k]} = \frac{t^k - 1}{t - 1}$ in general.

A contracting homotopy ψ for C is the $\mathbb{F}_p$-linear map given by

$$\begin{aligned} \psi(t^k \otimes \gamma_i(y) \otimes x^\eta) = {} & [k > 0][\eta = 0] \ t^{[k]} \otimes \gamma_i(y) \otimes x \\ & + [k = p - 1][\eta = 1] \ 1 \otimes \gamma_{i+1}(y) \otimes 1, \end{aligned} \tag{24}$$

where we use the Kronecker-Iverson notation $[b]$, which evaluates to 1 for Boolean expressions b having value true, and to 0 for those having value false, see [11].

Note that $\mathbb{F}_p[t_1,\ldots,t_n]/(t_1^p-1,\ldots,t_n^p-1) \cong \mathbb{F}_p[t_1](t_1^p-1)\otimes\cdots\otimes \mathbb{F}_p[t_n]/(t_n^p-1) \cong \mathbb{F}_pG_+^n$. Using tensor product formulae, we get a resolution $\mathcal{C}^{(n)}$ over $\mathbb{F}_pG_+^n$ of the form

$$(\mathbb{F}_pG_+^n \otimes \Gamma[y_1,\ldots,y_n]\otimes \Lambda_p[x_1,\ldots,x_n],\, d^{(n)}) \xrightarrow{\epsilon^{(n)}} \mathbb{F}_p \longrightarrow 0$$

using the fact that

$$\mathbb{F}_pG_+^n \otimes \Gamma_p[y_1,\ldots,y_n]\otimes \Lambda_p[x_1,\ldots,x_n] \cong \otimes^n\mathcal{C}.$$

Here $\epsilon^{(n)} = \otimes^n\epsilon$ and $d^{(n)}$ is the pullback of the tensor product differential

$$d^{\otimes^n} = \sum_{\nu=1}^{n} \underbrace{1\otimes\ldots\otimes 1}_{\nu-1}\otimes d\otimes \underbrace{1\otimes\ldots\otimes 1}_{n-\nu}. \tag{25}$$

and $\psi^{(n)}$ is the pullback of the tensor product homotopy given by

$$\psi^{\otimes^n} = \sum_{\nu=1}^{n} \underbrace{\pi\otimes\ldots\otimes \pi}_{\nu-1}\otimes \psi\otimes \underbrace{1\otimes\ldots\otimes 1}_{n-\nu}, \tag{26}$$

or by the reverse variant

$$\psi^{\otimes^n} = \sum_{\nu=1}^{n} \underbrace{1\otimes\ldots\otimes 1}_{\nu-1}\otimes \psi\otimes \underbrace{\pi\otimes\ldots\otimes \pi}_{n-\nu} \tag{27}$$

(or in fact, by any of the other permutations possible) where $\pi = \sigma\epsilon : \mathcal{C} \longrightarrow \mathcal{C}$. See [33, 2.3.2].

We will give a compact formula for the contracting homotopy $\psi^{(n)}$. Note first, that after fixing the position ν of the occurrence of ψ, we get a number of conditions for the appearances of non-zero terms by considering the positions $1 \le \mu < \nu$ where π occurs: $\pi(t_\mu^{a_\mu}\gamma_{i_\mu}(y_\mu)x_\mu^{\eta_\mu}) = [i_\mu = 0][\eta_\mu = 0]$ and hence it *kills* the element from the group algebra. Then we have to impose the conditions of ψ from (24) at position ν which can be associated to two sums. Hence, we immediately get

$$\psi^{(n)}(t^a\gamma_i(y)x^\eta) = \Sigma_0 + \Sigma_1 \tag{28}$$

where

$$\Sigma_0 = \sum_{\nu=1}^{n}\left(\prod_{\mu=1}^{\nu-1}[i_\mu=0][\eta_\mu=0]\right)[a_\nu>0][\eta_\nu=0]\sum_{\kappa=0}^{a_\nu-1} t^{\kappa e_\nu+\lambda_\nu(a)}\gamma_{\lambda_{\nu-1}(i)}(y)x^{e_\nu+\lambda_\nu(\eta)},$$

$$\Sigma_1 = \sum_{\nu=1}^{n}\left(\prod_{\mu=1}^{\nu-1}[i_\mu=0][\eta_\mu=0]\right)[a_\nu=p-1][\eta_\nu=1]\; t^{\lambda_\nu(a)}\gamma_{e_\nu+\lambda_{\nu-1}(i)}(y)x^{\lambda_\nu(\eta)}$$

and where we define

$$\lambda_\nu(a_1,\ldots,a_n) = (0,\ldots,0,a_{\nu+1},\ldots,a_n)$$

for $1 \le \nu \le n$ and we recall that e_ν is the νth standard basis vector in dimension n. Our programs use exactly these formulae and definitions we have just given.

Summing up we have a resolution $\mathcal{C}^{(n)}$ of $\mathbb{F}_p$ over $\mathbb{F}_p G_+^n$ and in fact, an SDR

$$\mathbb{F}_p \underset{\epsilon}{\overset{\sigma}{\rightleftarrows}} (\mathcal{C}^{(n)}, \psi^{(n)}).$$

Note that, in fact, $\mathcal{C}^{(n)}$ is a *minimal* resolution, i.e. the differential in $\mathbb{F}_p \otimes_{\mathbb{F}_p G_+^n} \mathcal{C}^{(n)}$ vanishes. This is easy to see from the explicit form of the differential. Thus, in fact,

$$H_*(\mathbb{F}_p G_+^n) \cong \mathbb{F}_p \otimes_{\mathbb{F}_p G_+^n} \mathcal{C}^{(n)} \cong \Gamma_p[y_1, \ldots, y_n] \otimes \Lambda_p[x_1, \ldots, x_n] \tag{29}$$

(as is well-known).

In the case of $p = 2$ this construction can be simplified.

3.6.4 Cartan's Little Resolution in Characteristic 2 In characteristic 2, we can define $\Gamma_2[x]$ for elements of any degree using the same multiplication formula as in the case of odd primes. Note then that if $|y| = 2$ and $|x| = 1$, we have

$$\Gamma_2[y] \otimes \Lambda_2[x] \cong \Gamma_2[x]. \tag{30}$$

An explicit algebra isomorphism is given by $\gamma_i(y) \otimes x^\eta \longmapsto \gamma_{2i+\eta}(x)$ as is easily verified. By tensoring, this generalizes to the n-variable case for $n \ge 1$. Thus, the underlying vector space for the minimal resolution of $\mathbb{F}_2$ over $\mathbb{F}_2(G_+)$ is $\mathcal{C} = \mathbb{F}_2(G_+) \otimes \Gamma_2[x]$ (G_+ is the underlying group structure of $\mathbb{F}_2$ in this case). The differential transfers over as

$$da = 0, \; d\gamma_i(x) = (t+1)\gamma_{i-1}(x) \tag{31}$$

for $a \in \mathbb{F}_2(G_+)$ and $\gamma_i(x)$ for $i \ge 1$. This follows from (23) and the fact that $t + 1 = t - 1 = t^{[2]}$.

As before, we obtain the resolution $\mathcal{C}^{(n)} = \mathbb{F}_2 G_+^n \otimes \Gamma[x_1, \ldots, x_n]$ of $\mathbb{F}_2$ over $\mathbb{F}_2(G_+)[n]$ by tensoring $\mathcal{C}$ with itself n times. Thus, we have the usual tensor product formula for the differential and we have the tensor product homotopy $\psi^{(n)}$ by formula (27). Note that in this case, we have $a_k = [a_k > 0]$. Since we will also have an explicit need of the formula in Sect. 6, we present it here.

$$\psi^{(n)}(t_1^{a_1} \ldots t_n^{a_n} \gamma_{j_1}(x_1) \ldots \gamma_{j_n}(x_n)) =$$

$$\sum_{k=1}^{n} \left(\prod_{p=k+1}^{n} [j_p = 0] \right) a_k t_1^{a_1} \ldots t_{k-1}^{a_{k-1}} \gamma_{j_1}(x_1) \ldots \gamma_{j_{k-1}}(x_{k-1}) \gamma_{j_k+1}(x_k). \tag{32}$$

3.6.5 Splitting Off of the Bar Construction Let $A = \mathbb{F}_p G_+^n$ in this section. The minimal resolutions given above split off of the bar construction in the terminology used in [33]. In other words, we have an SDR

$$\mathcal{C}^{(n)} \underset{f^{(n)}}{\overset{\nabla^{(n)}}{\rightleftarrows}} (B(A), \phi^{(n)}) \tag{33}$$

for each $n \geq 1$. The maps involved are given by the general formulae (17–22). But more can be said. Its not hard to see that $B(A)$ can be given an algebra structure $*$. The definition is inductive. In degree 0, $B_0(A)$ is just A and we take multiplication $*$ to be from this algebra. Inductively, on reduced elements $x, y \in \bar{B}(A)$, we take $x * y = s(\partial(x) * y + (-1)^{|x|} x * \partial(y))$. With this, $B(A)$ is a differential graded commutative algebra (this is true for any commutative algebra A). The recursive definition of $\nabla^{(n)}$ given by (18) makes it into a one-one map of differential graded algebras (see [4]). Thus, $\nabla^{(n)}$ is completely determined on the exterior algebra part of the minimal resolution by $\nabla^{(n)}(x_i) = [t_i - 1]$.

Still more can be said. Again, following Cartan in [4], one can define a *divided power* structure in $B(A)$ as follows. For p an odd prime and $x \in B(A)$, an *even degree reduced* element, define $\gamma_0(x) = x$ and inductively $\gamma_i(x) = s(\gamma_{i-1}(x) * \partial(x))$. With this definition, it is also not hard to see that the map $\nabla^{(n)}$ preserves divided powers, i.e. that $\nabla^{(n)}(\gamma_i(y_j)) = \gamma_i(\nabla^{(n)}(y_j))$.

When $p = 2$, the formula $\gamma_i(x) = s(\gamma_{i-1}(x) * \partial(x))$ is valid for x of any degree and it follows immediately that

$$A \otimes \Gamma_2[x_1, \ldots, x_n] \xrightarrow{\nabla^{(n)}} B(A)$$

is the unique multiplicative extension of the map

$$\nabla^{(n)}(\gamma_i(x_j)) = [t_j | \ldots | t_j] \ (i \text{ times}). \tag{34}$$

4 Main Theorems

Theorem 4.1. *Let G be a finite p-group and let A be its augmented group ring over $\mathbb{F}_p$. There is a free resolution*

$$(A \otimes \Gamma_p[y_1, \ldots, y_n] \otimes \Lambda_p[x_1, \ldots, x_n], d) \longrightarrow \mathbb{F}_p \longrightarrow 0 \tag{35}$$

of $\mathbb{F}_p$ over A. Furthermore, there is an SDR

$$(A \otimes \Gamma_p[y_1, \ldots, y_n] \otimes \Lambda_p[x_1, \ldots, x_n], d) \underset{f}{\overset{\nabla}{\rightleftarrows}} ((B(A), \partial), \phi). \tag{36}$$

If $p = 2$, this simplifies. In this case, there is a resolution

$$(A \otimes \Gamma_2[x_1, \ldots, x_n], d) \longrightarrow \mathbb{F}_2 \longrightarrow 0 \tag{37}$$

of $\mathbb{F}_2$ *over* A *and there is an SDR*

$$(A \otimes \Gamma_2[x_1, \ldots, x_n], d) \underset{f}{\overset{\nabla}{\rightleftarrows}} ((B(A), \partial), \phi) \tag{38}$$

These resolutions and strong deformation retracts can be constructed with algorithm 5.4.

Note that we do not claim that the differential is a derivation of the *obvious* algebra structure in either case.

One can iterate Wall's construction of twisted tensor product resolutions [38] to see why such resolutions exist, but that procedure requires making choices at various stages as explained below. In Sect. 7.1, we will prove however that up to conjugation by a chain isomorphism, any resolution obtained this way is given by our algorithm.

4.1 Twisted Tensor Product Resolutions

Given a group extension

$$1 \longrightarrow K \longrightarrow \widetilde{G} \longrightarrow G \longrightarrow 1$$

and two resolutions

$$\mathbb{F}_p K \otimes X \longrightarrow \mathbb{F}_p \longrightarrow 0, \ \mathbb{F}_p G \otimes Y \longrightarrow \mathbb{F}_p \longrightarrow 0,$$

the procedure in [38] constructs a differential on $\mathbb{F}_p\widetilde{G} \otimes X \otimes Y$, giving a resolution of $\mathbb{F}_p$ over $\mathbb{F}_p\widetilde{G}$, crucially using the fact that, as vector spaces, $\mathbb{F}_p\widetilde{G} \cong \mathbb{F}_p(K \times G)$. The construction is done in stages and several choices (which are abstractly known to be possible) have to be made. Now one can see how to get resolutions as above by induction. For $K = (\mathbb{F}_p, +)$, take the minimal resolution. If the theorem is true for all groups of order p^n and K' is of order p^{n+1}, then since K' is an extension of a group K of order p^n by $\mathbb{F}_p$, we can use the Wall construction on the resolution of the theorem over K and the minimal resolution over $\mathbb{F}_p$. Clearly the result is isomorphic to a resolution of the form of the theorem for K'.

Since iterating the construction above (essentially up through the lower central series) compounds the number of choices that have to by made considerably, it would be quite nice if there were a uniform procedure, suitable for programming, that could be given. In fact, the algorithm we mentioned in the introduction from [32] and which we implement in this paper does exactly that. We will discuss the relationship with the iterated twisted tensor product resolution in Sect. 7.1.

Finally, we mention that using the methods of this paper a stronger theorem actually holds. We have

Theorem 4.2. *Let G be a finite p-group and let A be its augmented group ring over $\mathbb{F}_p$. Let M be an A-module. There is a free resolution*

$$(A \otimes \Gamma_p[y_1, \ldots, y_n] \otimes \Lambda_p[x_1, \ldots, x_n] \otimes M, d) \longrightarrow \mathbb{F}_p \longrightarrow 0. \tag{39}$$

of M over A. Furthermore, there is an SDR

$$(A \otimes \Gamma_p[y_1, \ldots, y_n] \otimes \Lambda_p[x_1, \ldots, x_n] \otimes M, d) \underset{f}{\overset{\nabla}{\rightleftarrows}} ((B(A, M), \partial), \phi).$$

Note that this uses the two-sided bar construction [32, S3]. There is, of course, the obvious simplification for $p = 2$. We will however, only discuss the details of the version for $M = \mathbb{F}_p$ in this paper.

4.2 Complexes for Cohomology

By taking the A-linear dual of the onto map f from the theorems, we obtain an embedding of the linear dual over A of the small resolution into the dual $C(G)$ of the bar construction $B(A)$ (3.4.1).

Now note that for any A-module X, we have

$$X^* = \mathrm{Hom}_A(A \otimes X, \mathbb{F}_p) \cong \mathrm{Hom}_{\mathbb{F}_p}(X, \mathbb{F}_p).$$

Furthermore, it is well-known (e.g. [4]) that

$$\begin{aligned} &\mathrm{Hom}_{\mathbb{F}_p}(\Gamma_p[y_1, \ldots, y_n] \otimes \Lambda_p[x_1, \ldots, x_n], \mathbb{F}_p) \\ &\cong \mathbb{F}_p[w_1, \ldots, w_n] \otimes \Lambda_p[z_1, \ldots, z_n] \end{aligned}$$

and

$$\mathrm{Hom}_{\mathbb{F}_2}(\Gamma_2[x_1, \ldots, x_n], \mathbb{F}_2) \cong \mathbb{F}_2[z_1, \ldots, z_n].$$

as algebras (in fact, as Hopf-algebras), where $\{w_1, \ldots, w_n\}$ is dual to $\{y_1, \ldots, y_n\}$ and $\{z_1, \ldots, z_n\}$ is dual to $\{x_1, \ldots, x_n\}$. We do not claim that the corresponding differential δ_∞ is a derivation. In fact, generally this is false. There is however an algebra structure for which δ_∞ *is* a derivation. The discussion of this and its consequences is beyond the scope of the current paper however. The interested reader should see [26], [14]. We have the immediate

Corollary 4.3. *Let G be a finite p-group. If $p = 2$, let $X^* = \mathbb{F}_2[z_1, \ldots, z_n]$ otherwise let $X^* = \mathbb{F}_p[z_1, \ldots, z_n] \otimes \Lambda_p[w_1, \ldots, w_n]$. There is a differential δ_∞ on X and an embedding $X^* \hookrightarrow C(G)$ which is a chain homotopy equivalence. Hence $H^*(G) \cong H^*(X^*, \delta_\infty)$.*

As a consequence, given explicit cocycles in X^*, we can produce explicit functional cochains on the group G. Examples will be given in Sect. 8 below.

5 Explicit Algorithms

Let G be a finite p-group of order p^n and A be its augmented group ring over $\mathbb{F}_p$. We outline and detail the algorithms from [28], [33], [31], and [32] for constructing free resolutions. We then look specifically at the case of finite p-groups.

5.1 Perturbation Principle

The idea behind the algorithms is the following. Let A be an algebra over a field k and suppose there is free A resolution $\mathcal{B}(A,M) \xrightarrow{\epsilon} 0$ for all A-modules M where $\mathcal{B}(A,M) = A \otimes \bar{\mathcal{B}}(A,M)$ for some vector space $\bar{\mathcal{B}}(A,M)$. For example one can take $\mathcal{B}$ to be the bar construction. We will say that an algebra A is a *perturbation* of an algebra A_0 if there is a k-linear isomorphism $A \cong A_0$. We suppose that if A is a perturbation of A_0, there is a vector space isomorphism $\mathcal{B}(A,M) \cong \mathcal{B}(A_0,M)$, as is the case with the bar construction. Thus, $\mathcal{B}(A_0,M)$ supports two differentials. The first is the differential d, corresponding to A_0 and the second, d' is the pullback to $\mathcal{B}(A_0,M)$ of the differential on $\mathcal{B}(A,M)$. Thus, if there is an SDR

$$(X, d_X) \underset{f}{\overset{\nabla}{\rightleftarrows}} (\mathcal{B}(A_0,M), \phi),$$

we can use $t = d' - d$ as an initiator and see if the perturbation formulae (3.2) converge. If they do, we obtain a resolution over A as small as the given one over A_0. There are various contexts in which a given algebra may be realized as a perturbation of another algebra and we give two examples below.

5.2 Polynomial Groups Laws Revisited

Suppose that G is given in the form of Sect. 2.4, i.e. we have $G = \mathbb{F}_p{}^n$ (as sets) with groups law ρ that satisfies (1). In this case, we have generators $\{t_1, \ldots, t_n\}$ for G such that every element of G has a unique "normal form" $t_1^{a_1} \ldots t_n^{a_n}$ and the multiplication in G is given by

$$t_1^{a_1} \ldots t_n^{a_n} \cdot t_1^{b_1} \ldots t_n^{b_n} = t_1^{a_1+b_1+\mu_1} \ldots t_n^{a_n+b_n+\mu_n}$$

where $\mu_i = \mu_i(a_1, \ldots, a_{i-1}, b_1, \ldots, b_{i-1})$. On the same underlying set, we have the elementary abelian group law which we write as

$$t_1^{a_1} \ldots t_n^{a_n} + t_1^{b_1} \ldots t_n^{b_n} = t_1^{a_1+b_1} \ldots t_n^{a_n+b_n}.$$

As before write G_+^n for this group. We thus have two group laws on the same underlying set and so we have two different differentials (11) on $B(\mathbb{F}_p G_+^n)$, viz., the differential ∂ for G and the differential ∂^+ for G_+^n. We set $\mathcal{T} = \partial - \partial^+$. Thus, using the notation $t^a = t_1^{a_1}\ldots, t_n^{a_n}$, we have an initiator (3.2)

$$\mathcal{T}[t^{a_1}|\ldots|t^{a_k}] = \sum_{i=1}^{k-1}([t^{a_1}|\ldots|t^{a_i}\cdot t^{a_{i+1}}|\ldots|t^{a_k}] - [t^{a_1}|\ldots|t^{a_i}+t^{a_{i+1}}|\ldots|t^{a_k}]). \quad (40)$$

Provided the maps given in the perturbation lemma converge, we obtain a resolution of $\mathbb{F}_p$ over A. If G is not presented in the form above, it is more complicated to obtain an initiator. We describe this next.

5.3 Using the Isomorphisms of Sect. 2 for an Initiator

In general, since we are over a field, we have that for any filtration of the algebra A, A is a perturbation of $E_0(A)$ in the sense defined above. Let G be a finite p-group and $A = \mathbb{F}_pG$. By definition, we have $\mathbb{F}_pG_+^n \cong \mathbb{F}_p[t_1, \ldots, t_n]/(t_1^p - 1, \ldots, t_n^p - 1)$. But the latter algebra is clearly isomorphic to $\mathbb{F}_p[t_1, \ldots, t_n]/(t_1^p, \ldots, t_n^p)$ since $(t-1)^p = t^p - 1 \bmod p$. We have a sequence of vector space-isomorphisms

$$E_0(\mathcal{V}(\mathrm{gr}_pG)) \cong \mathcal{V}(\mathrm{gr}_pG) \cong E_0(\mathbb{F}_pG) \cong \mathbb{F}_pG \quad (41)$$

using the observation above and the isomorphism of Theorem 2.2. But we also have the p-modular Birkhoff-Poincaré-Witt theorem 2.1 that gives an isomorphism

$$\mathbb{F}_p[t_1, \ldots, t_n]/({t_1}^p, \ldots, {t_n}^p) \cong E_0(\mathcal{V}(\mathrm{gr}_pG)) \quad (42)$$

Putting all these maps together, we have an explicit realization of $A = \mathbb{F}_pG$ as a perturbation of $\mathbb{F}_pG_+^n$ by

$$\Xi : \mathbb{F}_pG_+^n \longrightarrow \mathbb{F}_pG \quad (43)$$

and we can form an intitator as described above. Again, provided the maps given in the perturbation lemma 3.2 converge, we obtain a resolution of $\mathbb{F}_p$ over A.

We will give an indication of why the maps in the perturbation lemma converge in these cases only for the case $p = 2$. The general case is similar. Also note that the isomorphism just given is quite similar to the situation in Sect. 5.2. In essence, by refining the mod-p lower central series to have cyclic factors and pulling back the generators to G, we obtain a set of generators $\{t_1, \ldots, t_n\}$ for the group G so that every element has the unique (normal) form $t_1^{a_1}\ldots t_n^{a_n}$ and clearly, the normal form of the product of two such elements satisfies (1).

5.4 The Algorithm

Note that in what follows, improvements for the special case $p = 2$ can easily derived by using 3.6.4.

INPUT: – A group G of prime power order p^n given in some computationally accessible form and its group algebra $A = \mathbb{F}_p G$ over the prime field $\mathbb{F}_p$ with p elements.

– The elementary abelian group $G_+^n = (\mathbb{F}_p^n, +)$ of order p^n, and its group algebra $A_+ = \mathbb{F}_p G_+^n$.

step 1. Construct an $\mathbb{F}_p$-isomorphism $\Xi : A_+ \longrightarrow A$. If G is given as a perturbation of G_+^n by a polynomial group law ρ as in 2.4, then set Ξ to be the identity on the underlying sets, which are equal and by means of 5.2 proceed with step 2. Otherwise by means of 5.3 construct Ξ as by determining the p-modular lower central series

$$G = Z_1 \geq Z_2 \geq \ldots \geq Z_m \geq \; Z_{m+1} = 1,$$

according to 2.1, refining it to have cyclic factors, exhibit a generator of each factor and pull them back to get a sequence $t_1, \ldots, t_n$ of generators for G. Ξ then maps the generator corresponding to the unit vector e_i in $G_+^n \cong (\mathbb{F}_p^n, +)$ to t_i.

step 2. Construct the bar constructions $(B(A_+), \partial_+)$ and $(B(A), \partial)$ and according to 3.4, form the $\mathbb{F}_p$-vector space isomorphism

$$\Theta : (B(A_+), \partial_+) \longrightarrow (B(A), \partial)$$

induced by Ξ and transfer the differential ∂ by $\partial' := \Theta^{-1} \partial \Theta$ as a second differential on $B(A_+)$. Define the initiator $\mathcal{T} := \partial_+ - \partial' : B(A_+) \longrightarrow B(A_+)$ for the transfer process in the perturbation lemma 3.2. Furthermore, construct the SDR

$$\mathbb{F}_p \underset{\epsilon}{\overset{s_{-1}}{\rightleftarrows}} (B(A_+), s).$$

step 3. Construct Cartan's little resolution

$$\mathcal{C} = (A_+ \otimes \Gamma_p[y_1, \ldots, y_n] \otimes \Lambda_p[x_1, \ldots, x_n], d)$$

according to Sect. 3.6.3 for the elementary abelian group G_+^n by constructing the divided power algebra $\Gamma_p[y]$ with n generators (3.6.1) and the exterior power algebra $\Lambda_p(x)$ generated by n generators, see 3.6.2.

step 4. Form the SDR (see 3.6.5)

$$\mathcal{C}^{(n)} \underset{f^{(n)}}{\overset{\nabla^{(n)}}{\rightleftarrows}} (B(A_+), \phi^{(n)}).$$

step 5. Use (4)–(7) in 3.2 to recursively define maps

$$\partial_k^+, \ \nabla_k^{(n)}, \ f_k^{(n)}, \text{ and } \phi_k^{(n)}$$

for $k = 1, \ldots$

OUTPUT: A resolution $(\mathcal{C}, \partial_\infty)$ of $\mathbb{F}_p$ over the group algebra $\mathbb{F}_pG$ and an SDR

$$(\mathcal{C}, \partial_\infty) \underset{f}{\overset{\nabla}{\rightleftarrows}} ((B(A_+), \partial), \phi)$$

as the limit of the constructions in step 4.

Note that the tensor products over $\mathbb{F}_p$ can be realized algorithmically by designing a data type for divided power algebras and exterior algebras, which allow arbitrary commutative rings as coefficient domains.

6 Details of the Perturbation for p = 2

As before, let $G_+^n = \mathbb{F}_2^n$ be the elementary abelian group 2^n and let $A_+ = \mathbb{F}_2 G_+^n$. We want to examine the SDR of the bar construction and the minimal resolution in more detail. By (30–32), we may write this as

$$(A_+ \otimes \Gamma_2[x_1, \ldots, x_n], d) \underset{f}{\overset{\nabla}{\rightleftarrows}} ((B(A_+), \partial), \phi) \tag{44}$$

where

$$\nabla(t_1^{a_1} \ldots t_n^{a_n} \gamma_{j_1}(x_1) \ldots \gamma_{j_n}(x_n)) = t_1^{a_1} \ldots t_n^{a_n} \gamma_{j_1}([t_1]) * \cdots * \gamma_{j_n}([t_n])) \tag{45}$$

and f and ϕ are generally given by (21) and (22).

Let $\mathcal{T}$ be the initiator from the last section. We need to see that $\mathcal{T}\phi$ is nilpotent in each degree. We will indicate why this is true in degree one. The higher degrees are similar.

In degree one we have

$$f([t_1^{a_1} \ldots t_n^{a_n}]) = \sum_{i=1}^{n} a_i t_{i+1}^{a_{i+1}} \ldots t_{n+1}^{a_{n+1}} \gamma_1(x_i) \tag{46}$$

Note that for a uniform formula, we have taken $t_{n+1} = 1$ here. See (32) above. Using this formula for f and (22) we have

$$\phi([t_1^{a_1} | \ldots | t_n^{a_n}]) = \sum_{i=1}^{n-1} a_i [t_{i+1}^{a_{i+1}} \ldots t_n^{a_n} \, | \, t_i]. \tag{47}$$

Assume now that we have a finite 2-group G and let $A = \mathbb{F}_2 G$. We assume that G is of the form in Sect. 2.4. So the group law ρ is a polynomial function that satisfies (1). We want to examine the perturbation formulae more closely in this situation.

Thus by (40) and (47), we immediately have

$$\mathcal{T}\phi([t_1^{a_1}\dots t_n^{a_n}]) = \sum_{i=1}^{n-2} a_i([t_{i+1}^{a_{i+1}}\dots t_n^{a_n}\cdot t_i] + [t_i t_{i+1}^{a_{i+1}}\dots t_n^{a_n}]). \tag{48}$$

Note that the last term in the sum vanished since t_n is central as follows from (1). Now note that in the terms above,

$$t_{i+1}^{a_{i+1}}\dots t_k^{a_n}\cdot t_i = t_i t_i^{\mu_i}\dots t_n^{\mu_n}$$

where $i \le n-2$ which again follows from (1). Thus, every term in the sum above is of the form $t_i t_i^{b_{i+1}}\dots t_n^{b_n}$ for some $b_j \in \{0,1\}$. We will say that such elements in G have *rank* i. Thus, it suffices to see that $\mathcal{T}\phi$ is nilpotent on elements of rank i for $i \le n-2$. The proof is by downward induction.

If $z = t_{n-2}t_{n-1}^{b_{n-1}}t_n^{b_n}$ is any rank $n-2$ element, then we have

$$\mathcal{T}\phi([z]) = [t_{n-1}^{b_{n-1}}t_n^{b_n}\cdot t_{n-2}] + [t_{n-2}t_{n-1}^{b_{n-1}}t_n^{b_n}].$$

but $t_{n-1}^{b_{n-1}}t_n^{b_n}\cdot t_{n-2} = t_{n-2}t_{n-1}^{b_{n-1}}t_n^{b_n+q}$ where $q = q(b_{n-1})$ and q is a polynomial over $\mathbb{F}_2$ with no constant term. In fact, from (1), it is clear that

$$q(b_{n-1}) = \mu_n(0,\dots,0,b_{n-1},0,\dots,0,1,0).$$

We thus have $\mathcal{T}\phi(z) = [t_{n-2}t_{n-1}^{b_{n-1}}t_n^{b_n+q}] + [z]$ and we need to consider cases.

If $b_{n-1} = 0$, then clearly $q = 0$ and the terms above cancel. If $b_{n-1} = 1$, then if $q = 0$, the terms cancel once more. If $q = 1$ in this case, we have $z = t_{n-2}t_{n-1}t_n^{b_n}$ and $\mathcal{T}\phi([z]) = [w] + [z]$ where $w = t_{n-2}t_{n-1}t_n^{b_n+1}$, but then $(\mathcal{T}\phi)^2([z]) = [t_{n-2}t_{n-1}t_n^{b_n+1+1}] + [w] + [w] + [z] = 0$. The proof for the inductive step is similar.

7 Some Observations and Consequences

7.1 Relationship to Twisted Tensor Product Resolutions

Consider again the twisted tensor product construction of Sect. 4.1. We have

Theorem 7.1. *Let G be a finite p-group and let $(X,d) \longrightarrow \mathbb{F}_p \longrightarrow 0$ be the twisted tensor product resolution over G from Sect. 4.1. We have that (X,d) is chain isomorphic to the resolution given by the perturbation lemma using the algorithm from Sect. 5.*

Proof. We use the uniqueness theorem from [1]. We have set up an explicit *transference problem* (see [1] for terminology), in this case, in Sect. 5. Since the twisted tensor product differential is a solution to this problem, the main theorem of [1, S5] shows that we have a chain homotopy *isomorphism* so that the differential obtained by iterating the construction in [38] is conjugate to the differential obtained via the perturbation formulae.

Note that, of course, by the comparison theorem, all such resolutions must be chain homotopy equivalent. This theorem gives a much stronger statement. In essence, up to the choices made in [38], the construction is exactly the same as the one given uniformly by the perturbation lemma.

7.2 An Explicit SDR and an Explicit Contracting Homotopy

Using the algorithms from Sect. 5, we obtain an SDR

$$(\mathcal{C}^{(n)}, \partial_\infty) \underset{f_\infty^{(n)}}{\overset{\nabla_\infty^{(n)}}{\rightleftarrows}} ((B(A), \partial), \phi_\infty^{(n)}). \tag{49}$$

but more can be said.

It turns out that this construction actually provides an explicit contracting homotopy for $(\mathcal{C}^{(n)}, \partial_\infty)$. This can be seen from the general

Lemma 7.2. *Let X be a resolution of R over A and suppose that we have an SDR*

$$X \underset{f}{\overset{\nabla}{\rightleftarrows}} (B(A), \phi).$$

Then $fs\nabla$ is a contracting homotopy for X.

7.3 2-Cocyles and Codes

For any group G, the special case of the cochain differential (16) for $i = 2$ is of some interest. A function $f : G \times G \longrightarrow \mathbb{F}_p$ that satisfies $\delta(f) = 0$ is said to be a 2-cocycle. I.e. f is a 2-cocycle if and only if

$$f(y, z) - f(xy, z) + f(x, yz) - f(x, y) = 0 \tag{50}$$

for all $x, y, z \in G$. As was pointed out in section 2, every p-group is built up inductively from $\mathbb{F}_p$ by a sequence of 2-cocycles.

Using classical methods involving the universal coefficient theorem, Schur multipliers, and transgression, methods for finding 2-cocycles representing 2-dimensional cohomology classes can be worked out in some cases. See [7], [8] and [17]. Also see [6]. Connections between combinatorial design theory and 2-cocycles has been pointed out in [15] (also see [16] for errata). Connections between coding theory and 2-cocycles have also been made in [18].

Evaluating functional 2-cocycles on all pairs of the group elements yields a matrix with entries from $\mathbb{F}_2$ in case of $p = 2$. In light of the connection between cocycles and combinatorics just mentioned, it seems useful to have a means to generate whole families of functional cocycles to examine for various properties. Our algorithm above actually can calculate explicit representative cocycles (and hence codes) directly for any finite p-group. An example is given in 8.2.4.

7.4 Universal Cochain

Let G be a finite p-group for which the group law is given by a polynomial as in (1). We want to examine Cor. 4.3 in more detail. First of all, we have an onto map $\bar{f}_\infty : \bar{B}(\mathbb{F}_p G) \longrightarrow X$ where $X = \Gamma_p[y_1, \ldots, y_n] \otimes \Lambda_p[x_1, \ldots, x_n]$ in case p is odd and $X = \Gamma_2[x_1, \ldots, x_n]$ when $p = 2$ by "reducing" the SDR from Theorem 4.1, i.e., by tensoring the objects and maps over $\mathbb{F}_p G$ with $\mathbb{F}_p$. We claim the

Lemma 7.3.

$$\bar{f}_\infty[t^{a_1}|\ldots|t^{a_k}] = \sum_{i,\epsilon} \lambda_{i,\epsilon}\gamma_i(y)x^\epsilon \tag{51}$$

where $a_\kappa = (a_{1,\kappa}, \ldots, a_{n,\kappa})$, $t^{a_\kappa} = t_1^{a_{1,\kappa}} \ldots t_n^{a_{n,\kappa}}$, $i = (i_1, \ldots, i_n)$, $\epsilon = (\epsilon_1, \ldots, \epsilon_n)$, $\gamma_i(y) = \gamma_{i_1}(y_1) \cdots \gamma_n(y_n)$, $x^\epsilon = x_1^{\epsilon_1} \ldots x_n^{\epsilon_n}$, *and*

$$\lambda_{i,\epsilon} = \lambda_{i,\epsilon}(a_1, \ldots, a_k) \tag{52}$$

is a polynomial in the coordinates $a_{\nu,\kappa}$*. As such, we may think of the expression (51) as lying in* $\mathbb{F}_p[a_{1,1}, \ldots, a_{n,i}]/(a_{\nu,\kappa}^p - a_{\nu,\kappa}) \otimes X$.

The proof is quite easy once one realizes that f_∞ is constructed from the maps f and ϕ in the SDR (33) and the initiator $\mathcal{T}$ from Sect. 5. using only arithmetic operations. Its clear that f, $\mathcal{T}$, and ϕ can be expressed in terms of polynomials, so the lemma follows. We will denote this polynomial expression by $f_\infty^{(i)}$ and call it *the universal i-cochain.* As an immediate consequence of all this, we have

Corollary 7.4. *Let* $\alpha = \sum_{j,\mu} \alpha_{j,\mu}\gamma_j(z)w^\mu$ *be any i-cochain in* $X^* = \mathbb{F}_p[z_1, \ldots, z_n] \otimes \Lambda_p[w_1, \ldots, w_n]$*. The corresponding functional i-cochain*

$$f_\infty^{(i)}(\alpha) : G^i \to \mathbb{F}_p$$

is given by

$$f_\infty^{(i)}(\alpha)(t^{a_1}, \ldots, t^{a_k}) = \sum \alpha_{i,\epsilon}\lambda_{i,\epsilon}(a_1, \ldots, a_k).$$

Obviously, if α *is a cocycle, so is* $f_\infty^{(i)}(\alpha)$*, and in this way, we obtain polynomial representatives for cohomology classes.*

Explicit examples of such universal cochains will be given in the next section.

8 Implementations and Computations

8.1 Implementations in AXIOM

The broad variety of algebraic data types required to realize and implement the the extensions of the algorithm from 5.4 seen in [32] in its most general setting is a topic of interest itself (see [5]). For the case of a given polynomial group law of the form (2.4) we have implemented the algorithm in the computer algebra system AXIOM (in arbitrary characteristic) and with focus on $p = 2$ and applications to coding theory. For the general case when the map Θ has to be considered, we also have written a program in the group theory system GAP as a preprocessing step when the finite p-group is given by power-commutator relations. The source code and some examples can be downloaded from the authors' homepages

`http://www.bangor.ac.uk/~mas019/`

or

`http://www.hd.shuttle.de/grabm/jg-top.html`.

8.1.1 Implementations for Arbitrary Characteristic The necessary structure for elementary abelian groups of order p^n, written multiplicatively, is given by the domain MWEA MultiplicativelyWrittenElementaryAbelian, and p-groups given by a polynomial group law are realized in PPGP PolynomialP-Group (see 2.4).[1] All kinds of graded, differential and augmented structures are provided, e.g. GRALALG GradedAugmentedLeftAlgebra. [2] The multiplicative structure of a monoid ring is given by the AXIOM domain constructor MRING MonoidRing, which was enhanced. [3] The bar construction (see 3.4) is implemented in the domain BAR BarConstruction [4]. Its basis over the group (monoid) ring has the infinite basis consisting of all lists of group elements.

We have constructed exterior algebras using the category EXTALGC ExteriorAlgebraCategory, and the domain EXTALG ExteriorAlgebra which requires the package EXMER ExMerge. [5] The algebra of divided powers is implemented in the domain DIVPOW DividedPowerAlgebra. [6] The full power and beauty of AXIOM is used to construct a domain for Cartan's little resolution (see 3.6) by combined use of the domain constructors ExteriorAlgebra MonoidRing, PrimeField, DividedPowerAlgebra and yields the constructor CLR CartanLittleResolution. [7] The functions (see 3.1.2) of the strong deformation retract for a p-group given by a polynomial perturbation of the elementary abelian group are given in the package SDRPG StrongDeformationRetractionPGroup. [8]

[1] The code is in `ppgp.spad`
[2] The code is in `graded.spad`.
[3] The code is in `mring.spad`.
[4] The code is in `bar.spad`
[5] The code is in `extalg.spad`.
[6] The code is in `divp.spad`
[7] The code is in `cartan.spad`
[8] The code is in `sdrpg.spad`

8.1.2 Implementations in Characteristic 2 In this case, the bar construction BarConstruction[9] is implemented using the constructor FreeModule[10] with coefficient ring R of category AssociativeAlgebra SingleIntegerMod 2 and with basis the type List ElementaryCommutativeGroup(2, n, R). The dimension n: SingleInteger is the first argument to the bar constructor, while R is either simply SingleIntegerMod 2 itself or an extension of it. This allows symbolic computation with generic group elements, e.g. SymbolicExponents[11]. The last argument to the bar constructor is the perturbation group law of type (Array R, Array R) $\longrightarrow$ Array R, which describes the group of order 2^n as a perturbation of the elementary abelian group. The implemented functions include t2 and t3 which implement the initiator $\mathcal{T}$ (48) in degrees 2 and 3, both on the basis type and on the bar construction. Similarly the contracting homotopy ϕ_+ of the strong deformation retract of Cartan's little resolution in its characteristic 2 variant is realized in degrees 1 and 2 by phi1 and phi2. Finally, the perturbation lemma iteration is done for degree 1 and 2 in the code for t_phi_iteration1 and t_phi_iteration2. A function basisOfDegree returns lists of basis elements as elements of the bar construction.

The necessary functions for

$$\mathcal{C}^{(n)} \underset{f^{(n)}}{\overset{\nabla^{(n)}}{\rightleftarrows}} (B(A), \phi^{(n)})$$

for the given 2-group (see (3.6.5)) are implemented in StrongDeformationRetraction2Group[12], which has the same first 3 arguments as the implementation of the corresponding bar construction. Moreover, the fourth argument is the group algebra and has to be given as S: AssociativeAlgebra R[13]. The embedding $\nabla^{(n)}$ is implemented as nabla for degrees 1, 2, and 3 or for exactly one variable in arbitrary degree. The projection $f^{(n)}$ is implemented as f1 and f2 for the degrees 1 and 2 and on basis elements as well as linearly extended on arbitrary bar elements. The new differential is implemented as d and realized by 5 functions: dOnBasisOfDegree2 and dOnBasisOfDegree3 are constants which store the results on basis elements of degree 2 and 3, setTableForDegree sets up the constant hashtable (a data structure set up for efficient storage and retrevial) dTable for these degrees, while d extracts the values from the hashtable. The new projection is implemented as f and gained as the result of the transfer problem. It is implemented for degree 1 and 2 in f_infinity1 and f_infinity2, while the function lambda is an internal help function.

The package DualComputations[14] has the same arguments as StrongDeformationRetraction2Group. Here Cartan's Little Resolution C is imple-

[9] This is coded in file `twococ2.as`.
[10] This is coded in file `freemod.as`.
[11] This is coded in file `pgroups.as`.
[12] This is coded in file `twococ2.as`.
[13] This is coded in file `algebra.as`.
[14] This is coded in file `twococ2.as`.

mented by DividedPowerBasisChar2(n, S, vx)[15]. The package exports the constant lists d1_dual and d2_dual with entries from C. They contain the values of the differential on the the (dual) basis of the cochain complex $\mathrm{Hom}_{\mathbb{F}_p}(\Gamma_2[x_1,...,x_n],\mathbb{F}_p)$ (see 4.2 and 3.4.1). In addition the function d_dual implements its differential both for C or the type of its canonical basis CB, defined to be DividedPowerBasisChar2(n, S, vx)[16]. To be able to use linear algebra to determine the kernels and images, the constants matrix_d1_dual and matrix_d2_dual return objects of Matrix S[17], where each row consists of the coefficients of the differential applied to a dual basis element.

The package CocycleComputations[18] has two arguments, the dimension n and the polynomial group law rho: (ASE, ASE) ⟶ ASE, where ASE abbreviates Array SymbolicExponents n[19]. It conveniently puts all pieces of computations together and exports cocycleRepresentatives and secondCohomologyGroup. The second function takes the polynomial group law (ASE, ASE) ⟶ ASE as its argument and returns the a 2-tupel. The first component is the kernel of the differential, the second component is the image of the differential for degree 2. This can be conveniently used as argument for the function cocycleRepresentatives, which returns a basis consisting of 2-cocycle representatives as elements of the type SE.

8.2 Upper Triangular Matrices

We present results of executing the algorithm in Sect. 5 first for the matrix groups $\mathrm{UT}_3(\mathcal{F}_2)$ and $\mathrm{UT}_4(\mathcal{F}_2)$ and then the group $\mathrm{UT}_3(\mathcal{F}_5)$.

Using spectral sequence methods, one could work out the ranks of the cohomology of the groups given in this section. Indeed, this provides a check of our results. In fact, even though we obtain a complex for computing the (co)homology of a group directly, it can be advantageous to use the spectral sequence associated to it when the number of generators is large and it is possible to interpret our results as giving a closed form expression for the corresponding differentials [32]. Since we obtain resolutions, we will however produce more than just cohomology. In addition, when we want to obtain results about (co)homology, we can easily get (co)cycle representatives from our complexes using the universal cochain from Sect. 7.4. This is illustrated in the last example.

8.2.1 $\mathbf{UT_3(\mathcal{F}_2)}$ This group G is dihedral and the group law may be written

$$(a_1,a_2,a_3)(b_1,b_2,b_3) = (a_1+b_1, a_2+b_2, a_3+b_3+a_1b_2).$$

[15] This is coded in file `divpow.as`.
[16] This is coded in file `divpow.as`.
[17] This is coded in file `matrix.as`.
[18] This is coded in file `twococ2.as`.
[19] This is coded in file `pgroups.as`.

Of course, it is straightforward to compute $H^*(G, \mathbb{F}_2)$ in this case using the Lyndon-Hochschild-Serre spectral sequence. The result is $H^*(G, \mathbb{F}_2) = R[z]$, the polynomial algebra in one variable z of degree 2 over the ring $R = \mathbb{F}_2[x, y]/(xy)$. Thus, H^2 has dimension 3.

It is also quite easy to compute $\bar{\partial}^*_{\infty,i}$ for $i = 1, 2$ in this case. We give only the non-zero differentials on the dual of the canonical basis:

$$x_3^* \longmapsto (x_1x_2)^* \tag{53}$$

$$(x_2x_3)^* \longmapsto (x_1\gamma_1(x_2))^* \tag{54}$$

$$(x_1x_3)^* \longmapsto (x_2\gamma_1(x_1))^* \tag{55}$$

$$(\gamma_2(x_3))^* \longmapsto (x_1\gamma_1(x_2))^* + (x_2\gamma_1(x_1))^*. \tag{56}$$

Thus, representatives of the three cohomology classes in dimension two are given by

$$\alpha_1 = (\gamma_2(x_1))^*, \quad \alpha_2 = (\gamma_2(x_2))^*, \quad \alpha_3 = (\gamma_2(x_3))^* + (x_1x_3)^* + (x_2x_3)^*.$$

The universal polynomial cochain is also easily computed. It is

$$\begin{aligned} f_\infty(a, b) = {} & (a_3b_3 + a_1a_3b_2 + a_1b_2b_3)\gamma_2(x_3) \\ & + a_2b_2\gamma_2(x_2) + a_1b_1\gamma_2(x_1) \\ & + a_1b_2x_1x_2 + a_1b_3x_1x_3 \end{aligned}$$

so three cocycle representatives $c_i : G \times G \longrightarrow \mathbb{F}_2$ are given by

$$\begin{aligned} c_1(a, b) &= a_1b_1 \\ c_2(a, b) &= a_2b_2 \\ c_3(a, b) &= a_3b_3 + a_1a_3b_2 + a_1b_2b_3 + a_1b_2 + a_1b_3. \end{aligned}$$

8.2.2 The universal Cochain for $\mathrm{UT}_4(\mathcal{F}_2)$ The interested reader can find further results for the example above as well as results for $\mathrm{UT}_4(\mathcal{F}_2)$ at the author's web sites (see Sect. 8.1), however we will list the universal cochain for $\mathrm{UT}_4(\mathcal{F}_2)$ in degree 2 here. It is

$$\begin{aligned}
& a_1b_1\gamma_2(x_1) + a_1b_2\gamma_1(x_1)\gamma_1(x_2) + a_1b_3\gamma_1(x_1)\gamma_1(x_3) + a_1b_4\gamma_1(x_1)\gamma_1(x_4) + \\
& a_1b_5\gamma_1(x_1)\gamma_1(x_5) + a_1b_6\gamma_1(x_1)\gamma_1(x_6) + a_2b_2\gamma_2(x_2) + a_2b_3\gamma_1(x_2)\gamma_1(x_3) + \\
& (a_2b_4 + a_1a_2b_2)\gamma_1(x_2)\gamma_1(x_4) + a_2b_5\gamma_1(x_2)\gamma_1(x_5) + \\
& (a_2b_6 + a_1a_2b_5)\gamma_1(x_2)\gamma_1(x_6) + a_3b_3\gamma_2(x_3) + \\
& (a_3b_4 + a_1b_2b_3 + a_1a_3b_2)\gamma_1(x_3)\gamma_1(x_4) + (a_3b_5 + a_2a_3b_3)\gamma_1(x_3)\gamma_1(x_5) + \\
& (a_3b_6 + a_1a_3b_5)\gamma_1(x_3)\gamma_1(x_6) + (a_1b_2b_4 + a_4b_4 + a_1a_4b_2)\gamma_2(x_4) + \\
& (a_4b_5 + a_2b_3b_4 + a_2a_4b_3)\gamma_1(x_4)\gamma_1(x_5) + (a_4b_6 + a_1a_4b_5 + a_1b_2b_3b_4 + \\
& a_1a_3b_2b_4 + a_3a_4b_4 + a_1a_4b_2b_3 + a_1a_3a_4b_2)\gamma_1(x_4)\gamma_1(x_6) + (a_2b_3b_5 + a_5b_5 + a_2a_5b_3)\gamma_2(x_5) + \\
& (a_5b_6 + a_3b_4b_5 + a_1b_2b_3b_5 + a_1a_3b_2b_5 + a_1a_5b_5 + a_3a_5b_4 + a_1a_5b_2b_3 + a_1a_2a_3b_2b_3 + \\
& a_1a_2b_2b_3 + a_1a_3a_5b_2)\gamma_1(x_5)\gamma_1(x_6) + (a_1b_5b_6 + a_3b_4b_6 + a_1b_2b_3b_6 + a_1a_3b_2b_6 + \\
& a_6b_6 + a_3a_4b_6 + a_1a_3b_4b_5 + a_1b_2b_3b_5 + a_1a_3b_2b_5 + a_1a_6b_5 + a_1a_3a_4b_5 + a_1b_2b_3b_4 + \\
& a_1a_3b_2b_4 + a_3a_6b_4 + a_3a_4b_4 + a_1a_6b_2b_3 + a_1a_3a_4b_2b_3 + a_1a_3a_6b_2 + a_1a_3a_4b_2)\gamma_2(x_6)
\end{aligned}$$

8.2.3 UT$_3(\mathcal{F}_5)$ The AXIOM program given in App. A can be used to output the resolution in section 8.2.3. These procedures rely heavily upon our constructed libraries described in Sect. 8.1.1.

The Differential in the Resolution up to Degree 4 We will give the resolution only up to degree 4; only non-zero differentials are shown. In degree 1, we have

$$\partial_\infty(x_1) = 4 + t_1, \quad \partial_\infty(x_2) = 4 + t_2, \quad \partial_\infty(x_3) = 4 + t_3,$$

in degree 2,

$$\begin{aligned}
\partial_\infty(x_1x_2) &= 4t_1t_2x_3 + (4 + t_1)x_2 + (1 + 4t_2)x_1, \\
\partial_\infty(x_1x_3) &= (4 + t_1)x_3 + (1 + 4t_3)x_1, \\
\partial_\infty(x_2x_3) &= (4 + t_2)x_3 + (1 + 4t_3)x_2,
\end{aligned}$$

$$\partial_\infty(\gamma_1(y_1)) = t_1{}^{[5]}x_1, \quad \partial_\infty(\gamma_1(y_2)) = t_2{}^{[5]}x_2, \quad \partial_\infty(\gamma_1(y_3)) = t_3{}^{[5]}x_3,$$

in degree 3,

$$\begin{aligned}
\partial_\infty(x_1x_2x_3) &= (4 + t_1)x_2x_3 + (1 + 4t_2)x_1x_3 + (4 + t_3)x_1x_2, \\
\partial_\infty(\gamma_1(y_1)x_1) &= (t_1 + 4)\gamma_1(y_1), \\
\partial_\infty(\gamma_1(y_1)x_2) &= (t_2 + 4)\gamma_1(y_1) + t_2\gamma_1(y_3) + p_1x_1x_3 + t_1{}^{[5]}x_1x_2, \\
\partial_\infty(\gamma_1(y_1)x_3) &= (t_3 + 4)\gamma_1(y_1) + t_1{}^{[5]}x_1x_3, \\
\partial_\infty(\gamma_1(y_2)x_1) &= (t_1 + 4)\gamma_1(y_2) + 4t_1\gamma_1(y_3) + p_2x_2x_3 + 4t_2{}^{[5]}x_1x_2, \\
\partial_\infty(\gamma_1(y_2)x_2) &= (t_2 + 4)\gamma_1(y_2), \\
\partial_\infty(\gamma_1(y_2)x_3) &= (t_3 + 4)\gamma_1(y_2) + t_2{}^{[5]}x_2x_3, \\
\partial_\infty(\gamma_1(y_3)x_1) &= (t_1 + 4)\gamma_1(y_3) + 4t_3{}^{[5]}x_1x_3, \\
\partial_\infty(\gamma_1(y_3)x_2) &= (t_2 + 4)\gamma_1(y_3) + 4t_3{}^{[5]}x_2x_3, \\
\partial_\infty(\gamma_1(y_3)x_3) &= (t_3 + 4)\gamma_1(y_3),
\end{aligned}$$

where

$$\begin{aligned}
p_1 &= 4t_1t_2 + 4t_1^3t_2 + 4t_1^3t_2t_3 + 4t_1^3t_2t_3^2 + 4t_1^2t_2 + 4t_1^2t_2t_3 + 4t_1^4t_2t_3^3 \\
&\quad + 4t_1^4t_2t_3^2 + 4t_1^4t_2t_3 + 4t_1^4t_2 \\
p_2 &= t_1t_2^3t_3^2 + t_1t_2^3t_3 + t_1t_2^3 + t_1t_2^4t_3^3 + t_1t_2^4t_3^2 + t_1t_2^4t_3 + t_1t_2^4 \\
&\quad + t_1t_2^2t_3 + t_1t_2^2 + t_1t_2
\end{aligned}$$

and in degree 4,

$$
\begin{aligned}
\partial_\infty(\gamma_1(y_1)x_1x_2) &= (4t_1t_2\gamma_1(y_1) + 4t_1t_2\gamma_1(y_3))x_3 + (t_1+4)\gamma_1(y_1)x_2 \\
&\quad +((4t_2+1)\gamma_1(y_1) + 4t_2\gamma_1(y_3))x_1 \\
\partial_\infty(\gamma_1(y_2)x_1x_2) &= (4t_1t_2\gamma_1(y_2) + t_1t_2\gamma_1(y_3))x_3 + ((t_1+4)\gamma_1(y_2) + 4t_1\gamma_1(y_3))x_2 \\
&\quad +(4t_2+1)\gamma_1(y_2)x_1 \\
\partial_\infty(\gamma_1(y_3)x_1x_2) &= 4t_1t_2\gamma_1(y_3)x_3 + (t_1+4)\gamma_1(y_3)x_2 + (4t_2+1)\gamma_1(y_3)x_1 + t_3^{[5]}x_1x_2x_3 \\
\partial_\infty(\gamma_1(y_1)x_1x_3) &= (t_1+4)\gamma_1(y_1)x_3 + (4t_3+1)\gamma_1(y_1)x_1 \\
\partial_\infty(\gamma_1(y_2)x_1x_3) &= ((t_1+4)\gamma_1(y_2) + 4t_1\gamma_1(y_3))x_3 + (4t_3+1)\gamma_1(y_2)x_1 + 4t_2^{[5]}x_1x_2x_3 \\
\partial_\infty(\gamma_1(y_3)x_1x_3) &= (t_1+4)\gamma_1(y_3)x_3 + (4t_3+1)\gamma_1(y_3)x_1 \\
\partial_\infty(\gamma_1(y_1)x_2x_3) &= ((t_2+4)\gamma_1(y_1) + t_2\gamma_1(y_3))x_3 + (4t_3+1)\gamma_1(y_1)x_2 + t_1^{[5]}x_1x_2x_3 \\
\partial_\infty(\gamma_1(y_2)x_2x_3) &= (t_2+4)\gamma_1(y_2)x_3 + (4t_3+1)\gamma_1(y_2)x_2 \\
\partial_\infty(\gamma_1(y_3)x_2x_3) &= (t_2+4)\gamma_1(y_3)x_3 + (4t_3+1)\gamma_1(y_3)x_2 \\
\partial_\infty(\gamma_2(y_1)) &= t_1^{[5]}\gamma_1(y_1)x_1 \\
\partial_\infty(\gamma_1(y_1)\gamma_1(y_2)) &= q_1\gamma_1(y_3)x_3 + (t_2^{[5]}\gamma_1(y_1) + q_2\gamma_1(y_3))x_2 + (t_1^{[5]}\gamma_1(y_2) + q_3)\gamma_1(y_3)x_1 \\
&\quad +q_4x_1x_2x_3 \\
\partial_\infty(\gamma_1(y_1)\gamma_1(y_3)) &= t_3^{[5]}\gamma_1(y_1)x_3 + t_1^{[5]}\gamma_1(y_3)x_1 \\
\partial_\infty(\gamma_2(y_2)) &= t_2^{[5]}\gamma_1(y_2)x_2 \\
\partial_\infty(\gamma_1(y_2)\gamma_1(y_3)) &= t_3^{[5]}\gamma_1(y_2)x_3 + t_2^{[5]}\gamma_1(y_3)x_2 \\
\partial_\infty(\gamma_2(y_3)) &= t_3^{[5]}\gamma_1(y_3)x_3,
\end{aligned}
$$

where

$$
\begin{aligned}
q_1 &= 4t_1^4 + 4t_1^4t_3 + 4t_1^4t_3^2 + 4t_1^4t_3^3 + 4t_1^3t_2^3 + 4t_1^3t_2^3t_3 + 4t_1^3t_2^3t_3^2 + 4t_1^2t_2^4 + 4t_1^2t_2^4t_3 \\
&\quad +4t_1^3 + 4t_1^3t_3 + 4t_1^3t_3^2 + 4t_1^2t_3 + 4t_1^2 + 4t_1^4t_2^4t_3^3 + 4t_1^4t_2^4t_3^2 + 4t_1^4t_2^4t_3 + 4t_1^4t_2^4 \\
q_2 &= t_1^3t_2^4 + t_1^2t_2^4 + t_1^4t_2^4 + t_2^4 + t_2^3 + t_2 + t_1^3t_2^2 + t_1^4t_2^3 + t_1^2t_2^3 + t_2^2 \\
q_3 &= 4t_1^4t_2^4 + 4t_1^3 + 4t_1^4t_2^2 + 4t_1^4 + 4t_1^3t_2^4 + 4t_1^3t_2^2 + 4t_1^2t_2^3 + 4t_1 + 4t_1^2 + 4t_1^4t_2^3 \\
q_4 &= t_1^2t_2^4t_3^2 + t_1^2t_2^4t_3 + t_1^2t_2^4 + t_1^2t_2t_3 + t_1^2t_2 + t_1^4t_2^2t_3^2 + t_1^4t_2^2t_3 + t_1^4t_2^2 + t_1^4t_2^3t_3 \\
&\quad +t_1^4t_2^3 + t_1^2t_2^3 + t_1^2t_2^2t_3^3 + t_1^2t_2^2t_3^2 + t_1^2t_2^2t_3 + t_1^2t_2^2 + t_1^4t_2t_3^3 + t_1^4t_2t_3^2 + t_1^4t_2t_3 \\
&\quad +t_1^4t_2 + t_1^4t_2^4 + t_1t_2^4t_3^3 + t_1t_2^4t_3^2 + t_1t_2^4t_3 + t_1t_2^4 + t_1t_2 + t_1^3t_2^2 + t_1^3t_2^3 + t_1^3t_2^3t_3 \\
&\quad +t_1^3t_2^3t_3^2 + t_1^3t_2^3t_3^3 + t_1t_2^3 + t_1t_2^3t_3 + t_1t_2^3t_3^2 + t_1t_2^2 + t_1t_2^2t_3 + t_1^3t_2 + t_1^3t_2t_3 + t_1^3t_2t_3^2 \\
&\quad +t_1^3t_2^4 + t_1^3t_2^4t_3.
\end{aligned}
$$

The Contracting Homotopy for the Resolution Only non-zero values are shown. Furthermore, we give the contracting homotopy only on elements that are needed to prove that the differential given above is indeed a resolution up to the given degree. This has been computed using Lemma 7.2.

$$\phi_\infty(t_1) = x_1, \quad \phi_\infty(t_2) = x_2, \quad \phi_\infty(t_3) = x_3$$

$$
\begin{aligned}
\phi_\infty(t_2x_1) &= 4x_1x_2, & \phi_\infty(t_3x_1) &= 4x_1x_3, & \phi_\infty(t_3x_2) &= 4x_2x_3, \\
\phi_\infty(t_1^4x_1) &= \gamma_1(y_1), & \phi_\infty(t_2^4x_2) &= \gamma_1(y_2), & \phi_\infty(t_3^4x_3) &= \gamma_1(y_3)
\end{aligned}
$$

$$
\begin{aligned}
&\phi_\infty(t_3x_1x_2) = x_1x_2x_3, \\
\phi_\infty(t_1\gamma_1(y_1)) &= \gamma_1(y_1)x_1, \ \phi_\infty(t_2\gamma_1(y_1)) = \gamma_1(y_1)x_2, \ \phi_\infty(t_3\gamma_1(y_1)) = \gamma_1(y_1)x_3, \\
\phi_\infty(t_2^4x_1x_2) &= 4\gamma_1(y_2)x_1, \phi_\infty(t_2\gamma_1(y_2)) = \gamma_1(y_2)x_2, \ \phi_\infty(t_3\gamma_1(y_2)) = \gamma_1(y_2)x_3, \\
\phi_\infty(t_3^4x_1x_3) &= 4\gamma_1(y_3)x_1, \ \phi_\infty(t_3^4x_2x_3) = 4\gamma_1(y_3)x_2, \phi_\infty(t_3\gamma_1(y_3)) = \gamma_1(y_3)x_3
\end{aligned}
$$

$$\begin{aligned}
\phi_\infty(t_2\gamma_1(y_1)x_1) &= 4\gamma_1(y_1)x_1x_2, & \phi_\infty(t_2\gamma_1(y_2)x_1) &= 4\gamma_1(y_2)x_1x_2,\\
\phi_\infty(t_3^4x_1x_2x_3) &= \gamma_1(y_3)x_1x_2, & \phi_\infty(t_3\gamma_1(y_1)x_1) &= 4\gamma_1(y_1)x_1x_3,\\
\phi_\infty(t_3\gamma_1(y_2)x_1) &= 4\gamma_1(y_2)x_1x_3, & \phi_\infty(t_3\gamma_1(y_3)x_1) &= 4\gamma_1(y_3)x_1x_3,\\
\phi_\infty(t_3\gamma_1(y_1)x_2) &= 4\gamma_1(y_1)x_2x_3, & \phi_\infty(t_3\gamma_1(y_2)x_2) &= 4\gamma_1(y_2)x_2x_3,\\
\phi_\infty(t_3\gamma_1(y_3)x_2) &= 4\gamma_1(y_3)x_2x_3, & \phi_\infty(t_1^4\gamma_1(y_1)x_1) &= \gamma_2(y_1),\\
\phi_\infty(t_2^4\gamma_1(y_1)x_2) &= \gamma_1(y_1)\gamma_1(y_2), & \phi_\infty(t_3^4\gamma_1(y_1)x_3) &= \gamma_1(y_1)\gamma_1(y_3),\\
\phi_\infty(t_2^4\gamma_1(y_2)x_2) &= \gamma_2(y_2), & \phi_\infty(t_3^4\gamma_1(y_2)x_3) &= \gamma_1(y_2)\gamma_1(y_3),\\
\phi_\infty(t_3^4\gamma_1(y_3)x_3) &= \gamma_2(y_3).
\end{aligned}$$

The Differential in the Reduced Complex up to Degree 4 By definition, the reduced complex is the one obtained by tensoring the resolution with $\mathcal{F}_5$ over the group ring $\mathcal{F}_5\mathrm{UT}_3(\mathcal{F}_5)$. It is a suitable complex for computing the homology of $\mathrm{UT}_3(\mathcal{F}_5)$. Again, only non-zero differentials are shown.

$$\partial_\infty(x_1x_2) = 4x_3$$

$$\partial_\infty(\gamma_1(y_1)x_2) = \gamma_1(y_3) \quad \partial_\infty(\gamma_1(y_2)x_1) = 4\gamma_1(y_3)$$

$$\begin{aligned}
\partial_\infty(\gamma_1(y_1)x_1x_2) &= (4\gamma_1(y_1) + 4\gamma_1(y_3))x_3 + 4\gamma_1(y_3)x_1\\
\partial_\infty(\gamma_1(y_2)x_1x_2) &= (4\gamma_1(y_2) + \gamma_1(y_3))x_3 + 4\gamma_1(y_3)x_2\\
\partial_\infty(\gamma_1(y_3)x_1x_2) &= 4\gamma_1(y_3)x_3\\
\partial_\infty(\gamma_1(y_2)x_1x_3) &= 4\gamma_1(y_3)x_3\\
\partial_\infty(\gamma_1(y_1)x_2x_3) &= \gamma_1(y_3)x_3\\
\partial_\infty(\gamma_1(y_1)\gamma_1(y_2)) &= 2\gamma_1(y_3)x_3
\end{aligned}$$

8.2.4 A Cocyclic Code with a Hadamard Property

Continuing the computation we find as representatives of a basis of the cohomology group $H^2(\mathrm{UT}_4(\mathbb{F}_2))$ the following 7 (abstract) cocycles.

$$\gamma_2(x_1)^*,\ \gamma_2(x_2)^*,\ \gamma_2(x_3)^*,\ (\gamma_1(x_1)\gamma_1(x_3))^*,\ \gamma_2(x_4)^* + (\gamma_1(x_1)\gamma_1(x_4))^* + (\gamma_1(x_2)\gamma_1(x_4))^*$$

$$\gamma_2(x_5)^* + (\gamma_1(x_3)\gamma_1(x_5))^* + (\gamma_1(x_2)\gamma_1(x_5))^*,\ (\gamma_1(x_2)\gamma_1(x_6))^* + (\gamma_1(x_3)\gamma_1(x_4))^* + (\gamma_1(x_4)\gamma_1(x_5))^*$$

Applying these functions to the universal cochain from Sect. 8.2.2 results in the following functional cocycle representatives of the second cohomology group:

$$\begin{aligned}
&\mu_1 = a_1b_1, \mu_2 = a_2b_2, \mu_3 = a_3b_3, \mu_4 = a_1b_3,\\
&\mu_5 = a_1b_2b_4 + a_4b_4 + a_2b_4 + a_1b_4 + a_1a_4b_2 + a_1a_2b_2,\\
&\mu_6 = a_2b_3b_5 + a_5b_5 + a_3b_5 + a_2b_5 + a_2a_5b_3 + a_2a_3b_3,\\
&\mu_7 = a_2b_6 + a_4b_5 + a_1a_2b_5 + a_2b_3b_4 + a_3b_4 + a_1b_2b_3 + a_2a_4b_3 + a_1a_3b_2.
\end{aligned}$$

Considering the cocycle $\mu = \sum_{i=1}^{7} \mu_i$ and evaluating this function on all pairs of group elements results in a 0-1-matrix H which satisfies the following combinatorial property

$$HH^t = (64([i = j] + [i = 32j]))_{1\le i,j\le 64}$$

– a generalization of the Hadamard property (cf. [18]). In particular the first 32 rows without column 1 determine a non-linear $(63, 32, 32)$ code.

8.3 The 2-Sylow Subgroup of $\mathrm{Sp}_4(F_{2^2})$

The symplectic group $\mathrm{Sp}_4(F_{2^2})$ is a sporadic simple group of order $979200 = 2^8 3^2 5^2 17$. The power-commutator presentation of its 2-Sylow subgroup was computed with the system GAP. [20] It is given by

$$\langle t_1, t_2, \ldots, t_8 | t_1^2 = t_8, (t_2, t_1) = t_5, (t_2, t_1) = t_6, (t_3, t_1) = t_8, (t_3, t_2) = t_7, (t_4, t_2) = t_5 \rangle.$$

Its mod-2 lower central series is $Z_1 = \mathrm{Sp}_4(F_{2^2}) > Z_2 = \langle t_5, t_6, t_7, t_8 \rangle > Z_3 = \{1\}$. This was refined to cyclic factors using GAP's functions `RightCoset` and `CanonicalRightCosetElement`.[21] This gave $t_1 t_8$ and did not change $t_i, i \geq 2$. Writing t_1 for $t_1 t_8$, which does not change the given relations, we have found the images under the vector space isomorphims Θ of the canonical generators $e_1, \ldots, e_n$ of the corresponding elementary abelian group of order 2^n. Using a *symbolic* collecting algorithm, which we have implemented[22], we can multiply two generic elements t^a and t^b to get the polynomial group law. We find that $t^a t^b = t^c$ where c is equal to

$$(b_1 + a_1, b_2 + a_2, b_3 + a_3, b_4 + a_4, b_5 + a_4 b_2 + a_2 b_1 + a_5, b_6 + a_3 b_1 + a_6, b_7 + a_3 b_2 + a_7, b_8 + a_4 b_1 + a_1 b_1 + a_8).$$

For a group of nilpotency class 2 the power-commutator relations are directly reflected. The results for the reduced differential computed with our algorithm are as follows. In degree 1 $\bar{\partial}_\infty$ is 0. The non-zero images of the canonical basis in degree 2 are

$$\begin{aligned}
\bar{\partial}_\infty(x_1 x_2) &= x_5, \\
\bar{\partial}_\infty(x_1 x_3) &= x_6, \\
\bar{\partial}_\infty(x_1 x_4) &= x_8, \\
\bar{\partial}_\infty(x_2 x_3) &= x_7, \\
\bar{\partial}_\infty(x_2 x_4) &= x_5, \\
\bar{\partial}_\infty(\gamma_1(y_1)) &= x_8,
\end{aligned}$$

while in degree 3 we have,

[20] `http://www-history-mcs.st-and.ac.uk/~gap/`

[21] See `sp4-4.gap` and `corrplie.gap`.

[22] The code is in `binstr.spad`.

$$
\begin{aligned}
\bar{\partial}_\infty(x_1x_2x_3) &= x_6x_7 + x_5x_7 + x_5x_6 + x_3x_5 + x_2x_6 + x_1x_7,\\
\bar{\partial}_\infty(x_1x_2x_4) &= x_4x_5 + x_2x_8 + x_1x_5,\\
\bar{\partial}_\infty(x_1x_2x_6) &= x_5x_6,\\
\bar{\partial}_\infty(x_1x_2x_7) &= x_5x_7,\\
\bar{\partial}_\infty(x_1x_2x_8) &= x_5x_8,\\
\bar{\partial}_\infty(x_1x_3x_4) &= x_6x_8 + x_4x_6 + x_3x_8,\\
\bar{\partial}_\infty(x_1x_3x_5) &= x_5x_6,\\
\bar{\partial}_\infty(x_1x_3x_7) &= x_6x_7,\\
\bar{\partial}_\infty(x_1x_3x_8) &= x_6x_8,\\
\bar{\partial}_\infty(x_1x_4x_5) &= x_5x_8,\\
\bar{\partial}_\infty(x_1x_4x_6) &= x_6x_8,\\
\bar{\partial}_\infty(x_1x_4x_7) &= x_7x_8,\\
\bar{\partial}_\infty(x_2x_3x_4) &= x_5x_7 + x_4x_7 + x_3x_5,\\
\bar{\partial}_\infty(x_2x_3x_5) &= x_5x_7,\\
\bar{\partial}_\infty(x_2x_3x_6) &= x_6x_7,\\
\bar{\partial}_\infty(x_2x_3x_8) &= x_7x_8,\\
\bar{\partial}_\infty(x_2x_4x_6) &= x_5x_6,\\
\bar{\partial}_\infty(x_2x_4x_7) &= x_5x_7,\\
\bar{\partial}_\infty(x_2x_4x_8) &= x_5x_8,\\
\bar{\partial}_\infty(\gamma_1(y_1)x_1) &= x_1x_8,\\
\bar{\partial}_\infty(\gamma_1(y_2)x_1) &= \gamma_1(y_5) + x_2x_5,\\
\bar{\partial}_\infty(\gamma_1(y_3)x_1) &= \gamma_1(y_6) + x_3x_6,\\
\bar{\partial}_\infty(\gamma_1(y_4)x_1) &= \gamma_1(y_8) + x_4x_8,\\
\bar{\partial}_\infty(\gamma_1(y_1)x_2) &= \gamma_1(y_5) + x_2x_8 + x_1x_5,\\
\bar{\partial}_\infty(\gamma_1(y_3)x_2) &= \gamma_1(y_7) + x_3x_7,\\
\bar{\partial}_\infty(\gamma_1(y_4)x_2) &= \gamma_1(y_5) + x_4x_5,\\
\bar{\partial}_\infty(\gamma_1(y_1)x_3) &= \gamma_1(y_6) + x_3x_8 + x_1x_6,\\
\bar{\partial}_\infty(\gamma_1(y_2)x_3) &= \gamma_1(y_7) + x_2x_7,\\
\bar{\partial}_\infty(\gamma_1(y_1)x_4) &= \gamma_1(y_8) + x_4x_8 + x_1x_8,\\
\bar{\partial}_\infty(\gamma_1(y_2)x_4) &= \gamma_1(y_5) + x_2x_5,\\
\bar{\partial}_\infty(\gamma_1(y_1)x_5) &= x_5x_8,\\
\bar{\partial}_\infty(\gamma_1(y_1)x_6) &= x_6x_8,\\
\bar{\partial}_\infty(\gamma_1(y_1)x_7) &= x_7x_8.
\end{aligned}
$$

A An AXIOM-Program for $\mathrm{UT}_3(\mathcal{F}_5)$

The first part defines the ingredients for this particular group. From the line starting with `The data structures` on, the program is generally applicable for other p-groups as well. It computes ∂_∞ (`dNew`) and ∇_∞ (`nablaNew`) as well as $\bar{\partial}_\infty$ (`dNewReduced`) and $\bar{\nabla}_\infty$ (`nablaNewReduced`) for the degrees 1 up to 4. For brevity, obvious parts are left out. Similar programs for computing the projection f_∞ and the contracting homotopy ϕ_∞ can be found on the internet (see 8.1).

```
)clear all
)spool ut3-5.out

-- load all the necessary code
)r loadall

-- the prime
p := 5
-- the field with p elements
F := PrimeField p
-- the dimension
n := 3

-- sets of variables
ly := [subscript('y,[i]) for i in 1..n]
lt := [subscript('t,[i]) for i in 1..n]
lx := [subscript('x,[i]) for i in 1..n]

-- the group law for the 3x3 upper triangular matrices
-- with 1's along the diagonal

rho:(List PF 5,List PF 5) -> List PF 5
rho(x,y) == [x.1+y.1,x.2+y.2,x.3+y.3+x.1*y.2]

-----------------------------------------------------------------------------------------
-- The data structures
-----------------------------------------------------------------------------------------
-- the group
G := PPGP(p, n, rho)

-- the elementary abelian group of order p^n
Gp := MultiplicativelyWrittenElementaryAbelian(p, n)

-- the group algebra of the elementary abelian group Gp
-- written multiplicatively
Ap := MonoidRing(F, Gp)

-- The divided power algera over Ap.  This is the algebra
-- Fp( (Z/pZ)^n ) x Gamma(y1,..,yn) where "x" denotes tensor
-- product over F, the prime field with p elements,
-- and Gamma is the algebra with infinitely
-- many generators g_i(j_j), i = 0,1,..., j = 1,..,n and
-- multiplication g_i(y_j)g_k(y_j) = [i+j,i]g_{i+j} where
-- [i+j,i] denotes the binomial coefficient mod p.
DP := DIVPOW(Ap, n, ly)

-- This is the Cartan "little resolution" over the group
-- ring of (Z/pZ)^n.  It is the differential graded augmented
-- algebra Z/pZ( (Z/pZ)^n ) x Gamma(y1,..,yn) x Lambda[x1,..,xn]
-- where Lambda denoted the exterior algebra and d is the classical
-- differential.  This includes the contracting homotopy.
C := CLR(p, n, lt, ly, lx)

--  This is the bar construction of Eilenberg and MacLane
--  for (Z/pZ)^n.
Bp := BAR(F, Gp)

--  This is the bar construction of Eilenberg and MacLane
--  for G.
B   := BAR(F, G)

--  This is the strong deformation retraction of the
--  Cartan little resolution into the bar construction.
--  It includes the inclusion, the retraction and the homotopy.
SDR := SDRPG(p,n,lt,ly,lx)

-----------------------------------------------------------------------------------------
-- The initiator
-----------------------------------------------------------------------------------------
--  The package PerturbationUtilites provides conversion functions (::, coerce)
--  to accomplish the isomorphisms Xi and Theta from the paper.
--  The actual initiator T:
tee : Bp -> Bp
tee(b) ==
  d1 :=  d(b :: B )$B :: Bp
  d2 :=  d(b)$Bp
  d1-d2

-- t composed with phi = homot
tphi : Bp -> Bp
tphi(b) == tee homot(b)$SDR

-----------------------------------------------------------------------------------------
-- The iterated transference process for the chain maps on Cartan's little resolution
-----------------------------------------------------------------------------------------

-----------------------------------------------------------------------------------------
-- The basis elements in Cartan's little resolution
-----------------------------------------------------------------------------------------
```

```
-- Now we go about constructing the degree 1, 2, 3, 4 components in the Cartan little
-- resolution. The canonical bases of Cartan's little resolutions for degrees 1,2,3,4
Cdegree1 := [monomial(gamma(r.divpow)$DP, r.extalg)$C for r in basisOfDegree(1)$CARTUTS(p,n)]
Cdegree2 := [monomial(gamma(r.divpow)$DP, r.extalg)$C for r in basisOfDegree(2)$CARTUTS(p,n)]
Cdegree3 := [monomial(gamma(r.divpow)$DP, r.extalg)$C for r in basisOfDegree(3)$CARTUTS(p,n)]
Cdegree4 := [monomial(gamma(r.divpow)$DP, r.extalg)$C for r in basisOfDegree(4)$CARTUTS(p,n)]


-------------------------------------------------------------------------------------
-- degree 1
-------------------------------------------------------------------------------------
tphiListCdegree1 : List List Bp := [];
zerosBp1 : List Bp := [0$Bp for i in 1..#Cdegree1];
nablaCdegree1 := [inc(c)$SDR for c in Cdegree1];

-- apply initiator to images of basis elements of degree 1
tnablaCdegree1 := [tee bp for bp in nablaCdegree1];
zerosBpCheck: Boolean := (nablaCdegree1 = zerosBp1);
if not zerosBpCheck then tphiListCdegree1 := [tnablaCdegree1];
while (not zerosBpCheck) repeat
   tphiIteration := [tphi bp for bp in tphiListCdegree1.1]
   zerosBpCheck := (tphiIteration = zerosBp1)
   if (not zerosBpCheck) then tphiListCdegree1 := cons(tphiIteration,tphiListCdegree1)

-------------------------------------------------------------------------------------
-- degree 2
-------------------------------------------------------------------------------------

...

-------------------------------------------------------------------------------------
-- degree 4
-------------------------------------------------------------------------------------

...

-------------------------------------------------------------------------------------
-- the new limit differential on Cartan's Little Resolution
-------------------------------------------------------------------------------------

-- the lists tphiListCdegree4 contain all the non-zero powers of tphi
-- now project the summands back (f = proj) to Cartan's Little Resolution
ftphiListCdegree1 := [ [proj(bp)$SDR for bp in LL] for LL in reverse tphiListCdegree1];
ftphiListCdegree2 := [ [proj(bp)$SDR for bp in LL] for LL in reverse tphiListCdegree2];
ftphiListCdegree3 := [ [proj(bp)$SDR for bp in LL] for LL in reverse tphiListCdegree3];
ftphiListCdegree4 := [ [proj(bp)$SDR for bp in LL] for LL in reverse tphiListCdegree4];

-- the lists dCdegree contain the applications of the given differential on C to the basis
-- elements, for uniform handling, also put them into ftphiListCdegree
dCdegree1 := [d c for c in Cdegree1];
dCdegree2 := [d c for c in Cdegree2];
dCdegree3 := [d c for c in Cdegree3];
dCdegree4 := [d c for c in Cdegree4];

ftphiListCdegree1 := cons(dCdegree1, ftphiListCdegree1);
ftphiListCdegree2 := cons(dCdegree2, ftphiListCdegree2);
ftphiListCdegree3 := cons(dCdegree3, ftphiListCdegree3);
ftphiListCdegree4 := cons(dCdegree4, ftphiListCdegree4);

-- the new differential is the sum of the given differential and the result of
-- the perturbation process (perturbation lemma)
sumListsC(lc: List C, lc': List C): List C == [c+c' for c in lc for c' in lc']
-- summing up all lists componentwise
dnewCdegree1 := reduce(sumListsC, ftphiListCdegree1, [0$C for i in 1..#Cdegree1])
dnewCdegree2 := reduce(sumListsC, ftphiListCdegree2, [0$C for i in 1..#Cdegree2])
dnewCdegree3 := reduce(sumListsC, ftphiListCdegree3, [0$C for i in 1..#Cdegree3])
dnewCdegree4 := reduce(sumListsC, ftphiListCdegree4, [0$C for i in 1..#Cdegree4])

-- some function for pretty output
--)r homolprt

-- function for pretty printing
O := OutputForm
say(str) == messagePrint(str)$OutputForm
form(c,yy) == print(hconcat [message("d ")$O,c::O,message(" = ")$O,yy::O])$O
form(f, x, fx) == print(hconcat [message(concat(f,"("))$O,x::O,message(") = ")$O,fx::O])$O

printChainMap(f, degree, basisList, fbasisList) ==
  say "--------------------------------------------------------------------------------"
  print(center [" The chain map "::O, f::O])$O
  print(center [" on the canonical basis elements of degree "::O, degree :: O])$O
  say "--------------------------------------------------------------------------------"
  for x in basisList for fx in fbasisList repeat form(f, x, fx)


printChainMap("dNew", 1, Cdegree1, dnewCdegree1)
printChainMap("dNew", 2, Cdegree2, dnewCdegree2)
printChainMap("dNew", 3, Cdegree3, dnewCdegree3)
printChainMap("dNew", 4, Cdegree4, dnewCdegree4)
```

```
------------------------------------------------------------------------------------
-- the new limit inclusion nabla from Cartan's Little Resolution to Bar Construction
------------------------------------------------------------------------------------

-- the lists tphiListCdegree contain all the non-zero powers of tphi
-- now once again apply homotopy phi
phitphiListCdegree1 := [ [homot(bp)$SDR for bp in LL] for LL in reverse tphiListCdegree1];
phitphiListCdegree2 := [ [homot(bp)$SDR for bp in LL] for LL in reverse tphiListCdegree2];
phitphiListCdegree3 := [ [homot(bp)$SDR for bp in LL] for LL in reverse tphiListCdegree3];
phitphiListCdegree4 := [ [homot(bp)$SDR for bp in LL] for LL in reverse tphiListCdegree4];

-- the lists nablaCdegree are already constructed,
-- for uniform handling, also put them into phitphiListCdegree

phitphiListCdegree1 := cons(nablaCdegree1, phitphiListCdegree1);
phitphiListCdegree2 := cons(nablaCdegree2, phitphiListCdegree2);
phitphiListCdegree3 := cons(nablaCdegree3, phitphiListCdegree3);
phitphiListCdegree4 := cons(nablaCdegree4, phitphiListCdegree4);

-- the new limit inclusion nablanew is the sum of the given inclusion and the result of
-- the perturbation process (perturbation lemma)
sumListsBp(lbp: List Bp, lbp': List Bp): List Bp== [bp+bp' for bp in lbp for bp' in lbp'];
-- summing up all lists componentwise
nablanewCdegree1 := reduce(sumListsBp, phitphiListCdegree1, [0$Bp for i in 1..#Cdegree1]);
nablanewCdegree2 := reduce(sumListsBp, phitphiListCdegree2, [0$Bp for i in 1..#Cdegree2]);
nablanewCdegree3 := reduce(sumListsBp, phitphiListCdegree3, [0$Bp for i in 1..#Cdegree3]);
nablanewCdegree4 := reduce(sumListsBp, phitphiListCdegree4, [0$Bp for i in 1..#Cdegree4]);

printChainMap("nablaNew", 1, Cdegree1, nablanewCdegree1)
printChainMap("nablaNew", 2, Cdegree2, nablanewCdegree2)
printChainMap("nablaNew", 3, Cdegree3, nablanewCdegree3)
printChainMap("nablaNew", 4, Cdegree4, nablanewCdegree4)


------------------------------------------------------------------------------------
-- the new limit inclusion nabla from Cartan's Little Resolution to Bar Construction
------------------------------------------------------------------------------------

-- the lists tphiListCdegree4 contain all the non-zero powers of tphi
-- now once again apply homotopy phi
phitphiListCdegree1 := [ [homot(bp)$SDR for bp in LL] for LL in reverse tphiListCdegree1];
phitphiListCdegree2 := [ [homot(bp)$SDR for bp in LL] for LL in reverse tphiListCdegree2];
phitphiListCdegree3 := [ [homot(bp)$SDR for bp in LL] for LL in reverse tphiListCdegree3];
phitphiListCdegree4 := [ [homot(bp)$SDR for bp in LL] for LL in reverse tphiListCdegree4];

-- the lists nablaCdegree are already constructed,
-- for uniform handling, also put them into phitphiListCdegree

phitphiListCdegree1 := cons(nablaCdegree1, phitphiListCdegree1);
phitphiListCdegree2 := cons(nablaCdegree2, phitphiListCdegree2);
phitphiListCdegree3 := cons(nablaCdegree3, phitphiListCdegree3);
phitphiListCdegree4 := cons(nablaCdegree4, phitphiListCdegree4);

-- the new limit inclusion nablanew is the sum of the given inclusion and the result of
-- the perturbation process (perturbation lemma)
sumListsBp(lbp: List Bp, lbp': List Bp): List Bp== [bp+bp' for bp in lbp for bp' in lbp']
-- summing up all lists componentwise
nablanewCdegree1 := reduce(sumListsBp, phitphiListCdegree1, [0$Bp for i in 1..#Cdegree1]);
nablanewCdegree2 := reduce(sumListsBp, phitphiListCdegree2, [0$Bp for i in 1..#Cdegree2]);
nablanewCdegree3 := reduce(sumListsBp, phitphiListCdegree3, [0$Bp for i in 1..#Cdegree3]);
nablanewCdegree4 := reduce(sumListsBp, phitphiListCdegree4, [0$Bp for i in 1..#Cdegree4]);


------------------------------------------------------------------------------------
-- reduction, i.e. tensoring with the ground field, i.e. summing up coefficients
------------------------------------------------------------------------------------

-- Here we tensor the resolution with Z/pZ over the group ring:
reductionC : C -> C
reductionC c ==
  brFc := basisRepresentationOverF(c)$C
  ans : C := 0
  -- note that we do that by simply 'forgetting' the group elements
  -- i.e. summing up their coefficients for the augmentation
  for rc in brFc repeat ans := ans + rc.f * gamma(rc.y)$DP * (rc.x::EAB::C)
  ans

-- Here we tensor the bar construction with Z/pZ over the group ring:
reductionBp : Bp ->Bp
reductionBp bp == reduce(+, [monomial(epsilon(c)::Ap, s)$Bp for c in coefficients bp
    for s in support bp],0$Bp)

say("---------------------------------------------------------------------------")
say("  The reduced complex ")
say("---------------------------------------------------------------------------")


dnewredCdegree1 := [reductionC dnewc for  dnewc in dnewCdegree1];
```

```
nablanewredCdegree1 := [reductionBp nablanewc for  nablanewc in nablanewCdegree1];

printChainMap("dNewReduced", 1, Cdegree1, dnewredCdegree1)
printChainMap("nablaNewReduced", 1, Cdegree1, nablanewredCdegree1)

...

printChainMap("nablaNewReduced", 4, Cdegree4, nablanewredCdegree4)

say("---------------------------------------------------------------------------")
say "-- End of homological computation (differential, inclusion) of given Group"
say("---------------------------------------------------------------------------")


)spool
```

References

1. Donald W. Barnes and Larry A. Lambe. A fixed point approach to homological perturbation theory. *Proc. Amer. Math. Soc.*, 112(3):881–892, 1991.
2. R. Brown. The twisted Eilenberg-Zilber theorem. In *Simposio di Topologia (Messina, 1964)*, pages 33–37. Edizioni Oderisi, Gubbio, 1965.
3. Henri Cartan and Samuel Eilenberg. *Homological algebra.* Princeton University Press, Princeton, N. J., 1956.
4. Henri Cartan, J. C. Moore, R. Thom, and J-P. Serre. *Algèbras d'Eilenberg-MacLane et homotopie. 2ieme ed., revue et corrig ee.* Secretariat mathematique (Hektograph), Paris, 1956.
5. Torsten Ekedahl, Johannes Grabmeier, and Larry Lambe. A Generic Language for algebraic computations, 2000. In preparation.
6. G. Ellis and I. Kholodna. Second cohomology of finite groups with trivial coefficients. *Homology, Homotopy & Appl.*, 1:163–168, 1999.
7. D. L. Flannery. Transgression and the calculation of cocyclic matrices. *Austral. J. Combin.*, 11:67–78, 1995.
8. D. L. Flannery. Calculation of cocyclic matrices. *J. Pure Appl. Algebra*, 112(2):181–190, 1996.
9. D. L. Flannery, K. J. Horadam, and W. de Launey. Cocyclic hadamard matrices and difference sets. *Discrete Appl. Math.*, 102:47–61, 2000.
10. D. L. Flannery and E. A. O'Brien. Computing 2-cocycles for central extensions and relative difference sets. *Comm. Algebra*, 28(4):1939–1955, 2000.
11. R.L. Graham, D.E. Knuth, and O. Patashnik. *Concrete Mathematics.* Addison-Wesley, Reading, Massachusetts, 1989.
12. V. K. A. M. Gugenheim. On the chain-complex of a fibration. *Illinois J. Math.*, 16:398–414, 1972.
13. V. K. A. M. Gugenheim and L. A. Lambe. Perturbation theory in differential homological algebra. I. *Illinois J. Math.*, 33(4):566–582, 1989.
14. V. K. A. M. Gugenheim, L. A. Lambe, and J. D. Stasheff. Perturbation theory in differential homological algebra. II. *Illinois J. Math.*, 35(3):357–373, 1991.
15. K. J. Horadam and W. de Launey. Cocyclic development of designs. *J. Algebraic Combin.*, 2(3):267–290, 1993.
16. K. J. Horadam and W. de Launey. Erratum: "Cocylic development of designs". *J. Algebraic Combin.*, 3(1):129, 1994.
17. K. J. Horadam and W. de Launey. Generation of cocyclic Hadamard matrices. In *Computational algebra and number theory (Sydney, 1992)*, volume 325 of *Math. Appl.*, pages 279–290. Kluwer Acad. Publ., Dordrecht, 1995.
18. K. J. Horadam and A. A. I. Perera. Codes from cocycles. In *Lecture Notes in Computer Science*, volume 1255, pages 151–163. Springer-Verlag, Berlin-Heidelberg-New York, 1997.
19. Johannes Huebschmann. The homotopy type of $F\psi^q$. The complex and symplectic cases. In *Applications of algebraic K-theory to algebraic geometry and number theory, Part I, II (Boulder, Colo., 1983)*, pages 487–518. Amer. Math. Soc., Providence, R.I., 1986.
20. Johannes Huebschmann. Perturbation theory and free resolutions for nilpotent groups of class 2. *J. Algebra*, 126(2):348–399, 1989.
21. Johannes Huebschmann and Tornike Kadeishvili. Small models for chain algebras. *Math. Z.*, 207(2):245–280, 1991.

22. Bertram Huppert and Norman Blackburn. *Finite groups. II.* Springer-Verlag, Berlin-New York, 1982. Grundlehren der Mathematischen Wissenschaften, Band 242.
23. N. Jacobson. Restricted Lie algebras of characteristic p. *Trans. Amer. Math. Soc.*, 50:15–25, 1941.
24. Richard D. Jenks and Robert S. Sutor. *Axiom. The scientific computation system.* Springer-Verlag, Berlin, Heidelberg, New York, 1992.
25. S. A. Jennings. The structure of the group ring of a p-group over a modular field. *Trans. Amer. Math. Soc.*, 50:175–185, 1941.
26. Leif Johansson and Larry Lambe. Transferring algebra structures up to homology equivalence. *Math. Scand.*, 88(2), 2001.
27. Leif Johansson, Larry Lambe, and Emil Sköldberg. On constructing resolutions over the polynomial algebra, 2000. Preprint.
28. Larry Lambe. Next generation computer algebra systems AXIOM and the scratchpad concept: applications to research in algebra. In *Analysis, algebra, and computers in mathematical research (Luleå, 1992)*, volume 156 of *Lecture Notes in Pure and Appl. Math.*, pages 201–222. Dekker, New York, 1994.
29. Larry Lambe. The 1996 Adams Lectures at Manchester University: New computational methods in algebra and topology, May 20 1996.
30. Larry Lambe and Jim Stasheff. Applications of perturbation theory to iterated fibrations. *Manuscripta Math.*, 58(3):363–376, 1987.
31. Larry A. Lambe. Resolutions via homological perturbation. *J. Pure Appl. Algebra*, 12:71–87, 1991.
32. Larry A. Lambe. Homological perturbation theory, Hochschild homology, and formal groups. In *Deformation theory and quantum groups with applications to mathematical physics (Amherst, MA, 1990)*, volume 134 of *Contemp. Math.*, pages 183–218. Amer. Math. Soc., Providence, RI, 1992.
33. Larry A. Lambe. Resolutions which split off of the bar construction. *J. Pure Appl. Algebra*, 84(3):311–329, 1993.
34. Saunders Mac Lane. *Homology.* Classics in Mathematics. Springer-Verlag, Berlin, 1995. Reprint of the 1975 edition.
35. J. Peter May. The cohomology of restricted Lie algebras and of Hopf algebras. *Bull. Amer. Math. Soc.*, 71:372–377, 1965.
36. D. Quillen. On the associated graded ring of a group ring. *J. Algebra*, 10:411–418, 1968.
37. Jean-Pierre Serre. *Lie algebras and Lie groups. 1964 lectures, given at Harvard University, 2nd ed.*, volume 1500 of *Lecture Notes in Mathematics.* Springer-Verlag, Berlin-Heidelberg-New York, 1992.
38. C.T.C. Wall. Resolutions for extensions of groups. *Proc. Phil. Soc.*, 57:251–255, 1961.

The Energy of a Graph: Old and New Results

Ivan Gutman

Faculty of Science, University of Kragujevac, P. O. Box 60
YU–34000 Kragujevac, Yugoslavia
e-mail: gutman@knez.uis.kg.ac.yu

Abstract Let G be a graph possessing n vertices and m edges. The energy of G, denoted by $E = E(G)$, is the sum of the absolute values of the eigenvalues of G. The connection between E and the total electron energy of a class of organic molecules is briefly outlined. Some (known) fundamental mathematical results on E are presented: the relation between $E(G)$ and the characteristic polynomial of G, lower and upper bounds for E, especially those depending on n and m, graphs extremal with respect to E, n-vertex graphs for which $E(G) > E(K_n)$. The characterization of the n-vertex graph(s) with maximal value of E is an open problem.

1 Introduction

In this article we are concerned with schlicht graphs, namely graphs without multiple, weighted or directed edges and without loops. Let G be such a graph. The number of its vertices and edges is denoted by n and m, respectively. Its vertices are labeled by $v_1, v_2, \ldots, v_n$.

The adjacency matrix $\mathbf{A} = \mathbf{A}(G)$ of G is a square matrix of order n whose (i, j)-entry is defined as

$$A_{ij} = \begin{cases} 1 & \text{if the vertices } v_i \text{ and } v_j \text{ are adjacent} \\ 0 & \text{otherwise .} \end{cases}$$

The eigenvalues of $\mathbf{A}(G)$ are said to be the eigenvalues of the graph G. A graph on n vertices has n eigenvalues (not all of which need to be distinct); these will be denoted by $x_1, x_2, \ldots, x_n$ and labeled in a non-decreasing manner: $x_1 \geq x_2 \geq \cdots \geq x_n$. The collection of all n eigenvalues of G forms the spectrum of G.

The characteristic polynomial of $\mathbf{A}(G)$ is said to be the characteristic polynomial of the graph G. Thus, the characteristic polynomial of G is a monic polynomial of degree n defined via

$$\phi(G, x) = \det[x\,\mathbf{I} - \mathbf{A}(G)]$$

where $\mathbf{I}$ stands for the unit matrix of order n.

The graph eigenvalues are just the zeros of the characteristic polynomial. The graph eigenvalues are necessarily real–valued numbers.

Spectral properties of graphs (including properties of the characteristic polynomial) have been extensively studied; for a detailed survey see [1].

2 A Quantum–Chemical Excursion

Within the Hückel molecular orbital (HMO) theory (for details see [2,3]), the total energy of the so-called π-electrons is calculated from the molecular orbital energy levels $E_1, E_2, \ldots, E_n$ by means of the formula

$$E_\pi = \sum_{j=1}^{n} g_j \, E_j$$

where g_j is the number of π-electrons in the jth molecular orbital. On the other hand, the HMO Hamiltonian operator $\hat{\mathbf{H}}$ is related to the adjacency matrix $\mathbf{A}$ of a pertinently constructed graph, the so–called molecular or Hückel graph, via

$$\hat{\mathbf{H}} = \alpha \, \mathbf{I} + \beta \, \mathbf{A} \tag{1}$$

where the parameters α and β are assumed to be constants; for details see [4,5,6]. Consequently,

$$E_j = \alpha + \beta \, x_j$$

and therefore

$$E_\pi = n_e \, \alpha + \beta \sum_{j=1}^{n} g_j \, x_j \tag{2}$$

with n_e denoting the number of π-electrons in the underlying molecule. The right–hand side summation in (2) uniquely depends on the molecular graph, provided the occupation numbers g_j are known. If one restricts the consideration to molecules in their ground states, then Eq. (2) becomes

$$\begin{aligned} E_\pi &= n_e \, \alpha + \left[2 \sum_{j=1}^{n_e/2} x_j \right] \beta && \text{if } n_e \text{ is even} \\ E_\pi &= n_e \, \alpha + \left[2 \sum_{j=1}^{(n_e-1)/2} x_j + x_{(n_e+1)/2} \right] \beta && \text{if } n_e \text{ is odd .} \end{aligned} \tag{3}$$

The non-trivial part of the HMO total π-electron energy is the expression in square brackets in Eqs. (3). This fact is usually stressed by using so–called β-units, i.e. by formally setting $\alpha = 0$, $\beta = 1$. The form of the expression (3) is rather awkward and is not suitable for mathematical analysis. However, in the almost ubiquitously encountered special case when

$$g_j = 2 \text{ whenever } x_j > 0 \qquad \& \qquad g_j = 0 \text{ whenever } x_j < 0$$

we get

$$E_\pi = n_e \, \alpha + 2 \beta \sum_{+} x_j$$

with $\sum_{+}$ indicating summation over all positive–valued graph eigenvalues.

Because all diagonal elements of **A** are zero,

$$\sum_{j=1}^{n} x_j = 0$$

implying

$$E_\pi = n_e\,\alpha + \beta \sum_{j=1}^{n} |x_j|$$

or, in β-units

$$E_\pi = \sum_{j=1}^{n} |x_j| \ . \tag{4}$$

3 The Energy of a Graph

Whereas the left–hand side of (4) is meaningful only if the underlying graph belongs to the restricted class of "Hückel graphs", its right–hand side is well–defined for all graphs. In addition, the right–hand side expression in Eq. (4) possesses a certain mathematical beauty; for instance, it is a symmetric function of the graph eigenvalues.

This motivated the present author [7] to put forward the following graph invariant:

Definition 3.1. If G is an n-vertex graph and $x_1, x_2, \ldots, x_n$ are its eigenvalues, then the energy of G is

$$E(G) = \sum_{j=1}^{n} |x_j| \ .$$

From Definition 3.1 it straightforwardly follows that

$$E(G) = 2 \sum_{+} x_j \ . \tag{5}$$

If misunderstanding is not possible, then we write E instead of $E(G)$.

The (mathematical) theory of the graph energy is nowadays relatively well elaborated, although some fundamental problems are still waiting to be solved. In this article we outline a few results on the energy of graphs, some long known and a few recently obtained. Needless to say that these results are applicable in chemistry – in the theory of total π-electron energy.

4 The Coulson Integral Formula

Theorem 4.1 (Coulson, 1940 [8]). *If G is a graph on n vertices, then*

$$E(G) = \frac{1}{\pi} \int_{-\infty}^{+\infty} \left[n - \frac{ix\,\phi'(G,ix)}{\phi(G,ix)} \right] dx$$

$$= \frac{1}{\pi} \int_{-\infty}^{+\infty} \left[n - x \frac{d}{dx} \log \phi(G,ix) \right] dx \tag{6}$$

where $\phi'(G,x) = (d/dx)\phi(G,x)$ and $i = \sqrt{-1}$.

In the above formula, as well as in what follows, $\int_{-\infty}^{+\infty} F(x)\,dx$ stands for the principal value of the respective integral, i. e., for

$$\lim_{t \to \infty} \int_{-t}^{+t} F(x)\,dx \ .$$

Proof. Denote by $\xi_1, \xi_2, \ldots, \xi_p$ the distinct eigenvalues of the graph G, and by μ_j the algebraic multiplicity of ξ_j,

$$\mu_1 + \mu_2 + \cdots + \mu_p = n \ . \tag{7}$$

Let z be a complex variable, $z = x + i\,y$. Then, because of

$$\phi(G,z) = \prod_{j=1}^{p} (z - \xi_j)^{\mu_j}$$

we have

$$\frac{\phi'(G,z)}{\phi(G,z)} = \sum_{j=1}^{p} \frac{\mu_j}{z - \xi_j} \ . \tag{8}$$

Hence $\phi'(G,z)/\phi(G,z)$ is an analytical function possessing only simple poles, which are just $\xi_1, \xi_2, \ldots, \xi_p$. Define

$$f(z) = z\,\phi'(G,z)/\phi(G,z) \ . \tag{9}$$

Then by applying to Eq. (8) the Cauchy formula,

$$\frac{1}{2\pi i} \oint_{\Gamma} f(z)\,dz = \sum_{+} \mu_j\,\xi_j = \sum_{+} x_j \tag{10}$$

where Γ is a contour shown in Fig. 1.

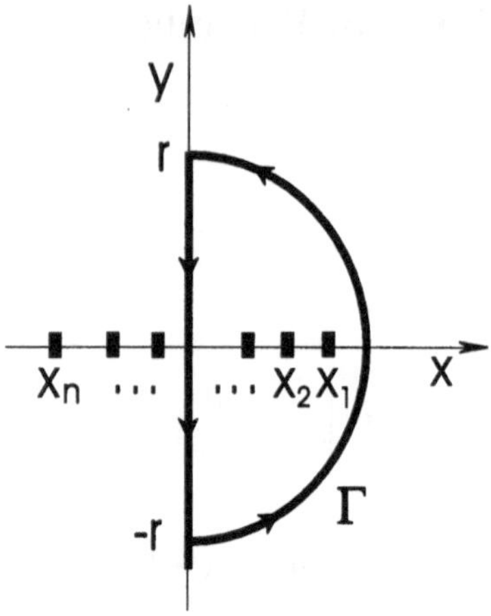

Figure1. For some $r > x_1$, the contour Γ goes along the y-axis from point $(0, r)$ to point $(0, -r)$ and then returns to $(0, r)$ along a semicircle with radius r

For n being a constant,

$$\frac{1}{2\pi i} \oint_\Gamma f(z)\, dz = \frac{1}{2\pi i} \oint_\Gamma [f(z) - n]\, dz \ . \tag{11}$$

Because of (7), (8) and (9) the integrand on the right–hand side of (11) has the property

$$\lim_{|z| \to \infty} [f(z) - n] = 0 \ .$$

In view of it, if $r \to \infty$ (see Fig. 1) then the integrand $f(z) - n$ vanishes everywhere on Γ, except on the y-axis. This change of Γ will, however, not affect the value of the contour integral itself. Thus for $r \to \infty$,

$$\frac{1}{2\pi i} \oint_\Gamma [f(z) - n]\, dz = \frac{1}{2\pi i} \int_{+\infty}^{-\infty} [f(iy) - n]\, d(iy) = \frac{1}{2\pi} \int_{-\infty}^{+\infty} [n - f(iy)]\, dy \ . \tag{12}$$

Formula (6) is a direct consequence of Eqs. (10)–(12) and Eq. (5).

The Coulson integral formula and its various modifications (see below) have important chemical applications. Namely, a theorem by Sachs [1,9] establishes the explicit dependence of the coefficients of the characteristic polynomial of a graph on the structure of this graph. The Coulson formula establishes the explicit dependence of the energy of a graph on the characteristic polynomial of this graph. By combining the Coulson integral formula with the Sachs theorem we gain insight into the dependence of the energy of a graph on the structure of this graph, hence a complete information on the dependence of the total π-electron energy of a molecule (as computed within the HMO model) on the structure of this molecule. More on this matter can be found elsewhere [5,10,11,12].

Corollary 4.2. *If G_1 and G_2 are two graphs with equal number of vertices, then*

$$E(G_1) - E(G_2) = \frac{1}{\pi} \int_{-\infty}^{+\infty} \log \frac{\phi(G_1, ix)}{\phi(G_2, ix)}\, dx\ .$$

Corollary 4.3. *If G is a graph on n vertices, then*

$$E(G) = \frac{1}{\pi} \int_{-\infty}^{+\infty} \frac{dx}{x^2} \log |x^n\, \phi(G, i/x)|\ .$$

Corollary 4.4 ([13]). *Let ζ be any real number greater than x_1. Then*

$$E(G) = \frac{2}{\pi} \sum_{k=1}^{\infty} (-1)^{k-1} \frac{M_{2k}}{(2k-1)\, \zeta^{2k-1}} + \frac{1}{\pi} \int_{-\zeta}^{+\zeta} \left[n - x \frac{d}{dx} \log \phi(G, ix) \right] dx \tag{13}$$

where $M_p = \sum_{j=1}^{n} (x_j)^p$ is the p-th spectral moment of the graph G.

The infinite summation on the right–hand side of (13) is, of course, convergent. Numerous authors attempted to express the energy in terms of spectral moments [14]–[19]. Corollary 4.4 may be understood as the final (negative) solution of this problem.

If B is a bipartite graph on n vertices, then its characteristic polynomial is of the form [1]

$$\phi(B, x) = x^n + \sum_{k \geq 1} (-1)^k\, b(B, k)\, x^{n-2k} \tag{14}$$

with $b(B, k) > 0$ for $k = 1, 2, \ldots, n_+$ and $b(B, k) = 0$ for $k > n_+$, where n_+ is the number of positive eigenvalues of B.

Recall that n_+ is also the number of negative eigenvalues of B, and therefore B has $n - 2\, n_+$ zero eigenvalues.

If F is an acyclic bipartite graph (= a forest), then $b(F, k) = m(F, k)$, where $m(F, k)$ is the number of k-matchings of F, that is the number of k-element independent edge sets of F [1].

Corollary 4.5 ([20]). *If B and F are a bipartite graph and a forest, respectively, then*

$$E(B) = \frac{1}{\pi} \int_{-\infty}^{+\infty} \frac{dx}{x^2} \log \left[1 + \sum_{k \geq 1} b(B,k)\, x^{2k} \right]$$

$$E(F) = \frac{1}{\pi} \int_{-\infty}^{+\infty} \frac{dx}{x^2} \log \left[1 + \sum_{k \geq 1} m(F,k)\, x^{2k} \right] .$$

Corollary 4.5 implies that the energy of a bipartite graph B is a monotonically increasing function of each of the parameters $b(B,1), b(B,2), \ldots, b(B,n_+)$, and that the energy of a forest F is a monotonically increasing function of each of the parameters $m(F,1), m(F,2), \ldots, m(F,n_+)$.

Corollary 4.6 ([20]). *(a) If for two bipartite graphs B_1 and B_2 (not necessarily with equal number of vertices), the relation*

$$b(B_1,k) \ \leq \ b(B_2,k) \tag{15}$$

is satisfied for all $k \geq 1$, then $E(B_1) \leq E(B_2)$. If, in addition, $b(B_1,k) < b(B_2,k)$ for at least one value of k, then $E(B_1) < E(B_2)$.
(b) If for two forests F_1 and F_2 (not necessarily with equal number of vertices), the relation

$$m(F_1,k) \ \leq \ m(F_2,k) \tag{16}$$

is satisfied for all $k \geq 1$, then $E(F_1) \leq E(F_2)$. If, in addition, $b(F_1,k) < b(F_2,k)$ for at least one value of k, then $E(F_1) < E(F_2)$.

Relations (15) and (16) have been established for numerous pairs of graphs [21]–[26], inferring inequalities between their energies. Of them we mention only the following [20,21]:

Let, as usual, $\overline{K_n}$, S_n and P_n denote the graph without edges, the star and the path, respectively, each on n vertices. If F_n is an n-vertex forest, different from $\overline{K_n}$ and P_n, then

$$E(\overline{K_n}) \ < \ E(F_n) \ < \ E(P_n) .$$

If T_n is an n-vertex tree, different from S_n and P_n, then

$$E(S_n) \ < \ E(T_n) \ < \ E(P_n) .$$

5 Bounds for the Energy

Theorem 5.1 ((McClelland, 1971 [27])). *If G is a graph with n vertices, m edges and adjacency matrix $\mathbf{A}$, then*

$$\sqrt{2\,m + n(n-1)\,|\det \mathbf{A}|^{2/n}} \;\le\; E(G) \;\le\; \sqrt{2\,m\,n}\,. \tag{17}$$

Proof. We recall a well known relation for graph eigenvalues:

$$\sum_{j=1}^{n} (x_j)^2 = 2\,m \tag{18}$$

and start with

$$E^2 = \left(\sum_{j=1}^{n} |x_j|\right)^2 = \sum_{j=1}^{n} |x_j|^2 + 2\sum_{j<k} |x_j|\,|x_k| = 2\,m + n(n-1)\,AM\{\,|x_j|\,|x_k|\,\} \tag{19}$$

where $AM\{\,|x_j|\,|x_k|\,\}$ indicates the arithmetic mean of the $(n^2-n)/2$ distinct terms $|x_j|\,|x_k|$, $j<k$. The geometric mean of the same terms is

$$GM\{\,|x_j|\,|x_k|\,\} = \left(\prod_{j<k} |x_j|\,|x_k|\right)^{2/(n^2-n)}$$

$$= \left(\prod_{j=1}^{n} |x_j|^{n-1}\right)^{2/(n^2-n)} = \left(\prod_{j=1}^{n} |x_j|\right)^{2/n} = |\det A|^{2/n}$$

where we have taken into account that $\prod_{j=1}^{n} x_j = \det \mathbf{A}$.

The lower bound is now a consequence of the fact that the geometric mean of non-negative numbers cannot exceed their arithmetic mean.

The variance of the numbers $|x_j|$, $j = 1, 2, \ldots, n$, is equal to:

$$Var\{\,|x_j|\,\} = AM\{\,|x_j|^2\,\} - [AM\{\,|x_j|\,\}]^2$$

$$= \frac{1}{n}\sum_{j=1}^{n} |x_j|^2 - \left(\frac{1}{n}\sum_{j=1}^{n} |x_j|\right)^2 = \frac{2\,m}{n} - \left(\frac{E}{n}\right)^2$$

and the upper bound follows from the fact that the variance is a non-negative quantity.

Corollary 5.2. *If* $\det \mathbf{A} \ne 0$, *then* $E(G) \;\ge\; \sqrt{2\,m + n(n-1)} \;\ge\; n$.

A large number of other bounds for the energy has been reported in the literature (see the subsequent section). Curiously, however, the following simple estimates have evaded attention until quite recently [28].

In addition to (18) we now need also the relation

$$\sum_{j<k} x_j\, x_k = -m \ .$$

Then, in analogy to Eq. (19),

$$E^2 = 2m + 2\sum_{j<k} |x_j|\,|x_k| \ \geq \ 2m + 2\left|\sum_{j<k} x_j\, x_k\right| = 2m + 2|-m| = 4m$$

implying

$$E \ \geq \ 2\sqrt{m} \ . \tag{20}$$

If G has isolated vertices (i. e., vertices to which no other vertex is adjacent), then each isolated vertex results in an eigenvalue equal to zero. Adding isolated vertices to G will thus change neither m nor E. In view of this consider, for a moment, graphs having m edges and no isolated vertices. The maximum number of vertices of such graphs is $2m$, which happens if $G = m\,K_2$, i. e., if the graph G consists of m isolated edges. For all other graphs, $n < 2m$. Bearing this in mind, we have $\sqrt{2mn} \ \leq \ \sqrt{2m\,(2m)} = 2m$ which combined with the upper bound (17) yields $E \ \leq \ 2m$. As explained above, the latter inequality holds also if G possesses isolated vertices.

By this we deduced:

Corollary 5.3. *If G is a graph containing m edges, then*

$$2\sqrt{m} \ \leq \ E(G) \ \leq \ 2m \ . \tag{21}$$

A more detailed analysis [28] reveals that the bounds (21) are sharp: $E(G) = 2\sqrt{m}$ holds if and only if G is a complete bipartite graph plus arbitrarily many isolated vertices, and $E(G) = 2m$ holds if and only if G consists of m isolated edges and of arbitrarily many isolated vertices. Thus, among lower and upper bounds for E, depending solely on m, those given by Eq. (21) are the best possible.

In some cases the lower bound (21) can be improved.

It is known (see [1]) that the greatest eigenvalue of a graph cannot be less than the average vertex degree $2m/n$. Therefore,

$$\frac{2m}{n} \ \leq \ x_1 \ \leq \ \sum_{+} x_j = \frac{1}{2}\,E$$

resulting in

$$E \ \geq \ \frac{4m}{n} \ . \tag{22}$$

The estimate (22) is better than the lower bound in (21) if $n^2/4 < m \leq n(n-1)/2$. Besides, equality in (22) occurs if and only if G is a complete multipartite graph [28].

At this point it is natural to ask about bounds for E, depending solely on n. Then, of course, the consideration has to be restricted to graphs without isolated vertices. In addition to Corollary 5.2 we have:

Theorem 5.4. *If G is an n-vertex graph without isolated vertices, then*

$$E(G) \geq 2\sqrt{n-1} \tag{23}$$

with equality if and only if G is the n-vertex star.

Proof. Case (a): G is connected.
If G is a connected graph, then $m \geq n-1$ and (23) follows from (20). Equality is attained if G is the complete bipartite graph with $m = n-1$, which is the star.

Case (b): G is disconnected.
Let G be composed of $p > 1$ components, $G_1, G_2, \ldots, G_p$, having $n_1, n_2, \ldots, n_p$ vertices, respectively, $n_1 + n_2 + \cdots + n_p = n$. Note that since G has no isolated vertices, for all values of j it must be $n_j \geq 2$ and therefore $\sqrt{n_j - 1} \geq 1$. Then $E(G) = E(G_1) + E(G_2) + \cdots + E(G_p)$ and (what just has been proven), $E(G_j) \geq 2\sqrt{n_j - 1}$ for $j = 1, 2, \ldots, p$. Consequently,

$$\begin{aligned}
E(G) &\geq 2\left(\sqrt{n_1 - 1} + \sqrt{n_2 - 1} + \cdots \sqrt{n_p - 1}\right) \\
&= 2\sqrt{\left(\sqrt{n_1 - 1} + \sqrt{n_2 - 1} + \cdots \sqrt{n_p - 1}\right)^2} \\
&= 2\sqrt{(n_1 - 1) + (n_2 - 1) + \cdots + (n_p - 1) + 2\sum_{j<k} \sqrt{n_j - 1}\sqrt{n_k - 1}} \\
&\geq 2\sqrt{n - p + p(p-1)} = 2\sqrt{n - 1 + (p-1)^2}
\end{aligned}$$

because there are $p(p-1)/2$ summands of the form $\sqrt{n_j - 1}\sqrt{n_k - 1}$, each being greater than or equal to unity.

We thus arrived at a stronger result than needed: For a graph G with p components, $p \geq 1$, and without isolated vertices,

$$E(G) \geq 2\sqrt{n - 1 + (p-1)^2}$$

from which Theorem 5.4 follows immediately.

Another way to formulate Theorem 5.4 is:

Corollary 5.5. *Among n-vertex graphs without isolated vertices, the star has minimal energy.*

Clearly, among n-vertex graphs, the edgeless graph $\overline{K_n}$ has minimal energy, equal to zero.

What is missing from Theorem 5.4 is a sharp upper bound for the energy of a graph, depending solely on the number n of vertices. Finding of such a bound is related to the problem of identifying the n-vertex graph(s) with maximal energy.

6 More Bounds for the Energy

Let $U = 2m - n\,|\det \mathbf{A}|^{2/n}$. Then the McClelland bounds (17) can be rewritten as

$$0 \leq 2mn - E^2 \leq (n-1)U \ .$$

These bounds were improved [29]. For all graphs:

$$U \leq 2mn - E^2 \leq (n-1)U$$

whereas for bipartite graphs

$$2U \leq 2mn - E^2 \leq (n-2)U \ .$$

Let n_+ and n_- be the number of positive and negative eigenvalues of the graph G. Then [30]

$$E(G) \leq \sqrt{\frac{8\,m\,n_+\,n_-}{n_+ + n_-}} \ .$$

If t and q are, respectively, the number of triangles and quadrangles in the graph G, and if D is the sum of squares of its vertex degrees, then [30]

$$12\,t\,\sqrt{n_+\,n_-[(n_+ + n_-)(8\,q + 2\,D - 2\,m) - 4\,m^2]} \leq E \leq$$

$$\sqrt{12(n_+ + n_-) + (n_+ - n_-)^2 \left(\frac{4\,q + D - m}{m} - \frac{9\,t^2}{m^2}\right)} \ .$$

Let B be a bipartite graph with n vertices and m edges. Let $b(B,2)$ and $b(B,3)$ be quantities defined via Eq. (14), and recall that $b(B,1) = m$. Introduce the auxiliary quantity E_T:

$$E_T = 2\sqrt{m + \sqrt{\frac{n(n-2)\,b(B,2)}{2}}} \ .$$

Then [31]

$$E(B) \leq E_T \leq \sqrt{2\,m\,n}$$

and furthermore [32]

$$E(B) \leq 2\left[\sqrt{\frac{3n(n-2)(n-4)\,b(B,3)}{4}} - \sqrt{\frac{8\,m^3}{n} + \frac{3\,m\,E_T}{2}}\right]^{1/3} \ .$$

Additional bounds for E, applicable to all graphs are reported in [33,34]. The bounds valid for various special classes of graphs are too numerous to be considered here; for some of them see [35].

7 Searching for the Graph(s) with Maximal Energy: Hyperenergetic Graphs

Various empirical and statistical studies (usually performed on graphs of chemical interest, which are connected and possess relatively few edges), point towards a simple regularity: The energy of a graph increases with the increase of the number of vertices and edges. A famous approximation, quantifying the above regularity, is the McClelland formula [27]

$$E \approx a\sqrt{2mn} \quad ; \quad a \approx 0.9$$

which was found to be chemically quite satisfactory (for details see [35]).

A naive extension of this rule to all graphs resulted in the conjecture [7] that among n vertex graphs, the complete graph K_n has maximal energy. It was soon shown (first by Chris Godsil in the early 1980s) that there exist graphs whose energy exceeds $E(K_n)$.

Because the spectrum of K_n consists of the numbers $n-1$ and -1 ($n-1$ times) [1], it follows that $E(K_n) = 2n-2$.

Definition 7.1. A graph G, such that $E(G) > 2n-2$, is said to be hyperenergetic.

The fact that hyperenergetic graphs exist has been known for a long time (C. D. Godsil, unpublished, [36]), but a systematic construction of such graphs was achieved only quite recently.

Proposition 7.2 ([37]). *The line graph of the complete graph on n vertices is hyperenergetic if $n \geq 5$.*

Proof. The complete graph K_n is regular of degree $n-1$. The line graph of K_n has $n(n-1)/2$ vertices. It is known [1,38] that the characteristic polynomial of a regular graph R and of its line graph $L(R)$ are related as

$$\phi(L(R), x) = (x+2)^{n(r-2)/2}\,\phi(R, x-r+2)$$

where n and r are the number of vertices and the degree of R. Bearing in mind that $\phi(K_n, x) = (x-n+1)(x+1)^{n-1}$ we calculate

$$E(L(K_n)) = 2n^2 - 6n$$

which for $n \geq 5$ is greater than

$$E(K_{n(n-1)/2}) = n^2 - n - 2 \, .$$

Proposition 7.2 provides hyperenergetic graphs with $n(n-1)/2$ vertices for $n \geq 5$, that is hyperenergetic graphs with $10, 15, 21, 28, 36, \ldots$ vertices. In what follows we construct hyperenergetic graphs for all $n = 9, 10, 11, \ldots$.

Definition 7.3. Let K_n be the complete graph on n vertices, $n \geq 3$. Let v be a vertex of K_n and let e_i , $i = 1, \ldots, r$, $1 \leq r \leq n-1$, be edges of K_n, all incident to v. The graph obtained by deleting e_i , $i = 1, \ldots, r$, from K_n will be denoted by K_n^r. Besides, for $r = 0$, $K_n^r \equiv K_n$.

By performing a number of pertinent transformations on the determinant of $x\mathbf{I} - \mathbf{A}(K_n^r)$ (whose details are given elsewhere [39,40]), we arrive at

Proposition 7.4. *For* $0 \leq r \leq n-1$,

$$\phi(K_n^r, x) = (x+1)^{n-3}[x^3 - (n-3)\,x^2 - (2n-r-3)\,x + (r-1)(n-1-r)] \ .$$

From Proposition 7.4 one straightforwardly deduces the spectrum of K_n^r, and from it the respective energy:

Proposition 7.5. *For* $n \geq 4$ *and* $2 \leq r \leq n-2$,

$$E(K_n^r) = n - 3 + x_1 + x_2 - x_3$$

where x_i , $i = 1, 2, 3$, *are the three solutions of the cubic equation*

$$x^3 - (n-3)\,x^2 - (2n-r-3)\,x + (r-1)(n-1-r) = 0$$

of which x_1, x_2 *are positive and* x_3 *is negative.*

The inequality $E(K_n^r) > E(K_n)$ will be satisfied if and only if

$$x_1 + x_2 - x_3 > n + 1 \ . \tag{24}$$

Determining the conditions under which (24) is obeyed is an elementary (yet somewhat tedious) exercise from algebra. Skipping its details we state our main result:

Theorem 7.6. *The energy of the graph* K_n^r *exceeds the energy of* K_n *for:* $r = 2$ *and* $n \geq 10$; $r = 3$ *and* $n \geq 9$; $r = 4$ *and* $n \geq 9$; $r = 5$ *and* $n \geq 10$; $r \geq 6$ *and* $n \geq r + 4$.

Corollary 7.7. *Each choice of the parameters* n *and* r, *specified in Theorem 7.6, provides a hyperenergetic graph.*

Corollary 7.8. *There exist hyperenergetic graphs on* n *vertices, for every* $n \geq 9$.

There are hyperenergetic graphs also with 8 vertices. For instance, let the edges f_1, f_2, f_3, f_4 lie on a quadrangle of K_8. Then $E(K_8 - f_1 - f_2 - f_3 - f_4) > E(K_8)$.

By designing hyperenergetic graphs we are still far from characterizing graphs with maximal energy. Computer–aided searches of such graphs were performed [28,36] and the maximum–energy graphs up to $n = 12$ reported. Their inspection, however, gives no clue about the structure of such graphs in the general case.

Another approach to the same problem was done in [41], where edges were added one–by–one, in a random manner, starting with $\overline{K_n}$ (for which $m = 0$) and ending with K_n (for which $m = n(n-1)/2$). This construction produces labeled (n, m)-graphs uniformly at random.

By repeating the construction many times, a statistical regularity emerges: if $n \geq 9$ then by varying m between 0 and $n(n-1)/2$ the expectation value of the energy of a random (n, m)-graph first increases, attains a maximum E_{max} at some $m = m_{max}$ and then decreases. The following approximate behavior was found for $9 \leq n \leq 30$ [41]:

$$E_{max} \approx A\, n^p \qquad A = 0.733 \pm 0.007 \qquad p = 1.390 \pm 0.004$$

$$m_{max} \approx B\, n^q \qquad B = 0.47 \pm 0.03 \qquad q = 1.87 \pm 0.02\ .$$

These results give a hint where one should look for graphs with maximum energy, but the structure of these graphs remains almost completely obscure.

In the author's opinion, finding and characterizing the maximum–energy graphs is the most challenging open problem in the, now more than half a century old, theory of graph energy. At this stage the solving of the problem deserves to be attacked both by means of computer–aided combinatorial search and optimization algorithms, and by the proof techniques of algebraic graph theory.

It is easy to prove that for v being any vertex of the graph G, $E(G-v) \leq E(G)$. Another hard–to–crack problem would be to characterize the graphs G and their edges e for which $E(G-e) \leq E(G)$.

References

1. Cvetković, D., Doob, M., Sachs, H.: Spectra of Graphs – Theory and Application. Academic Press, New York, 1980; 2nd revised ed.: Barth, Heidelberg, 1995
2. Streitwieser, A.: Molecular Orbital Theory for Organic Chemists. Wiley, New York, 1961
3. Coulson, C. A., O'Leary, B., Mallion, R. B.: Hückel Theory for Organic Chemists. Academic Press, London, 1978
4. Günthard, H. H., Primas, H.: Zusammenhang von Graphentheorie und MO–Theorie von Molekeln mit Systemen konjugierter Bindungen. Helv. Chim. Acta **39** (1956) 1645–1653

5. Gutman, I., Polansky, O. E.: Mathematical Concepts in Organic Chemistry. Springer–Verlag, Berlin, 1986
6. Dias, J. R.: Molecular Orbital Calculations Using Chemical Graph Theory. Springer–Verlag, Berlin, 1993
7. Gutman, I.: The energy of a graph. Ber. Math.–Statist. Sekt. Forschungszentrum Graz **103** (1978) 1–22
8. Coulson, C. A.: On the calculation of the energy in unsaturated hydrocarbon molecules. Proc. Cambridge Phil. Soc. **36** (1940) 201–203
9. Sachs, H.: Beziehungen zwischen den in einem Graphen entaltenen Kreisen und seinem charakteristischen Polynom. Publ. Math. (Debrecen) **11** (1964) 119–134
10. Gutman, I., Trinajstić, N.: Graph theory and molecular orbitals. XV. The Hückel rule. J. Chem. Phys. **64** (1976) 4921–4925
11. Gutman, I.: A class of approximate topological formulas for total π-electron energy. J. Chem. Phys. **66** (1977) 1652–1655
12. Gutman, I.: Proof of the Hückel rule. Chem. Phys. Lett. **46** (1977) 169–171
13. Gutman, I.: Remark on the moment expansion of total π-electron energy. Theor. Chim. Acta **83** (1992) 313–318
14. Hall, G. G.: The bond orders of alternant hydrocarbon molecules. Proc. Roy. Soc. London **A 229** (1955) 251–259
15. Stepanov, N. F., Tatevskii, V. M.: Decomposing the energy of π-electrons in terms of bonds in the simplest variant of the molecular orbital method (in Russian). Zh. Strukt. Khim. **2** (1961) 204–208
16. Gutman, I., Trinajstić, N.: Graph theory and molecular orbitals. Total π-electron energy of alternant hydrocarbons. Chem. Phys. Lett. **17** (1972) 535–538
17. Gutman, I., Trinajstić, N.: Graph theory and molecular orbitals. The loop rule. Chem. Phys. Lett. **20** (1973) 257–260
18. Y. Jiang, Y., Tang, A., Hoffmann, R.: Evaluation of moments and their application to Hückel molecular orbital theory. Theor. Chim. Acta **65** (1984) 255–265
19. Schmalz, T. G., Živković, T., Klein, D. J.: Cluster expansion of the Hückel molecular orbital energy of acyclics: Application to pi resonance theory. Stud. Phys. Theor. Chem. **54** (1988) 173–190
20. Gutman, I.: Acyclic systems with extremal Hückel π-electron energy. Theor. Chim. Acta **45** (1977) 79–87
21. Gutman, I.: Partial ordering of forests according to their characteristic polynomials. in: Hajnal, A., Sós, V. T. (Eds.), Combinatorics, North–Holland, Amsterdam, 1978, pp. 429–436
22. Zhang, F.: Two theorems of comparison of bipartite graphs by their energy. Kexue Tongbao **28** (1983) 726–730
23. Zhang, F., Lai, Z.: Three theorems of comparison of trees by their energy. Science Exploration **3** (1983) 12–19
24. Gutman, I., Zhang, F.: On a quasiordering of bipartite graphs. Publ. Inst. Math. (Beograd) **40** (1986) 11–15
25. Gutman, I., Zhang, F.: On the ordering of graphs with respect to their matching numbers. Discr. Appl. Math. **15** (1986) 25–33
26. Zhang, Y., Zhang, F., Gutman, I.: On the ordering of bipartite graphs with respect to their characteristic polynomials. Coll. Sci. Papers Fac. Sci. Kragujevac **9** (1988) 9–20
27. McCelland, B. J.: Properties of the latent roots of a matrix: The estimation of π-electron energies. J. Chem. Phys. **54** (1971) 640–643

28. Caporossi, G., Cvetković, D., Gutman, I., Hansen, P.: Variable neighborhood search for extremal graphs. 2. Finding graphs with extremal energy. J. Chem. Inf. Comput. Sci. **39** (1999) 984–996
29. Gutman, I. Bounds for total π-electron energy. Chem. Phys. Lett. **24** (1974) 283–285
30. Gutman, I.: Bounds for total π-electron energy of conjugated hydrocarbons. Z. Phys. Chem. (Leipzig) **266** (1985) 59–64
31. Türker, L.: An upper bound for total π-electron energy of alternant hydrocarbons. Commun. Math. Chem. (MATCH) **16** (1984) 83–94
32. Gutman, I., Türker, L., Dias J. R.: Another upper bound for total π-electron energy of alternant hydrocarbons. Commun. Math. Chem. (MATCH) **19** (1986) 147–161
33. Gutman, I.: Bounds for Hückel total π-electron energy. Croat. Chem. Acta **51** (1978) 299–306
34. D. A. Bochvar, D. A., Stankevich, I. V.: Approximate formulas for some characteristics of the electronic structure of molecules. 1. Total electron energy (in Russian). Zh. Strukt. Khim. **21** (1980) 61–66
35. Gutman, I.: Total π-electron energy of benzenoid hydrocarbons. Topics Curr. Chem. **162** (1992) 29–63
36. Cvetković, D., Gutman, I.: The computer system GRAPH: A useful tool in chemical graph theory. J. Comput. Chem. **7** (1986) 640–644
37. Walikar, H. B., Ramane, H. S., Hampiholi, P. R.: On the energy of a graph. in: Balakrishnan, R., Mulder, H. M., Vijayakumar, A. (Eds.), Graph Connections, Allied Publishers, New Delhi, 1999, pp. 120–123
38. Sachs, H.: Über Teiler, Faktoren und charakteristische Polynome von Graphen, Teil II. Wiss. Z. TH Ilmenau **13** (1967) 405–412
39. Gutman, I.: Hyperenergetic molecular graphs. J. Serb. Chem. Soc. **64** (1999) 199–205
40. Gutman, I., Pavlović, L.: The energy of some graphs with large number of edges. Bull. Acad. Serbe Sci. Arts (Cl. Math. Natur.) **118** (1999) 35–50.
41. Gutman, I., Soldatović, T., Vidović, D.: The energy of a graph and its size dependence. A Monte Carlo approach. Chem. Phys. Lett. **297** (1998) 428–432

Bethe Ansatz: Quasi-particles, Spectral Parameters and the Hypothesis of Strings

M. Kuźma, B. Lulek, and T. Lulek

Institute of Physics, Pedagogical University
ul. Rejtana 16A, 35-359 Rzeszów, Poland

Abstract Some combinatoric and physical aspects of two forms of the Bethe Ansatz formalisms have been presented. The first approach expressed in the terms of quasi-momenta p_α leads to a set of transcendental equations whose solutions are very tedious to find. Therefore for decreasing these difficulties the second approach is formulated consisting in the transformation of Bethe equations to a combinatoric set of equations for the spectral parameters. In the asymptotic limit the Bethe Ansatz solutions satisfy the hypothesis of strings.

1 Introduction

The one-dimensional finite Heisenberg magnetic ring [1] is one of the fundamental models in Quantum Mechanics. The exact solution of the eigenproblem of this model was given by Bethe [2] and is called the "Bethe Ansatz". A lot of later papers were based on this solution and gave rise to important results in mathematics and physics. The ground state of the antiferromagnetic system was studied by Hulthen [3] and Cloiseau and Pearson [4]. Orbach [5] and Baxter [6] have generalized the Bethe-Hulthen method on an anisotropic case. Bound states (the case of nonreal quasi-momenta) were studied by Ovchinnikov [7]. Takahashi [8] invastigated the thermodynamics of the one-dimensional Heisenberg model.

The Bethe's hypothesis was applied in the several problems and models of modern theoretical physics. First Yang and Yang [9] investigated the thermodynamics of the one-dimensional Bose gas using BA and then application of BA in statistical physics became very popular [10,11]. Another applications should be mentioned like Kondo model of magnetic impurities [12,13], Hubbard model of itinerant electrons [14,15], solitons [16], Hofstadter butterfly [17]. etc.

Bethe Ansatz can be formulated in two mathematical formalisms. In the approach of Bethe's [2], the eigenstates are solutions of a set of the transcendental equations and they are parametrized by pseudomomenta p_α and phases $\phi_{\alpha\alpha'}$. The second approach transfers this set to the combinatoric set of equations for the spectral parameters λ [18,19]. It is the so called algebraic Bethe Ansatz. This form of BA bases on the quantum method of the inverse problem [20,21]. Solutions of this set of equations parametrise eigenfunctions of the Heisenberg magnetic Hamiltonian. These solutions are also parametrised by combinatoric objects, so called rigged configurations [22,24,25].

In Sec. 2 of this paper we introduce briefly these two descriptions of BA, and discuss the genesis of the pseudo-particles (quasi-particles). In Sec. 3 the structure of configuration space of pseudo-particles is presented. The difference between this space and the classical configuration space for particles in quantum mechanics is stressed. Based on the properties of this space the model of "nearest-neighbors" (n-n) is introduced in Sec. 4. In Sec. 5 the properties of the spectral parameters and the hypothesis of strings are presented.

2 Algebraic form of the Bethe Ansatz and quasiparticles

We consider the finite one-dimensional Heisenberg model with the nearest-neighbor Hamiltonian for spin 1/2. Let N be a number of nodes on a chain. Then the Hamiltonian of the system is

$$\hat{H} = J \sum_{j \in \tilde{N}} \left(\hat{\boldsymbol{s}}_j \hat{\boldsymbol{s}}_{j+1} - \frac{1}{4} \right).$$

The periodic boundary conditions require that $\hat{\boldsymbol{s}}_{N+1} = \hat{\boldsymbol{s}}_N$. Bethe's solution of the eigenvalue problem of (2) is given by eigenfunctions ψ:

$$\psi = \sum_{1 \le j_1 \le j_2 \le \ldots \le j_r \le N} a(j_1, j_2, \ldots, j_r) \cdot \hat{\boldsymbol{s}}^-_{j_1} \hat{\boldsymbol{s}}^-_{j_2} \ldots \hat{\boldsymbol{s}}^-_{j_r} |0\rangle ,$$

where $a(j_1, j_2, \ldots, j_r)$ are amplitudes of Ising states $\hat{\boldsymbol{s}}^-_{j_1} \hat{\boldsymbol{s}}^-_{j_2} \ldots \hat{\boldsymbol{s}}^-_{j_r} |0\rangle$, j_α, $\alpha = 1, 2, \ldots, r$ are positions of the reversed z-spin projection in the chain, r – the number of the reversed spins. The amplitudes are expressed by quasi-momenta $p = \{p_\alpha | \alpha \in \tilde{r}\}$ and phases $\phi_{\alpha\alpha'}$:

$$a(j_1, j_2, \ldots, j_r) = \sum_{\pi \in \Sigma_r} A_\pi \exp\left(\mathrm{i} \sum_{\alpha \in \tilde{r}} p_{\pi(\alpha)} j_\alpha \right),$$

where amplitudes A_π are

$$A_\pi = \exp\left(\mathrm{i} \sum_{\substack{\alpha' > \alpha \\ \pi(\alpha') < \pi(\alpha)}} \phi_{\alpha\alpha'} \right)$$

and π are permutations of 1, 2, ..., r.

The energy eigenvalue of the system is given by the formula:

$$E = -J \sum_{\alpha=1}^{r} (1 - \cos p_\alpha).$$

Quasi-momenta p_α and phases $\phi_{\alpha\alpha'}$ should satisfy the relations:

$$2\cot(\phi_{\alpha\alpha'}/2) = \cot(p_\alpha/2) - \cot(p_{\alpha'}/2)$$

and

$$\mathrm{e}^{\mathrm{i}p_\alpha N} = \prod_{\alpha\neq\alpha'} \exp(\mathrm{i}\phi_{\alpha\alpha'}), \qquad \alpha = 1, 2, \ldots, r.$$

Introducing parameters $\lambda_1, \lambda_2, \ldots, \lambda_r$, which are defined by $\lambda_\alpha = \cot(p_\alpha/2)$, we get the algebraic form of Bethe equations [8,19]:

$$\frac{(\lambda_\alpha + \mathrm{i})^N}{(\lambda_\alpha - \mathrm{i})^N} = \prod_{\alpha'\neq\alpha} \frac{(\lambda_\alpha - \lambda'_\alpha + 2\mathrm{i})}{(\lambda_\alpha - \lambda'_\alpha - 2\mathrm{i})}, \qquad \alpha \in \tilde{r},$$

which gives the energy:

$$E = \sum_{\alpha=1}^{r} \left(-\frac{2J}{\lambda_\alpha^2 + 1} \right).$$

The reflection conditions (2) in terms of the parameters λ take the form:

$$\lambda_\alpha - \lambda_{\alpha'} = 2\cot(\phi_{\alpha\alpha'}/2)\ .$$

Eqs. (2)-(2) form an algebraic BA.

According to formula (2), (2) and (2), each eigenstate of the original Heisenberg Hamiltonian with fixed M can be treated as an ensemble of pseudoparticles with some, possibly complex, pseudomomenta p_α, which move freely up to mutual scattering. The scatterings are two body interactions. The pseudoparticles have several properties which are distinct from those for the obvious particles in physics:

1. The scattering of two quasiparticles preserves momenta, but generates an exchange of phases of the free motion.
2. They have a hard core what means that they cannot be placed together on the same node of the chain. Therefore the space of the scattering is very special, as described by the reflection conditions (2).
3. Pseudoparticles are really indistinguishable (the denotation of them works only between scatterings, after a scattering one can not distinguish between scattered particles).
4. Amplitudes A_π of quasiparticles are in a relation with a statistics of particles. Since A_π are neither 1 nor $(-1)^\pi$, these "particles" have no definite statistics.
5. Pseudomomenta do not fulfill the cyclic Born-Karman boundary conditions. Instead of these, the following are valid:

$$p_\alpha - \frac{1}{N} \sum_{\alpha'\neq\alpha} \phi_{\alpha\alpha'} = 2\pi \frac{n_\alpha}{N},$$

where $n_\alpha \in \mathbb{Z}/N\mathbb{Z} \equiv \tilde{N}$ and $\mathbb{Z}/N\mathbb{Z}$ is a set of quantum numbers (the winding numbers). This means that the quantized observable is not a pure momentum, but a combined effect of free motion and a correction arising from scattering.

6. The set of pseudomomenta contains not only real, but also complex values $p_\alpha = p'_\alpha + \mathrm{i}p''_\alpha$.

3 Structure of the configuration space of pseudoparticles

Let

$$\tilde{N} = \{j = 1, 2, \ldots, N\}$$

be a set of all nodes of the Heisenberg linear chain, and let

$$\tilde{2} = \{+, -\}$$

denote the set of all projections of the single-node spin $s = 1/2$. $\tilde{N}$ forms a regular orbit of the cyclic group C_N and therefore we refer to it as to the crystal. Assuming that we have r particles (r inversed spins) in a crystal the classical configuration space is formed by the Cartesian product

$$\tilde{N}^{\times r} = \{(j_1, j_2, \ldots, j_r) | j_\alpha \in \tilde{N}, \alpha \in \tilde{r}\},$$

where $\tilde{r} = \{\alpha = 1, 2, \ldots, r\}$ and $(j_1, j_2, \ldots, j_r)$ is a point in this space.

Eq. (3) is not the space of positions of the Heisenberg chain because of the properties of quasi-particles listed in previous section. First, due to the "hard core" property we have to eliminate the fat diagonal $D(r)$ from the space (3). Secondly, we have to consider the indistinguishability property by introducing an action of the Pauli group Σ_N. The set of orbits of this action [26,27] forms a true configuration space $Q^{(r)}$ of all magnetic configurations with r spin deviations

$$Q^{(r)} = \left(\tilde{N}^{\times r} \backslash D(r)\right) / \Sigma_r.$$

These orbits can be presented by their representatives: $(j_1, j_2, \ldots, j_r)$ for $1 \leq j_1 \leq j_2 \leq \ldots \leq j_r \leq N$. Thus

$$Q^{(r)} = \{(j_1, j_2, \ldots, j_r) | 1 \leq j_1 \leq j_2 \leq \ldots \leq j_r \leq N\}.$$

The set of all magnetic configurations of the crystal $\tilde{N}$ is formed by the set of mappings f

$$\tilde{2}^{\tilde{N}} = \{f : \tilde{N} \to \tilde{2}\}.$$

Therefore the space of all quantum states of the magnet is given by the linear closure of the set $\tilde{2}^{\tilde{N}}$ over the field $\mathbb{C}$ of complex numbers,

$$\mathcal{H} = \mathrm{lc}_{\mathbb{C}}\, \tilde{2}^{\tilde{N}},$$

where $\tilde{2}^{\tilde{N}}$ is an orthonormal basis in $\mathcal{H}$. The space of all magnetic configurations with r spin deviations is given by

$$\mathcal{H}_r = \mathrm{lc}_{\mathbb{C}}\, Q^{(r)}.$$

The space $\mathcal{H}_r$ is invariant under the action of the Heisenberg Hamiltonian $\hat{H}$ (2). Therefore the BA solutions for eigenvectors of $\hat{H}$ can be presented in the form

$$|\psi\rangle = \sum_{f \in Q^{(r)}} a_f \, |f\rangle ,$$

where $f \in Q^{(r)}$ is a magnetic configuration

$$|f\rangle = |j_1, j_2 \ldots, j_r\rangle , \qquad 1 \leq j_2 \leq j_2 \leq \ldots \leq j_r \leq N.$$

Thus, one can interpret $Q^{(r)}$ as the classical configuration space for the system of r pseudoparticles on the crystal $\tilde{N}$, and $\mathcal{H}_r$ is the corresponding space for the system quantized along the Schroedinger picture of Quantum Mechanics.

4 "Nearest Neighbour" model

Let f' be a nearest neighbor (n-n) of the configuration $f \in Q^{(r)}$. An f' is, by the definition, a n-n of f if the distribution of spin deviations in the crystal $\tilde{N}$ for f' differs from that for f by a single jump of a single deviation by one lattice constant. Then the action of the Hamiltonian (2) on f yields

$$\frac{2}{J} \hat{H} \, |f\rangle = \sum_{f' \in Q_f^{(r)}} (|f'\rangle - |f\rangle),$$

where $Q_f^{(r)}$ is the set of all n-n configurations of f.
The relation (4) determines the dynamics of the system and describes the motion laws of pseudo-particles. This relation can be considered as a "projection" of the Heisenberg model dynamics on a space with r deviations. One can notice too that matrix elements of (4) are integers what allows to treat the problem by the combinatoric methods.

5 Hypothesis of strings

Each Bethe solution is determined by a set of r values of the spectral parameters λ. This set, due to a physical reasons, is decomposed into subsets called strings.

A string is described by its length l associated with a number of particles bounded into a single object being just a string [25]. Let m_l be the number of strings of the length l, so that

$$r = \sum_{l=1}^{\infty} l \cdot m_l$$

defines a partition ν of r. The partition ν is called a string configuration. The length of string is associated with its spin by the formula

$$s = \frac{l-1}{2}, \qquad (\text{or } l = 2s+1).$$

The Bethe Ansatz hypothesis of strings consists in observation that in the asymptotic limit $N \to \infty$ spectral parameters achieve the form

$$\lambda_m^{l\nu} = \lambda^{l\nu} + im + O(\mathrm{e}^{-\delta N}),$$

where $\delta > 0$, $\lambda^{l\nu}$ is real, m is an element of the set $\{-s, -s+1, \ldots, s\}$, and $O(\mathrm{e}^{-\delta N})$ is a remainder which vanishes exponentially.

Each string is classified by a quantum number L called admissible rigging

$$0 < L < P_L,$$

where $P_L = N - 2Q_L$ is the boundary and Q_L is the number of cells in the first L columns of the diagram ν. Denoting by $Z(\nu)$ the number of admissible riggings of the configuration ν one can obtain

$$Z(\nu) = \binom{P_L + m_l}{m_l}.$$

The number $Z(\nu)$ determines the dimension of a irreducible representation Δ^λ of the symmetric group Σ_N

$$[\Delta^\lambda] = \sum_{\nu \to r} Z(\nu).$$

The dimension (5) is the number of states with spin s and the number r.

6 Conclusions

The solutions of the Bethe Ansatz are parametrized by a set of r numbers of pseudomomenta p_α or by spectral parameters λ_α. The first form of solutions allows us to treat the magnetic states as a pseudo-particles with several properties which distinguish them from ordinary particles in Quantum Mechanics. The introduction of the spectral parameters to the description of the Bethe solution allows us to apply combinatoric methods for a calculation of the problem, making it to be easier and preserve a good classification of states.

References

1. W. Heisenberg, Z. Phys. **49**, 619-636 (1928).
2. H. Bethe, Z. Phys. **71**, 205-226 (1931).
3. L. Hulthen, Arkiv Mat., Astron. Fysik **26A**, No 11, 1-106 (1938).
4. J. Des Cloizeaux and J.J. Pearson, Phys. Rev. **128**, No 5, 2131-2135 (1958).
5. R. Orbach, Phys. Rev. **112**, No 2, 309-316 (1958).
6. R.J.Baxter, Ann. Phys. **70**, No 2, 323-327 (1972).
7. A.A. Ovchinnikov, Zh. Eksp. Teor. Fiz. **56**, No 4, 1154-1365 (1969).
8. M. Takahaski, Progress of Theoretical Physics **46**, No 2 (1971).
9. C.N. Yang, C.P. Yang, Phys. Rev. **150**, No 1, 321-327 (1966); 150, No 1, 327-339 (1966).
10. R. Baxter, Exactly Solvable Models in Statistical Mechanics, New York, Academic Press, 1982.
11. C.N. Yang, M.L. Ge (Eds) Braid Group, Knot Theory and Statistical Mechanics, World Sci., Sinapore 1989 (vol. I) and 1994 (vol. II).
12. A.N. Kirillov, LOMI **131**, 88-105 (1983).
13. P. Schlottmann, Phys.Rev.Lett. **21**, 1697-1700 (1983).
14. F. Woynarovich and H.P. Edde, J. Phys. A: Math. Gen. **20**, L443-449 (1987).
15. M.M. Canchesi, A. Avella, F. Manchini, Europhys. Lett. **44**, 328-334 (1987).
16. L. Fadeev and L. Takhtajan, Hamiltonian Methods in the Theory of Solitons, Springer Berlin 1987.
17. A.V. Zabrodin in: T. Lulek, W. Florek, S. Walcerz (Eds), Symmetry and Structural Properties of Condensed Matter, World Sci., Singapore 1995, pp. 197-204.
18. L.A. Takhtadzhyan, L.D. Faddeev, Usp. Mat. Nauk **34**, No 5, 13-63 (1979).
19. L.D. Faddeev, L.A. Takhtadzhyan, Zapiski Nauchnykh Seminarov Leningradskogo Otdeleniya Matematicheskogo Instituta **109**, pp.134-178 (1981).
20. L.D. Faddeev, Soviet Scientific Reviews, Harvard Academic, London, pp. 107-155 (1980).
21. K.K. Sklyanin, L.A. Takhtadzhyan and L.D. Faddeev, Teor. Mat. Fiz. **40**, No 2, 194-220 (1979).
22. L.A. Takhtadzhyan and L.D. Faddeev, J. Sov. Math. **24**, No 2 (1984).
23. P.P. Kulish and N.Yu. Reshetikhin, J. Phys. A **16**, 591-596 (1983).
24. A.N. Kirillov, in Automorphic Functions and Number Theory, II, J. Sov. Math. **36**, No 1 (1987).
25. S.V. Kerov, A.N. Kirillov and N.Yu. Reshetikhin, Zapiski Nauchnykh Seminarov Leningradskogo Otdeleniya Matematicheskogo Instituta AN SSSR **155**, pp.50-64 (1986).
26. T. Lulek, The duality of Weyl and Bethe Ansatz for finite Heisenberg chain, to appear in: B. Lulek, T. Lulek, A. Wal, Symmetry and Structural Properties of Condensed Matter, World Sci., Singapore 1999.
27. T. Lulek, Molecular Physics Reports **23**, 56-61 (1999).

Ordering the Affine Symmetric Group

Alain Lascoux

C.N.R.S., Institut Gaspard Monge, Université de Marne-la-Vallée,
5 Bd Descartes, Champs sur Marne, 77454 Marne La Vallée Cedex 2 FRANCE
Alain.Lascoux@univ-mlv.fr

Abstract We review several descriptions of the affine symmetric group. We explicit the basis of its Bruhat order.

1 Introduction

For a group defined by generators, one has a notion of *Cayley graph* representing the succesive generation of elements of the group by product of generators. This is specially interesting in the case of a Coxeter group $(W, \mathcal{S})$, $\mathcal{S}$ being the set of generators, because one has a length function which filters the elements of W according to the minimum length of the words in the generators of W which expresses them (words of minimum length are called *reduced decompositions*).

The Cayley graph of W naturally provides an ordering (the *weak order*, cf. [1] for more details) of W: $w' \le w$ iff any reduced decomposition of w' is the left factor of at least one reduced decomposition of w, in other words, iff there is a path from w' to w in the Cayley graph of W.

However, if one takes the group algebra of W instead of W, and a reduced decomposition $[i, j, \dots, h] := s_i s_j \cdots s_h$ of an element w, then one wants the order to be such that any expression of the type $(1 + s_i)(1 + s_j) \cdots (1 + s_h)$ has leading term (with respect to the order) w.

In that case, it is clear that the expansion of $(1 + s_i) \cdots (1 + s_h)$ involves only the elements obtained by evaluating in W the subwords of $[i, j, \dots, h]$. One easily checks that the set of such elements is independent of the choice of a reduced decomposition of w (though the set of subwords is not).

Let us write $w' \le w$ (for the *Bruhat order*) iff there exists a decomposition of w' which is a subword of a reduced decomposition of w. Let us also write $[1, w]$ for the interval $\{w' \,|\, 1 \le w' \le w\}$, 1 being the identity element of W.

The above definition of the Bruhat order [3] is not very efficient, since one has much too many subwords of a reduced decomposition of w compared to the cardinality of $[1, w]$.

Ehresmann[5] found another definition, in the case of the symmetric group, by using all the projections $W \xrightarrow{p_i} W/W_i$ onto all the quotient spaces W/W_i (W_i= maximal parabolic, i.e. group generated by the elements of $\mathcal{S} \setminus s_i$). Ehresmann's definition is

$$w' \le w \;\Leftrightarrow\; p_i(w') \le p_i(w) \;,\; i = 1, 2, \dots \tag{1}$$

Now, it makes sense because one has an easy description of the Bruhat order on W/W_i and of the projection. In the case of the symmetric group, W/W_i, as an ordered set, can be identified with the space of all partitions, the diagram of which are included in a fixed rectangle (the order on partitions being inclusion of diagrams).

This description has been generalized by Deodhar[4] to all Coxeter groups.

In [13], M.P. Schützenberger and I gave another description. It uses a notion which is valid for any ordered set X. One defines the basis $\mathcal{B}$ of X to be the following subset:

$$b \in \mathcal{B} \Leftrightarrow \exists x \in X,\ b \text{ is minimum in the complement of } [\leq x] . \quad (2)$$

Given $x \in X$, let furthermore $\mathcal{B}(x)$ denote the set $\{b \in \mathcal{B},\, b \leq x\}$. We have shown in [13] that in the case X is finite, then

$$x \leq y \Leftrightarrow \mathcal{B}(x) \subseteq \mathcal{B}(y) . \quad (3)$$

Now, it happens that in the case of a Coxeter group W the set $\mathcal{B}(w)$ is "small" compared to $[1, w]$ or to the set of subwords of a reduced decomposition of $w \in W$.

The bases of the symmetric group and of the hyperoctahedral group, as well as the map $w \mapsto \mathcal{B}(w)$ are given in [13]. Geck and Kim [9] have determined the bases of all finite Coxeter groups.

The construction of bases passes to the quotient spaces W/W_i (and, as we have said, it is sufficient to treat only these spaces, because of (1)). For example, for the symmetric group $\mathfrak{S}_n$, its quotient $\mathfrak{S}_n/\mathfrak{S}_m \times \mathfrak{S}_r$, $n = m + r$, can be identified with the set of partitions contained in m^r, and its basis consists of all (non void) rectangular partitions i^j: $i \leq m$, $j \leq r$. The elements of the basis have in fact a geometrical meaning, being in correspondence with determinantal varieties [11].

Now, the notion of a basis can be extended to infinite sets X, when X satisfy some mild hypotheses.

Let us take an affine Coxeter group. Then one still has a length function, which allow to filter the group according to length. Let W^ℓ be the set of elements of length $\leq \ell$. If $\ell(w) \leq \ell$, then the full interval $[1, w]$ is contained in W^ℓ. Let $\mathcal{B}^\ell$ be the basis of W^ℓ.

One has that $\mathcal{B} = \cup \mathcal{B}^\ell$ and one can check that

$$w' \leq w \Leftrightarrow \mathcal{B}(w') \subseteq \mathcal{B}(w) . \quad (4)$$

In the following, we shall describe the basis of the affine symmetric group $\widetilde{\mathfrak{S}}_n$ and its Bruhat order.

2 The Affine Symmetric Group

Let n be an integer and $\widetilde{\mathfrak{S}}_n$ be the quotient of the infinite symmetric group (generated by s_i, $i \in \mathbb{Z}$ under the relations $s_i = s_{i+n}$), $i \in \mathbb{Z}$. In other words,

$\widetilde{\mathfrak{S}}_n$ is generated by $s_0, \dots s_{n-1}$ satisfying the braid relations (with $s_n := s_0$)

$$s_i s_{i+1} s_i = s_{i+1} s_i s_{i+1} \ , \ s_i s_j = s_j s_i \ , \ |i-j| \ncong 1 \text{ mod n} \ , \ 0 \le \text{i}, \text{j} \le \text{n}-1. \quad (5)$$

The group $\widetilde{\mathfrak{S}}_n$ can be thought as acting on infinite vectors $v = [\dots, v_i, \dots]$, $i \in \mathbb{Z}$: the generator s_i transposes in v components v_{i+kn}, v_{i+1+kn} for all $k \in \mathbb{Z}$ simultaneously. Usually, one takes vectors v such that $v_{i+n} = v_i + n$, $i \in \mathbb{Z}$. Associating to the identity element the "vacuum vector"$\triangleright : \triangleright_i = i$, $\forall i \in \mathbb{Z}$, then the orbit of $\triangleright$ is in bijection with $\widetilde{\mathfrak{S}}_n$ and one can code any element $\sigma \in \widetilde{\mathfrak{S}}_n$ by the image $v = \triangleright\sigma$ of $\triangleright$ under σ, or equivalently by the *window* $[v_1, \dots, v_n]$.

In other words, elements of $\widetilde{\mathfrak{S}}_n$ can be coded by vectors in $\mathbb{Z}^n$, starting from $[1, 2, \dots, n]$, with the action

$$\begin{aligned} [\cdots v_i, v_{i+1} \cdots] \, s_i &= [\cdots v_{i+1}, v_i \cdots] \, , \quad 1 \le i \le n-1 \\ [v_1 \, , \, \dots \, , \, v_n] \, s_0 &= [v_n - n \, , \, \dots \, , \, v_1 + n] \, . \end{aligned} \quad (6)$$

One has in fact many other choices to represent $\widetilde{\mathfrak{S}}_n$ by infinite vectors satisfying periodicity conditions. One can also modify the definition of "transposing two adjacent components".

For example, let $c(\triangleright)$ be the infinite vector with all components equal to 0, and let $c(\sigma)$ be the image of $c(\triangleright)$ under the action

$$[\cdots c_j, c_{j+1} \cdots] \, s_i = [\cdots c_{j+1}{+}1, c_j \cdots] \, , 1 \le i \le n{-}1, c_i \le c_{i+1}. \forall j \cong i \, mod \, n \, . \quad (7)$$

This action preserves equalities $c_j = c_{j+n}$, and once more, one can restrict vectors to their components $1, \dots, n$. In that set-up, one has

$$[c_1 \, , \, \dots \, , \, c_n] \, s_0 = [c_n \, , \, \dots \, , \, c_1 + 1] \, , c_n \le c_1 \, . \quad (8)$$

Now, it is easy to check that the mapping $\widetilde{\mathfrak{S}}_n \ni \sigma \mapsto [c_1, \dots, c_n] \in \mathbb{N}^n$ is a bijection. It can be easily described without passing through a reduced decomposition of σ (cf. [2]).

By restriction, it induces a bijection from $\mathfrak{S}_n$ onto the set of vectors such that $c_1 \le n-1, \ \dots, c_n \le 0$. Such a vector is called the *Lehmer code* of the corresponding permutation σ [10] and encodes the "inversions"of σ. This is still true in the infinite case (cf. [2], [6]). Let us call *scalar weight* of $c(\sigma)$ the sum $c_1 + \cdots + c_n$. Then the above action shows that the scalar weight increases by 1 if length increases, and thus the scalar weight of $c(\sigma)$ is equal to the length $\ell(\sigma)$. Shi [15] gives another description of the length function, directly from the window of σ.

3 Quotient of $\widetilde{\mathfrak{S}}_n$ modulo $\mathfrak{S}_n$

Since all s_i's play a symmetrical rôle, all quotient spaces of $\widetilde{\mathfrak{S}}_n$ modulo a maximal parabolic are isomorphic. Let us choose the finite symmetric group $\mathfrak{S}_n$, generated by $s_1, \dots, s_{n-1}$, as a maximal parabolic.

The unique element of minimum length of a class $\sigma\,\mathfrak{S}_n$ is the unique element ζ of the class such that

$$\ell(\zeta\, s_i) < \ell(\zeta) \Leftrightarrow i = 0\,.$$

Therefore, minimum elements are exactly those elements with (strictly) increasing window $[v_1, \ldots, v_n]$: $v_1 < v_1 < \cdots < v_n$.

From (8) one deduces that codes of such elements are (weakly) increasing vectors $[c_1, c_2, \ldots, c_n]$ such that $c_1 = 0$, that is, are partitions of length $\leq n-1$.

Now, there is a third coding of $\widetilde{\mathfrak{S}}_n/\mathfrak{S}_n$ by infinite vectors u, with periodicity $u_{j+n} = u_j - 1$, $\forall j \in \mathbb{Z}$. It amounts to take origin vector such that $[u_1, \ldots, u_n] = [0, \ldots, 0]$, and act by

$$\begin{aligned} [\cdots u_i, u_{i+1} \cdots] &\xrightarrow{s_i} [\cdots u_{i+1}\,,\, u_i \cdots]\;, 1 \leq i \leq n-1\,, \\ [u_1, \ldots, u_n] &\xrightarrow{s_0} [u_n + 1, \ldots, u_1 - 1]\,. \end{aligned} \tag{9}$$

From a vector u in $\mathbb{Z}^n$ such that $u_1 + \cdots + u_n = 0$ (call *weight* such a vector), one can recover an element of $\mathfrak{S}_n\backslash\widetilde{\mathfrak{S}}_n$ as follows (the quotient is now on the left): one has to sort increasingly the vector $[1 + nu_1, \ldots, n + nu_n]$.

For example, $n = 3$, $u = [7, -4, -3]$ gives the vector $[22, -10, -6] = [1 + 3 \times 7, 2 - 3 \times 4, 3 - 3 \times 3]$ which is sorted into $[-10, -6, 22]$, and is an element of $\widetilde{\mathfrak{S}}_3$ of minimum length (of code $[0, 1, 19]$) in its coset modulo $\mathfrak{S}_3$.

B. Leclerc has shown me that it is more convenient, in the theory of roots of affine Weyl groups, to add an "imaginary root"and use "extended weights"which are vectors in $\mathbb{Z}^{n+1}$: one takes for origin $[0, \ldots, 0] \in \mathbb{Z}^{n+1}$ and the action is

$$\begin{aligned} [\cdots w_i, w_{i+1} \cdots] &\xrightarrow{s_i} [\cdots w_{i+1}\,,\, w_i \cdots]\;, 1 \leq i \leq n-1\,, \\ [w_1, \ldots, w_n, w_{n+1}] &\xrightarrow{s_0} [w_n + 1, \ldots, w_1 - 1, w_1 - w_n + w_{n+1} - 1]\,. \end{aligned} \tag{10}$$

A weight $u \in \mathbb{Z}^n$ already has an unnecessary component, since the sum of components is zero : one can write $u = x_1\alpha_1 + \cdots + x_{n-1}\alpha_{n-1}$, the α_i being the simple roots $[1, -1, 0, \ldots, 0], \ldots, [0, \ldots, 0, 1, -1]$.

However, it happens that going to $\mathbb{Z}^{n+1}$ simplifies some constructions, the extra component having combinatorial interpretations.

4 n-Cores and Bruhat order

None of these codings furnishes a simple description of the Bruhat order. Fortunately, Fock spaces and crystal graphs indicate a fourth one ([14], [12], [7]). Now one will represent elements of $\widetilde{\mathfrak{S}}_n/\mathfrak{S}_n$ by partitions which are n-*cores*, that is partitions such that one cannot erase a border strip (also called outer ribbon) of length n from their diagram (cf. [8]).

One numbers diagonals of diagrams with the numbers (which are called *colours*) $0, 1, \ldots, n-1$ periodically, the main diagonal being of colour 0.

Now, take the diagram of a an n-core such that no corner box is of colour i. Then the action of s_i, $i = 0, \ldots, n-1$ on this diagram is to add all possible boxes which are of colour i, in such a way that the resulting object is still the diagram of a partition. If there is at least one corner box of colour i, then the action consists in erasing all peripheral boxes of colour i such that one still gets a diagram of a partition.

It is an easy combinatorics to check that from an n-core one still gets an n-core, and that the operation is involutive. Now, one does not try to check the braid relations, because one takes the classics again ([8] p.78) and one sees that one can use, instead of an n-core, still another object which is called *abacus*.

Let $\diamondsuit$ be a (decreasing) partition extended to infinity on the right by 0's: $\diamondsuit = [\diamondsuit_1, \diamondsuit_2, \ldots, \diamondsuit_k, 0, 0, \ldots]$. Let $\mathcal{A}$ be the set of numbers $\{\diamondsuit_1 - 0, \diamondsuit_2 - 1, \diamondsuit_3 - 2, \ldots\}$ interpreted as balls in an abacus of width n and bi-infinite height, with places numbered consecutively as follows:

$$\begin{array}{cccc|c} \vdots & \cdots & \vdots & \vdots & \textit{level} \\ 1-2n & \cdots & -n-1 & -n & -1 \\ 1-n & \cdots & -1 & 0 & 0 \\ 1 & \cdots & n-1 & n & 1 \\ n+1 & \cdots & 2n-1 & 2n & 2 \\ \vdots & \cdots & \vdots & \vdots & \vdots \end{array}$$

Putting balls in positions belonging to $\mathcal{A}$ gives by definition the *n-abacus* of the partition $\diamondsuit$.

That $\diamondsuit$ be an n-core is equivalent to the fact that balls are vertically packed upwards ([8] p.80).

Reading levels of bottom balls, one gets a vector in $u \in \mathbb{Z}^n$ such that $u_1 + \cdots + u_n = 0$, that is, one gets a weight. This construction is, indeed, a bijection between weights in $\mathbb{Z}^n$ and n-cores or elements of $\mathfrak{S}_n \backslash \widetilde{\mathfrak{S}}_n$ or of $\widetilde{\mathfrak{S}}_n / \mathfrak{S}_n$.

For example, the 5-core $[4, 2^3, 1^4]$ gives by subtraction the infinite vector

$$[4, 1, 0, -1, -3, -4, -5, -6, -8, -9, \ldots]$$

and the abacus

$$\begin{array}{ccccc|c} & & & & & \textit{level} \\ -14 & -13 & -12 & -11 & -10 & -2 \\ -9 & -8 & \cdot & -6 & -5 & -1 \\ -4 & -3 & \cdot & -1 & 0 & 0 \\ 1 & \cdot & \cdot & 4 & \cdot & 1 \end{array}$$

the weight $[1, 0, -2, 1, 0]$, the code $[0, 1, 2, 2, 4]$ and the element $[-7, 2, 5, 6, 9]$ of $\widetilde{\mathfrak{S}}_5 / \mathfrak{S}_5$.

For each of the different objects that have been given, the description of the action of the simple generators s_i allows to prove that the above constructions are indeed bijections and that the elementary operations represent the generators of $\widetilde{\mathfrak{S}}_n$.

Now each of the objects that we have displayed will provide informations about the corresponding element of the affine group, information which would sometimes not have been easy to get in another description, though all bijections here are elementary. For example, it is amusing to look at what becomes the addition of weights (plain addition of vectors) in terms of n-cores, or of codes.

Here is some example computed with ACE [16], where we give a chain of elements of $\widetilde{\mathfrak{S}}_4/\mathfrak{S}_4$, with their code, the corresponding extended weight in $\mathbb{Z}^5$ and the 4-core with its colours. The reader should beware that simple transpositions act from the left on codes and on group elements, and from the right on weights and cores.

Code	$[0,1,2,7]$		$[0,2,2,7]$	
Group element	$[-6,1,4,11]$		$[-7,2,4,11]$	
Ext. weight	$[0,-2,2,0,-4]$		$[-2,0,2,0,-4]$	
Core	$[7,4,2,2,1,1,1]$		$[7,4,2,2,2,1,1,1]$	
Diagram	2 3 0 1 2 2 3 3 0 1 2 0 1 2 3 0 1 2	$\xrightarrow{s_1}$	1 2 3 0 1 1 2 2 3 3 0 1 2 0 1 2 3 0 1 2	$\xrightarrow{s_3}$

$[0,2,2,8]$		$[0,2,2,9]$		$[0,2,3,9]$
$[-7,2,3,12]$		$[-8,2,3,13]$		$[-9,2,4,13]$
$[-2,0,0,2,-4]$		$[3,0,0,-3,-9]$		$[3,0,-3,0,-9]$
$[8,5,2,2,2,1,1,1]$		$[9,6,3,2,2,2,1,1,1]$		$[9,6,3,3,2,2,2,1,1,1]$
1 2 3 0 1 1 2 2 3 3 0 1 2 3 0 1 2 3 0 1 2 3	$\xrightarrow{s_0}$	0 1 2 3 0 0 1 1 2 2 3 0 3 0 1 2 3 0 0 1 2 3 0 1 2 3 0	$\xrightarrow{s_3}$	3 0 1 2 3 3 0 0 1 1 2 3 2 3 0 3 0 1 2 3 0 0 1 2 3 0 1 2 3 0

For what concerns the Bruhat order, we shall see that we need n-cores. Codes, which are also (increasing) partitions for minimum elements in their coset modulo $\mathfrak{S}_n$, would not give (by inclusion) the proper order, but only a sub-order. Björner and Brenti ([2], th 6.3 and th.6.5) obtained a criterium describing the Bruhat order in terms of monotonous functions; it is equivalent to the following proposition.

Let us denote by C the morphism from $\widetilde{\mathfrak{S}}_n/\mathfrak{S}_n$ (identified with the set of elements of $\widetilde{\mathfrak{S}}_n$ whicht are of minimum length in their coset) onto n-cores.

Proposition 4.1. *Let $\sigma, \omega \in \widetilde{\mathfrak{S}}_n/\mathfrak{S}_n$. Then $\sigma \le \omega$ iff the diagram of $C(\sigma)$ is contained in the diagram of $C(\omega)$.*

Proof. Take any corner of the diagram of $C(\omega)$ and let i be its colour. Then $s_i\,\omega < \omega$ and one knows that ([3], [4])

$$\sigma \le \omega \;\Leftrightarrow min(s_i\sigma, \sigma) \le s_i\omega$$

Now, we have to compare two n-cores with less boxes, and as in the finite case, we are finally reduced to know how to compare two elements of length differing by 1. □

5 Basis of the Bruhat order

When one needs to store huge sets of elements of the affine group, together with the information about the Bruhat order, then it is not efficient to also store the associated partitions (for an element in $\widetilde{\mathfrak{S}}_n$, we need n partitions, because we have n different projections $\widetilde{\mathfrak{S}}_n \mapsto \widetilde{\mathfrak{S}}_n/\mathfrak{S}_n$). It is well illustrated by Geck and Kim [9], in the finite case, that one has instead to use the basis of the order, as defined above. In this section we shall determine the basis of $\widetilde{\mathfrak{S}}_n/\mathfrak{S}_n$.

Let us first introduce some special n-cores.

Given two strictly positive numbers p, q such that $p + q = n$, a (p, q)-*staircase* is a partition E such that the stairs of its diagram are all of heigth p and width q, except for the top stair which is of width $\le q$ and the bottom stair which is of heigth $\le p$.

In other words, $E = [E_1, E_2, \ldots, E_k, 0, \ldots]$, the non zero differences $E_i - E_{i+1}$ are all equal to q, except possibly for the last one $E_k - 0$ which is such that $1 \le E_k \le q$, and the partition conjugate to E (obtained by transposing the axes) satisfies the same conditions with p instead of q.

Let us notice that all the corners of a staircase have the same colour, and that, given (p,q) and a point in the Cartesian quadrant $\mathbb{N} \times \mathbb{N}$, there exists one and only one (p, q)-staircase having this point as a corner.

Let $\mathcal{D}$ be the diagram of an n-core, and $\triangle$ one of its corners. $\triangledown$ the maximal box of $\mathcal{D}$ on the first diagonal of the same colour above (resp. below) the diagonal containing $\triangle$ (if it exists). Define the *left-rectrix* (resp. *right-rectrix*) of $\mathcal{D}$ of *pivot* $\triangle$ to be the staircase having $\triangle$ and $\bigtriangledown$ as corners. n-Cores which are too small to have two diagonals of the same colour are in fact staircases. Let us say in that case that they have only one rectrix which coincide with themselves.

For the 4-core $[5,4,3,2,1,1,1]$, here is an example of a left-rectrix (figured by $\heartsuit$'s):

$$
\begin{array}{lcl cl}
2 & & 2 & & 2 \\
3 & & 3 & & 3 \\
0 & & \bigtriangledown & & \heartsuit \\
1\,2 & \mapsto & 1\;2 & \mapsto & \heartsuit\,2 \\
2\,3\,0 & & 2\;3\;\triangle & & \heartsuit\,\heartsuit\,\heartsuit \\
3\,0\,1\,2 & & 3\;0\,1\;2 & & \heartsuit\,\heartsuit\,\heartsuit\,2 \\
0\,1\,2\,3\,0 & & 0\;1\,2\;3\,0 & & \heartsuit\,\heartsuit\,\heartsuit\,\heartsuit\,\heartsuit
\end{array}
$$

Recall that in the finite symmetric group case, cosets representatives are diagrams contained in a rectangle; rectrices of a diagram $\mathcal{D}$ are the maximal rectangular partitions contained in it [13]. Describing Bruhat order on a finite symmetric group essentially reduces to decomposing partitions into maximal rectangles.

Lemma 5.1. *Let $\mathcal{D}$ be the diagram of an n-core. Then $\mathcal{D}$ is the supremum (with respect to inclusion of diagrams) of its rectrices.*

Proof. Since $\mathcal{D}$ is contained in the supremum of any family of diagrams such that each corner of $\mathcal{D}$ is contained into at least one of them, it is sufficient (and immediate) to check that each rectrix of $\mathcal{D}$ is contained entirely in $\mathcal{D}$. □

In order to determine the basis of $\widetilde{\mathfrak{S}}_n/\mathfrak{S}_n$, we need some more properties of staircases.

Lemma 5.2. *Let $\mathcal{D}$ and $\mathcal{F}$ be diagrams of n-cores such that $\mathcal{D}$ is not included into $\mathcal{F}$. Then there exists at least one staircase $E \subset \mathcal{D}$ such that $E \not\subset \mathcal{F}$.*

Proof. Take any corner $\clubsuit$ of $\mathcal{D}$ which does not belong to $\mathcal{F}$. Then any staircase with corner $\clubsuit$ and included into $\mathcal{D}$ will do. □

Lemma 5.3. *Let E be a (p,q)-staircase, having at least one internal corner $\clubsuit$ (i.e. a corner which is not on the first column, nor on the bottom row). Let $\bigtriangledown$, $\triangle$ be the west and south neighbour boxes of $\clubsuit$, and let $E_{\bigtriangledown}$ be the $(n-1,1)$-staircase of corner $\bigtriangledown$ and $E_{\triangle}$ be the $(1,n-1)$-staircase of corner $\triangle$. Let moreover F be the partition which is the supremum of $E_{\bigtriangledown}$ and of $E_{\triangle}$ (it is an n-core).*

Then E is minimum in the complementary of the interval $[\emptyset, F]$.

Proof. Suppose that an n-core E' is strictly contained in E and not in F. Take a box of E' which does not belong to F and a staircase E'' contained in E' of corner this box. Then E'' is not included in F. Therefore it is sufficient to prove that each staircase E'' strictly contained in E is also contained in F. If $E'' \subset E$ contains a box which is an internal corner of E, then this box is a corner of E'' and $E'' = E$. Therefore E'' avoids all internal corners of E, in particular $\clubsuit$. But is is clear that the part of E'' which is right of $\clubsuit$ is contained in $E_{\triangle}$, and that the part which is left of $\clubsuit$ is contained in $E_{\triangledown}$. This implies that $E'' \subseteq F$. □

For example, take the $(2,2)$-staircase ⁂ $:= [7,5,5,3,3,1,1]$, choosing the corner pointed by $\clubsuit$, its diagonal being figured by $\spadesuit$'s:

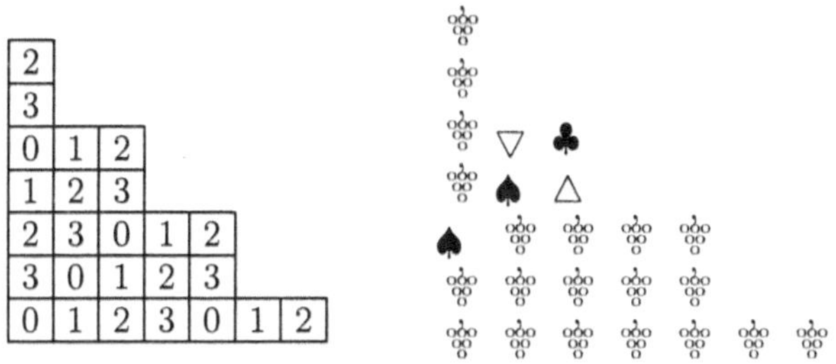

Then any 4-core strictly contained in ⁂ is also contained in the 4-core

$F = [12,9,6,3,\ 2,1,1,1] =$ 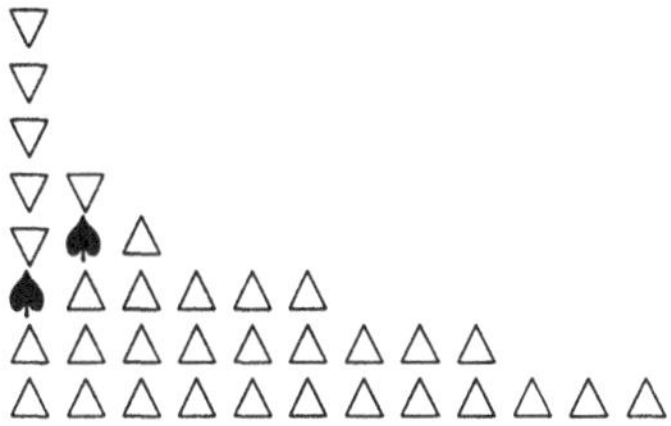

We have left aside the cores having no internal corner, that is *hook-partitions*. A case-by-case analysis which offers no difficulty gives the following classification.

Lemma 5.4. *Partitions*

$$[1];\ [2], \ldots [n-1];\ [1^2], \ldots, [1^{n-1}];\ [n,1], [2,1^{n-1}]$$

belong to the basis, but not the other hooks with $n+1$ boxes: $[3,1^{n-2}], \ldots, [n-1,1^2]$, *though they are (p,q)-staircases.*

For example, for $n = 5$, the basis contains the hooks $[1]; [2], [3], [4]; [1^2], [1^3], [1^4]; [5,1], [2,1^3]$, but not $[3,1^3]$ nor $[4,1^2]$. Indeed, $[3,1,1,1]$ is the supremum of $[3]$ and $[1,1,1,1]$, and $[4,1,1]$ is the supremum of $[4]$ and $[1,1,1]$.

In summary, one has the following characterization of the basis of $\widetilde{\mathfrak{S}}_n/\mathfrak{S}_n$.

Theorem 5.5. *The basis of $\widetilde{\mathfrak{S}}_n/\mathfrak{S}_n$ for the Bruhat order consists in all (p,q)-staircases, $p+q=n$, excepted the hooks $[3\,1^{n-2}],\ldots,[n-1,1,1]$.*

We have now all the necessary tools to handle the affine symmetric group together with its Bruhat order. In particular, elements will be represented by the decomposition of their associated cores into maximal sub-staircases.

For example, the 5-core  $=[6,6,5,4,3^4,2^4,1^4]$, corresponding to the group element $[-15,2,4,11,13]$, with code $[0,3,3,7,8]$ and length 21, is decomposed into the following three staircases

0		0				
1		1				
2		2				
3		3				
4 0		4 0				
0 1		0 1				
1 2		1 2				
2 3		2 3				
3 4 0	=	3 4 0	$\cup$		$\cup$	
4 0 1		4 0 1				
0 1 2		0 1 2		0 1 2		
1 2 3		1 2 3		1 2 3		
2 3 4 0		2 3 4 0		2 3 4		2 3 4
3 4 0 1 2		3 4 0 1		3 4 0 1 2		3 4 0
4 0 1 2 3 4		4 0 1 2		4 0 1 2 3		4 0 1 2 3 4
0 1 2 3 4 0		0 1 2 3		0 1 2 3 4		0 1 2 3 4 0

For a 5-core to be bigger than  with respect to the Bruhat order, it is necessary and sufficient to be bigger than each of the above three staircases.

As in the case of a finite symmetric group, one has to reduce the numbers of projections $\widetilde{\mathfrak{S}}_n \mapsto \widetilde{\mathfrak{S}}_n/\mathfrak{S}_n$ or $\widetilde{\mathfrak{S}}_n \mapsto \mathfrak{S}_n\backslash\widetilde{\mathfrak{S}}_n/\mathfrak{S}_n$ to get an efficient coding [13]. We shall not go into these technicalities here. Let us just mention that in the case of a finite symmetric group, we have used such a "decomposition" of a permutation into rectangular partitions to compute certain Kazhdan-Lusztig polynomials [11].

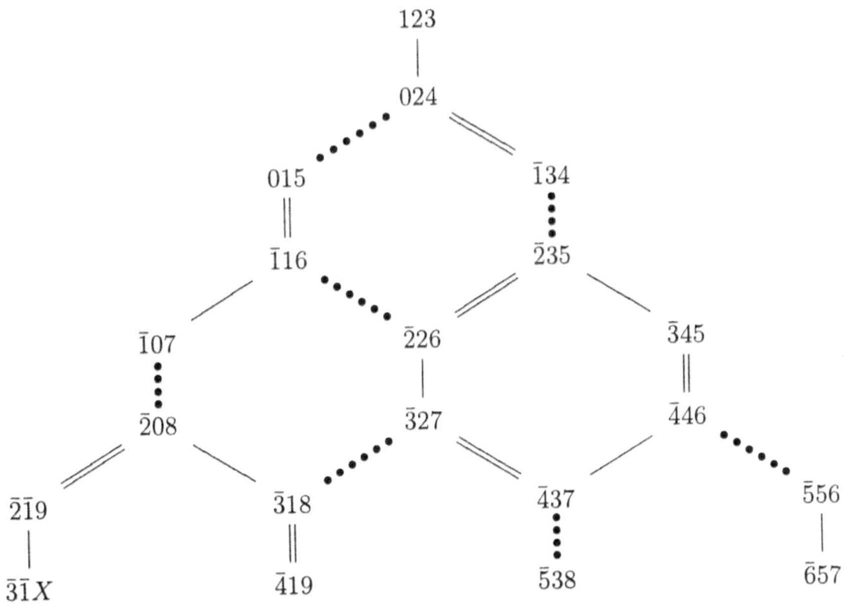

Cosets representatives for n=3

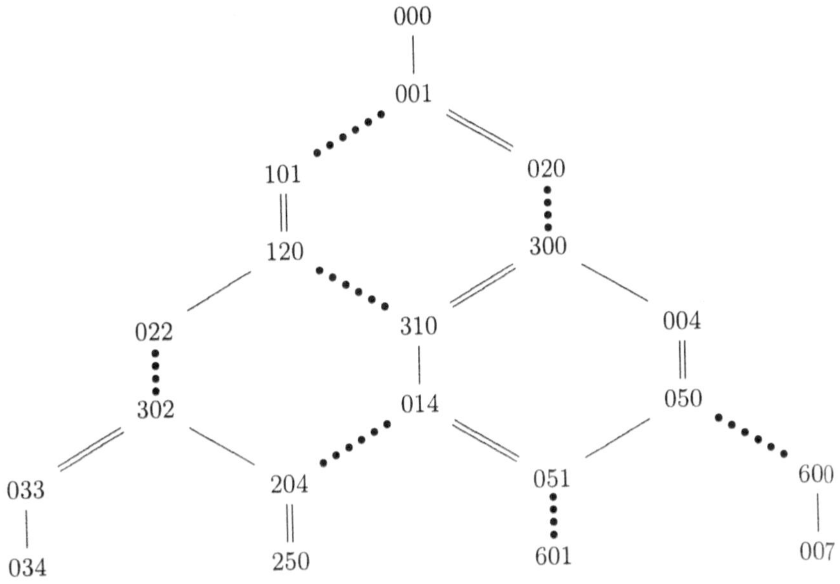

Codes

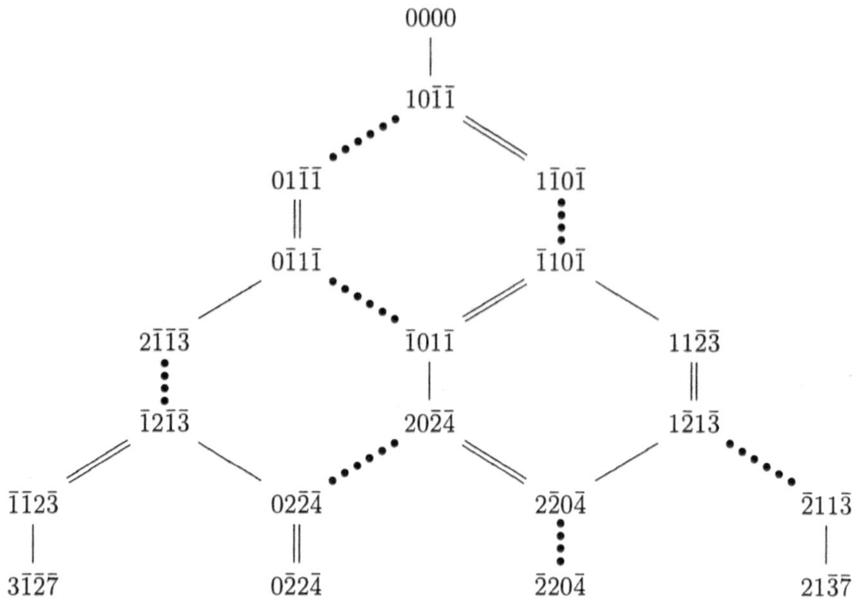

Extended Weights

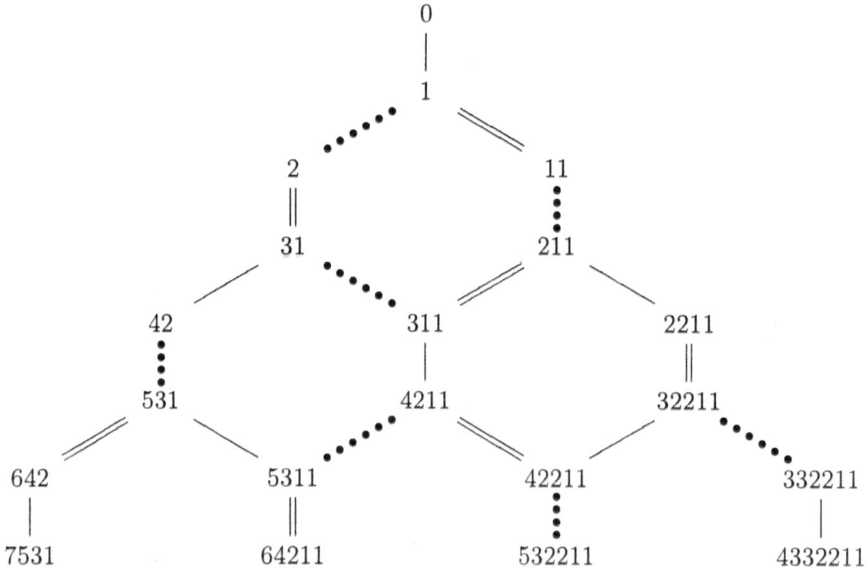

3-Cores

References

1. A. Björner, *Orderings of Coxeter groups*, Combinatorics and Algebra (Boulder, 1983), Contemp. Math. **34** (1984) 155-162.
2. A. Björner, F. Brenti, *Affine permutations of type A*, Electronic J. Comb. **3**# R18 (1996).
3. N. Bourbaki, *Groupes et Algèbres de Lie*, Fasc *34*, Hermann, Paris (1968).
4. V.V. Deodhar, *Some characterizations of Bruhat ordering on a Coxeter group*, Invent. Math. **39** (1977) 187-198.
5. C. Ehresmann, *Sur la topologie de certains espaces homogènes*, Ann. Math. **35** (1934) 396-443.
6. H. Eriksson, *Computational and combinatorial aspects of Coxeter groups*, thesis, KTH, Stockholm (1994).
7. O. Foda, B. Leclerc, M. Okado, J-Y Thibon, T. Welsh, *Branching functions of $A_{n-1}^{(1)}$ and Jantzen-Seitz problem for Ariki-Koike algebras*, Adv. in Maths **141** (1999) 322-365.
8. G. James, A. Kerber, *The Representation Theory of the Symmetric Group*, Encyclopedia of Mathematics, Addisson-Wesley, Reading MA, (1991).
9. M. Geck, S. Kim, *Bases for the Bruhat-Chevalley order on all finite Coxeter groups*, J. of Algebra **197** (1997) 278–310.
10. A. Kerber, *Algebraic Combinatorics via finite group actions*, Wissenschaftsverlag, Mannheim (1991).
11. A. Lascoux, , *Ordonner le groupe symétrique: pourquoi utiliser l'algèbre de Iwahori-Hecke ?*, Congrès International Berlin 1998, Documenta Mathematica, extra volume ICM 1998, III (1998) 355–364.
12. A. Lascoux, B. Leclerc et J.Y. Thibon, *Crystal graphs and q-analogues of weight multiplicities for root systems of type A_n*, Letters in Math. Physics, **35** (1995) 359–374.
13. A. Lascoux, M.P. Schützenberger, *Treillis et bases des groupes de Coxeter*, Electronic J. of Comb. **3** # R27 (1996).
14. K.C. Misra, T. Miwa, *Crystal base of the basic representation of $U_q(\widehat{sl}_n)$*, Commun. Math. Phys., **134**, 1990, p.79-88.
15. Jian-Yi Shi, *The Kazhdan-Lusztig cells in certain affine Weyl Groups*, Springer L.N. **1179** (1986).
16. S. Veigneau, *ACE, an algebraic environment for the computer algebra system MAPLE*, `http://phalanstere.univ-mlv.fr/~ace` (1998).

Constructing Objects up to Isomorphism, Simple 9-Designs with Small Parameters

R. Laue

Universität Bayreuth, Lehrstuhl II für Mathematik (Informatik)
Germany

Dedicated to Professor Adalbert Kerber on the occasion of his 60th birthday

Abstract Group actions are reviewed as a tool for classifying combinatorial objects up to isomorphism. The objective is a general theory for constructing representatives of isomorphism types. Homomorphisms of group actions allow to reduce problem sizes step by step. In particular, classifying by stabilizer type, i.e. the automorphism group of the objects, is generalized to using only sufficiently large subgroups of stabilizers. So, less knowledge of the full subgroup lattice of the classifying group is needed. For single steps in the homomorphism decomposition, isomorphism problems are transformed into double coset problems in groups. New lower bounds are given for the number of long double cosets such that corresponding bounds for the number of objects with trivial automorphism group can be derived.

The theory is illustrated by an account of recent work on the construction of t-designs including new results. Based on a computer search by DISCRETA several simple 8-designs and the first simple 9-designs with small parameters are presented. The automorphism group is $ASL(3,3)$ acting on 27 and 28 points. There are many isomorphism types in each case. The number of isomorphism types is determined in the smaller cases. By relating the isomorphism types of design extensions to double cosets designs with small automorphism groups are also accessible. There result more than 10^{16} isomorphism types of 8-$(28,14,\lambda')$ designs from each 8-$(27,13,\lambda)$ design. There are exactly 131,210,855,332,052,182,104 isomorphism types of 7-$(25,9,45)$ designs obtained from extending all the 7-$(24,8,5)$ designs with automorphism group $PSL(2,23)$ by all the 7-$(24,9,40)$ designs with automorphism group $PGL(2,23)$. Most of these designs have a trivial automorphism group. Iterating forming extensions then results in more than 10^{62} isomorphism types of 7-$(26,10,342)$ designs.

keywords: Group actions, isomorphism problem, double coset, t-design, Kramer-Mesner method.

1 Introduction

In mathematics, a natural aim is to describe the objects that are considered. Ideally, a Fundamental Theorem would fully determine some infinite series and maybe some finitely many additional sporadic objects comprising all cases. This has been achieved in algebra for finite fields, finite abelian groups, finite simple groups etc. The results are used to derive further classifications from these.

In combinatorics, the objects usually have a less regular structure to allow such a comprehensive theorem. So, often the weaker aim to only count the objects of a fixed size is pursued. There are ingenious solutions for many cases, some relying on a fairly general method. A prominent example is Pólya's and Redfield's theory of counting [47,13,27].

In applications, there is a need of not only knowing abstractly the existence of some number of objects but to really have the objects. This is obvious if the isomers of a molecule are searched to look for those which may have a specified spectrum. Also, block-designs can of course be applied in the planning of experiments in agriculture only if they are explicitly known. A code can be used only if it is at hand.

The development of powerful and cheap computers in the last decade now allows to solve such construction problems in many interesting cases. It is even possible to find constructive solutions where an efficient counting method is not available.

It should therefore be a natural task to excerpt from the different algorithmic approaches the common aspects. Just as a theory of counting allows to tackle various problems in a similar way, a theory of construction should give general rules applicable in a larger variety of problems.

This has been a motivation for several papers by A. Kerber and the author and some books by A. Kerber [27]. These all rely on implementations of the algorithms and got an important stimulus from practical experiences. Many aspects of applications to the construction of isomers, of groups and of codes have already sufficiently been explained in some specialized papers and some review articles, see [37,26,38] and the references there. Here we add some material that resulted from the search for t-designs with "large"t on small point sets, where large means $t \geq 3$. The t-designs are combinatorial objects defined on a point set V of v points. We only consider simple t-designs $\mathcal{D}$ which consist of a collection of k-element subsets, called blocks, of V, such that each t-element subset of V lies in exactly the same number λ of blocks. The numeric parameters of $\mathcal{D}$ are listed as t-(v, k, λ). Usually, constructing t-designs and solving isomorphism problems for t-designs are difficult. We use a group-action approach for solving these problems by algebraic means. It is important to notice that isomorphism problems sometimes are easier to solve if information about the way of construction is used. So, we follow up this idea and thus avoid to solve the general isomorphism problem.

We first give a summary of group theoretic methods which form the abstract background. They are collected out of several recent papers. Then, the use of these methods in the search for t-designs is explained. On one hand, prescribed automorphism groups are used to deal with whole orbits of these groups instead of the individual elements like t-sets, k-sets or even designs. On the other hand, these groups yield a powerful tool for isomorphism classification. This had already been developed in the recent papers on 6-, 7-, and 8-designs constructed with help of a computer by our system DISCRETA. In this paper we continue with the first simple 9-designs on small point sets and then consider the isomorphism problem for design extensions. We use double cosets which often correspond to the isomorphism types. So, for the first time huge numbers of isomorphism types can be determined. Most of these designs have a trivial automorphism group. The following tables illustrate these results.

The new and earlier results on t-designs with $t \geq 8$ and $v \leq 40$ and big automorphism group are summarized in the following table. All results concerning $ASL(3,3)$ are new.

Simple 8- and 9-designs

Parameters	Group	Size of KM-matrix	Number of isomorphism types
8-(27,11,432)	$ASL(3,3)$	31 × 121	1
8-(27,12,1296)	$ASL(3,3)$	31 × 154	4336
8-(27,12,1932)	$ASL(3,3)$	31 × 154	2110899
8-(27,13,3204)	$ASL(3,3)$	31 × 176	538218
8-(27,13,3240)	$ASL(3,3)$	31 × 176	618421
8-(27,13,4608)	$ASL(3,3)$	31 × 176	$\geq$ 200000000
8-(27,13,5076)	$ASL(3,3)$	31 × 176	many
8-(27,13,5148)	$ASL(3,3)$	31 × 176	many
8-(28,13,5832)	$ASL(3,3)+$	48 × 330	$\geq$ 5000000000
8-(28,13,7080)	$ASL(3,3)+$	48 × 330	many
8-(28,13,7128)	$ASL(3,3)+$	48 × 330	many
8-(28,14,10680)	$ASL(3,3)+$	48 × 352	$\geq$ 1
8-(28,14,10800)	$ASL(3,3)+$	48 × 352	$\geq$ 1
8-(28,14,14040)	$ASL(3,3)+$	48 × 352	$\geq$ 1
8-(28,14,15360)	$ASL(3,3)+$	48 × 352	$\geq$ 1
8-(28,14,16920)	$ASL(3,3)+$	48 × 352	$\geq$ 1
8-(28,14,17160)	$ASL(3,3)+$	48 × 352	$\geq$ 1
8-(28,14,18600)	$ASL(3,3)+$	48 × 352	$\geq$ 1
8-(31,10,93)	$PSL(3,5)$	42 × 174	138
8-(31,10,100)	$PSL(3,5)$	42 × 174	1658
8-(36,11,1260)	$Sp(6,2)_{36}$	79 × 694	$\geq$ 1
8-(40,11,1440)	$PSL(4,3)$	53 × 569	$\geq$ 150000000
9-(28,14,3204)	$ASL(3,3)+$		538218
9-(28,14,3240)	$ASL(3,3)+$		618421
9-(28,14,4608)	$ASL(3,3)+$		$\geq$ 200000000
9-(28,14,5076)	$ASL(3,3)+$		many
9-(28,14,5148)	$ASL(3,3)+$		many

The parameter list on 27 points and the group $ASL(3,3)$ is complete.

It is remarkable that up to now no 8-design with an automorphism group $PGL(2,q)$ has been found. Also the values of λ for 8-designs are large compared to those of the 7-designs found by prescribing some $PGL(2,q)$, [4]. Since small values of λ are of interest, we also list the few known parameter sets of 6- and 7-designs with $\lambda \leq 10$, omitting derived designs.

Simple 6- and 7-designs

Parameter set	Constructed by	No. of isomorphism types
6-(14,7,4)	$C_{13}+$	2
6-(19,7,4)	$Hol(C_{17})++$	1
6-(19,7,6)	$Hol(C_{19})$	3
6-(22,7,8)	large set recursion	
6-(28,7,6)	$PSU(3,9)$	$\geq$ 10
6-(32,7,6)	$PSL(2,31)$	$\geq$ 18
7-(24,8,4)	$PSL(2,23)$	1
7-(24,8,5)	$PSL(2,23)$	138
7-(24,8,6)	$PSL(2,23)$	$\geq$ 132
7-(24,8,7)	$PSL(2,23)$	$\geq$ 126
7-(24,8,8)	$PSL(2,23)$	$\geq$ 63
7-(26,8,6)	$PGL(2,25)$	$\geq$ 7
7-(33,8,10)	$P\Gamma L(2,32)$	4996426

Without restrictions on λ there are about 400 parameter sets of 7-designs and about 1100 parameter sets of 6-designs with up to 40 points in the database of DISCRETA now.

Using the theory of design extensions presented in this paper, we obtain the following table of lower bounds for the number of isomorphism types of 7-designs.

Extensions of Designs

No.	Parameters	Group	Number of isomorphism types	Parameters	Number of isomorphism types
1	7-(24, 9, 48)	$PGL(2,23)$	$\geq$ 2827		
1	7-(24,10,240)	$PGL(2,23)$	$\geq$ 91	7-(25,10,288)	$\geq$ 23786911342165204970
2	7-(24, 9, 64)	$PGL(2,23)$	$\geq$ 15335		
2	7-(24,10,320)	$PGL(2,23)$	$\geq$ 2	7-(25,10,384)	$\geq$ 5161263324118902274
3	7-(24,8, 5)	$PSL(2,23)$	138		
3	7-(24,9,40)	$PGL(2,23)$	113	7-(25,9,45)	131210855332052182104
4	7-(24,8, 6)	$PSL(2,23)$	$\geq$ 132		
4	7-(24,9,48)	$PGL(2,23)$	$\geq$ 2827	7-(25,9,54)	$\geq$ 125594891886632282241
5	7-(24,8, 8)	$PSL(2,23)$	$\geq$ 63		
5	7-(24,9,64)	$PGL(2,23)$	$\geq$ 15335	7-(25,9,72)	$\geq$ 7951408124200930620
6	7-(26,8, 6)	$PGL(2,25)$	7		
6	7-(26,9,54)	$P\Gamma L(2,25)$	3989	7-(27,9,60)	$\geq$ 121802772685441446018
7	7-(26,12, 5796)	$P\Gamma L(2,25)$	$\geq$ 1		
7	7-(26,13,13524)	$\geq A\Gamma L(1,25)$	$\geq$ 1	7-(27,13,19320)	$\geq$ 1
8	7-(27,10, 540)	$P\Gamma L(2,25)+$	$\geq$ 1		
8	7-(27,11,2295)	$AGL(3,3)$	$\geq$ 105	7-(28,11,2835)	$\geq$ 5754099169659337180
9	7-(27,11, 810)	$ASL(3,3)$	1188		
9	7-(27,12,2592)	$AGL(3,3)$	33	7-(28,12,3402)	$\geq$ 281311515186391173924
10	7-(27,11,2025)	$AGL(3,3)$	57		
10	7-(27,12,6480)	$AGL(3,3)$	$\geq$ 500	7-(28,12,8505)	$\geq$ 3374317419594730345500
11	7-(28,13,10080)	$Sp(6,2)$	1		
11	7-(28,14,21600)	$\geq Sp(6,2)_1$	$\geq$ 1	7-(29,14,31680)	$\geq$ 1
12	7-(25, 9, 54)	Id	$\geq$ 11106724087393318560		
12	7-(25,10,288)	Id	$\geq$ 23786900753834023916	7-(26,10,342)	$\geq 10^{62}$

The system DISCRETA is freely available from our web-page
http://www.mathe2.uni-bayreuth.de/ discreta/
which also contains an account of the presently known t-designs for $t > 5$, Steiner 5-designs, and further information. The author thanks the DISCRETA research group, in particular Anton Betten and Alfred Wassermann, for their support and Axel Kohnert for computing numbers of double cosets with his system SYMMETRICA.

2 Definitions and Notations

If a group G acts on a set Ω and Δ is a subset of Ω then $N_G(\Delta) = \{g \in G | \{\delta^g | \delta \in \Delta\} = \Delta\}$ is the normalizer of Δ in G. This generalizes the notion of the normalizer in the special case of the conjugation action of a group on its lattice of subgroups. This normalizer acts on Δ and the kernel of this action is $C_G(\Delta)$, the centralizer of Δ in G. We prefer this notion to the set-wise and point-wise stabilizers if Δ consists of more than one point. As usual, G_ω is also used to denote the stabilizer of ω in G. For a group element $g \in G$ the set of fixed points is $C_\Omega(g) = \{\omega \in \Omega | \omega^g = \omega\}$.

We assume throughout the paper that $V = \{1, 2, \cdots, v\}$ is a set of natural numbers. Any subset of size k is called a k-set. The symmetric group on V is denoted by S_V.

The right cosets Ag of a subgroup A of a group G form the set $A \backslash G$ and similarly the left cosets gA form the set A/G. If A and B are two subgroups of the group G then $A \backslash G / B = \{AgB | g \in G\}$ is the set of double cosets of A and B in G.

3 Group Actions

An important strategy is to transform an isomorphism problem from some family of objects into a group theoretic problem. The following basic results often allow this transfer.

Theorem 3.1 (Fundamental Lemma). *Let a group G act transitively on a set Ω and $\omega \in \Omega$. Then the mapping $\phi : \Omega \mapsto N_G(\omega) \backslash G$ such that $\phi(\omega^g) = N_G(\omega)g$ is a bijection.*

The action of G on Ω is replaced by right-multiplication on the set of right cosets of $N_G(\omega)$ in G. Restricting the acting group to a subgroup gives a description of the orbits of that subgroup.

Theorem 3.2 (Split of Orbits). *Let U be a subgroup of G where G acts transitively on Ω. Then*

$$\omega^{gU} \mapsto N_G(\omega)gU$$

defines a bijection between the set of U-orbits on Ω and the set of double cosets $N_G(\omega) \backslash G / U$.

There is another situation which leads to double cosets.

Theorem 3.3 (Glueing Lemma). *Let a group G_1 be a group of automorphisms of some object ω_1 and a group G_2 be a group of automorphisms of some object ω_2. Let $f : \omega_1 \mapsto \omega_2$ be a fixed isomorphism. Then each isomorphism is obtained by composing f with some automorphism α of ω_2, such that the set of all isomorphisms is described by a group;*

$$Iso(\omega_1, \omega_2) = f Aut(\omega_2).$$

$G_1 \times G_2$ acts on $Iso(\omega_1, \omega_2)$ by

$$f\alpha^{(g_1,g_2)} = g_1^{-1} f \alpha g_2 = f(f^{-1} g_1^{-1} f)\alpha g_2$$

for $(g_1, g_2) \in G_1 \times G_2$. Thus, the orbits of $G_1 \times G_2$ are in bijection to the double cosets

$$(f^{-1} G_1 f)\alpha G_2$$

in $Aut(\omega_2)$.

This Lemma appears in different applications independently in the literature. An early instance can be found in Ph. Hall's lecture series in Göttingen in 1939, [24], where ω_1 is a factor group of one group and ω_2 a subgroup of another group. The different ways of identifying ω_1 with ω_2 have to be classified with respect to equivalence under two groups of automorphisms acting on ω_1 and ω_2, respectively. Such identifications had already earlier been carried out by Lunn and Senior [43] for a classification of subdirect products of groups. It thus may have been known to these authors before. Other group constructions like semi-direct products, central amalgamations etc. are considered by the present author in [31], [38] [26]. In Chemistry, Ruch et al. [50] identified places on a skeleton of a molecule with ligands that should be distributed to these places. This plays an important role in mathematical generators for isomers like the early Dendral [41] and Molgen [22]. We will give a new application to the construction of t-designs below.

Algorithms for solving double coset problems are presented in several papers, we mention [14][32][17] [42][52][53]. In many cases the number of double cosets is very big. Then one can at least count them by combinatorial methods, like the Cauchy Frobenius Lemma. We refer to Kerber's book [27]. So, from an implementation of Redfield's cap-product by H. Fripertinger in A. Kohnert's system SYMMETRICA we obtained the following numbers of double cosets that we will use in our section on designs.

Double Cosets	Number
$PSL(2,23)\backslash S_{24}/PSL(2,23)$	16,828,376,982,435,832
$PSL(2,23)\backslash S_{24}/PGL(2,23)$	8,414,188,491,217,916
$PGL(2,23)\backslash S_{24}/PGL(2,23)$	4,207,094,330,061,055
$PGL(2,25)\backslash S_{26}/PGL(2,25)$	1,657,180,580,754,274,540
$P\Gamma L(2,25)\backslash S_{26}/P\Gamma L(2,25)$	414,295,145,235,066,413
$PGL(2,25)\backslash S_{26}/P\Gamma L(2,25)$	828,590,290,377,152,694
$AGL(1,25)+\backslash S_{26}/P\Gamma L(2,25)$	10,771,673,642,332,865,588
$AGL(3,3)\backslash S_{27}/AGL(3,3)$	118,397,102,441,920,363
$ASL(3,3)\backslash S_{27}/AGL(3,3)$	236,794,204,702,349,473
$ASL(3,3)\backslash S_{27}/ASL(3,3)$	473,588,409,404,698,946
$AGL(3,3)\backslash S_{27}/P\Gamma L(2,25)+$	1,150,819,833,931,867,436
$Sp(6,2)_{28}\backslash S_{28}/Sp(6,2)_{28}$	144,708,746,195,525,184

In applications most of the orbits of a group are long orbits. The elements in such orbits then have a trivial stabilizer. Usually, it is difficult to determine the number of these orbits. We give at least a lower bound for this number. Orbits different from long orbits are called short orbits.

Lemma 3.4. *Let a group G act on a set Ω and let each $g \in G$, $g \neq id$, have at most c fixed points. Then there are at most $a_s = 2 \cdot c/|G|$ short orbits. The number a_l of long orbits is at least $|\Omega|/|G| - (1-1/|G|)c$. If the total number of orbits is a then we have for the number a_s of short orbits*

$$a_s \leq 2 \cdot a - 2\frac{|\Omega|}{|G|}.$$

Proof Let $\Omega' = \bigcup_{g \neq id} C_\Omega(g)$, where the union runs over all $g \in G$ different from the identity. Then all short orbits are formed from elements in Ω'. So, counting these orbits by the Cauchy-Frobenius Lemma would require to know the number of fixed points for each group element, including the identity. We use the crude bound $|C_{\Omega'}(id)| = |\Omega'| \leq \sum_{g \neq id} |C_\Omega(g)|$. Then we get for the number a_s of short orbits

$$\begin{aligned} a_s = \frac{1}{|G|}\sum_{g \in G} |C_{\Omega'}(g)| &= \frac{1}{|G|}(|C_{\Omega'}(id)| + \sum_{g \neq id} |C_\Omega(g)|) \\ \leq \frac{2}{|G|}\sum_{g \neq id} |C_\Omega(g)| &\leq \frac{2c}{|G|}. \end{aligned}$$

For the number a_l of long orbits we get

$$\begin{aligned} a_l &= \frac{1}{|G|}\sum_{g \in G} |C_\Omega(g)| - \frac{1}{|G|}\sum_{g \in G} |C_{\Omega'}(g)| = \frac{1}{|G|}(|C_\Omega(id)| - |C_{\Omega'}(id))| \\ &= \frac{1}{|G|}(|\Omega| - |\Omega'|) \geq \frac{1}{|G|}(|\Omega| - \sum_{g \neq id} |C_\Omega(g)|) \\ &\geq \frac{1}{|G|}(|\Omega| - \sum_{g \neq id} c) = \frac{1}{|G|}(|\Omega| - (|G| - 1)c). \end{aligned}$$

From this we obtain the first inequality. The second one is obtained from a combination of the above arguments with the Cauchy-Frobenius Lemma. So, we multiply

$$a = \frac{1}{|G|}(|\Omega| + \sum_{g \neq id} |C_\Omega(g)|)$$

by 2 and use the above bound

$$a_s \leq \frac{2}{|G|}\sum_{g \neq id} |C_\Omega(g)|$$

to get

$$2a - a_s \geq \frac{2|\Omega|}{|G|}$$

which is equivalent to the claimed inequality.

Example. $PGL(2,23)$ has $a = 83$ orbits on 8-sets such that the second equality gives $a_s \leq 44$. Actually, there exist exactly 39 short orbits in this situation. So, the bound seems to be reasonably good. But for our purpose to find designs with trivial automorphism group it is not good enough.

We obtain a sharper bound by considering only points which are fixed by some subgroup of prime order. We are interested in the action of $PGL(2,p)$, p an odd prime, by multiplication from the right on the set of right cosets of $PGL(2,p)$ in S_{p+1}. In other words, we investigate the double cosets $PGL(2,p)\backslash S_{p+1}/PGL(2,p)$.

Theorem 3.5. *$PGL(2,p)$ has at least*

$$\frac{1}{(p+1)p(p-1)}\Big\{ (p-2)! - \Big\{\tfrac{p(p-1)}{4(p+1)} \textstyle\sum_{d|p+1,d\ prime}(d-1)d^{(p+1)/d}((p+1)/d)! + (p+1) + \tfrac{p(p+1)}{4(p-1)} \textstyle\sum_{d|p-1,d\ prime}(d-1)d^{(p-1)/d}((p-1)/d)!\Big\}\Big\}$$

long double cosets in S_{p+1}.

Proof Let $G = PGL(2,p)$ and let $U \leq G \leq S_{p+1}$. Then $G\pi U = G\pi$ for some $\pi \in S_{p+1}$ if and only if $\pi U \pi^{-1} \leq G$. For some fixed $U' \leq G$ the elements π conjugating U onto U' form a coset of $N_{S_{p+1}}(U)$. So, there are $|N_{S_{p+1}}(U)|$ such elements. We have to multiply this number by the number of choices for U' and then divide by $|G|$, because these elements fall into cosets of G. Then the number of cosets fixed by U is determined. Now, U has $G : N_G(U)$ conjugates in G each of which has this number of fixed points. Lastly we have to sum these numbers of fixed points over all subgroups U of some prime order. Subtracting this number from the number of cosets of G in S_{p+1}, which is $(p-2)!$, gives a lower bound for the number of cosets which are not fixed under any non-trivial element of G. All these cosets then form double cosets consisting of $|G|$ cosets such that dividing by $|G|$ gives a lower bound for the number of long double cosets.

We use some well known results on subgroups of $PGL(2,p)$, as can be found in [25]. The elements of $G = PGL(2,p)$ and so also the subgroups U of prime order have at most 2 fixed points.

The fixed point free subgroups U lie in some cyclic subgroup C of order $p+1$ from a single conjugacy class and we have $N_G(U) = N_G(C)$ of order $2(p+1)$. A generator of U has $(p+1)/d$ cycles of length d if $|U| = d$. The centralizer of U then has order $d^{(p+1)/d}((p+1)/d)!$ [27] and the normalizer induces in addition an automorphism group of order $d-1$ on U. So,

$$|N_{S_{p+1}}(U)| = d^{(p+1)/d}((p+1)/d)! \cdot (d-1)$$

in this case. We have $|G : N_G(U)| = p(p-1)/2$ as the number of choices for U and as well for U'. The number of cosets $G\pi$ fixed by such subgroups U of order d is

$$\begin{aligned}\{\frac{p(p-1)}{2}\}^2|N_{S_{p+1}}(U)|\frac{1}{|G|} &= \{\frac{p(p-1)}{2}\}^2 d^{\frac{p+1}{d}}\frac{p+1}{d}!(d-1)\frac{1}{(p+1)p(p-1)} \\ &= \frac{p(p-1)}{4(p+1)}(d-1)d^{(p+1)/d}((p+1)/d)!\end{aligned}$$

If there is only one fixed point then $|U| = p$ and $|N_{S_{p+1}}(U)| = p(p-1)$. Then we have $|G : N_G(U)| = p+1$ and

$$(p+1)^2 p(p-1)\frac{1}{(p+1)p(p-1)} = p+1$$

cosets $G\pi$ are fixed by such subgroups U.

If U of order d has two fixed points then $d|(p-1)$. We have

$$|N_{S_{p+1}}(U)| = 2d^{(p-1)/d}((p-1)/d)! \cdot (d-1).$$

U is contained in the cyclic subgroup of order $p-1$ of a dihedral subgroup D of G of order $2(p-1)$ and which lies in a single conjugacy class. Then $N_G(U) = D$ such that the number of cosets $G\pi$ fixed by such subgroups U of order d is

$$\begin{aligned}\{\frac{p(p+1)}{2}\}^2|N_{S_{p+1}}(U)|\frac{1}{|G|} &= \{\frac{p(p+1)}{2}\}^2 2d^{\frac{p-1}{d}}\frac{p-1}{d}!(d-1)\frac{1}{(p+1)p(p-1)} \\ &= \frac{p(p+1)}{4(p-1)}(d-1)d^{(p-1)/d}((p-1)/d)!\end{aligned}$$

In case of $q = 23$ we obtain that the subgroup $PGL(2,23)$ of S_{24} has at least 4,207,092,457,345,954 long double cosets compared to a total number of 4,207,094,330,061,055 double cosets. So, only a very small part of the set of all double cosets is small. By splitting a long double coset in

$$PGL(2,23)\backslash S_{24}/PGL(2,23)$$

into double cosets in

$$PGL(2,23)\backslash S_{24}/PSL(2,23)$$

or in

$$PSL(2,23)\backslash S_{24}/PSL(2,23)$$

we obtain 2 resp. 4 long double cosets. Thus, also for these cases we easily obtain an even much larger number of long double cosets.

Since many isomorphism problems can be transformed into double coset problems, results on the number of long double cosets correspond to results on the number of objects with trivial automorphism group from a large scale of structures. We give an application to t-designs in the next section. For t-designs no easy way was known before to obtain bigger examples with trivial automorphism group.

A great many of instances for glueings arise from creating objects from smaller ones by adding new features or forming extensions. We use the notion of homomorphisms of group actions for a formal setting.

Definition 3.6 (Homomorphism of group actions). *Let G_1 be a group acting on a set Ω_1 and G_2 be a group acting on a set Ω_2. A pair $\sigma = (\sigma_\Omega, \sigma_G)$ of mappings, where σ_Ω maps Ω_1 into Ω_2 and $\sigma_G : G_1 \to G_2$ is a group homomorphism, is a homomorphism of group actions if σ is compatible with both actions, i.e. for all $g \in G_1$ and all $\omega \in \Omega_1$*

$$(\omega^g)^{\sigma_\Omega} = \omega^{\sigma_\Omega g^{\sigma_G}}.$$

If both components of σ are surjective σ is an epimorphism, if both components are bijective σ is an isomorphism.

If σ_G is not surjective orbits of the image group can be determined by the Split of Orbits Lemma from G_2-orbits. So, we further on will restrict to the case of surjective σ_G. Then, the action of G_2 can be replaced by an appropriate action of G_1 on Ω_2. We will thus simplify the notation by these assumptions.

Theorem 3.7 (Homomorphism Principle). *Let a group G act on two sets Ω_1 and Ω_2 and let $\sigma : \Omega_1 \longrightarrow \Omega_2$ be compatible with both group actions. Then the preimage sets of two elements of Ω_2 from the same G-orbit intersect the same G-orbits on Ω_1. If $\sigma(\omega) = \sigma(\omega')$ for two elements $\omega, \omega' \in \Omega_1$ then any $g \in G$ with $\omega^g = \omega'$ must lie in the stabilizer of $\sigma(\omega)$.*

By Theorem 3.7 a set of orbit representatives from the G-orbits on Ω_1 can be obtained by first determining orbit representatives from the G-orbits on Ω_2, together with their stabilizers, and then determining representatives from the stabilizer-orbits on the preimage sets of the representatives from Ω_2.

If G acts trivially on Ω_2 then the image points are *invariants*. This is a widely used method to show that two preimage points are from different orbits.

If the group acts non-trivially on Ω_2 then the stabilizers are much smaller than G. As well the preimage sets are small compared to Ω_1. So, the problem size is drastically reduced. In many cases, the stabilizers are even trivial, such that the full preimage sets can be taken as sets of representatives. Then an explicit listing can be avoided.

There are many important examples of homomorphisms, usually when there is an induced group action [34]. In computational group theory, the SOGOS system [33] made use of this. In combinatorics, multigraphs can be mapped to simple graphs setting each edge multiplicity to 1, see [12], directed graphs can be reduced to undirected graphs, labelings of edges or vertices may be omitted etc. In each case the isomorphism types of objects are just the orbits of the symmetric group on the set of vertices acting induced on the set of objects by renaming the vertices of each object. This induced action of the symmetric group is compatible with the simplifications to simple graphs. So, these simplifications are homomorphisms of the group action. When applying the homomorphism principle, mostly no group action has to be considered when the simple graphs are extended to multigraphs or directed graphs. We only have to notice that most simple graphs have a trivial automorphism group such that the stabilizer of the object simple graph is trivial. So, in most cases the set of full preimages of the simplification consists of pairwise non-isomorphic objects, i. e. all multigraphs that are reduced to the same simple graph with a trivial automorphism group are pairwise non-isomorphic.

A very useful homomorphism of group actions is given by mapping each object onto its stabilizer, see also [49]. So, again the action on some set Ω is transported into an internal group action, this time the conjugation on the subgroup lattice.

Corollary 3.8. *If Δ is the set of objects in Ω with full stabilizer U then the orbits of $N_G(U)$ on Δ are the intersections of G-orbits with Δ. Each orbit of $N_G(U)$ on Δ has length $N_G(U) : U$.*

We emphasize an important special case.

Corollary 3.9. *If a subgroup U is equal to its normalizer in the acting group G then all objects with stabilizer U lie in pairwise different G-orbits. In particular, if U is a maximal not normal subgroup of G then all objects fixed by U and not fixed by G lie in pairwise different G-orbits.*

Generally, an orbit of $N_G(U)$ on the set of objects with stabilizer U corresponds to that part of the G-orbit that has the same stabilizer U. Each of these orbits of $N_G(U)$ of course is in bijection to the right cosets of U.

Usually, it is much easier to determine the fixed points of a subgroup U than to find only those objects with full stabilizer U. If all minimal overgroups are known, then one can compute their fixed points as well and subtract them from the set of fixed points of U. Then there remain those with full stabilizer U. In the finite case, the number of orbits then can be determined by first computing the number of fixed points of U by the principle of exclusion-inclusion, equivalent to Möbius inversion on the subgroup lattice, and then dividing by the index of U in its normalizer. This can be done by a matrix computation, see Burnside's [15] or Kerber's book [27]. From the mere counting point of view this approach has been investigated by several authors, among the first are M. Klin [28], D.L. Stockmeyer [54]. There is also a long series of papers of D.E. White on this topic, see for example [58].

We emphasize that our point of view is constructive. So, we are interested in a set of representatives for the isomorphism types of objects. Since the action on sets of mappings is of importance for many constructions of combinatorial objects, we repeat the explicit formula for this case from [35].

Theorem 3.10 (Orbits of Mappings). *Let a group G act on a set X and let Y be another set. For any subgroup U of G the set $(Y^X)_U$ of mappings fixed by U is given by*

$$(Y^X)_U = \prod_{B\ U\text{-orbit}} \bigcup_{y \in Y} \{y\}^B$$

where the union denotes all mappings which are constant on the orbit B and the product of sets of mappings defined on disjoint sets is just the cartesian product. Then for each U a system of representatives from the orbits of $N_G(U)/U$ on

$$\prod_{B\ U\text{-orbit}} \bigcup_{y \in Y} \{y\}^B \setminus \bigcup_{U <_{max} V \leq G} \prod_{B\ V\text{-orbit}} \bigcup_{y \in Y} \{y\}^B$$

give a full system of representatives from the G-orbits with stabilizers from the conjugacy class of U.

The main problem in applying the Möbius inversion is the requirement that all overgroups of the subgroup U must be known. In some situations we need less information.

Definition 3.11 (Control of Fusion). *Let a group G act on a set Ω, let U be a subgroup of G, and let Δ be a subset of Ω. Then U controls the G-fusion on Δ if for each $\delta_1, \delta_2 \in \Delta$ and a $g \in G$ with $\delta_1^g = \delta_2$ there exists some $u \in U$ such that $\delta_1^u = \delta_2$.*

The homomorphism principle 3.7 describes one occurrence of such a situation. There Δ is just the preimage set of some point in Ω_2 and U is its stabilizer. Here we exhibit another case.

Theorem 3.12 (Localization). *Let a group G act on a set Ω, let U be a subgroup of G, and let Δ be a set of fixed points of U. If for all $\delta \in \Delta$ the stabilizer $N_G(\delta)$ controls the G-fusion on $\{U^g | g \in G,\ U^g \leq N_G(\delta)\}$ where the action is the conjugation then $N_G(U)$ controls the G-fusion on Δ.*

Thus, a control of fusion within the subgroup lattice yields a control of fusion on some exterior action of the group.

Corollary 3.13. *Let U be a subgroup of a group G where G is acting on a set Ω. If U is the unique subgroup of some isomorphism type in the stabilizer $N_G(\delta)$ of each point δ fixed by U then $N_G(U)$ controls the G-fusion on the set of fixed points of U.*

Another instance results from Sylow's Theorem.

Corollary 3.14. *Let U be a largest p-subgroup of each $N_G(\delta)$ where δ is fixed by U. Then $N_G(U)$ controls the G-fusion on the set of fixed points of U.*

If we interpret the approach as guessing the automorphism group of the objects we require that our guess at least covers a Sylow subgroup of the full automorphism group.

If U is even larger than a Sylow p-subgroup P of a stabilizer then U will have less fixed points, in general. Then we want to reduce the group that controls G-fusion on this smaller set also.

Theorem 3.15 (Reduction). *Let U be a subgroup of a group G where G is acting on a set Ω such that U contains a Sylow p-subgroup P of the stabilizer of each of its fixed points. Then $N_G(P) \cap N_G(U)$ controls the G fusion on the set of those fixed points that are not fixed by any proper overgroup $< U^h, U >$ of U for $h \in N_G(P)$.*

While the Moebius inversion above requires the knowledge of all minimal overgroups of U we here can construct the specific overgroups whose fixed points have to be taken out of consideration. On the remaining set of fixed points of U the smaller subgroup $D = N_G(P) \cap N_G(U)$ controls G-fusion. We remark that $DU = N_G(U)$ in this case. The objects taken out then are fixed points of a known larger group $V =< U^h, U >$ for which we can proceed in the same way. Of course V then still contains the Sylow subgroup P of the full stabilizers such that $N_G(P)$ still controls fusion on the set of fixed points of V. So, we can apply the same technique again. But some of these groups V may lie in the same conjugacy class. So, this problem has to be solved first. If U is contained in V and V^g for some $g \in G$ then also $U^{g^{-1}}$ is contained in V. Instead of deciding the conjugacy of overgroups of U one can determine those conjugates of U that are contained in the same overgroup V. The inverses of the conjugating elements then will produce the conjugates of V containing U. This approach can be realized with the help of group theoretic computer packages. Presently, the design construction program DISCRETA uses GAP [20] for some group theoretic computations needed to solve isomorphism problems for designs with some prescribed automorphism group automatically.

In many cases, if U is sufficiently large, each of the overgroups constructed has no fixed points. So, then again $N_G(U)$ controls the G-fusion on the set of fixed points of U. But it should be warned that even then U need not be the full stabilizer of its fixed points. This occurs for example if $U = PSL(2,11)$ is prescribed as an automorphism group of a 5-$(12,6,1)$ design, a Witt design with full automorphism group M_{12}. There are two such designs which are interchanged by the normalizer $PGL(2,11)$ of U. Prescribing $PSL(2,23)$ as an automorphism group of a 5-$(24,8,1)$ design, the big Witt design, as well results in two solutions which are interchanged by $PGL(2,23)$. The Mathieu groups, which are the full automorphism groups of these designs, are not obtained in this way.

It is a strong feature of this approach that in many cases one can decide that all objects admitting a certain group of automorphisms all must be pairwise non-isomorphic without even knowing the objects.

As an example consider $PGL(2,p)$ for some prime p. This subgroup of S_{p+1} contains a Sylow p-subgroup P of S_{p+1} and even the normalizer of P. So, all objects fixed by $PGL(2,p)$ lie in pairwise different orbits under S_{p+1}. The smallest subgroup for which this argument holds in this case is the holomorph of P acting with an additional fixed point. For any overgroup U of P any overgroup of the holomorph of P controls the S_{p+1} fusion on the set of fixed points of U. In particular $PGL(2,p)$ controls the S_{p+1} fusion on the set of all fixed points of $PSL(2,p)$.

4 Iterative Constructions

The homomorphism principle is well suited for an iteration. So, a problem is simplified in several steps. The solution strategy starts with the simplest version and step by step tries to lift the solutions to the preimage spaces. In our aim to construct objects in each step some kind of extension occurs, depending on the actual structure.

We want to discuss some general aspects of such extensions and consider a single extension step. So, suppose an object ω and another object δ are the building parts of a new object γ, which is an extension of ω by δ. For the same pair (ω, δ) there will be several extensions, in general. These have to be classified up to isomorphism.

Usually, forming γ means to identify some structure S_1 derived from ω with a respective structure S_2 derived from δ. An automorphism of ω preserving S_1 can be applied to ω without changing the isomorphism type of the extension. The same holds for automorphisms of δ preserving S_2. So, we frequently are led to a situation where the Glueing Lemma applies. A more detailed approach may even use prescribed stabilizers to single out objects with certain automorphisms.

The building parts will be considered as distinguished parts of the extension, and classifying these objects will only solve the isomorphism problem up to these substructures being distinguished. By selecting canonical representatives from these orbits one will obtain only *semi-canonical* representatives. Omitting the distinction of the substructures used in the construction may cause that different semicanonical representatives become isomorphic.

We proceed in two sub-steps. Firstly, we classify triples (ω, δ, γ) and from these classes we, secondly, form the classes of objects γ.

Theorem 4.1 (Iteration Step). *Let a group A and a group B act faithfully on the space $\Omega \times \Delta \times \Gamma$ such that the projection onto $\Omega \times \Delta$ is compatible with the group action of A and the projection onto Γ is compatible with the group action of B. For each triple (ω, δ, γ) let $N_A((\omega, \delta, \gamma)) = N_B((\omega, \delta, \gamma))$. Then representatives for the B-orbits on Γ and their stabilizers in B can be obtained by the following steps.*

- *For each representative (ω, δ) from an A-orbit and its stabilizer compute representatives from the orbits of the stabilizer on the set of extensions (ω, δ, γ), where γ varies, together with their stabilizers $N_A((\omega, \delta, \gamma))$. Declare all such γ as candidates for representatives.*
- *Run through the candidates (ω, δ, γ) and do:*
  Declare γ to be a representative, determine all $(\omega', \delta', \gamma)$ for this γ.
  For each $(\omega', \delta', \gamma)$ decide whether there exists some $b \in B$ such that $(\omega, \delta, \gamma)^b = (\omega', \delta', \gamma)$.
  If such an element $b \in B$ exists then enlarge $N_A((\omega, \delta, \gamma))$ by the coset $N_A((\omega, \delta, \gamma))b$.
  Determine the representative $(\omega', \delta', \gamma)^a$ of its A-orbit and test whether $\gamma^a = \gamma$. If the test is negative then γ^a is removed from the set of candidates.

The condition that $N_A((\omega, \delta, \gamma)) = N_B((\omega, \delta, \gamma))$ is fulfilled if both normalizers act faithfully on the object (ω, δ, γ) and all its automorphisms are contained in A and in B.

The proof of Theorem 4.1 is straightforward, using the homomorphism principle twice. The two projections are the homomorphisms needed. The most interesting part is the determination of $N_B(\gamma)$. Here the bijection between an orbit and the set of right cosets of a stabilizer is used.

An important special case of Theorem 4.1 is the Leiterspiel (snakes and ladders) by B. Schmalz[51] see also [37] which computes double coset representatives in this way.

Further examples are provided by semi-direct products of groups where homomorphisms from a factor group into the automorphism group of the normal subgroups have to be classified, see [31], [36], [38]. A fast graph generator relying on these principles is described in [21]. In a generator for isomers, ligands have to be placed onto places of a skeleton [50], [26], and below we will form extensions of t-design.

There is a variation called *Canonocal Generation* by B.D. McKay [44]. In some way a unique orbit of δ in the resulting object γ under $Aut(\gamma)$ must be distinguished. Then all extensions are rejected in which γ does not lie in the distinguished orbit. For details the reader is referred to [44].

A technique called *Orderly Generation* was introduced by R. Read [48], I. Faradzev [19]. If there exists a total ordering $\leq$ on $\Omega \times \Delta$ and each γ with $\gamma \leq \gamma^B$ is an extension of some (ω, δ) such that $(\omega, \delta) \leq (\omega, \delta)^A$ then only smallest elements of all orbits need to be extended and a test for $\gamma \leq \gamma^B$ suffices for these extensions.

In addition, invariants may be used to reduce the computation time. Each first appearance of a new value of the invariant indicates that a new isomorphism type has been found. A comparison of such a value with the previous values can be obtained in constant time using a good hash function.

The requirements for orderly generation often can be fulfilled in combinatorial construction problems. Here, in an extension step, often some set is extended by just one element. We consider a fairly general version but explicitly fix the action.

Suppose a group G acts on a finite set X. We impose on X an ordering $<$ such that also the set 2^X of all subsets of X is lexicographically ordered. Each orbit S^G for some $S \in 2^X$ contains a lexicographically minimal element S_0 which we denote as the canonical representative with respect to $<$. In short we say $S \in canon_<(2^X, G)$ iff $S \leq S^G$. Then we have the following fundamental lemma [21].

Theorem 4.2 (Orderly Generation). *If* $S \in canon_<(2^X, G)$, $T \subset S$, *and* $T < S$ *then also* $T \in canon_<(2^X, G)$.

Proof: Let $S = T_1 \cup T_2$ and $T_1 < S$ but T_1 not a canonical representative. Then there exists some $g \in G$ such that $T_1^g < T_1$. If $T_1^g = \{x_1, \cdots, x_t\}$ where $x_1 < x_2 < \cdots < x_t$ then for some $i \leq t$ we have $T_1 = \{x_1, \cdots, x_{i-1}, x_i', \cdots, x_t'\}$ and $x_i < x_i'$. Since $S^g = T_1^g \cup T_2^g \supseteq \{x_1, \cdots, x_i\}$, we obtain $S^g < T_1 < S$ contradicting the hypothesis on S.

Thus, we only have to enlarge representatives T of smaller cardinality by elements x which are larger than each element in T to obtain candidates for representatives of greater cardinality. This approach can be refined by noticing that there are some further elements y larger than each element in T which can be excluded as x.

Lemma 4.3 (Semi-canonicity). *Let $T = \{x_1, \cdots, x_t\}$ be canonical, where $x_1 < x_2 < \cdots < x_t$. Then for $y \in x^{N_G(\{x_1,\cdots,x_i\})}$ for $x_i < x < x_{i+1}$ and $i < t$ the set $T \cup \{y\}$ is not in $canon_<(2^X, G)$. If $i = t$ then if y is not minimal in its orbit under $N_G(T)$ the set $T \cup \{y\}$ is not in $canon_<(2^X, G)$.*

Proof: Let $y = x^g$ for some $g \in N_G(\{x_1, \cdots, x_i\})$ and some x with $x_i < x < x_{i+1}, \quad i < t$. Then

$$(\{x_1, \cdots, x_i\} \cup \{x\}) < \{x_1, \cdots, x_i\}^g \cup \{x^g\} \leq T \cup \{y\}$$

such that the subset $\{x_1, \cdots, x_i, y\}$ of T is smaller than T but not canonical. Therefore by Theorem 4.2 also $T \cup \{y\}$ is not canonical. The second case is obvious.

The candidates obtained after removing the cases of the preceding lemma are semi-canonical [46].

A test for minimality for each remaining candidate S now has to decide whether there exists some $g \in G$ such that $S^g < S$.

Often the required solutions have to fulfill some constraints. Checking these constraints is usually much faster than a canonicity check. So, a sieving with respect to the constraints will save time. One may even delay a canonicity check to the end of several extension steps hoping that after sieving only a few candidates remain. Now, if a candidate S is not minimal in its orbit then already its predecessor may not have been minimal also. In the light of Theorem 4.2 it is therefore useful to determine the first extension step where this non-canonicity could have been detected. Then all further extensions of this candidate must also be rejected. Depending on the selectivity of the additional constraints a delicate balance of steps with constraint checking only and steps with canonicity check combined with tracing back to the earliest detection point is needed for a fast strategy.

5 Groups and Designs

In this section, the theory shall be illustrated by a task from combinatorics. We apply the theorems of the preceding section to the problem of constructing t-(v, k, λ) designs up to isomorphism. The problems are first to find such designs for some large t but small v and then to solve the isomorphism problem for these designs. A successful strategy has been to prescribe a large group A of automorphisms and reduce the question of which k-sets should be taken as blocks to the question of which A-orbits on k-sets should be combined to form the set of blocks. So, a big group will reduce the problem size considerably.

It remains to find t-designs with only trivial automorphisms and it is to be expected that most of the designs will be of this type. But there are some parameter sets where this expectation is wrong. So, it is known that there is only one isomorphism type of designs for each of the parameter sets 2-$(7,3,1)$, 3-$(8,4,1)$, 3-$(10,4,1)$, 5-$(12,6,1)$, 5-$(24,8,1)$, each with a big automorphism group, $PGL(2,3)$, $AGL(3,4)$, $P\Gamma L(2,9)$ M_{12}, and M_{24}, respectively. It is not clear whether there are only finitely many such cases. On the other hand, our results below will confirm that most t-designs will have a trivial automorphism group.

We start by constructing t-designs with a prescribed group of automorphisms from k-orbits of that group. For collecting k-orbits requires to have these orbits. The preceding section provides at least three ways to approach this problem.

- Orderly generation
- Homomorphism principle à la *Leiterspiel (snakes and ladders)* [51]
- Prescribed stabilizers

While the use of orderly generation on a high level description is sufficiently explained in the preceding section the other two topics need some explanation. The Leiterspiel is an example for Theorem 4.1. It proceeds from orbits on $(k-1)$-sets to orbits on k-sets in two steps. In the first step sequences are classified consisting in the first entry of a $(k-1)$-set and in the second entry of a single point not contained in the first entry. Iterating these two up and down steps results in a growing amount of information to be stored to find out at which step representatives from previously different orbits fuse into one orbit.

The prescribed stabilizer method is useful for determining k-orbits with non-trivial stabilizers directly. So, from a knowledge of the subgroup lattice of the prescribed automorphism group one starts with a set of representatives from those conjugacy classes of subgroups that may occur as a stabilizer of a k-set. Notice that a subgroup may only fix a k-set if the sizes of point orbits may be added up to k. For illustration we explain an example from [39].

The only non-trivial subgroups of $PSL(2,23)$ that leave a 10-set invariant are subgroups of order 2. So, one can conclude by the Cauchy-Frobenius Lemma or some direct argument that there are exactly 66 orbits with stabilizers of order 2 and 290 orbits with trivial stabilizers. Since $PSL(2,23)$ is 3-homogeneous, each 10-orbit is a 3-design. The 3-designs formed by the orbits with stabilizer of order 2 have only half as many blocks as those with a trivial stabilizer. Each pair of these smaller designs then forms a design with the same number of blocks as the bigger designs. Thus, by grouping the smaller designs into pairs we get 33 designs with the same size as the remaining 290 designs. So, all 10-sets are partitioned into 323 3-designs with the same parameters. Such a partition is called a large set. Large sets are important because there are some famous iterative constructions of infinite families of t-designs that need large sets as a starting point.

In more general situations, a Möbius inversion as mentioned after Corollary 2 in the preceding chapter can determine those subgroups that are the full stabilizers of k-sets.

A t-design consists of a selection of k-sets as blocks such that each t-set is contained in exactly λ blocks. Constructing a t-design with a prescribed group of automorphisms A means to select appropriate A-orbits on k-sets. If a t-set T is contained in a blocks from an orbit K^A of k-sets then also each T^α for $\alpha \in A$ is also contained in a blocks from this orbit. So, only orbit representatives need to be considered. Kramer and Mesner [29] formalized this approach by a matrix equation. The matrix contains a row for each t-orbit and a column for each k-orbit. The entry for row T^A and column K^A is the number of k-sets in K^A that contain T. Then selecting columns such that each T is contained in exactly λ k-sets from these columns amounts to solving a diophantine system of equations with a $0-1$ vector and right hand side a column vector with constant entry λ.

Though solving this problem implies solving the binary packing problem which is NP-complete, there are several algorithms which are successful at least for moderately sized problems. For very small λ one can use backtracking [44] or some clever tabu search [45,11]. These programs constructed the largest known Steiner 5-systems on up to 244 points. That approach is also successful when there are only a few rows and many columns. For larger values of λ a version of the LLL-basis reduction algorithm is applied, see [30,57]. The software package DISCRETA developed in Bayreuth by A. Betten, A. Wassermann and the author contains implementations of these algorithms. A graphical user interface allows an easy handling of them. The system led to many new results some of which are listed in the introduction.

The new 8-designs with automorphism group $ASL(3,3)$ were found as solutions of the Kramer-Mesner system of diophantine equations. There were many solutions and the number of isomorphism types is obtained using Theorem 3.8. Using DISCRETA we find that $AGL(3,3)$ is not admitted as a group of automorphisms of any of these designs. So, the normalizer $AGL(3,3)$ of $ASL(3,3)$ has orbits of length 2 on the set of these designs. Since $AGL(3,3)$ is a maximal subgroup of S_{27} and $ASL(3,3)$ is a maximal subgroup of A_{27}, see [40], then $ASL(3,3)$ is the full automorphism group of these designs. The number of isomorphism types can thus be obtained by dividing the number of solutions by 2. We have determined this number for the smaller cases in this way.

This argument was already applied by Schmalz to the classification of t-designs. The special case Corollary 3.9 was later used to show that all 4,996,426 7-$(33, 8, 10)$ designs with automorphism group $P\Gamma L(2,32)$ are pairwise non-isomorphic.

The known Steiner 5-designs can also be classified by this approach. There are no 5-$(p+1,6,1)$ designs admitting $PGL(2,p)$ by a result of Denniston [18]. So, all such designs found by prescribing $PSL(2,p)$ are grouped into isomorphic pairs under $PGL(2,p)$ and these are the isomorphism types. Thus, the number of isomorphism types in this case is just half the number of solutions. This could be applied to obtain the exact number of isomorphism types for $p = 11, 23, 41, 71$ and lower bounds for further incomplete sets of solutions. In particular, there are exactly 3 isomorphism types of 5-$(84,6,1)$ designs consisting of orbits with trivial stabilizer only and group $PSL(2,83)$ [11].

The localization technique Theorem 3.15 was applied to classify the 8-$(31,10,\lambda)$ designs with prescribed group $PSL(3,5)$. It can also be used to solve the problems given in Kramer and Mesner's paper [29] mentioned above. There subgroups of the holomorph of C_{13} containing C_{13} had been prescribed as groups of automorphisms. For the full holomorph which is the normalizer of C_{13} in S_{13} all designs found are pairwise non isomorphic. Thus, we find 28 isomorphism types of 2-$(13,5,45)$ designs with this automorphism group. For the unique subgroup of index 2 there exist 890 designs allowing this group. After removing the 28 designs of the overgroup which we already considered the remaining designs fall into orbits of length 2 under the action of the holomorph which controls the S_{13} fusion. So, there result 431 new isomorphism types. Similarly, for the unique subgroup of index 3 we obtain from 24643 designs admitting this group 8205 new isomorphism types. The subgroup D_{13} admits more than 21,030,000 solutions such that with this automorphism group there exist more than 3,500,000 further isomorphism types.

We now proceed with an analysis of a well known construction and applications of it to 7- and 8-designs.

The extension method of van Leijenhorst [56] and Tran van Trung [55] builds a new design from two given designs with some appropriate parameter sets. The construction can be explained in the following way. From any t-(v,k,λ) design $\mathcal{D}$ one can obtain two smaller designs. A point x is fixed and the blocks are classified into those that contain x and those that do not contain x. Then $\{B\backslash\{x\} | B \in \mathcal{D}\}$ is a $(t-1)$-$(v-1,k-1,\lambda)$ design, the derived design at x, and $\{B | B \in \mathcal{D}, x \notin B\}$ is a $(t-1)$-$(v-1,k,\lambda)(v-k)/(k-t+1)$ design, the residual design at x. Of course it looks promising to reverse this process. Then two given $(t-1)$-designs $\mathcal{D}_1$ with the parameters of a derived and $\mathcal{D}_2$ with the parameters of a residual design should be combined to a t-design. The construction simply has to add a new point to each block of $\mathcal{D}_1$, obtaining $\mathcal{D}_1 * \{v\}$, and then forms $\mathcal{D} = \mathcal{D}_1 * \{v\} \cup \mathcal{D}_2$. Unfortunately, only very rarely $\mathcal{D}$ is a t-design, as in the case of Alltop's Theorem [1]. But, as van Leijenhorst and Tran van Trung noticed the result is at least a $(t-1)$-$(v,k,\lambda+\lambda(v-k)/(k-t+1))$ design.

We will take a closer look at this construction. So, suppose the new point added is v. Then any $(t-1)$-set T' not containing v is contained in λ blocks from $\mathcal{D}_1 * \{v\}$ and $\lambda(v-k)/(k-t+1)$ blocks from $\mathcal{D}_2$. A $(t-1)$-set T' containing v is only contained in blocks from $\mathcal{D}_1 * \{v\}$. So, after removing v from T' and each block in $\mathcal{D}_1 * \{v\}$ we obtain the number of blocks from $\mathcal{D}_1$ containing a $(t-2)$-set. Therefore, $\mathcal{D}$ is a $(t-1)$-design if this number is equal to $\lambda + \lambda(v-k)/(k-t+1)$. But this holds because the $(t-1)$-design $\mathcal{D}_1$ is also a $(t-2)$-design with just this parameter.

We notice, that we know more about $\mathcal{D}$. Each t-set containing v is contained in exactly λ blocks. Only those t-sets not containing v may be contained in a different number of blocks.

Another important aspect results from the fact that in the construction both designs $\mathcal{D}_1$ and $\mathcal{D}_2$ can be replaced by any other design with the same parameters. So, even when starting with only two designs we can replace them by isomorphic copies to get a large number of extensions. Of course, many of them will be isomorphic but one can also obtain non-isomorphic designs in this way. We want to determine the isomorphism types in important cases. So, if $\mathcal{D}_1$ and $\mathcal{D}_2$ are replaced by isomorphic copies then one can apply a permutation to the point set such that at least one of $\mathcal{D}_1 * \{v\}$ or $\mathcal{D}_2$ is in its original form. We therefore assume that only $\mathcal{D}_2$ is replaced by an isomorphic copy. Then we formally have applied a permutation π on the set of points $\{1, \cdots, v-1\}$ to the elements in the blocks of $\mathcal{D}_2$. We denote the result by $\mathcal{D}_2^\pi$ and get the extension $\mathcal{D}(\pi) = \mathcal{D}_1 * \{v\} \cup \mathcal{D}_2^\pi$. In this situation the Glueing Lemma 3.3 applies.

Theorem 5.1. *Let $\mathcal{D}_1$ be a $(t-1)$-$(v-1, k-1, \lambda)$ design with automorphism group A_1 and $\mathcal{D}_2$ be a $(t-1)$-$(v-1, k, \lambda(v-k)/(k-t+1))$ design with automorphism group A_2, where the point set in each case is $V' = \{1, \cdots, v-1\}$. Then there exists an isomorphism*

$$\phi : \mathcal{D}(\pi_1) \mapsto \mathcal{D}(\pi_2)$$

for permutations π_1, π_2 on V' such that ϕ fixes v if and only if

$$A_1 \pi_1 A_2 = A_1 \pi_2 A_2.$$

For the proof notice that any isomorphism ϕ fixing v has to map the derived design of $\mathcal{D}(\pi_1)$ at v onto the derived design of $\mathcal{D}(\pi_2)$ at v. The restriction to V' is an automorphism α_1 of $\mathcal{D}_1$. Similarly, the residual designs are mapped one onto the other such that $\pi_1^{-1}\phi\pi_2$ restricted to V' is an automorphism α_2 of $\mathcal{D}_2$. Thus, $\pi_1^{-1}\alpha_2\pi_2 = \alpha_1$ and $\pi_2 = \alpha_1\pi_1^{-1}\alpha_2$. On the other hand, if π_1 and π_2 lie in the same double coset modulo A_1 and A_2 then $\pi_2 = \alpha_1\pi_1^{-1}\alpha_2$, for some α_i in A_i, and $\alpha_1\pi_1^{-1}\alpha_2$ extended by the fixed point v maps $\mathcal{D}(\pi_1)$ onto $\mathcal{D}(\pi_2)$.

The group of all permutations fixing the new point v acts on the set of design extensions and its orbits are refinements of the general isomorphism classes of designs. So, we can obtain the general isomorphism classes if we can decide which of the special extension classes belong to the same general class. This can be done by the following result.

Theorem 5.2. *Let $\mathcal{D}_1$ and $\mathcal{D}_2$ be as before with automorphism groups A_1 and A_2 respectively. Let $\mathcal{D}(\pi_1)$ and $\mathcal{D}(\pi_2)$ be two design extensions of $\mathcal{D}_1$ and $\mathcal{D}_2$. Suppose, $\phi_1 : \mathcal{D} \mapsto \mathcal{D}(\pi_1)$ and $\phi_2 : \mathcal{D} \mapsto \mathcal{D}(\pi_2)$ are two isomorphisms from a design $\mathcal{D}$ to $\mathcal{D}(\pi_1)$ and $\mathcal{D}(\pi_2)$. Then there exists an isomorphism $\phi : \mathcal{D}(\pi_1) \mapsto \mathcal{D}(\pi_2)$ fixing some point x if and only if there exists an automorphism α of $\mathcal{D}$ mapping the point $\phi_1^{-1}(x)$ to the point $\phi_2^{-1}(x)$. In particular,*

$$Aut(\mathcal{D}(\pi_1)))_v = A_1 \cap A_2^{\pi_1}$$

The number of isomorphism types of extensions is at least

$$\frac{1}{v}|A_1 \backslash S_v / A_2|$$

For the proof notice that for any such α the composition of isomorphisms $\phi_1^{-1}\alpha\phi_2$ is an isomorphism ϕ fixing v. On the other hand, given such a ϕ we obtain α by $\alpha = \phi_1\phi\phi_2^{-1}$. The special case $\phi_1 = \phi_2$ yields the description of the stabilizer of v in the automorphism group $Aut(\mathcal{D}(\pi_1))$. From the description of the special isomorphism classes by means of double cosets in the Glueing Lemma 3.3 we obtain the lower bound.

More generally, if there are n_1 t-$(v-1, k-1, \lambda)$ designs with automorphism group A_1 and n_2 t-$(v-1, k, \lambda')$ designs with automorphism group A_2 and m double cosets with stabilizer order up to l then there exist at least $n_1 \times n_2 \times m/v$ isomorphism types of t-$(v, k, \lambda + \lambda')$ designs with automorphism group order up to $l \times v$.

The pairs of designs for which the extension method can be applied can be obtained from any t-design. One only has to take a derived and a residual design and then can combine them again twisted by a renaming of the points of one of the two designs. Thus, from only one t-design there results a large number of $(t-1)$-designs.

In particular each of the new 8-$(27, 13, \lambda)$ designs with automorphism group $ASL(3,3)$ gives by Alltop's construction a 9-$(28, 14, \lambda)$ design $\mathcal{D}$ with automorphism group $ASL(3,3)+$. We form the derived and the residual design of $\mathcal{D}$ with respect to some point. Then we can apply the extension construction to these two designs with a twisting of the residual design. The different twistings using the described procedure then yield

$$|SL(3,3)\backslash S_{27}/ASL(3,3)|/28 \geq 16913871764453533$$

new 8-$(28, 14, \lambda + \lambda')$ designs with various groups of automorphisms.

We now look for situations where it can be shown that the new point v must be fixed by all isomorphisms between any extensions of two given designs. Then the double cosets above are in bijection to the isomorphism types. Also the stabilizers of v are the full automorphism groups of the designs obtained by extension and instead of the lower bound we have an exact number of isomorphism types.

Lemma 5.3. *For a t-subset T of X let $a_i(T)$ denote the number of blocks of $\mathcal{D}_i$ containing T, for $i = 1, 2$. If for each point p there exists a T containing p such that for all permutations π $\lambda \neq a_1(T) + a_2(T^\pi)$ then in each extension the new point x is the only point such that every t-subset containing x lies in exactly λ blocks.*

Proof The t-subsets that contain x are contained in exactly the blocks that result from adding x to the blocks of $\mathcal{D}_1$. Thus those t-subsets lie in exactly λ blocks. So, any isomorphism α of any extension $\mathcal{D}_\pi$ mapping x to a point $p \neq x$ will have to map the set of blocks containing some t-subset T with $x \in T$ onto the set of blocks containing T^α where $p \in T^\alpha$. Thus, both sets of blocks must have the same cardinality. Now, T is contained in exactly λ blocks and $T' = T^\alpha$ is contained in $a_1(T')$ blocks of $\mathcal{D}_1$. The remaining blocks containing T' are from $\mathcal{D}_\pi$ such that the renaming π of the points in $\mathcal{D}_2$ causes these blocks to contain T'. So, for this π the existence of α would imply $\lambda = a_1(T') + a_2(T'^\pi)$ contrary to our assumption.

Of course it is not feasible to run through all permutations π to check whether the assumptions of the Lemma are satisfied. So, we look for sufficient conditions that are easier to check and still give the conclusion of the Lemma. First, we have orbits of the automorphism groups A_1 of $\mathcal{D}_1$ and A_2 of $\mathcal{D}_2$ on the set of all t-subsets. All T from such an orbit are contained in the same number of blocks of the respective design. A permutation π maps an orbit $T_i^{A_1}$ into several orbits $T_j^{A_2}$. Let

$$a_{ij} = |\{S : S \in T_i^{A_1}, S^\pi \in T_j^{A_2}\}|$$

where i and j run through the orbit numbers. Then

$$\sum_j a_{ij} = |T_i^{A_1}|$$

and

$$\sum_i a_{ij} = |T_j^{A_2}|.$$

If the condition $a_1(T) + a_2(T^\pi) = \lambda$ is violated then α cannot exist. Therefore all a_{ij} where $a_1(T_i^{A_1}) + a_2(a_j^{A_2}) \neq \lambda$ are zero. If the remaining system of Diophantine equations has no solutions then α cannot exist. So, this set of equations yields a sufficient condition to conclude that all isomorphism types of extensions of two particular designs are in bijection to the double cosets of A_1 and A_2 in S_{v-1}.

A special situation occurs when a prescribed automorphism group is transitive on the set of points.

A very prominent example is formed by the smallest 6-designs. These designs have parameters 6-$(14, 7, 4)$ and are constructed by Alltop's Theorem from a 5-$(13, 6, 4)$ design, [30]. The automorphism group of the 5-design is C_{13} and there exist exactly 24 solutions of the Kramer-Mesner system of equations. Thus the isomorphism types are given by the orbits of the normalizer $Hol(C_{13})$ of C_{13} in S_{13} on the set of points which have sizes 1 and 12. So, there are exactly 2 isomorphism types of 5-$(13, 6, 4)$ designs with automorphism group C_{13}. By an argument of Kreher and Radziszowski in [30] the isomorphism types of the extensions often can also be determined in such a situation.

In Alltop's construction, the blocks of the residual design that by which the derived design is extended are uniquely determined by the derived design. So, all automorphisms of $\mathcal{D}$ extend to the extended design $\mathcal{D}^+$. Therefore,

$$Aut(\mathcal{D}) = Aut(\mathcal{D}^+)_v.$$

Taking the derived designs at other points thus give designs whose automorphism groups are the corresponding other point stabilizers.

Theorem 5.4. *Let A be the full automorphism group of t-(v, k, λ) designs where $v = 2k + 1$ and t is even or $\lambda = \frac{1}{2}\binom{v-t}{k-t}$. If A acts transitively on the point set but has no transitive extension then two Alltop extensions $\mathcal{D}_1^+$ and $\mathcal{D}_2^+$ of t-(v, k, λ) designs $\mathcal{D}_1$ and $\mathcal{D}_2$ with full automorphism group A are isomorphic if and only if $\mathcal{D}_1$ and $\mathcal{D}_2$ are isomorphic.*

The proof is immediate from the fact that the full automorphism group of an Alltop extension here either is transitive or has the new point as a fixed point. So, if A cannot be transitively extended then the derived design at the new point is not isomorphic to any other derived design and thus characterizes the isomorphism type of the Alltop extension.

In case of the 5-$(13, 6, 4)$ design the extended design still has C_{13} as its full automorphism group with an additional fixed point 14. So, all other points form just one orbit and have a trivial stabilizer. Therefore the other derived designs have trivial automorphism groups. In particular, different isomorphism types of 5-$(13, 6, 4)$ designs with automorphism group C_{13} extend to different isomorphism types of 6-$(14, 7, 4)$ designs. The new 9-$(28, 14, \lambda)$ designs are obtained from 8-$(27, 13, \lambda)$ designs by using Alltop's construction. Here, the automorphism group $ASL(3, 3)$ acts transitively but cannot be transitively extended. Otherwise, the extended group would have to be at least 2-transitive and there is even no primitive group on 28 points different from the alternating and the full symmetric group containing $ASL(3, 3)$, see for example [16]. So, different isomorphism types of 8-$(27, 13, \lambda)$ designs with automorphism group $ASL(3,3)$ extend to different isomorphism types of 9-$(28,14,\lambda)$ designs.

There are many further situations where the automorphism group is transitive and Alltop's construction applies. So, it is sufficient in these cases to verify that the group is not the stabilizer of a point in a primitive group. Then the Theorem allows to determine the isomorphism types of Alltop extensions from the isomorphism types of the given designs.

We now again consider the general situation of extensions and assume a transitive automorphism group on the design with the larger block size.

Theorem 5.5. *Let A_1 be the automorphism group of a $(t-1)$-$(v-1,k-1,\lambda)$ design $\mathcal{D}_1$ and A_2 the automorphism group of a $(t-1)$-$(v-1,k,\lambda(v-k)/(k-t+1))$ design $\mathcal{D}_2$ both defined on a point set V. Let A_2 act transitively on the set of $v-1$ points and let none of the extensions be a t-(v,k,λ) design. Then the isomorphism types of extensions of $\mathcal{D}_1$ and $\mathcal{D}_2$ to a $(t-1)$-$(v,k,\lambda(v-t+1)/(k-t+1))$ design are in bijection to the double cosets $A_1 \backslash Sym(V)/A_2$. The automorphism group of an extension of $\mathcal{D}_1$ by $\mathcal{D}_2^\pi$ for some permutation π is $A_1 \cap A_2{}^\pi$.*

Proof By the Lemma it suffices to show that for each point of the point set different from the added point there exists a t-subset T such that this T is not contained in λ blocks of the extension design. Since A_2 is transitive on these points, each orbit of A_2 on t-subsets contains a block containing a designated point. So, we only have to find one orbit T^{A_2} such that $a_1(T) + a_2(T)$ is different from λ. This means that the extension is not a t-design, as assumed in the Theorem.

Corollary 5.6. *If for a prescribed automorphism group A there exist n_1 designs with parameter set $(t-1)$-$(v-1,k-1,\lambda)$ and n_2 designs with parameter set $(t-1)$-$(v,k,\lambda(v-k)/(k-t+1))$ then under the assumptions of the last Theorem there exist $n_1 \cdot n_2 \cdot |A \backslash Sym(V)/A|$ isomorphism types of extensions.*

In the last Theorem, it suffices to find only one t-orbit of A_2 such that any t-subset T from this orbit lies in strictly more than λ blocks of $\mathcal{D}_2$. Then one can also conclude that none of the extensions will be a t-design. This holds for example for each of the 113 7-$(24,9,40)$ designs with automorphism group $PGL(2,23)$ as can be verified by DISCRETA. So, by the Corollary forming the extensions with any of the 138 7-$(24,8,5)$ designs with automorphism group $PSL(2,23)$ yields as many isomorphism types of 7-$(25,9,45)$ designs as there are double cosets of these automorphism groups in S_{24}. Thus, we obtain in this way exactly

$$113 \times 138 \times 8{,}414{,}188{,}491{,}217{,}916 = 131{,}210{,}855{,}332{,}052{,}182{,}104$$

isomorphism types of 7-$(25, 9, 45)$ designs.

In the introduction we have given a table on extensions of designs with results obtained from this approach. We discuss some entries of that table.

- The lower bound for the number of 7-$(26, 10, 342)$ designs in the last row is obtained by multiplying the numbers of designs with the parameters 7-$(25, 9, 54)$ and 7-$(25, 10, 288)$ in that row by the number of double cosets of the identity in S_{25}, i. e. 25!, and then dividing by 26, which is the maximal number of designs that may be isomorphic after making the new point an ordinary point. It is likely, that all these designs are pairwise non-isomorphic such that the last division is superfluous.
- The new point v is distinguished in some extensions in the following cases: Row No. 1: the third and fourth 7-$(24, 10, 240)$ designs from the list in [26], Row No. 3: all extensions, Row No. 4: all of 10 extensions tested, Row No. 5: all of 15 extensions tested, Row No. 6: each of 21 7-$(26, 9, 54)$ designs, Row No. 8: at least five 7-$(27, 11, 2295)$ designs, Row No. 9: the ninth 7-$(27, 11, 2295)$ design from the list in [26], Row No.10: all of 500 extensions tested.
- Row number 8 shows an example of totally different automorphism groups.
- Row number 7 is interesting, because the 7-$(26, 13, 13524)$ design results from first taking the residual design of a 7-$(26, 12, 5796)$ design and then extending that by Alltop's construction. Thus, here the existence of only one design suffices for the Tran van Trung-van Leijenhorst construction. Taking the residual design reduces the automorphism group to the stabilizer of a point, in this case $A\Gamma L(1, 25)$.
- Row 11 with the group $Sp(6, 2)$ acting on 28 points is similar. There are some further values of λ for which there exists a 7-$(26, 13, \lambda)$ design with automorphism group $Sp(6, 2)$. For each of them the same construction can be applied. In each of these cases we cannot give the number of isomorphism types of extensions.
- Row 12 uses the results from row 1 and row 4. So, in this case we are supposed to see all kinds of subgroups of $PGL(2, 23)$ as automorphism groups. We then can combine any such pair and form their double cosets in S_{25}. This number of double cosets then has to be multiplied with the number of solutions belonging to these automorphism groups. So, this illustrates the theory given above.

References

1. W. O. ALLTOP: Extending t-designs. *J. Comb. Theory(A)* **18** (1975), 177-186.
2. A. BETTEN, A. KERBER, R. LAUE, A. WASSERMANN: Es gibt 7-Designs mit kleinen Parametern! *Bayreuther Math. Schr.* **49** (1995), 213.

3. A. BETTEN, A. KERBER, A. KOHNERT, R. LAUE, A. WASSERMANN: The discovery of simple 7-designs with automorphism group $P\Gamma L(2,32)$. *Proc of AAECC 11, Springer* LN in Computer Science **948** (1995), 131–145.
4. A. BETTEN, R. LAUE, A. WASSERMANN: Simple 7-Designs With Small Parameters. *J.Comb. Designs* **7** (1999), 79–94.
5. A. BETTEN, R. LAUE, A. WASSERMANN: Some simple 7-designs. *Geometry, Combinatorial Designs and Related Structures, Proceedings of the First Pythagorean Conference.* J. W. P. Hirschfeld, S. S. Magliveras, M. J. de Resmini eds. Cambridge University Press, LMS Lecture Notes **245** (1997), 15-25.
6. A. BETTEN, R. LAUE, A. WASSERMANN: Simple 6- and 7-designs on 19 to 33 points. *Congressus Numerantium* **123** (1997), 149-160.
7. A. BETTEN, M. C. KLIN, R. LAUE, A. WASSERMANN: Graphical t-Designs via polynomial Kramer-Mesner matrices. *Discrete Mathematics* **197/198** (1999), 83–109.
8. A. BETTEN, A. KERBER, R. LAUE, A. WASSERMANN: Simple 8-designs with small parameters. *Designs, Codes and Cryptography* **15** (1998), 5-27.
9. A. BETTEN, R. LAUE, A. WASSERMANN: Simple 8-(40,11,1440) designs. *Discrete Applied Mathematics* **95** (1999), 109-114.
10. A. BETTEN, R. LAUE, A. WASSERMANN: A Steiner 5-Design on 36 Points. *Designs, Codes and Cryptography* **17** (1999), 181-186.
11. A. BETTEN, S. MOLODTSOV, R. LAUE, A. WASSERMANN: Steiner systems with automorphism group $PSL(2,71)$, $PSL(2,83)$ and $P\Sigma L(2,3^5)$. *J. of Geometry* **67** (2000), 35–41.
12. J. BIEGHOLDT: Computerunterstützte Berechnung von Multigraphen mittels Homomorphieprinzip. Diplomarbeit Universität Bayreuth, may 28, 1995.
13. N. G. DE BRUIJN: Pólya's theory of counting. *Applied Combinatorial Mathematics* BECKENBACH ED.) Wiley, New York, 1964.
14. H. BROWN, L. HJELMELAND, L. MASINTER: Constructive graph labeling using double cosets. *Discrete Math.* **7** (1974), 1-30.
15. W. BURNSIDE: Theory of groups of finite order. Dover Publ. , New York, 1955, reprint of 2nd. edition 1911.
16. L. G. CHOUINARD II, J. JAJCAY, S. S. MAGLIVERAS: Finite groups and designs. *The CRC Handbook of Combinatorial Designs,* C. J. COLBOURN, J. H. ED., CRC Press 1996, 587-615.
17. J. H. CONWAY: An Algorithm for double coset enumeration? *Proceedings of the London Math. Soc. Symp on Computational Group Theory* (1984), ed. M. Atkinson, Academic Press, 33-37
18. R. H. F. DENNISTON: The problem of the higher values of t. *Annals of Discrete Mathemtics* **7** (1980), 65-70.
19. I. A. FARADZHEV: Generation of nonisomorphic graphs with a given degree sequence (russian). In Algorithmic Studies in Combinatorics, Ed. Nauka, Moscow(1978), 11-19.
20. THE GAP-TEAM Groups, algorithms, and programming. Version 4. *Lehrstuhl D für Mathematik, RWTH Aachen, Germany and School of Mathematical and Computational Sciences, U. St.Andrews, Scotland, 1997.*
21. T. GRÜNER, R. LAUE, M. MERINGER: Algorithms for group actions applied to graph generation. *Groups and Computation II, Workshop on Groups and Computation,* june7–10, 1995,113–123, L. Finkelstein, W. M. Kantor, ed., *DIMACS* **28**, AMS 1997.

22. T. GRÜNER, A. KERBER, R. LAUE, M. MERINGER MOLGEN 4.0. *MATCH* **37** (1998), 205–208.
23. R. HAGER, A. KERBER, R. LAUE, D. MOSER, W. WEBER: Construction of orbit representatives. *Bayreuther Math. Schr.* **35** (1991), 157–169.
24. P. HALL: On groups of automorphisms. *J. Reine Angew. Math.* **182** (1940), 194-204.
25. B. HUPPERT: *Endliche Gruppen I.* Springer Grundlehren der mathematischen Wissenschaften Bd **134**, (1967).
26. A. KERBER, R. LAUE: Group actions, double cosets, and homomorphisms: unifying concepts for the constructive theory of discrete structures. *Acta Applicanda Mathematicae* **52**(1998), 63–90.
27. A. KERBER: *Algebraic combinatorics via finite group actions.* BI-Wissenschaftsverlag Mannheim, 1991, 2.nd ed. Springer 1999.
28. M. H. KLIN: On the number of graphs for whicha given permutation group is the automorphism group. *Kibernetika* **6** (1970), 131–137.
29. E. S. KRAMER, D. M. MESNER: t-designs on hypergraphs. *Discrete Math.* **15** (1976), 263–296.
30. D. L. KREHER, S. P. RADZISZOWSKI: The existence of simple 6-(14,7,4) designs. *J. Comb. Theor.* **A** (1986), 237-243.
31. R. LAUE: Zur Konstruktion und Klassifikation endlicher auflösbarer Gruppen. *Bayreuther Math. Schr.* **9** (1982),309 Seiten.
32. R. LAUE: Computing double coset representatives for the generation of solvable groups. *Proceedings EUROCAM'82, Marseille 1982, Springer LN in Computer Science* **144** (1982), 65–70.
33. R. LAUE, J. NEUBÜSER, U. SCHOENWAELDER: Algorithms for finite soluble groups and the SOGOS system. *Computational Group Theory, M. D. Atkinson ed.,Academic Press, London* (1984),105–135.
34. R. LAUE: Abbildungen und Algorithmen. *Seminaire lotharingien de combinatoire, 14^e^ session, Burg Feuerstein,*(1986),115–132.
35. R. LAUE: Eine konstruktive Version des Lemmas von Burnside. *Bayreuther Math. Schr.* **28** (1989), 111-125.
36. R. LAUE: Konstruktionen von Gruppen, Graphen, etc. *Darstellungstheorietage Mai 1992, Akademie gemeinnütziger Wissenschaften zu Erfurt. Sitzungsberichte der Mathematisch-Naturwissenschaftlichen Klasse, Bd.* **4** (1992), 53–65.
37. R. LAUE: Construction of combinatorial objects -A tutorial. *Bayreuther Math. Schr.* **43** (1993), 53–96.
38. R. LAUE: Construction of groups and the constructive approach to group actions. *Symmetry and Structural Properties of Condensed Matter (Zajaczkowo 1994).* T. Lulek, W. Florek, S. Walcerz ed., World Sci., Singapore (1995), 404–416.
39. R. LAUE, S. S. MAGLIVERAS, A. WASSERMANN: New large sets of t-designs. To appear in *J. Combinat. Designs.*
40. M. W. LIEBECK, C. E. PRAEGER, J. SAXL: A classification of the maximal subgroups of the finite alternating and symmetric groups. *J. Algebra* **111** (1987), 365-383.
41. R. K. LINDSAY, B. G. BUCHANAN, E. A. FEIGENBAUM, J. LEDERBERG: Applications of artificial intelligence for organic chemistry: The Dendral Project. McGraw-Hill, New York(1980).

42. S. LINTON: Double coset enumeration. *J. Symbolic Computation* **12** (1991), 415-426.
43. A. C. LUNN, J. K. SENIOR: A method of determining all the solvable groups of given order and its application to the orders 16p and 32p. *Amer. J. Math.* **56** (1934), 319-327.
44. B. D. MCKAY: Isomorph-free exhaustive generation. *J. Algorithms* **26** (1998), 306-324.
45. R. MATHON: Searching for spreads and packings. *Geometry, Combinatorial Designs and Related Structures, Proceedings of the First Pythagorean Conference.* J. W. P. Hirschfeld, S. S. Magliveras, M. J. de Resmini eds. Cambridge University Press, LMS Lecture Notes **245** (1997), 161-176.
46. S. G. MOLODTSOV: Computer-Aided generation of molecular graphs. *Commun. in Math. Chem. (MATCH)* **30** (1994), 213-224.
47. G. PÓLYA: Kombinatorische Anzahlbestimmung für Gruppen, Graphen und chemische Verbindungen. *Acta Math.* **68** (1937), 145-254.
48. R. C. READ: Everyone a winner. *Ann. Discr. Math.* **2** (1978), 107-120.
49. G.-C. ROTA, D. A. SMITH: Enumeration under group action. *Annali Scuola Normale Superiore-Pica. Classe de Scienze* **(4)4**, (1977), 637-646.
50. E. RUCH, W. HÄSSELBARTH, B. RICHTER: Doppelnebenklassen als Klassenbegriff und Nomenklaturprinzip für Isomere und ihre Abzählung. *Theor. Chim. Acta (Berlin)* **19** (1970), 288-300.
51. B. SCHMALZ: Verwendung von Untergruppenleitern zur Bestimmung von Doppelnebenklassen. *Bayreuther Math. Schr.* **31** (1990), 109-143.
52. B. SCHMALZ: The t-Designs with prescribed automorphism group, new simple 6-designs. *J. Combinatorial Designs* **1** (1993),125-170.
53. M. C. SLATTERY: Computing double cosets in soluble groups. To appear in *J. Symbolic Computation* **11** (1999).
54. P. K. STOCKMEYER: Enumeration of graphs with prescribed automorphism group. *Ann Arbor*, 1971.
55. TRAN VAN TRUNG: On the construction of t-designs and the existence of some new infinite series of simple 5-designs. *Arch. Math.* **47** (1986), 187-192.
56. D. C. VAN LEIJENHORST: Orbits on the projective line. *J. Comb. Theory A* **31** (1981), 146–154.
57. A. WASSERMANN: Finding simple t-designs with enumeration techniques. *J. Comb. Designs* **6** (1998), 79-90.
58. D. E. WHITE: Classifying patterns by automorphiom group: an operator theoretic approach. *Discrete Math.* **13** (1975), 277–295.

Group Actions and Classification of Quantum States of the Heisenberg Model of Magnetism

Tadeusz Lulek

Institute of Physics, Pedagogical University
ul. Rejtana 16A, 35-359 Rzeszów, Poland

Abstract The kinematics and dynamics of the Heisenberg model of magnetism is reviewed from the point of view of combinatorics. The general scheme of the duality of Weyl is presented at two levels: (i) the total space of all quantum states of the magnet, (ii) the subspace with the definite number of spin deviations. The role of dual actions of appropriate groups is emphasised and the corresponding quantum numbers are pointed out.

1 Introduction

We aim to review here some important aspects of the Heisenberg model of magnetism from the point of view of combinatorics. The main emphasis will be put onto various group actions, along the general scheme of the duality of Weyl. From the physical point of view, the Heisenberg model of magnetism [1] consists in the assumption of localised elementary carriers of crystalline magnetization: they are ionic magnetic moments, each placed at a node of the crystal, and represented by an angular momentum s referred to as the single-node spin. The angular momentum s might be either the intrinsic spin of an ion (e.g. in the 3d–metals), or its total angular momentum (e.g. from the partially filled 4f shells of the rare earth). An alternative to Heisenberg is the band model [2], in which magnetism is a collective effect, resulting from delocalised electrons, interacting through the whole metallic crystal.

Thus from the structural point of view, the Heisenberg model can be looked at as a composite system of, say, N elementary subsystems (nodes of the crystal), with a definite structure given by the crystal arrangement of ions. We assume for simplicity that each magnetic ion in a crystal is characterised by the same spin s. Our elementary subsystems are thus, by definition, identical, but can be distinguished by their localisation in the crystal.

Quantum-mechanical description involves the linear unitary structure of both the total space $\mathcal{H}$ of all quantum states of the magnet, and the single node space h of a constituent subsystem – a magnetic ion. The latter is the $(2s+1)$-dimensional carrier space of the irreducible representation $D^{(s)}$ of the group SU(2) – the universal cover of the group SO(3,$\mathbb{R}$) of three-dimensional rotations. SO(3,$\mathbb{R}$) is the symmetry group for the isotropic (XXX) Heisenberg model, in which all interactions between spins are spherically symmetric. It is also an approximate symmetry group in anisotropic models (XXY or XYZ).

These model assumptions yield the tensor product structure of the space $\mathcal{H}$, namely

$$\mathcal{H} = \otimes_{j \in \tilde{N}} h_j$$

with h_j being a faithful copy of the space h, associated with the j-th node of the crystal.

In this way, the Heisenberg model of magnetism can be seen as an ideal example of the duality of Weyl [3] between the symmetric group Σ_N of all permutations of some elementary identical objects (single-node spins) and the unitary group SU(2) of the symmetry of a single object (spherical symmetry). The Weyl duality was introduced in early stages of formulation of Quantum Mechanics but soon was obscured by Pauli exclusion principle [4] which admitted only one-dimensional representations of the symmetric group Σ_N as physically sensible (fermions and bosons). It is worth to mention at this point that the Pauli exclusion principle applies to *free* elementary particles and, in particular, does *not* apply to the localised spins of the Heisenberg model. Thus in the latter case all composite quantum states, corresponding to multidimensional parastatistic, are physically realizable.

It is natural to consider the set

$$\tilde{s} = \{|sm\rangle | m = -s, -s+1, \ldots, s\}$$

of a standard basis of angular momentum, so that $|sm\rangle$ is a quantum state in the single-node space h, with m being the z-projection of the spin s. If we introduce, moreover, the set

$$\tilde{N} = \{j = 1, 2, \ldots, N\}$$

of labels of all nodes of the magnetic crystal, then, from the mathematical point of view, we are left with the classical combinatorial problem of "balls" $\tilde{N}$ and "boxes" $\tilde{s}$, or, in another phrasing, with "colouring" of elements of the set $\tilde{N}$ by the "palette" $\tilde{s}$. Physically, it corresponds to a distribution of elementary magnetic momenta over the crystal lattice.

2 Kinematics of the Heisenberg Model: a Linearised Version of the Problem of Balls and Boxes

The sets $\tilde{s}$ ("boxes") and $\tilde{N}$ ("bals"), given respectively by Eqs. (1) and (1), can be considered as a "Troian horse", which introduces combinatorics into the midst of quantum theory of magnetism. Each mapping $f : \tilde{N} \to \tilde{s}$ has the meaning of a distribution of elementary magnetic momenta $\tilde{s}$ over the crystal $\tilde{N}$, and will be refered herefrom to as a *magnetic configuration*. The set

$$\tilde{s}^{\tilde{N}} = \{f : \tilde{N} \to \tilde{s}\}$$

of all magnetic configurations spans the space $\mathcal{H}$ of all quantum states of the magnet. More precisely, this set is an orthonormal basis in the (unitary) space $\mathcal{H}$, so that

$$|\tilde{s}^{\tilde{N}}| = \dim\mathcal{H} = (2s+1)^N ,$$

and

$$\mathcal{H} = lc_{\mathbb{C}}\tilde{s}^{\tilde{N}} ,$$

i.e. $\mathcal{H}$ is the linear closure of the set $\tilde{s}^{\tilde{N}}$ of all magnetic configurations over the field $\mathbb{C}$ of complex numbers. In Dirac notation, elements of $\mathcal{H}$, i.e. possible quantum states of the magnet, are denoted by $|\Psi\rangle$. In particular, the state corresponding to a magnetic configuration $f \in \tilde{s}$ is denoted by $|f\rangle$, or more explicitly,

$$|f\rangle = |m_1, m_2, \ldots m_N\rangle ,$$

with $m_j = f(j)$ being the spin-projection of the j-th node. A magnetic configuration f can thus be looked at as a word of the length N in the alphabet $\tilde{s}$.

The space $\mathcal{H}$, defined in terms of the set $\tilde{s}^{\tilde{N}}$, determines the quantum kinematics of the model. The dynamics is defined by the Hamiltonian – a Hermitean operator $\hat{H} \in \mathrm{End}\mathcal{H}$, expressed in terms of a bilinear form of Cartesian components of the spin operators $\hat{\boldsymbol{s}}_j = (\hat{s}_j^x, \hat{s}_j^y, \hat{s}_j^z)$ for each magnetic node $j \in \tilde{N}$. This Hamiltonian describes interactions between localised spins. Some general physical speculations yield a model assumption that the predominant part of these interactions exhibits spherical symmetry. It can be written in the form

$$\hat{H} = \sum_{j_1, j_2 \in \tilde{N}} J_{j_1 j_2} \hat{\boldsymbol{s}}_{j_1} \cdot \hat{\boldsymbol{s}}_{j_2} ,$$

where $J_{j_1 j_2} = J_{j_2 j_1}$ are real parameters, called exchange integrals (between nodes j_1 and j_2; usually one puts $J_{jj} = 0$). Eq. (2) provides the most general form for the isotropic (XXX) billinear interaction model. One consider also anisotropic bilinear models (XXY and XYZ), in which exchange parameters are indexed as $J_{j_1 j_2}^{\alpha_1 \alpha_2}$, with $\alpha_1, \alpha_2 \in \{x, y, z\}$ but we shall not use such models in the sequal.

The Hamiltonian $\hat{H}$ has also, by the model definition, the symmetry of the group G, describing geometric symmetry of the spatial distribution of the set $\tilde{N}$. It can be, for example, the translational symmetry of the "periodic" crystal, that is a crystal with standard periodic Born-Karman conditions of the solid state physics. Clearly, we have

$$G \subset \Sigma_N ,$$

i.e. G is a subgroup of the symmetric group Σ_N on the set $\tilde{N}$.

From physical reasons, it is sometimes sufficient to consider a simplified model, with the only non-vanishing interactions between nearest neighbours, with the Hamiltonian

$$H_{nn} = J \sum_{j \in \tilde{N}} \sum_{\delta} \hat{s}_j \cdot \hat{s}_{j+\delta} \,,$$

where $j + \delta$ stands for a nearest neighbour (n.n.) of the node j, and the sum over δ extends over all n.n. The Hamiltonian (2) is thus determined by a single parameter J.

If the magnet is, at the time $t = 0$, in a state $|\Psi; 0\rangle \in \mathcal{H}$, then at the time $t > 0$ it reaches another state $|\Psi; t\rangle$, according to the formula

$$|\Psi; t\rangle = \mathrm{e}^{(\mathrm{i}/\hbar)\hat{H}t} |\Psi; 0\rangle \,,$$

where $\hbar$ is the Planck constant, and $\hat{H}$ is the Hamiltonian of the magnet. The main problem in physics consists in finding all quantum stationary states of the magnet, together with their energies. In other words, we are looking for a complete solution of the eigenproblem of the Hamiltonian $\hat{H}$, which is written in a form

$$\hat{H} |\Psi_{Ea}\rangle = E |\Psi_{Ea}\rangle \,.$$

The set of all eigenvalues E of $\hat{H}$ forms the spectrum of energy of the system, whereas the set

$$b = \{|\Psi_{Ea}\rangle\}$$

of all (linearly independent and orthonormal) eigenvectors forms a basis in the space $\mathcal{H}$, so that

$$\mathcal{H} = lc_{\mathbb{C}} b \,, \quad |b| = (2s+1)^N \,.$$

The eigenproblem (2) is solved exactly for an arbitrary N only in the case of a one-dimensional crystal (a magnetic linear chain) by the so called Bethe Ansatz [5]. The corresponding two-and three dimensional problems are still waiting for their solutions. The framework described above allows us to treat the one-, two-, three-dimensional Heisenberg models on equal feating, from a unique point of view of a quantum-mechanical system which is composed from identical subsystems, and its both kinematics and dynamics is described along the duality of Weyl. We have to mention, however, that the viewpoint presented here is far from exhausting the subject "combinatorics and the Heisenberg model". In particular, it leaves aside the specific valuable results for the one-dimensional case, related to the so called Bethe Ansatz [5], and associated very interesting statistical properties of magnetic materials, related to exactly solvable models and integrable systems (see. e.g., a review and collection of papers in [6]. The relevant mathematics exhibits also a rich combinatorial content related to the string hipothesis, the charge of Young tableaux, rigged configurations etc. [7]-[14]. All these subjects are left beside the present article, but are pursued in some other contributions to the ALCOMA Symposium.

3 The First Duality of Weyl: Symmetric vs. Unitary Group

We define now two actions, $P : \Sigma_N \times \mathcal{H} \to \mathcal{H}$ and $Q : \mathrm{U}(2) \times \mathcal{H} \to \mathcal{H}$, of the symmetric group Σ_N of all permutations of nodes of the magnet, and of the unitary group U(2) of linear unitary transformations of the single-object space h related to the rotational symmetry of the model, with both groups acting on the total space $\mathcal{H}$ of all quantum states of the magnet. The corresponding formulas read

$$P(\sigma) = \begin{pmatrix} f \\ f \circ \sigma^{-1} \end{pmatrix} \quad \sigma \in \Sigma_N ,$$

$$Q(a) = \begin{pmatrix} f \\ af \end{pmatrix} \quad a \in \mathrm{U}(2) ,$$

where $f \in \tilde{s}^{\tilde{N}}$, $f \circ \sigma^{-1}$ is the composition of mappings $f : \tilde{N} \to \tilde{s}$ and $\sigma^{-1} : \tilde{N} \to \tilde{N}$, af reads for the multilinear extension of the defining action of $a \in \mathrm{U}(2)$ in h to the N-th tensor power $\mathcal{H}$ (cf. Eq. (1)). Note that P is treated as the linear (not only permutational) representation in the space $\mathcal{H}$, even if one can restrict it to the set $\tilde{s}^{\tilde{N}}$.

Clearly, the action P permutes the nodes of the crystal, whereas the action Q rotates simultaneously all single-particle spins. These actions are considered as mutually dual, in a spirit of terminology of balls (P emerging from $\tilde{N}$) and boxes (Q emerging from h), both acting on the total space $\mathcal{H}$.

The main mathematical observation is that both actions mutually commute, i.e.

$$P(\sigma)Q(a) = Q(a)P(\sigma) = \begin{pmatrix} f \\ af \circ \sigma^{-1} \end{pmatrix} , \quad f \in \tilde{s}^{\tilde{N}} , \quad a \in \mathrm{U}(2) , \quad \sigma \in \Sigma_N .$$

The physical meaning of this fact is expressed in terms of the Heisenberg uncertainty principle of Quantum Mechanics: there exists a basis in $\mathcal{H}$ which is irreducible under both dual groups simultaneously. Moreover, for $s = 1/2$, this basis is complete. In the sequel we consider this case in some more detail. Thus, for $s = 1/2$, $\mathcal{H}$ decomposes into mutually orthogonal subspaces

$$\mathcal{H} = \sum_{\lambda} \oplus \mathcal{H}^{\lambda}$$

labelled by standard bipartitions λ of N, i.e. $\lambda = (\lambda_1, \lambda_2)$ such that $\lambda_1 \geq \lambda_2$ and

$$\lambda_1 + \lambda_2 = N \quad , \lambda_1 - \lambda_2 = 2S .$$

Each bipartition λ is the label of the irreducible representation Δ^{λ} of the group Σ_N and, at the same time, of the irreducible representation D^S of the group U(2), corresponding to the total spin S of the magnet. Thus we have

$$P|_{\mathcal{H}^{\lambda}} \cong (2S+1)\Delta^{\lambda} ,$$

and

$$Q|_{\mathcal{H}^\lambda} \cong (\dim \Delta^\lambda)\, D^S \ ,$$

where $\dim\Delta^\lambda$ is given by the standard hooklength formula. In other words, the subspace $\mathcal{H}^\lambda$ is the carrier space of $(2S+1)$ copies of a standard carriers of the irreducible reresentation Δ^λ, and, at the same time, the carrier space of $(\dim\Delta^\lambda)$ copies of the irreducible representation D^S.

At the level of bases, the Weyl duality states that there exists a basis of $\mathcal{H}$ of the form

$$b_{\text{Weyl}} = \{|\lambda\, y_Y y_W\rangle\} \ ,$$

where λ runs over all bipartitions of N of the form (3), y_Y runs over all standard Young tableaux (i.e. the Young diagram λ, filled with the elements of the set $\tilde{N}$, strictly increasingly in both rows and columns), and y_W runs over all semistandard Weyl tableaux (i.e. the same diagram λ, filled with elements of the dual set $\tilde{2}$, nondecreasingly in rows, and strictly increasingly in columns).

Clearly, the duality of Weyl implies the sum rule

$$2^N = \sum_\lambda (2S+1)\dim\Delta^\lambda \ ,$$

where the sum runs over all standard bipartitions λ of N.

In the quantum theory of identical particles, one can interpret $\tilde{N}$ as the set of labels of particles, and h as the single-particle space. Then the partition λ plays formally the role of a statistics [15]. This comparison has , however, a limited application, since, e.g., introduction of fermions ($\lambda = 1^N$) requires $\dim h \to \infty$.

The Weyl basis b_{Weyl}, given by Eq. (3), is not the required basis b of eigenstates of the Heisenberg Hamltonian H (cf. Eq. (2)), since, clearly, Σ_N is *not* the symmetry group of the model. Still, b_{Weyl} is "better"than the intial basis $\tilde{2}^{\tilde{N}}$ of all magnetic configurations since the group U(2) in the Weyl scheme expresses the spherical symmetry, and thus its irreducible representations $D^{(S)}$ yield exact quantum numbers of the model, namely the total spin S of the magnet. More explicitly, we introduce the operator

$$\hat{\boldsymbol{S}} = \sum_{j\in\tilde{N}} \hat{\boldsymbol{s}}_j$$

of the total spin of the magnet as the generator of the representation Q of the Lie algebra SU(2) in the space $\mathcal{H}$. Then we have

$$\left(\hat{\boldsymbol{S}}\right)^2 |\lambda\, y_Y y_W\rangle = S(S+1)|\lambda\, y_Y y_W\rangle \ ,$$

where S is related to λ by Eq. (3). For the z-projection $\hat{S}^z$ of the operator $\hat{\boldsymbol{S}}$ we have

$$\hat{S}^z|\lambda\, y_Y y_W\rangle = M|\lambda\, y_Y y_W\rangle \ ,$$

where M is the magnetisation of the crystal in the quantum state $|\lambda\, y_Y y_W\rangle$. This magnetisation is given by the formula

$$M = (c_+ - c_-)/2 \ ,$$

where c_+ and c_- are the components of the weight vector

$$C(y_W) = (c_+, c_-)$$

of the Weyl tableau y_W (thus c_i is the number of occurances of the element $i \in \tilde{2}$ in y_W). In this isotropic model, S and M are exact quantum numbers which results from the spherical symmetry.

For $s > 1/2$, the dual groups should be Σ_N and U($2s$+1) to assure completeness of classification of basis states in $\mathcal{H}$ by the exclusive use of only irreducible representations of both groups. Then Eqs. (3) and (3) still hold, but the sum runs over all partitions λ of N into not more than $2s + 1$ parts, and the relations between spin operators and generators of the Lie algebra u($2s$+1) become more elaborated than the relatively simple formulas (3)-(3), or (3)-(3). From the physical point of view, the unitary group U($2s$+1) is the full quantum symmetry group of the single-object space h, which does *not* describe the symmetry of the Heisenberg model, at least with the bilinear Hamiltonians described in Sec. 2. Such larger settings are sometimes related to as "generalized Heisenberg models" [7,8], and we do not consider them here.

4 The Second Duality of Weyl: Spin Deviations as Indistinguishable Hard-Core Particles

The magnetization M is an exact quantum number already for each magnetic configuration f given by Eq. (2), namely

$$M = \sum_{j \in \tilde{N}} m_j \ .$$

This observation allows us to exploit the purely permutational structure of the action P in more detail. Clearly, the magnetization M is invariant under the permutations of the group Σ_N. Moreover, for $s = 1/2$, M classifies orbits of the action P. Each such orbit is characterised by the stabiliser

$$\Sigma^{\mu} = \Sigma_{N-r} \times \Sigma_r \ ,$$

i.e. by a Young subgroup of Σ_N, with Σ_{N-r} and Σ_r acting on the nodes occupied by the spins $+1/2$ and $-1/2$, respectively.

Here,

$$\mu = \{N - r, r\}$$

is a partition of N, with

$$r = \frac{N}{2} - M\,.$$

We call the magnetic configuration

$$|0\rangle = |++\cdots+\rangle$$

the *saturation*, or *vacuum* state, and refer to r as to the number of *spin deviations* from the state $|0\rangle$. Then

$$P = \sum_{r=0}^{N} R^{\Sigma_N:\Sigma^\mu}$$

with μ given by Eq. (4), is the decomposition of the *permutational* representation P into transitive representation $R^{\Sigma_N:\Sigma^\mu}$, with the stabiliser Σ^μ. Each term in the r.h.s. of Eq. (4) defines an orbit of the group Σ_N on the set $\tilde{2}^{\tilde{N}}$. We denote this orbit by $Q^{(r)}$, so that

$$|Q^{(r)}| = \binom{N}{r}\,.$$

The fact that M, and thus also r, is an exact quantum number, implies that the linear (unitary) space

$$\mathcal{H}^{(r)} = lc_{\mathbb{C}}Q^{(r)}\,,$$

spanned on each orbit $Q^{(r)}$ of Σ_N, is invariant under the action of the isotropic Heisenberg Hamiltonian $\hat{H}$. The dynamics in the space $\mathcal{H}^{(r)}$ consists in hopping of spin deviations between nodes, coupled by appropriate exchange parameters $J_{j_1 j_2}$. In particular, for the n.n. models, a hopping occurs between nearest neighbours only. Such a picture allows us to treat each space $\mathcal{H}^{(r)}$ separately, as the quantum space for the system of r pseudoparticles (spin deviations), which move on a finite lattice $\tilde{N}$. The laws of motion are implemented by the Hamiltonian $\hat{H}$. Moreover, in order to adjust to the initial Heisenberg model of the magnet, one has to impose (i) a statistical constraint that pseudoparticles are indistinguishable, and (ii) a kinematical constraint that (for the single-node spin $s = 1/2$) they have hard cores.

The statistical limitation follows from the fact that during the process of forming a magnetic configuration one takes, say, $N - r$ copies of the spin projection $+1/2$, and r copies of $-1/2$. Clearly, copies of the same species are indistinguishable. The kinematical constraint follows from the fact that, for the spin $s = 1/2$, two spin deviations cannot occupy the same node.

In this way, we arrive at another physical system: r indistinguishable hard-core pseudoparticles, each of them moving on the crystal $\tilde{N}$. It gives rise to another duality of Weyl [16,17], with the set

$$\tilde{r} = \{\alpha = 1, 2, \ldots, r\}$$

of labels of constituent pseudoparticles, and the crystal $\tilde{N}$ in the role of the "classical" configuration space of a single pseudoparticle (in the meaning of classical mechanics). The corresponding dual groups are the symmetric group Σ_r on the set $\tilde{r}$, responsible for indistinguishability of pseudoparticles and referred to as the Pauli group, and, say, the geometric symmetry group G of the spatial distribution of nodes of the crystal $\tilde{N}$. Again, we form the set

$$\tilde{N}^{\tilde{r}} = \{\varphi : \tilde{r} \to \tilde{N}\} ,$$

and consider another combinatorial problem of "balls" $\tilde{r}$ as pseudoparticles, and "boxes" $\tilde{N}$ as the single-particle classical configuration space. Thus, we define two action, $A : \Sigma_r \times \tilde{N}^{\tilde{r}} \to \tilde{N}^{\tilde{r}}$, and $B : G \times \tilde{N}^{\tilde{r}} \to \tilde{N}^{\tilde{r}}$, by standard formulas

$$A(\pi) = \begin{pmatrix} \varphi \\ \varphi \circ \pi^{-1} \end{pmatrix} , \quad \varphi \in \tilde{N}^{\tilde{r}} , \quad \pi \in \Sigma_r$$

$$B(g) = \begin{pmatrix} (\varphi_1, \ldots, \varphi_r) \\ (g\varphi_1, \ldots, g\varphi_r) \end{pmatrix} , \quad (\varphi_1, \ldots, \varphi_r) = \varphi \in \tilde{N}^{\tilde{r}} , \quad g \in G ,$$

where

$$g = \begin{pmatrix} j \\ g(j) \end{pmatrix} , \quad j \in \tilde{N}$$

is a permutation of nodes of the crystal $\tilde{N}$ under $g \in G$. Commutativity of theses two actions, i.e.

$$A(\pi)B(g) = B(g)A(\pi) , \quad \pi \in \Sigma_r , \quad g \in G ,$$

gives rise to a combinatorial version of the duality of Weyl [16].

The set $\tilde{N}^{\tilde{r}}$ has a natural interpretation of the classical configuration space for the system of r distinguishable and mutually penetrable particles, each moving on the crystal $\tilde{N}$, with no restriction for the number of particles occupying a single node simultaneously. The statistical requirement of the indistinguishability of spin deviations as pseudoparticles yields that a position of the whole system is not given by a configuration $\varphi \in \tilde{N}^{\tilde{r}}$, but rather by the orbit of the Pauli group Σ_r, generated from φ. Thus the set $\tilde{N}^{\tilde{r}}/\Sigma_r$ of all orbits of the Pauli group serves as the classical configuration space of the system of r indistinguishable, but mutually penetrable particles. This set decomposes into strata of the action A according to the formula

$$\tilde{N}^{\tilde{r}}/\Sigma_r = \bigcup_{\gamma \vdash r} S(A, \gamma) ,$$

where

$$\gamma = (\gamma_1, \gamma_2, \ldots, \gamma_r) , \quad \sum_{i=1}^{t} \gamma_i = r$$

is the corresponding epikernel. The epikernel γ describes occupation of the nodes of the crystal $\tilde{N}$, such that $\gamma_i \geq \gamma_{i+1} > 0$ are numbers of particles, occupying a single node. Now, the kinematic restriction of the hard core implies

$$Q^{(r)} = S\left(A, 1^r\right) ,$$

so that the classical configuration space of our r pseudoparticles coincides with the generic stratum of the action A of the Pauli group, with the epikernel given by the partition 1^r. The other strata, $S(A,\gamma)$ with $\gamma \neq 1^N$, form the fat diagonal $D(r)$ of the set $\tilde{N}^{\tilde{r}}$, so that

$$Q^{(r)} = \left(\tilde{N}^{\tilde{r}} \setminus D(r)\right) / \Sigma_r .$$

The dual action B of the geometric symmetry group G on the set $\tilde{N}^{\tilde{r}}$ is not too interesting in our case, since it corresponds to distinguishable particles. Then, when, e.g. the group G acts transitively and freely on the crystal $\tilde{N}$ (i.e. when $\tilde{N}$ is a regular orbit of G), the action B is free. Much more information is contained in the coarsened action $B' : G \times \tilde{N}^{\tilde{r}}/\Sigma_r \to \tilde{N}^{\tilde{r}}/\Sigma_r$ of the group G on the set of orbits of the Pauli group Σ_r. In particular, the restriction $B'|_{Q^{(r)}}$ describes the geometric symmetry of the classical configuration space $Q^{(r)}$ of the system of our r pseudoparticles.

Decomposition of the action $B'|_{Q^{(r)}}$ into strata depends upon a particular group G, and related geometric distribution of nodes of the crystal $\tilde{N}$. In the case when $G = C_N$, the cyclic group, $\tilde{N}$ is a ring, and then

$$B'|_{Q^{(r)}} = \bigcup_{\kappa \in K(N)} S\left(B'|_{Q^{(r)}}, \kappa\right) ,$$

where $K(N)$ is the lattice of all divisors of N. Each stratum $S'\left(B'|_{Q^{(r)}}, \kappa\right)$ is the set of all orbits with the stabiliser $C_\kappa \triangleleft C_N$. Each orbit in this stratum consists of

$$\overline{\kappa} = \frac{N}{\kappa}$$

elements, so that $\overline{\kappa} \in K(N)$ is the comlementary divisor to κ in the lattice $K(N)$. In particular, the case $\kappa = 1$ corresponds to the generic stratum, i.e. regular orbits, whereas the other extreme case $\kappa = N$ gives invariants of the action B' (two single-element orbits, $|++\ldots+\rangle$ and $|--\ldots-\rangle$ in the case of $s = 1/2$). Below, we give same examples of orbits of the action B' for $N = 6$:

$-+-+++$	$-++-++$	$-+-+-+$	$------$
$+-+-++$	$+-++-+$	$+-+-+-$	
$++-+-+$	$++-++-$		
$+++-+-$			
$-+++-+$			
$+-+++-$			
$\kappa = 1, \overline{\kappa} = 6$	$\kappa = 2, \overline{\kappa} = 3$	$\kappa = 3, \overline{\kappa} = 2$	$\kappa = 6, \overline{\kappa} = 1$

The second duality of Weyl allows us to use explicitly the geometric symmetry G of the Heisenberg model by means of study of properties of the action $B'|_{Q^{(r)}}$. Decomposition of this action into irreducible representations yields exact quantum numbers, in addition to already introduced S and M. We demonstrate it for the case when $\tilde{N}$ is a ring, so that $G = C_N$. The irreducible representations of the cyclic group C_N are given by

$$\Gamma_k(j) = e^{2\pi i k j/N} , \quad j \in C_N \cong \tilde{N} ,$$

and the label k has the physical meaning of a quasimomentum. The range of the quasimomentum k is called the Brillouin zone B, and given by the set

$$B = \left\{ k = 0, \pm 1, \pm 2, \ldots, \begin{cases} \pm(N/2-1),\ N/2 \text{ for } N \text{ even} \\ \pm(N-1)/2, \text{ for } N \text{ odd} \end{cases} \right\} .$$

The quasimomentum k is the exact quantum number of the Heisenberg model, resulting from its translational symmetry.

In solid state physics, when one deals with crystals with the translational symmetry, then the quasimomentum $\boldsymbol{k}$ (which can be a two- or three-dimensional vector) is the constant of motion, which ranges over an appropriate Brillouin zone B. As a rule, the spectrum of energy of the crystal can be organised into *bands*, so that the dependence $E_n(\boldsymbol{k})$ is smooth enough to define what is called the dispersion low for a band n. If one is interested in the limit $N \to \infty$, then the Brillouin zone B becomes a compact manifold, whereas the band index remains discrete. One tacitly assumes that the quasimomentum $\boldsymbol{k}$ ranges over the whole area of B. There are possible, however, some effects of a finite size of N, namely *rarefied bands* [17]-[19], with $\boldsymbol{k}$ ranging only over some rarefied subset of B. In the case of a finite N, these effects can be transparently demonstrated by an analysis of the action $B'|_{Q^{(r)}}$, and we show it here for the one-dimensional case of a linear ring [18,19].

Each orbit of the action $B'|_{Q^{(r)}}$ with the stabiliser C_κ spans a subspace in $\mathcal{H}^{(r)}$ of dimension $\overline{\kappa} = N/\kappa$. This space is clearly invariant under C_N, and decomposes into irreducible subspaces with a definite quasimomentum k according to the decomposition

$$R^{N:\kappa} = \sum_{k \in B/\kappa} \oplus \Gamma_k$$

of the corresponding transitive representation $R^{N:\kappa}$ on this orbit into irreducible representations Γ_k. The sum in (4) runs over the set

$$B/\kappa = \{ k \in B | k/\kappa \in B \} ,$$

referred to as the κ-tuply rarefied Brillouin zone. In particular, $B/1 = B$, and $B/N = \{0\}$. The example $N = 6$ mentioned above yields $B = \{0, \pm 1, \pm 2, 3\}$, $B/2 = \{0, \pm 2\}$, $B/3 = \{0, 3\}$, $B/6 = \{0\}$.

Determination of the energy structure of the Heisenberg model is a highly non-trivial problem of spectral theory. In particular, the problem of existence of energy bands is still open, even in the one-dimensional case when the Bethe Ansatz offers an exact solution. The structure (4) of the action $B'|_{Q^{(r)}}$ provides, at least, hints for existence of rarefied bands.

5 Conclusions

We have presented here some notions and problems of the Heisenberg model of magnetism from the point of view of combinatorics. The principal frame of the model can be adequately stated in terms of the duality of Weyl, and the skeleton of the latter is the old problem of "balls"and "boxes", i.e. nodes of the crystal and single-node spin projection. This duality allows us to relate the quantum numbers S and M of the total spin of the magnet to appropriate Young and Weyl tableaux. Restriction of considerations to a fixed M, or, equivalently, to a fixed number r of spin deviations from the saturation configuration, allows us to introduce the second duality of Weyl, with a new version of "balls and boxes"as, respectively, spin deviations treated as indistinguishable hard-core particles, and nodes of the crystal. Both approaches relate exact quantum numbers with appropriate, mutually dual actions of respective groups. In particular, the coarsened action of the geometric symmetry group yields an insight into the band structure of the magnet.

References

1. W. Heisenberg, Z. Phys. **49**, 619 (1928)
2. D.C. Mattis, The Theory of Magnetism, Springer Berlin 1981.
3. H. Weyl, The Theory of Groups and Quantum Mechanics, Princeton 1931.
4. W. Pauli, Z. Phys. **31**, 765 (1925)
5. H. Bethe, Z. Phys. **71**, 205 (1931) (English translation in [6])
6. D.C. Mattis, An Encyclopedia of Exactly Solved Models in One Dimension, World Sci., Singapore 1993, pp. 689-716.
7. L.A. Takhtajan, Phys. Lett. **87A**, 479 (1982)
8. S.V. Kerov, A.N. Kirillov, and N.Yu. Reshetikhin, J. Sov. Math. **41**, 916 (1988)
9. A.N. Kirillov and N.Yu. Reshetikhin, J. Sov. Math. **41**, 925 (1998).
10. A.N. Kirillov and N.Yu. Reshetikhin, Lett. Math. Phys. **12**, 199 (1986).
11. A.N. Kirillov, Zap. Nauchn. Sem. Leningrad. Otdel. Mat. Inst. Steklov. (LOMI) **131**, 88 (1983).
12. A. Lascoux, B. Leclerc, and J.-Y. Thibon, J. Math. Phys. **38**, 1041 (1997).
13. S. Dasmahapatra and O. Foda, Int. J. Mod. Phys. **A13**, 501 (1998).
14. A. Schilling and S.O. Warnaar, Commun. Math. Phys. **202**, 359 (1999).
15. B. Lulek and T. Lulek, Rep. Math. Phys. **38**, 277 (1996)
16. B. Lulek and T. Lulek, J. Phys. A.: Math. Gen. **29**, 4687 (1996)
17. B. Lulek and T. Lulek, Rep. Math. Phys. **38**, 267 (1996)
18. B. Lulek, Acta Phys. Polon. **B22**, 371 (1991)
19. B. Lulek, J. Phys: Condens. Matter **4**, 8737 (1992)

The Combinatorics of the Character Theory for some Group Extensions

Alun Morris[1] and Mohammed Almestady[2]

[1] University of Wales, Aberystwyth, Ceredigion SY23 3BZ, Wales, U.K.
[2] King Abdul Aziz University, P.O.Box 7732, Jeddah 21472, Saudi Arabia

Abstract Combinatorial methods are presented for calculating the Fischer matrices of the generalized symmetric group and one of its covering groups.

1 Introduction

The group extensions which will be considered in this lecture are wreath products or semi-direct products of groups, in particular, such products which involve the symmetric groups S_n and their covering groups $\tilde{S}_n$.

These groups arise naturally in many contexts and their representations and characters have been extensively studied with a number of different approaches adopted. The earliest work dates back to W. Specht's thesis written under the supervision of I. Schur in the early thirties. This work was published in a series of papers, the first of these [20] dealt with the wreath product $G \wr S_n$ and in [21], he considered the more general case $G \wr H$. Later, he dealt with the special case of the hyperoctahedral group $\mathbb{Z}_2 \wr S_n$, where $\mathbb{Z}_2$ is the cyclic group of order 2. This is the Weyl group of type B_n or the semi-direct product $\mathbb{Z}_2^n \rtimes S_n$. In the early forties, soon after R. Brauer and G. de B. Robinson's early stimulating work on the modular representations of symmetric groups, M. Osima [16] showed that the wreath product $\mathbb{Z}_m \wr S_n$ or the semi-direct product $\mathbb{Z}_m^n \rtimes S_n$, the *generalized symmetric groups*, arise naturally in this context and independently of Specht's earlier work gave a full account of the representation theory (and also applied this work to the modular representations of the symmetric groups).

It was in the middle sixties that Adalbert Kerber [11,12] realised the true significance of wreath products through his early work on the modular representations of symmetric groups. He was the first in [11] to apply A. H. Clifford's approach [8] to the representations of groups with a non-trivial normal subgroup to the representations of wreath products. In two major contributions [13] which are often cited he further modified and developed this work and additionally took up the difficult problem of explicitly calculating the characters of these groups for the wreath products $\mathbb{Z}_m \wr S_n$ and $S_m \wr S_n$.

Various group extensions and their representations were of crucial importance in the classification of simple groups and in the computation of their character tables. It was in this context that B. Fischer [5,6] presented a powerful and interesting method to compute the irreducible characters of an extension of a group N by a group G under certain conditions. This essentially reduces the calculation to knowledge of the character table of G and some of its subgroups, and also the determination of a certain matrix, now called a *Fischer matrix*, corresponding to each class of conjugate elements of G. R. List and I. Mahmoud [15] applied this theory to a wreath product $G \wr S_n$ and have given an explicit formula for the Fischer matrix in this case. In fact, through some interesting structural results, they essentially show that it is only necessary to calculate the Fischer matrix corresponding to the identity class of S_n which is parameterized by the partition (1^n). However, as the few explicit results they present for $\mathbb{Z}_2 \wr S_n$ for $n = 2, 3, 4$ show, the explicit calculation of these Fischer matrices presents a formidable problem.

This is the problem taken up by M. Almestady [1] in his recent thesis. There are well known combinatorial methods for calculating the irreducible characters of symmetric groups and their Young subgroups. These methods involve many fundamental combinatorial objects, Young tableau, hooks, etc. In particular, there is a powerful recursion formula for calculating the characters of S_n, the Murnaghan-Nakayama formula. This in turn leads to an explicit formula which gives the value of the irreducible character corresponding to the partition λ of n at an element in a class corresponding to the partition ρ of n.

It turns out that similar approach can be successfully applied to the calculation of Fischer matrices for the wreath product $G \wr S_n$ which are even more explicit in the case of the generalized symmetric groups $\mathbb{Z}_m \wr S_n$ or $\mathbb{Z}_m^n \rtimes S_n$. A Murnaghan-Nakayama type formula has been obtained and also an explicit formula for a general element of the Fischer matrix. For this purpose, suitable combinatorial objects have to be introduced, these are generalized magic squares and generalized Young tableaux and tabloids. A description of this approach will be given here - the details will appear in [2].

If $\tilde{S}_n$ denotes one of the covering groups of S_n, then $\mathbb{Z}_m^n \rtimes \tilde{S}_n$ is one of the eight possible covering groups of $\mathbb{Z}_m^n \rtimes S_n$. The construction of Fischer matrices for these groups was also considered in [1] and has been developed further in [3]. Although the problem is considerably more complicated in this case, it turns out that the Fischer matrices are the same as those for $\mathbb{Z}_m^n \rtimes S_n$. We explain here why this is so and indicate how they can be determined in the cases where they differ.

The structure of this paper is as follows. In Sect.2, all the combinatorial requirements are presented and developed, for example, Young tabloids and their various generalizations, magic squares and generalized magic squares. A theorem is stated which connects these two concepts. In Sect.3, the representation and character theory of group extensions is briefly described with an emphasis on semi-direct products. This section culminates with the definition of Fischer matrices. In Sect.4, after presenting some essential background material on the generalised symmetric group $\mathbb{Z}_m^n \rtimes S_n$, all the relevant results about their Fischer matrices are obtained; this will review the contribution of R. List and M. Mahmoud [15] and give the main results of [2]. The main theorems that are needed for the explicit computation of the Fischer matrices are given, for example, a Murnaghan-Nakayama formula is stated in Theorem 4.6. Examples are given to illustrate the computation in particular cases. In Sect.5, a covering group $\tilde{S}_n$ of S_n is defined in terms of generators and relations and the required information about the projective representations and characters is given; for the basic information on projective representations of finite groups required in this context we refer to [7]. In Sect.6, generators and relations are also given for the group $\mathbb{Z}_m^n \rtimes \tilde{S}_n$ which is one of the covering groups of $\mathbb{Z}_m^n \rtimes S_n$. In addition, some of the information about these groups which are required later is given. In Sect.7, the problem of calculating the Fischer matrices of $\mathbb{Z}_m^n \rtimes \tilde{S}_n$ is considered; this is where it is shown that in many cases, the computation reduces to that in the earlier case, however, other cases turn out to be more complex and present an interesting problem.

2 Combinatorial Background

2.1 Magic Matrices

Let $\mathbb{Z}^+$ be the set of non-negative integers and let $\mathbb{N}$ be the set of natural numbers. If $n \in \mathbb{N}$, then $\mathbf{a} = (a_1, \ldots, a_m)$ is a *weak composition* of n if $n = a_1 + \cdots + a_m$, where $a_i \in \mathbb{Z}^+, 1 \le i \le m$. If a composition $\mathbf{a}$ has m summands, that is, $\mathbf{a}$ has m parts, $\mathbf{a}$ is called a *weak m-composition* of n. Thus if $A(n, m)$ is the set of all weak m-compositions of n, then

$$A(n,m) = \left\{\mathbf{a} = (a_1, \ldots, a_m) |\ a_i \in \mathbb{Z}^+, 1 \le i \le m, a_1 + \cdots + a_m = n\right\}.$$

Example 2.1.

$$\begin{aligned} A(4,3) = \{&(4,0,0),(0,4,0),(0,0,4),(3,1,0),(3,0,1),(1,3,0),(1,0,3),\\ &(0,3,1),(0,1,3),(2,2,0),(2,0,2),(0,2,2),(2,1,1),(1,2,1),(1,1,2)\} \end{aligned}$$

Now, let $\mathbf{a} = (a_1, \ldots, a_m), \mathbf{k} = (k_1, \ldots, k_m) \in A(n, m)$. Let

$$\mathcal{R}_{\mathbf{a}} = \{R = (r_{ij}) \in M_m(\mathbb{Z}^+)|\ \mathbf{r}_i = (r_{i1}, \ldots, r_{im}) \in A(a_i, m)\ 1 \le i \le m\},$$

$$\mathcal{C}_{\mathbf{k}} = \{R = (r_{ij}) \in M_m(\mathbb{Z}^+)|\ \mathbf{c}_j = (r_{1j}, \ldots, r_{mj}) \in A(k_j, m)\ 1 \le j \le m\}$$

and

$$\mathcal{R}_{\mathbf{a},\mathbf{k}} = \mathcal{R}_{\mathbf{a}} \cap \mathcal{C}_{\mathbf{k}}.$$

The matrices $R \in \mathcal{R}_{\mathbf{a},\mathbf{k}}$ are generalisations of magic squares which correspond to the special case where $\mathbf{a} = \mathbf{k} = (a, \ldots, a) \in A(ma, m)$. In view of the above, we shall refer to the elements of $\mathcal{R}_{\mathbf{a},\mathbf{k}}$ as *magic matrices.*

Example 2.2. If $n = 5, \mathbf{a} = (3, 0, 2), \mathbf{k} = (2, 1, 2)$, then

$$\mathcal{R}_{\mathbf{a},\mathbf{k}} = \left\{ \begin{pmatrix} 2\,1\,0 \\ 0\,0\,0 \\ 0\,0\,2 \end{pmatrix}, \begin{pmatrix} 2\,0\,1 \\ 0\,0\,0 \\ 0\,1\,1 \end{pmatrix}, \begin{pmatrix} 1\,0\,2 \\ 0\,0\,0 \\ 1\,1\,0 \end{pmatrix}, \begin{pmatrix} 0\,1\,2 \\ 0\,0\,0 \\ 2\,0\,0 \end{pmatrix}, \begin{pmatrix} 1\,1\,1 \\ 0\,0\,0 \\ 1\,0\,1 \end{pmatrix} \right\}.$$

However, for application later in this paper, all of the above needs to be generalised. If $m, n \in \mathbb{N}$ the a_i and k_i $1 \le i \le m$ are replaced by vectors $\mathbf{a}_i = (a_{i1}, \ldots, a_{in})$ and $\mathbf{k}_i = (k_{i1}, \ldots, k_{in})$ respectively, where $a_{ij}, k_{ij} \in \mathbb{Z}^+$. Let $\mathbf{n} = (n_1, \ldots, n_n)$, where $n_i \in \mathbb{Z}^+, \mathbf{a} = (\mathbf{a}_1, \ldots, \mathbf{a}_m), \mathbf{k} = (\mathbf{k}_1, \ldots, \mathbf{k}_m)$, $\mathbf{a}_i = (a_{i1}, \ldots, a_{in}), 1 \le i \le m$, $\mathbf{k}_i = (k_{i1}, \ldots, k_{in}), 1 \le i \le m$, where $a_{ij}, k_{ij} \in \mathbb{Z}^+$ and $(a_{1j}, \ldots, a_{mj}), (k_{1j}, \ldots, k_{mj}) \in A(n_j, m)$, $1 \le j \le n$. For $1 \le l \le n$, let

$$\mathcal{R}_{\mathbf{a},n_l} = \{R = (r^l_{ij}) \in M_m(\mathbb{Z}^+) |\ \mathbf{r}^l_i = (r^l_{i1}, \ldots, r^l_{im}) \in A(a_{il}, m), 1 \le i \le m\},$$

and

$$\mathcal{C}_{\mathbf{k},n_l} = \{R = (r^l_{ij}) \in M_m(\mathbb{Z}^+) |\ \mathbf{c}^l_j = (r^l_{1j}, \ldots, r^l_{mj}) \in A(k_{jl}, m),\ 1 \le j \le m\}.$$

Then, put

$$\mathcal{R}_{\mathbf{a},\mathbf{k},n_l} = \mathcal{R}_{\mathbf{a},n_l} \cap \mathcal{C}_{\mathbf{k},n_l}.$$

and then let

$$\mathcal{R}_{\mathbf{a},\mathbf{k},\mathbf{n}} = \left(\mathcal{R}_{\mathbf{a},\mathbf{k},n_1} | \ldots | \mathcal{R}_{\mathbf{a},\mathbf{k},n_n}\right).$$

Hence, the matrices in $\mathcal{R}_{\mathbf{a},\mathbf{k},\mathbf{n}}$ are $m \times mn$ matrices partitioned into n blocks of $m \times m$ matrices. However, if $n_i \neq 0$, then the corresponding component is excluded.

Example 2.3. If $\mathbf{n} = (2,4,2,2)$, $\mathbf{a} = (\mathbf{a}_1, \mathbf{a}_2, \mathbf{a}_3, \mathbf{a}_4, \mathbf{a}_5), \mathbf{k} = (\mathbf{k}_1, \mathbf{k}_2, \mathbf{k}_3, \mathbf{k}_4, \mathbf{k}_5)$, where $\mathbf{a}_1 = (1,1,1,0)$, $\mathbf{a}_2 = (0,0,0,2)$, $\mathbf{a}_3 = (0,0,1,0)$, $\mathbf{a}_4 = (1,1,0,0)$, $\mathbf{a}_5 = (0,2,0,0)$; $\mathbf{k}_1 = (1,1,0,1)$, $\mathbf{k}_2 = (1,0,0,0)$, $\mathbf{k}_3 = (0,0,1,0)$, $\mathbf{k}_4 = (0,2,0,1)$, $\mathbf{k}_5 = (0,1,1,0)$. Then, we list one of the 28 elements of $\mathcal{R}_{\mathbf{a},\mathbf{k},\mathbf{n}}$.

$$\left(\begin{array}{ccccc|ccccc|ccccc|ccccc}
1&0&0&0&0&1&0&0&0&0&0&0&1&0&0&0&0&0&0&0\\
0&0&0&0&0&0&0&0&0&0&0&0&0&0&0&1&0&0&1&0\\
0&0&0&0&0&0&0&0&0&0&0&0&0&0&1&0&0&0&0&0\\
0&1&0&0&0&0&0&0&1&0&0&0&0&0&0&0&0&0&0&0\\
0&0&0&0&0&0&0&0&1&1&0&0&0&0&0&0&0&0&0&0
\end{array}\right)$$

2.2 Generalised Tabloids

We now describe the other combinatorial objects which are required and which turn out to be closely related to the above.

Let $\mathbf{a} = (a_1, \ldots, a_m), \mathbf{k} = (k_1, \ldots, k_m) \in A(n, m)$. Then $\mathbf{a}$ can be represented graphically by its *Young diagram* $Y_{\mathbf{a}}$ having a_i squares in its *ith* row $1 \le i \le m$. A *Young tableau* of *shape* $\mathbf{a}$ and *content* $\mathbf{k}$ is the Young diagram $Y_{\mathbf{a}}$ with its squares filled by the integers $1, \ldots, m$ with the integer i appearing k_i times $1 \le i \le m$. Let $\mathcal{Y}_{\mathbf{a},\mathbf{k}}$ be the set all Young tableaux of shape $\mathbf{a}$ and content $\mathbf{k}$. If $t_1, t_2 \in \mathcal{Y}_{\mathbf{a},\mathbf{k}}$, then t_1 and t_2 are row equivalent, written $t_1 \sim t_2$, if the rows of t_1 and t_2 contain the same elements, that is,

t_2 is obtained from t_1 by reordering the elements in each row. If $t \in \mathcal{Y}_{\mathbf{a},\mathbf{k}}$ the corresponding equivalence class is denoted by $\mathbf{t}$ and is called a *tabloid.* Each tabloid can be represented by a Young tableau in which the numbers in each row are non-decreasing. We will not distinguish between a tabloid and such a representative element except that the numbers are bold $\mathbf{1}, \mathbf{2}, \ldots$. Let $\mathcal{L}_{\mathbf{a},\mathbf{k}}$ denote the set of tabloids of shape $\mathbf{a}$ and content $\mathbf{k}$.

Example 2.4. If $\mathbf{a} = (3,0,2), \mathbf{k} = (2,1,2)$, then

$$\mathcal{L}_{\mathbf{a},\mathbf{k}} = \left\{ \begin{array}{|c|c|c|} \hline 1 & 1 & 2 \\ \hline \multicolumn{3}{c}{} \\ \hline 3 & 3 \\ \cline{1-2} \end{array}, \begin{array}{|c|c|c|} \hline 1 & 1 & 3 \\ \hline \multicolumn{3}{c}{} \\ \hline 2 & 3 \\ \cline{1-2} \end{array}, \begin{array}{|c|c|c|} \hline 1 & 3 & 3 \\ \hline \multicolumn{3}{c}{} \\ \hline 1 & 2 \\ \cline{1-2} \end{array}, \begin{array}{|c|c|c|} \hline 2 & 3 & 3 \\ \hline \multicolumn{3}{c}{} \\ \hline 1 & 1 \\ \cline{1-2} \end{array}, \begin{array}{|c|c|c|} \hline 1 & 2 & 3 \\ \hline \multicolumn{3}{c}{} \\ \hline 1 & 3 \\ \cline{1-2} \end{array} \right\}$$

Now, let $\mathbf{t} \in \mathcal{L}_{\mathbf{a},\mathbf{k}}$. Suppose that in the i-th row of $\mathbf{t}$, there are r_{ij} copies of j, $1 \le i,j \le m$. Then $\sum_{j=1}^{m} r_{ij} = a_i$, $1 \le i \le m$ and $\sum_{i=1}^{m} r_{ij} = k_j$, $1 \le j \le m$. Thus, if $\mathbf{a}_i^{\mathbf{t}} = (r_{i1}, \ldots, r_{im})$, $1 \le i \le m$ is the i-th row of the matrix $R_{\mathbf{t}} = (r_{ij})$ then $R_{\mathbf{t}} \in \mathcal{R}_{\mathbf{a},\mathbf{k}}$. The converse is also true in that, given a matrix $R \in \mathcal{R}_{\mathbf{a},\mathbf{k}}$, then the corresponding tabloid in $\mathcal{L}_{\mathbf{a},\mathbf{k}}$ is easily constructed and is uniquely defined. Thus, we have the following theorem.

Theorem 2.5. *There is a 1-1 correspondence between the elements of* $\mathcal{R}_{\mathbf{a},\mathbf{k}}$ *and* $\mathcal{L}_{\mathbf{a},\mathbf{k}}$.

Hence, in order to find the magic matrices in the set $\mathcal{R}_{\mathbf{a},\mathbf{k}}$ it is only necessary to find the tabloids in the set $\mathcal{L}_{\mathbf{a},\mathbf{k}}$ which in practice turns out to be a considerably easier task. Tabloids carry the same information as magic matrices in a far efficient manner. This relationship will be used later to considerably simplify some calculations involving magic matrices.

This work is now generalised. As before, if $m, n \in \mathbb{N}$, let $\mathbf{n} = (n_1, \ldots, n_n)$, where $n_l \in \mathbb{Z}^+$ and $\sum_{l=1}^{n} l n_l = n$. Let $\mathbf{a} = (\mathbf{a}_1, \ldots, \mathbf{a}_m), \mathbf{k} = (\mathbf{k}_1, \ldots, \mathbf{k}_m)$, where $\mathbf{a}_i = (a_{i1}, \ldots, a_{in}), 1 \le i \le m$, $\mathbf{k}_i = (k_{i1}, \ldots, k_{in}), 1 \le i \le m$, and where $a_{ij}, k_{ij} \in \mathbb{Z}^+$ with $(a_{1j}, \ldots, a_{mj}), (k_{1j}, \ldots, k_{mj}) \in A(n_j, m), 1 \le j \le n$. In addition, let $a_i = \sum_{j=1}^{n} j a_{ij}$, $k_i = \sum_{j=1}^{n} j k_{ij}$. We now define *generalised tabloids.* In these, the squares in Young tableaux and tabloids are replaced by *ribbons,* that is, by rectangular boxes consisting of a number of square cells. A ribbon consisting of l square cells is called an *l-ribbon.* (These are the same as the l-ribbons of [14] except that our ribbons are in a single row).

A *generalised tabloid* $\mathbf{t}$ of shape $\mathbf{a}$ and content $\mathbf{k}$ corresponding to each $(\mathbf{a}, \mathbf{k}, \mathbf{n})$ is a diagram consisting of m rows made up of n_l l-ribbons, $1 \le l \le n$, such that there are a_{il} l-ribbons in the i-th row $1 \le i \le m$. These ribbons are filled with the integers $1, \ldots, m$ with each l-ribbon filled with l copies of the same integer such that k_{il} of the l-ribbons are filled with the integer i, $1 \le i \le m$. Furthermore, the numbers in each row are non-decreasing. The set of all generalised tabloids corresponding to $(\mathbf{a}, \mathbf{k}, \mathbf{n})$ is denoted by $\mathcal{L}_{\mathbf{a},\mathbf{k},\mathbf{n}}$.

The importance of generalised tabloids in this context comes through the following theorem.

Theorem 2.6. *There is a* $1-1$ *correspondence between the sets* $\mathcal{R}_{\mathbf{a},\mathbf{k},\mathbf{n}}$ *and* $\mathcal{L}_{\mathbf{a},\mathbf{k},\mathbf{n}}$.

Example 2.7. Using the same $\mathbf{a}, \mathbf{k}, \mathbf{n}$ as in Example 3 , then corresponding to the tabloid

1	1 1	3 3 3
1 1 1 1	4 4 4 4	
5 5 5		
2	4 4	
4 4	5 5	

is the unique matrix given in Example 3.

This example shows that not only are the tabloids far easier to enumerate, but also, they carry the same information in a more efficient manner.

3 Group Extensions and Fischer Matrices

Let $H = N.G$ be the group extension of the group N by the group G; thus $N \lhd H$ and $H/N \cong G$. Let $\theta \in Irr(N)$, where $Irr(N)$ is the set of irreducible characters of N. An action of H on $Irr(N)$ is defined by

$$\theta^h(n) = \theta(hnh^{-1})$$

for $h \in H, n \in N$. Then $I_\theta = \{h \in H | \theta^h = \theta\}$ is the *inertia group* of θ. Clearly, $N \lhd I_\theta$, and the factor group $\bar{I_\theta} = I_\theta/N$ is called the *inertia factor* of I_θ. A character $\theta \in Irr(N)$ is said to be *extended* to a character $\tilde{\theta} \in Irr(\bar{I_\theta})$ if $\tilde{\theta} = \theta_N$.

The extensions that we shall consider are *split extensions*, that is, H contains a subgroup $G' \cong G$ such that $H = N.G'$ and $N \cap G' = \{e\}$. The group H is then also called a *semi-direct product* of N and G, written as $H = N \rtimes G$, where G has been identified with G'. Then, every element $h \in H$ can be uniquely expressed as $h = ng$, where $n \in N$ and $g \in G$ and where multiplication in H is $(n_1g_1)(n_2g_2) = n_1n_2^{g_1}g_1g_2$, where $n^g = gng^{-1}$. If in addition, N is an abelian subgroup of H and $\theta \in Irr(N)$ has the property that $\theta^h = \theta$ for all $h \in H$, then a well known theorem due to Mackey (see, for example [8]) shows that the extended character $\tilde{\theta} \in Irr(H)$ is defined by

$$\tilde{\theta}(ng) = \theta(n)$$

for all $n \in N$ and $g \in G$.

Fischer [5] has presented a method for determining the irreducible characters of H if each $\theta \in Irr(N)$ can be extended to a $\tilde{\theta} \in Irr(I_\theta)$. In this method, certain matrices now called *Fischer matrices*, are crucial. These matrices are now briefly described mainly following List and Mahmoud [15].

Let C be a class of conjugate elements of G. Let $C_1, C_2, \ldots, C_t$ be the classes of H which map onto the class C under the homomorphism $H \to G \cong H/N$.

Let $\mathcal{L}_1, \mathcal{L}_2, ..., \mathcal{L}_r$ be the classes of $\bar{I}_\theta$ which fuse to $\mathcal{C}$. If $\theta \in Irr(N)$ let $\tilde{\theta} \in Irr(\bar{I}_\theta)$ be an extension of θ to I_θ. Then by Clifford's theorem [8], every irreducible character of H is of the form $(\tilde{\theta}.\beta)^H$, where $\beta \in Irr(I_\theta)$ is such that $N \subseteq ker(\beta)$ (β also denotes the corresponding irreducible character of $Irr(\bar{I}_\theta)$). The evaluation of $(\tilde{\theta}.\beta)^H$ on an element $h \in H$ which maps onto an element in the class $\mathcal{C}$ involves a matrix $F_\theta^{\mathcal{C}}$, which is called the *Fischer matrix* of θ at the class $\mathcal{C}$. For the evaluation of the irreducible characters of H it is necessary to determine a Fischer matrix $(\tilde{\theta}.\beta)^H$ for each class $\mathcal{C}$ of G and for each θ which is a representative of the orbit of H acting on $Irr(N)$.

More precisely, the Fischer matrices are defined as follows. If $h \in H$ is mapped onto an element in the class $\mathcal{C}$, let $l_i \in \mathcal{L}_i, 1 \leq i \leq r$, let $\mathcal{L}_{ji}, 1 \leq j \leq s$ be the classes of I_θ which map to $\mathcal{L}_i$ under the homomorphism I_θ to $\bar{I}_\theta$ and let $l_{ji} \in \mathcal{L}_{ji}$, $1 \leq j \leq s$, then

$$(\tilde{\theta}.\beta)^H(h) = \sum_{i=1}^{r}\sum_{k=1}^{s} \frac{|C_H(h)|}{|C_{I_\theta}(l_{ki})|}(\tilde{\theta}.\beta)(l_{ki}) \tag{1}$$

$$= \sum_{i=1}^{r}\left(\sum_{k=1}^{s} \frac{|C_H(h)|}{|C_{I_\theta}(l_{ki})|}\tilde{\theta}(l_{ki})\right)\beta(l_i), \tag{2}$$

since $\beta(l_{ki}) = \beta(l_i)$ as $N \subseteq ker(\beta)$. Then, the Fischer sub-matrix $F_\theta^{\mathcal{C}}$ corresponding to θ is the $r \times t$ matrix with element in the ith row corresponding to the class $\mathcal{C}_j$ given by

$$\left(\sum_{k=1}^{s} \frac{|C_H(h)|}{|C_{I_\theta}(l_{ki})|}\tilde{\theta}(l_{ki})\right). \tag{3}$$

Let $\theta_1, \ldots, \theta_p$ be representatives of the orbits of H acting on $Irr(N)$ such that $\bar{I}_{\theta_k}$ contains a conjugate of h, $1 \leq k \leq p$. Then the Fischer matrix $\mathcal{F}^{\mathcal{C}}$ is the matrix

$$\mathcal{F}^{\mathcal{C}} = \begin{pmatrix} \mathcal{F}_{\theta_1}^{\mathcal{C}} \\ \mathcal{F}_{\theta_2}^{\mathcal{C}} \\ \vdots \\ \mathcal{F}_{\theta_p}^{\mathcal{C}} \end{pmatrix}. \tag{4}$$

4 The Generalised Symmetric Group $\mathbb{Z}_m^n \rtimes S_n$

The above is now applied to the special case where N is the abelian group $N = \mathbb{Z}_m \times \cdots \times \mathbb{Z}_m (n\ copies)$ of order m^n, where $\mathbb{Z}_m$ is the cyclic group of order m and G is the symmetric group S_n of order $n!$. This group is the *wreath product* $\mathbb{Z}_m \wr S_n$ of order $m^n n!$; in the sequel, this is denoted by B_n^m (when $m = 2$, we have the Weyl group of type B_n).

The classes of S_n are parametrised by the partitions $(1^{n_1}2^{n_2}\ldots n^{n_n})$ of n, where $n_i \geq 0$, $1 \leq i \leq n$. The cyclic group $\mathbb{Z}_m$ has m irreducible characters which are denoted by $\chi_1, \chi_2, \dots, \chi_m$. Then,

$$\{\chi_1^{k_1}\chi_2^{k_2}\cdots\chi_m^{k_m} \mid \mathbf{k} = (k_1, \dots, k_m) \in A(n,m)\}$$

are representatives of the orbits of B_n^m acting on $Irr(N)$. Following List and Mahmoud [15], we will denote the Fischer matrix corresponding to the class $(1^{n_1}2^{n_2}\dots n^{n_n})$ by $\mathcal{F}(\mathbb{Z}_m; 1^{n_1}2^{n_2}\dots n^{n_n})$ or $\mathcal{F}(m; 1^{n_1}2^{n_2}\dots n^{n_n})$. (In fact, List and Mahmoud had a suitable finite group in place of $\mathbb{Z}_m$, most of what follows is true in the more general context but we will concentrate on this special case here.) They proved two results

$$\mathcal{F}(m; k^a) = \mathcal{F}(m; 1^a),\ a, k \geq 1 \tag{5}$$

$$\mathcal{F}(m; 1^{n_1}2^{n_2}\dots n^{n_n}) = \mathcal{F}(m; 1^{n_1}) \otimes \mathcal{F}(m; 2^{n_2}) \otimes \cdots \otimes \mathcal{F}(m; n^{n_n}) \tag{6}$$

which reduces the calculation of Fischer matrices to determining the matrices $\mathcal{F}(m; 1^n)$ for $n \in \mathbb{Z}^+$. Our initial aim will be to develop a combinatorial method for determining the Fischer matrices in this case; however we shall show that these combinatorial methods can be extended to the general case and are more useful in practice than using (6).

In order to do this, it is necessary to give some detailed information (see, for example, [9]) about the group B_n^m, in particular about its classes of conjugate elements. Let ζ be a primitive m-th root of unity. The elements of B_n^m permute the set $\{1, 2, \dots, n\}$ and multiply each of the elements of this set by a power of ζ. Thus the elements of B_n^m are of the form

$$\rho = \begin{pmatrix} 1 & 2 & \dots & n \\ \zeta^{q_1}b_1 & \zeta^{q_2}b_2 & \dots & \zeta^{q_n}b_n \end{pmatrix},$$

where $\{b_1, b_2, \dots, b_n\}$ is a permutation of the set $\{1, 2, \dots, n\}$ and $1 \leq q_i \leq m$, $1 \leq i \leq n$. The classes of conjugate elements of B_n^m correspond to m-partitions of n

$$(1^{a_{11}}2^{a_{12}}\dots n^{a_{1n}}; 1^{a_{21}}2^{a_{22}}\dots n^{a_{2n}}; \dots; 1^{a_{m1}}2^{a_{m2}}\dots n^{a_{mn}}),$$

where $\sum_{i=1}^n a_{ij} = n_j\ 1 \leq j \leq m$. It is these classes of B_n^m which fuse to the class $(1^{n_1}2^{n_2}\dots n^{n_n})$ of S_n under the natural homomorphism $B_n^m \to S_n$. The order of this class is

$$\frac{m^n n!}{\prod_{p,q} a_{pq}!(qm)^{a_{pq}}} \tag{7}$$

We now revert to the special case (1^n). This is considerably simpler than the general case and it will illustrate the general case which is considered later. If $\theta = \chi_1^{k_1}\chi_2^{k_2}\cdots\chi_m^{k_m}$, where $\sum_{i=1}^m k_i = n$, then the inertia group of θ is $I_\theta = B_{k_1}^m \times \cdots \times B_{k_m}^m$ and the corresponding inertia factor is the Young subgroup of S_n, $\bar{I}_\theta = S_{k_1} \times \cdots \times S_{k_m}$. In this case, the classes of B_n^m which map onto (1^n) are of the form $(1^{a_1}; \dots; 1^{a_m})$, where $\sum_{i=1}^m a_i = n$. Also, there is only one class of $\bar{I}_\theta$, namely $(1^{k_1} : \dots : 1^{k_m})$, which fuses to the class (1^n)

(here and later, the : are used to separate the classes of the components of the direct products). Thus, the classes of I_θ which map onto this class will correspond to the m-partitions of n of the form

$$((1^{r_{11}}; 1^{r_{21}}; \ldots; 1^{r_{m1}}) : \ldots : (1^{r_{1m}}; 1^{r_{2m}}; \ldots; 1^{r_{mm}}))\,, \tag{8}$$

such that $\sum_{i=1}^{m} r_{ij} = k_j, 1 \le j \le m$ and $\sum_{j=1}^{m} r_{ij} = a_i, 1 \le i \le m$. Thus, $\mathbf{a} = (a_1, \ldots, a_m), \mathbf{k} = (k_1, \ldots, k_m) \in A(n, m)$ and if $R = (r_{ij})$, then $R \in \mathcal{R}_{\mathbf{a},\mathbf{k}}$ (which was defined in Sect.2). There is therefore an entry in $\mathcal{F}(m, 1^n)$ corresponding to a row indexed by $\mathbf{k} \in A(n, m)$ and a column indexed by $\mathbf{a} \in A(n, m)$; we denote this element by $f_n(\mathbf{a}, \mathbf{k})$.

We now calculate this element using (3). Formula (7) can be used to compute the order of the relevant centralizers, thus we need only compute the extended character $\tilde{\theta}$. For this, we need to be more precise about the values of the characters of the cyclic group $\mathbb{Z}_m$. Let x be a generator of $\mathbb{Z}_m$, then we put $\chi_i(x^{j-1}) = \zeta^{(i-1)(j-1)}$, $1 \le i, j \le m$. Then, the value of $\tilde{\theta}$ on an element in the class corresponding to the m-partition (8) is

$$\prod_{i=1}^{m} \prod_{j=1}^{m} (\chi_i(x^{j-1}))^{r_{ij}} = \zeta^{\left(\sum_{i=1}^{m} \sum_{j=1}^{m} (i-1)(j-1) r_{ij}\right)} \tag{9}$$

This results in the following theorem.

Theorem 4.1. *If* $\mathbf{a}, \mathbf{k} \in A(n, m)$, *then the entry* $f_n(\mathbf{a}, \mathbf{k})$ *in* $\mathcal{F}(m; 1^n)$ *is given by*

$$f_n(\mathbf{a}, \mathbf{k}) = \sum_{R \in \mathcal{R}_{\mathbf{a},\mathbf{k}}} \left(\prod_{i=1}^{m} \binom{a_i}{\mathbf{r_i}} \right) \left(\zeta^{\left(\sum_{i=1}^{m} \sum_{j=1}^{m} (i-1)(j-1) r_{ij}\right)} \right), \tag{10}$$

where $\mathbf{r_i} = (r_{i1}, \ldots, r_{im})$ *is the i-th row of the matrix* R.

Remark 4.2. As tabloids are easier to enumerate than magic matrices, using the one-one correspondence established in Theorem 5, the summation in the formula for $f_n(\mathbf{a}, \mathbf{k})$ can be taken over all tabloids $\mathbf{t} \in \mathcal{L}_{\mathbf{a},\mathbf{k}}$.

The following example shows how to apply these results in a simple case.

Example 4.3. If $\mathbf{a} = (2, 1, 1), \mathbf{k} = (1, 2, 1)$, then

$$\mathcal{L}_{\mathbf{a},\mathbf{k}} = \left\{ \begin{array}{ll} 1 & 2 \\ 2 & \\ 3 & \end{array}, \begin{array}{ll} 1 & 2 \\ 3 & \\ 2 & \end{array}, \begin{array}{ll} 1 & 3 \\ 2 & \\ 2 & \end{array}, \begin{array}{ll} 2 & 2 \\ 1 & \\ 3 & \end{array}, \begin{array}{ll} 2 & 2 \\ 3 & \\ 1 & \end{array}, \begin{array}{ll} 2 & 3 \\ 1 & \\ 2 & \end{array}, \begin{array}{ll} 2 & 3 \\ 2 & \\ 1 & \end{array} \right\}$$

and

$$
\begin{aligned}
&f_4((2,1,1),(1,2,1)) = \\
&\binom{2}{1,1,0}\binom{1}{0,1,0}\binom{1}{0,0,1}\omega^{0+1+4}+\binom{2}{1,1,0}\binom{1}{0,0,1}\binom{1}{0,1,0}\omega^{0+2+2}+\\
&\binom{2}{1,0,1}\binom{1}{0,1,0}\binom{1}{0,1,0}\omega^{0+1+2}+\binom{2}{0,2,0}\binom{1}{1,0,0}\binom{1}{0,0,1}\omega^{0+0+4}+\\
&\binom{2}{0,2,0}\binom{1}{0,0,1}\binom{1}{1,0,0}\omega^{0+2+0}+\binom{2}{0,1,1}\binom{1}{1,0,0}\binom{1}{0,1,0}\omega^{0+0+2}+\\
&\binom{2}{0,1,1}\binom{1}{0,1,0}\binom{1}{1,0,0}\omega^{0+1+0}\\
&= 2\omega^5 + 2\omega^4 + 2\omega^3 + \omega^4 + \omega^2 + 2\omega^2 + 2\omega \\
&= -3.
\end{aligned}
$$

where ω is a primitive cube root of unity.

However, by means of the multinomial theorem (see, for example [23]), an elementary (but complicated) calculation gives a more direct combinatorial formula for the calculation of $f_n(\mathbf{a}, \mathbf{k})$ as given in the following corollary to the above theorem.

Corollary 4.4. *If* $\mathbf{a}, \mathbf{k} \in A(n, m)$, *then the entry* $f_n(\mathbf{a}, \mathbf{k})$ *in* $\mathcal{F}(m; 1^n)$ *is the coefficient of* $x_1^{k_1} x_2^{k_2} \dots x_m^{k_m}$ *in*

$$\prod_{i=1}^{m}\left(\sum_{j=1}^{m}(\zeta^{(i-1)(j-1)}x_j)^{a_i}\right).$$

As an application of this corollary, we show how the Fischer matrix $\mathcal{F}(3; 1^4)$ may be calculated.

Example 4.5. Let ω denote a primitive cube root of unity. Then

$$
\begin{aligned}
&\sum_{k_1+k_2+k_3=4} f_4((3,0,1),(k_1,k_2,k_3))x_1^{k_1}x_2^{k_2}x_3^{k_3} \\
&= (x_1+x_2+x_3)^3(x_1+\omega^2 x_2+\omega x_3) \\
&= x_1^4 + (3+\omega^2)x_1^3x_2 + (3+\omega)x_1^3x_3 - 3\omega x_1^2x_2^2 + 3x_1^2x_2x_3 - 3\omega^2x_1^2x_3^2 \\
&\quad +(1+3\omega^2)x_1x_2^3 + 3\omega^2x_1x_2^2x_3 + 3\omega x_1x_2x_3^2 + (1+3\omega)x_1x_3^3 \\
&\quad +\omega^2x_2^4 + \omega(1+3\omega)x_2^3x_3 - 3x_2^2x_3^2 + \omega(3+\omega)x_2x_3^3 + \omega x_3^4.
\end{aligned}
$$

This gives the elements in the column indexed by $(1^3 3)$ in the table for $\mathcal{F}(3; 1^4)$. The remaining elements in that table are obtained using similar calculations.

However, this is not the most efficient way of calculating such tables. We now present the Murnaghan-Nakayama formula mentioned in the Introduction. This results in a more efficient recursive method for the calculation. Hence, the entries of the Fischer matrix $\mathcal{F}(m;1^n)$ can be calculated from the Fischer matrix $\mathcal{F}(m;1^{n-1})$ as in the following theorem.

Theorem 4.6. *For each* $1 \leq i \leq m$,

$$f_n((a_1,\ldots,a_m),(k_1,\ldots,k_m)) = \sum_{j=1}^{m} \zeta^{(i-1)(j-1)} f_{n-1}((a_1,\ldots,a_i-1,\ldots,a_m),(k_1,\ldots,k_j-1,\ldots,k_m)),$$

where $f_{n-1}((a_1,\ldots,a_i-1,\ldots,a_m),(k_1,\ldots,k_j-1,\ldots,k_m))$ *is the entry in* $\mathcal{F}(m;1^{n-1})$ *which is in the column indexed by* $(1^{a_1}\ldots i^{a_i-1}\ldots m^{a_m})$ *and the row indexed by* $(1^{k_1}\ldots j^{k_j-1}\ldots m^{k_m})$.

We now prove the same sequence of results in the general case corresponding to the class $(1^{n_1}2^{n_2}\ldots n^{n_n})$ of S_n. The conjugacy classes of B_n^m which map onto the class $(1^{n_1}2^{n_2}\ldots n^{n_n})$ of S_n correspond to the m-partitions

$$(1^{a_{11}}2^{a_{12}}\ldots n^{a_{1n}};1^{a_{21}}2^{a_{22}}\ldots n^{a_{2n}};\ldots;1^{a_{m1}}2^{a_{m2}}\ldots n^{a_{mn}}) \qquad (11)$$

where $\sum_{i=1}^m a_{ij} = n_j$, $1 \leq j \leq n$. If $\theta = \chi_1^{k_1}\chi_2^{k_2}\cdots\chi_m^{k_m}$, as in the previous case, then $I_\theta = B_{k_1}^m \times\cdots\times B_{k_m}^m$ and the corresponding inertia factor is $\bar{I}_\theta = S_{k_1}\times\cdots\times S_{k_m}$. The classes of $\bar{I}_\theta$ which fuse to $(1^{n_1}2^{n_2}\ldots n^{n_n})$ correspond to the partitions

$$(1^{k_{11}}2^{k_{12}}\ldots n^{k_{1n}} : 1^{k_{21}}2^{k_{22}}\ldots n^{k_{2n}} : \ldots : 1^{k_{m1}}2^{k_{m2}}\ldots n^{k_{mn}})$$

where $\sum_{i=1}^m k_{ij} = n_j$, $1 \leq j \leq n$ and where $\sum_{j=1}^n jk_{ij} = k_i$, $1 \leq i \leq m$. Thus, the classes of I_θ which map onto this class correspond to the m-partitions of n of the form

$$\Big((1^{r_{11}^1}2^{r_{11}^2}\ldots n^{r_{11}^n});\ldots;(1^{r_{m1}^1}2^{r_{m1}^2}\ldots n^{r_{m1}^n}) : \ldots : (1^{r_{1m}^1}2^{r_{1m}^2}\ldots n^{r_{1m}^n});\ldots(1^{r_{mm}^1}2^{r_{mm}^2}\ldots n^{r_{mm}^n})\Big) \qquad (12)$$

where

$$\sum_{j=1}^m r_{ij}^l = a_{il}, (1 \leq i \leq m, 1 \leq l \leq n),$$

$$\sum_{i=1}^m r_{ij}^l = k_{il}, (1 \leq j \leq m, 1 \leq l \leq n).$$

Now, let $\mathbf{a_i} = (a_{i1}, a_{i2},\ldots,a_{in}), 1 \leq i \leq m$ and $\mathbf{k_j} = (k_{j1},k_{j2},\ldots,k_{jn}), 1 \leq j \leq m$. Then the Fischer sub-matrix corresponding to the character θ has its columns indexed by $1^{\mathbf{a}_1}2^{\mathbf{a}_2}\dots m^{\mathbf{a}_m}$ and its rows indexed by $1^{\mathbf{k}_1}2^{\mathbf{k}_2}\dots m^{\mathbf{k}_m}$. The general entry of this matrix is denoted by

$$f_{(n_1,n_2,\dots,n_n)}((\mathbf{a}_1,\mathbf{a}_2,\dots,\mathbf{a}_m),(\mathbf{k}_1,\mathbf{k}_2,\dots,\mathbf{k}_m))$$

which will be abbreviated to $f_{\mathbf{n}}(\mathbf{a},\mathbf{k})$. This element is now computed. The value on the extended character $\tilde{\theta}$ on the class (12) is

$$\prod_{i=1}^{m}\prod_{j=1}^{m}(\chi_i(x^{j-1}))^{(\sum_{l=1}^{n} r_{ij}^l)} = \left(\zeta^{(\sum_{i=1}^{m}\sum_{j=1}^{m}(i-1)(j-1)\sum_{l=1}^{n} r_{ij}^l)}\right). \tag{13}$$

Then, by means of (7) the order of the centralizer of an element in the class (11) is

$$m^{(\sum_{i=1}^{m}\sum_{j=1}^{n} a_{ij})}\prod_{l=1}^{n}\left(l^{(\sum_{i=1}^{m} a_{il})}\prod_{i=1}^{m} a_{il}!\right)$$

and the order of the centralizer of an element in the class (12) is

$$m^{(\sum_{i=1}^{m}\sum_{j=1}^{m}\sum_{l=1}^{n} r_{ij}^l)}\prod_{l=1}^{n}\left(l^{(\sum_{i=1}^{m}\sum_{j=1}^{m} r_{ij}^l)}\prod_{i=1}^{m}\prod_{j=1}^{m} r_{ij}^l!\right).$$

By taking the quotient of these two orders and the value of the extended character $\tilde{\theta}$, we find that there is an element of the form

$$\left(\prod_{i=1}^{m}\prod_{l=1}^{n}\binom{a_{il}}{r_{i1}^l,\dots,r_{im}^l}\right)\left(\zeta^{(\sum_{i=1}^{m}\sum_{j=1}^{m}(i-1)(j-1)\sum_{l=1}^{n} r_{ij}^l)}\right)$$

corresponding to every matrix $R \in \mathbf{R}_{\mathbf{a},\mathbf{k},\mathbf{n}}$. Thus, we have proved the following theorem.

Theorem 4.7. *If* $\mathbf{n} = (n_1, n_2, \dots, n_n), \mathbf{a} = (\mathbf{a}_1, \mathbf{a}_2, \dots, \mathbf{a}_m)$, $\mathbf{k} = (\mathbf{k}_1, \mathbf{k}_2, \dots, \mathbf{k}_m)$, *the entry* $f_{\mathbf{n}}(\mathbf{a},\mathbf{k})$ *of the Fischer matrix* $\mathcal{F}(m; 1^{n_1}2^{n_2}\dots n^{n_n})$ *is*

$$f_{\mathbf{n}}(\mathbf{a},\mathbf{k}) = \sum_{R\in\mathcal{R}_{\mathbf{a},\mathbf{k},\mathbf{n}}}\left(\prod_{i=1}^{m}\prod_{l=1}^{n}\binom{a_{il}}{\mathbf{r}_{\mathbf{i}}^{\mathbf{l}}}\right)\left(\zeta^{(\sum_{i=1}^{m}\sum_{j=1}^{m}(i-1)(j-1)\sum_{l=1}^{n} r_{ij}^l)}\right)$$

where $\mathbf{r}_{\mathbf{i}}^{\mathbf{l}} = (r_{i1}^l, \dots, r_{im}^l)$ *is the i-th row of the l-th component of the matrix* R.

As in the earlier case, using the one-one correspondence established in Theorem 6, the actual enumeration of the generalised magic matrices involved in the summation is done by means of the corresponding generalised tabloids. Thus, if $\mathbf{t} \in \mathcal{L}_{\mathbf{a},\mathbf{k},\mathbf{n}}$, then a_{il} denotes the number of l-ribbons in the i-th row and k_{jl} denotes the number of l-ribbons filled by j in $\mathbf{t}$.

Example 4.8. We determine the entry

$$f_{(2,4,2)}(((1,0,2),(0,2,0),(1,2,0)),((2,1,1),(0,2,1),(0,1,0)))$$

in the Fischer matrix $\mathcal{F}(3;1^2 2^4 3^2)$. We see that only four tabloids are possible, namely

$$\begin{array}{l} \fbox{1}\fbox{1 1 1}\fbox{2 2 2} \\ \fbox{1 1}\fbox{2 2} \\ \fbox{1}\fbox{2 2}\fbox{3 3} \end{array} \quad \begin{array}{l} \fbox{1}\fbox{1 1 1}\fbox{2 2 2} \\ \fbox{1 1}\fbox{3 3} \\ \fbox{1}\fbox{2 2}\fbox{2 2} \end{array} \quad \begin{array}{l} \fbox{1}\fbox{1 1 1}\fbox{2 2 2} \\ \fbox{1 1}\fbox{2 2} \\ \fbox{1}\fbox{1 1}\fbox{3 3} \end{array} \quad \begin{array}{l} \fbox{1}\fbox{1 1 1}\fbox{2 2 2} \\ \fbox{1 1}\fbox{2 2} \\ \fbox{1}\fbox{1 1}\fbox{2 2} \end{array}$$

This results in the four elements

$$\binom{1}{1,0,0}\binom{2}{1,1,0}\binom{2}{1,1,0}\binom{1}{1,0,0}\binom{2}{0,1,1}\omega^{1+2+4}$$

$$\binom{1}{1,0,0}\binom{2}{1,1,0}\binom{2}{1,0,1}\binom{1}{1,0,0}\binom{2}{0,2,0}\omega^{2+4}$$

$$\binom{1}{1,0,0}\binom{2}{1,1,0}\binom{2}{0,2,0}\binom{1}{1,0,0}\binom{2}{1,0,1}\omega^{2+4}$$

$$\binom{1}{1,0,0}\binom{2}{1,1,0}\binom{2}{0,1,1}\binom{1}{1,0,0}\binom{2}{1,1,0}\omega^{1+2+2},$$

which implies that

$$\begin{aligned} f_{(2,4,2)}(((1,0,2),(0,2,0),(1,2,0)),((2,1,1),(0,2,1),(0,1,0))) &= 8(\omega^7+\omega^6+\omega^5) \\ &= 0. \end{aligned}$$

There are a number of special cases which are of interest which we state in the following corollaries (in addition, Theorem 8 and (5) could be obtained as corollaries of this more general result).

Corollary 4.9. *The Fischer matrix $\mathcal{F}(m;n)$ is the character table of the cyclic group $\mathbb{Z}_m$.*

The next corollary will be of particular interest when this work is applied to the projective representations of $\mathbb{Z}_m^n \rtimes S_n$.

Corollary 4.10. *If $(\lambda_1,\ldots,\lambda_r)$ is a partition of n with r distinct parts, that is, $\lambda_1 > \ldots > \lambda_r > 0$, then the Fischer matrix $\mathcal{F}(m;(\lambda_1,\ldots,\lambda_r))$ is the $m^r \times m^r$ matrix with general element*

$$\zeta^{\sum_{l=1}^{r}(i_l-1)(j_l-1)},$$

where the λ_l-ribbon appears in the i_l-th row and is filled with j_l, $1 \le l \le r$ in the generalised tabloid corresponding to that element.

5 The Covering Group $\tilde{S}_n$ of the Symmetric Group S_n and its Spin Representations

The background information on projective representations, Schur multipliers, twisted products, etc., may be found in [7]. The symmetric group S_n has a presentation in terms of generators and relations

$$\begin{aligned} S_n = < s_1, s_2, \ldots, s_{n-1} | s_i^2 = 1,\ 1 \le i \le n-1; \\ (s_i s_{i+1})^3 = 1,\ 1 \le i \le n-2; \\ (s_i s_j)^2 = 1, |i-j| \ge 2,\ 1 \le i,j \le n-1 >, \end{aligned}$$

where s_i is the transposition $(i\ i+1)$. Then Schur [19,7] showed that S_n has *Schur multiplier* $\mathbb{Z}_2$ if $n \ge 4$ and trivial otherwise; thus it has two covering groups; as the representations of one are easily obtained from the other, we concentrate on one of these, namely $\tilde{S}_n$, which has a presentation in terms of generators and relations

$$\begin{aligned} \tilde{S}_n = < t_1, t_2, \ldots, t_{n-1}, z | z^2 = 1, z t_i = t_i z, t_i^2 = 1,\ 1 \le i \le n-1; \\ (t_i t_{i+1})^3 = 1,\ 1 \le i \le n-2; \\ (t_i t_j)^2 = z, |i-j| \ge 2,\ 1 \le i,j \le n-1 >, \end{aligned}$$

If $\phi_n : \tilde{S}_n \to S_n$ is defined by $\phi_n(t_i) = s_i$, $1 \le i \le n-1$ and $\phi_n(z) = 1$, then we have the exact sequence

$$1 \longrightarrow Z \longrightarrow \tilde{S}_n \xrightarrow{\phi_n} S_n \longrightarrow 1.$$

If $(i_1 i_2 \ldots i_m)$ is a cycle in S_n of length m and if $[i_1 i_2 \ldots i_m] \in \phi_n^{-1}(i_1 i_2 \ldots i_m)$, then $[i_1 i_2 \ldots i_m]$ and $[j_1 j_2 \ldots j_p]$ commute unless m and p are both even, in which case

$$[i_1 i_2 \ldots i_m][j_1 j_2 \ldots j_p] = z[j_1 j_2 \ldots j_p][i_1 i_2 \ldots i_m]. \tag{14}$$

If $\sigma \in S_n$, then let $\phi_n^{-1}\{\sigma\} = \{\tilde{\sigma}, z\tilde{\sigma}\}$, then these two elements may not be conjugate in $\tilde{S}_n$. Let λ denote the conjugacy class of S_n corresponding to the partition λ of n. If $\tilde{\lambda} = \phi_n^{-1}(\lambda)$, then either $\tilde{\lambda}$ is a conjugacy class of $\tilde{S}_n$ or splits into two classes $\tilde{\lambda}^{\pm}$. Then Schur [19] showed that $\tilde{\lambda}$ is a *splitting class* if $\lambda \in OP(n)$ or $\lambda \in DP^-(n)$, where $OP(n)$ is the set of all partitions of n into odd parts and $DP^-(n)(DP^+(n))$ is the set of all partitions of n into distinct parts, with $n - l(\lambda)$ odd(even).

The projective representations of S_n which are not projectively equivalent to the linear representations correspond to linear representations T of $\tilde{S}_n$ which are non-trivial on the centre, namely, $T(z) = -1$. These representations of $\tilde{S}_n$ are called *spin representations* and the corresponding characters are called *spin characters*. The spin characters have the property that they are non-zero only on the splitting classes. Spin representations are associate relative to a subgroup of index 2 in $\tilde{S}_n$, in this case relative to the subgroup $\tilde{A}_n = \phi_n^{-1}(A_n)$, where A_n is the alternating group. The self-associate representations will be indicated by (SA) and the associate representations by (NSA).

Schur [19] showed that the irreducible self-associate spin representations are labelled by $DP^+(n)$ and the pairs of non-equivalent irreducible associate spin representation by $DP^-(n)$. Thus, the irreducible spin representations of $\tilde{S}_n$ are labelled by the distinct partitions of n, $DP(n)$. If $\lambda \in DP(n)$, then denote the corresponding spin representation by T_λ and if $\lambda \in DP^-(n)$ the corresponding associate representations by $T_\lambda^\pm$. The corresponding spin characters are denoted by χ_λ and $\chi_\lambda^\pm$. Schur [19] and others (see, [7]) have showed how these may be calculated and have given precise formulae for some important cases. We shall denote an element in the class λ in S_n also by λ and the corresponding element in $\tilde{\lambda}$ by $\tilde{\lambda}$. In the case of spin(projective) characters, the usual tensor product of representations is replaced by *twisted tensor products*, and direct products of groups by *twisted products* [7]. We explain how this applies in the case of $\tilde{S}_n$.

If S_k and S_{n-k} are the symmetric groups acting on $\{1,\dots,k\}$ and $\{k+1,\dots,n\}$ respectively, then $\tilde{S}_k = < z_1, t_1, \dots, t_{k-1} >$ and $\tilde{S}_{n-k} = < z_2, t_{k+1}, \dots, t_{n-1} >$ are the respective covering groups, where z_1, z_2 are the central elements. If $\tilde{S}_{(k,n-k)} = \phi_n^{-1}(S_k \times S_{n-k})$, then $\tilde{S}_{(k,n-k)}$ is the subgroup of $\tilde{S}_n$ generated by $\{z, t_1, \dots, t_{k-1}, t_{k+1}, \dots, t_{n-1}\}$. Then, the twisted product $\tilde{S}_k \hat{\otimes} \tilde{S}_{n-k}$ can be identified with $\tilde{S}_{(k,n-k)}$ via the isomorphism ψ defined by

$$\psi(t_i) = \begin{cases} (t_i, 1)Z, \ 1 \le i \le k-1 \\ (1, t_i)Z, \ k+1 \le i \le n-1, \end{cases}$$

where $Z = \{(1,1), (z_1, z_2)\}$. Similarly, if $(k_1, \dots, k_m)$ is a partition of n, the above can be extended to that case, that is, there is an isomorphism of $\tilde{S}_{(k_1,\dots,k_m)}$ with $\tilde{S}_{k_1} \hat{\otimes} \cdots \hat{\otimes} \tilde{S}_{k_m}$, which thus gives an embedding of $\tilde{S}_{k_1} \hat{\otimes} \cdots \hat{\otimes} \tilde{S}_{k_m}$ in $\tilde{S}_n$. In [7], there is a table which gives the character values of the twisted tensor product of two representations. This is extended to cover the general case $\tilde{S}_{k_1} \hat{\otimes} \cdots \hat{\otimes} \tilde{S}_{k_m}$. Let χ_i, $1 \le i \le m$ be irreducible spin characters of $\tilde{S}_{k_i}$, then Table 1 summarises the position in this case.

Table1.

$\chi_1, \dots, \chi_r$ $\chi_{r+1}, \dots, \chi_m$	SA SA	SA NSA ($m - r$ *odd*)	SA NSA ($m - r$ *even*)
$g_i \in A_{n_i}$, $1 \le i \le m$	$\prod_{i=1}^m \chi_i(g_i)$	$2^{\lfloor (m-r)/2 \rfloor} \prod_{i=1}^m \chi_i(g_i)$	$2^{\lfloor (m-r)/2 \rfloor} \prod_{i=1}^m \chi_i(g_i)$
$g_i \in A_{n_i}$, $1 \le i \le r$ $g_j \notin A_{n_j}$, $r+1 \le i \le m$	0	$(2i)^{\lfloor (m-r)/2 \rfloor} \prod_{i=1}^r \Delta\chi_i(g_i) \times$ $\prod_{j=r+1}^m \chi_j(g_j)$	0

6 A Covering Group $\tilde{B}_n^m$ of B_n^m and its Spin Representations

The group B_n^m has a presentation

$$\begin{aligned} B_n^m = < s_i,\ 1 \le i \le n-1,\ w_j,\ 1 \le j \le n | s_i^2 = 1, w_j^m = 1; \\ (s_i s_{i+1})^3 = 1,\ 1 \le i \le n-2,\ s_i w_i = w_{i+1} s_i,\ s_i w_j = w_j s_i, j \ne i, i+1 \\ (s_i s_j)^2 = 1, |i-j| \ge 2,\ 1 \le i,j \le n-1 \\ w_i w_j = w_j w_i, i \ne j,\ 2 \le i \le n-1 > . \end{aligned} \tag{15}$$

Comparing this with the presentation of S_n given earlier, we see the natural embedding of S_n in B_n^m, (w can be interpreted as the mapping which takes 1 onto $\zeta 1$, with $\{2,3,\ldots,n\}$ fixed, and $w_j = s_{j-1}s_{j-2}\cdots s_1 w s_1 \cdots s_{j-2}s_{j-1}$ for $1 \le j \le n$ with w_j being interpreted as the mapping which takes j onto ζj, with the other letters fixed.

The Schur multiplier of B_n^m was obtained in [4]

$$H^2(B_n^m, \mathbb{C}^*) = \begin{cases} \mathbb{Z}_2 = \{\gamma\} & \text{if } m \text{ is odd}, n \ge 4, \\ \mathbb{Z}_2 \times \mathbb{Z}_2 \times \mathbb{Z}_2 = \{(\gamma,\lambda,\mu)\} & \text{if } m \text{ is even}, n \ge 4, \\ \mathbb{Z}_2 \times \mathbb{Z}_2 = \{(\lambda,\mu)\} & \text{if } m \text{ is even}, n = 3, \\ \mathbb{Z}_2 = \{\mu\} & \text{if } m \text{ is even}, n = 2, \\ \{1\} & \text{otherwise}, \end{cases} \tag{16}$$

where $\gamma = \lambda = \mu = \pm 1$.

This means that if m is even B_n^m has eight factor sets $\{(\gamma,\lambda,\mu)|\gamma^2 = \lambda^2 = \mu^2 = 1\}$ and two factor sets if m is odd, $\{(\gamma)|\gamma^2 = 1\}$. We consider only one of these, namely, the one corresponding to the factor set $(-1,1,1)$ if m is even and the factor set (-1) if m is odd, for which it can be shown that the corresponding covering group denoted by $\tilde{B}_n^m$ is the semi-direct product $\mathbb{Z}_m \rtimes \tilde{S}_n$, and for which therefore the concept of Fischer matrices is relevant. The group $\tilde{B}_n^m$ has a presentation

$$\begin{aligned} \tilde{B}_n^m = < t_i,\ 1 \le i \le n-1,\ u_j,\ 1 \le j \le n | t_i^2 = 1, u_j^m = 1 \\ (t_i t_{i+1})^3 = 1,\ 1 \le i \le n-2,\ t_i u_i = u_{i+1} t_i,\ t_i u_j = u_j t_i, j \ne i, i+1 \\ (t_i t_j)^2 = -1, |i-j| \ge 2,\ 1 \le i,j \le n-1, \\ u_i u_j = u_j u_i, i \ne j,\ 2 \le i \le n-1 >, \end{aligned} \tag{17}$$

The group B_n^m has a total of $2m$ characters of degree 1, but the only relevant one in this case is η, which is defined by $\eta(s_i) = -1,\ 1 \le i \le n-1,\ \eta(w_j) = 1, 1 \le j \le n$; clearly $ker(\eta) = \mathbb{Z}_m^n \rtimes A_n =: A_n^m$ and it is relative to this normal subgroup of B_n^m that we will be working, for example, when considering representations they will be associate relative to A_n^m.

The splitting classes for projective representations of B_n^m for all factor sets were given by Read [17] (who in [18] was the first to determine all the irreducible projective representations of B_n^m for all factor sets). For the factor set $(-1,1,1)$, the splitting classes of B_n^m (or of $\tilde{B}_n^m$) interpreted in the notation of this paper are classes of the m-partition form $(OP, OP, \dots, OP)$ and $(DP, DP, \dots, DP)$, with the number of even parts odd. For convenience we will denote this latter m-partition by $(DP, DP, \dots, DP)^-$.

The irreducible spin characters of $\tilde{B}_n^m$ for this factor set correspond to the m-partitions $(\lambda^1, \lambda^2, \dots, \lambda^m) \in (DP, DP, \dots, DP)$, where λ^i is a partition of k_i with distinct parts $1 \le i \le m$, $\sum_{i=1}^m k_i = n$. These characters are denoted by $\chi_{(\lambda^1,\lambda^2,\dots,\lambda^m)}$. If $n-l$ is even, then $\chi_{(\lambda^1,\lambda^2,\dots,\lambda^m)}$ is (SA), but if $n-l$ is odd, it is (NSA); the two spin characters are denoted by $\chi^{\pm}_{(\lambda^1,\lambda^2,\dots,\lambda^m)}$, here l is the total number of parts in $(\lambda^1, \lambda^2, \dots, \lambda^m)$. The character and representation theory of $\tilde{B}_n^m$ is well understood, starting with the work of Read [18] culminating in the work of Jones [10], who built on the authoritative account by Stembridge [24] on the projective representations of hyperoctahedral groups which is the special case $m = 2$.

In this case also, the irreducible spin characters are determined by inducing from an appropriate subgroup. If $\mathbf{k} = (k_1, \dots, k_m) \in A(n,m)$ then that group is $\tilde{B}_{k_1}^m \tilde{\otimes} \cdots \tilde{\otimes} \tilde{B}_{k_m}^m$ which is isomorphic to a subgroup $\tilde{B}^m_{(k_1,\dots,k_m)}$ of $\tilde{B}_n^m$.

7 Fischer Matrices for $\tilde{B}_n^m$.

In a certain sense, the position is considerably more complicated here. We are now dealing with spin characters and also the earlier direct product of subgroups is replaced by the twisted product which needs to be embedded in the group. On the other hand, not all classes are splitting classes and therefore, there are less Fischer matrices to be computed. It turns out, that in 'most' of the relevant cases, the Fischer matrices are identical with those in the ordinary linear case and with the remaining case, it should be possible to provide an explicit form for the Fischer matrix. We can only outline the position here, the details will appear in [3], which will develop further what has already appeared in [1].

The Fischer matrix corresponding to the class $(1^{n_1}, \dots, n^{n_n})$ of $\tilde{S}_n$ will be denoted by $\tilde{\mathcal{F}}(m; 1^{n_1}, \dots, n^{n_n})$. As mentioned above, not all these classes are relevant now; in order to determine which classes need to be considered the groups involved in the induction formula (3) must be closely analysed.

If $\theta = \chi_1^{k_1}\chi_2^{k_2}\cdots\chi_m^{k_m}$, is a representative of an orbit of $\tilde{B}_n^m$ acting on the irreducible characters of $\mathbb{Z}_m^n$, then the inertia group is now $I_\theta = \tilde{B}_{k_1}^m \tilde{\otimes} \cdots \tilde{\otimes} \tilde{B}_{k_m}^m$ and the corresponding inertia factor is $\bar{I}_\theta = \tilde{S}_{k_1} \tilde{\otimes} \cdots \tilde{\otimes} \tilde{S}_{k_m}$. Thus it is the splitting classes of these groups which will determine which classes of $\tilde{S}_n$ are relevant here.

As we saw in Sect.6, the splitting classes of $\tilde{B}_n^m$ correspond to m-partitions of the form $(OP, OP, \dots, OP)$ and $(DP, DP, \dots, DP)^-$. In order to determine the classes of $\tilde{S}_{k_1} \tilde{\otimes} \cdots \tilde{\otimes} \tilde{S}_{k_m}$ which map to a class of $\tilde{S}_n$, it is necessary to analyse Table 1 in Sect.5 and also see exactly when the spin characters of $\tilde{S}_n$

are non-zero. Since, if $\lambda^i \in DP^+$, χ_{λ^i} is a (SA) and if $\lambda^i \in DP^-$, $\chi_{\lambda^i}{}^{\pm}$ is a (NSA), it can be seen that the splitting classes of $\tilde{S}_{k_1}\tilde{\otimes}\cdots\tilde{\otimes}\tilde{S}_{k_m}$ correspond to the partitions

$$(i)\ (OP(k_1) : OP(k_2) : \ldots : OP(k_m))$$

$$(ii)\ (DP(k_1) : DP(k_2) : \ldots : DP(k_m))$$

such that $n-l$ is odd, where l is the total number of parts in the m-partition $\lambda^1, \lambda^2, \ldots, \lambda^m$.

Thus, the relevant classes of $\tilde{S}_n$ are (i) (OP) (ii) $(DP^{(m)})$, where $(DP^{(m)})$ is the set of all partitions of n with no part repeated more than m times, (if $m = 1$, then $(DP^{(1)}) = (DP)$ and if $m = 2$, then $(DP^{(2)}) =$ $\{(\lambda_1^2, \ldots, \lambda_r^2, \mu_1, \ldots, \mu_s) | \lambda_1 > \cdots > \lambda_r > 0,\ \mu_1 > \cdots > \mu_s > 0,\ \lambda_i \neq \mu_j\}$. Thus, there are Fischer matrices corresponding to these classes only in $\tilde{S}_n$.

There is also another fundamental difference in this case; great deal of care has to be exercised in applying (3) which defines Fischer matrices in the general case. In the first place, $\tilde{S}_{k_1}\tilde{\otimes}\cdots\tilde{\otimes}\tilde{S}_{k_m}$ is not a subgroup of $\tilde{S}_n$, we must deal with its isomorphic image $\tilde{S}_{(k_1,\ldots,k_m)}$. For example, the fact that 'cycles' do not commute in $\tilde{S}_n$, means that in 'straightening out' the elements in going from $\tilde{S}_{(k_1,\ldots,k_m)}$ to $\tilde{S}_n$, (14) needs to be applied and whenever cycles of even order are interchanged a factor of -1 occurs. Furthermore, if w_{kl} is an element of S_n which interchanges a cycle of length k with a cycle of length l, then $\epsilon(w_{kl}) = (-1)^{kl}$, where ϵ is the sign character of S_n; that is, $\epsilon(w_{kl}) = -1$ if and only if both k and l are even.

The result is that formula (4) which defines a Fischer matrix needs to be modifed in this case to compensate for these differences with a factor $\psi(l_{ki})$ added to give

$$\left(\sum_{k=1}^{s} \frac{|C_H(h)|}{|C_{I_\theta}(l_{ki})|} \tilde{\theta}(l_{ki}) \psi(l_{ki}) \right). \tag{18}$$

It turns out, that in the first case where only (OP)'s are involved, there are no difficulties, here, the first row of Table 2 shows that all the elements now are in their respective alternating groups, only products of characters are involved (that is, no difference characters are involved). Also, all the classes involved in the ordinary case are splitting classes. Thus the Fischer matrix here has the same shape as in the ordinary case, and so we have the following theorem.

Theorem 7.1.

$$\tilde{\mathcal{F}}(m; OP) = \mathcal{F}(m; OP)$$

There is also, a special case of the second case which also results in the same Fischer matrix, namely, when the class of $\tilde{S}_n$ has all its part distinct, that is, of the form (DP). If (DP) is a class of $\tilde{S}_n$, then the classes of $\tilde{B}_n^m$ which map onto this class are all of the form$(DP; DP; \ldots; DP)$ and hence are all splitting classes of $\tilde{B}_n^m$. The conjugacy classes of $\tilde{S}_{k_1}\tilde{\otimes}\cdots\tilde{\otimes}\tilde{S}_{k_m}$ which fuse to (DP) are of the form $(DP:DP:\ldots:DP)^-$. The classes of $\tilde{B}_{k_1}\tilde{\otimes}\cdots\tilde{\otimes}\tilde{B}_{k_m}$ which maps onto one of these classes of $\tilde{S}_{k_1}\tilde{\otimes}\cdots\tilde{\otimes}\tilde{S}_{k_m}$ turn out to be uniquely determined, indeed just as in the ordinary case. Thus, the resulting Fischer matrix is of the same shape as in that case and as there is a unique element in each position in that matrix, and we have the following theorem.

Theorem 7.2.

$$\tilde{\mathcal{F}}(m;DP) = \mathcal{F}(m;DP)$$

However, the general case is far more complex. For the special case $m = 2$, a complete explicit answer is available based on the work of Stembridge [24] and will appear in [3]. We will illustrate what occurs with two examples. The first is for the case $m = 2$.

Example 7.3. We compute the Fischer matrix $\tilde{\mathcal{F}}(2;1^22^234)$ of $\tilde{B}_{13}$. This is a 4×4 matrix, since the splitting classes of $\tilde{B}_{13}$ which map to the class (1^22^234) are $(1234;12), (124;123), (123;124)$ and $(12;1234)$. Furthermore, there are four tabloids corresponding to each element in the Fischer matrix as the only choice available are the fillings corresponding to the $\mathbf{1},\mathbf{2}$. For example, we compute the element

$$\tilde{f}_{(2,2,1,1)}(((1,1,0,1),(1,1,1,0)),((1,1,0,1),(1,1,1,0))).$$

The four tabloids are

1	1 1	1 1 1 1	2	1 1	1 1 1 1	1	2 2	1 1 1 1	2	2 2	1 1 1 1
2	2 2	2 2 2	1	2 2	2 2 2	2	1 1	2 2 2	1	1 1	2 2 2

Applying the formula for the extended characters and the straightening formula now gives $(-1)^0(-1)^3 = -1$, $(-1)^1(-1)^2 = -1$, $(-1)^3(-1)^2 = -1$ and $(-1)^4(-1)^1 = -1$ respectively; giving a total of -4. Similarly, the remainder of Table 2 for $\tilde{\mathcal{F}}(2;1^22^234)$ can be computed.

Table2.

Type	$1^{10}2^3$	1^72^6	1^62^7	1^32^{10}
$1^{10}2^3$	4	-4	-4	4
1^72^6	-4	-4	4	4
1^62^7	-4	4	-4	4
1^32^{10}	4	4	4	4

As a further more complex example, we consider the following.

Example 7.4. In the Fischer matrix $\tilde{\mathcal{F}}(3;1^22^3)$, the entry $\tilde{f}_{(2,3)}(((1,1),(0,1),(1,1)),((1,1),(1,1),(0,1))) = 9\omega$, since the 12 corresponding tabloids produce the following contributions

1 1 | 1 2 2 3 3 | 2	1 1 | 1 3 3 2 2 | 2	2 2 | 1 1 1 3 3 | 2	2 2 | 1 3 3 1 1 | 2	3 3 | 1 1 1 2 2 | 2	3 3 | 1 2 2 1 1 | 2
ω^7	$-\omega^6$	$-\omega^6$	ω^4	ω^4	$-\omega^3$
1 1 | 2 2 2 3 3 | 1	1 1 | 2 3 3 2 2 | 1	2 2 | 2 1 1 3 3 | 1	2 2 | 2 3 3 1 1 | 1	3 3 | 2 1 1 2 2 | 1	3 3 | 2 2 2 1 1 | 1
$-\omega^5$	ω^4	ω^4	$-\omega^2$	$-\omega^2$	ω^7

References

1. M.O.Almestady, *Fischer matrices for generalized symmetric groups - a combinatorial approach*, Ph.D. thesis, University of Wales, Aberystwyth (1998).
2. M.O.Almestady, A.O.Morris, *Fischer matrices for generalized symmetric groups - a combinatorial approach*, (to appear).
3. M.O.Almestady, A.O.Morris, *Fischer matrices for projective representations of generalized symmetric groups*, (to appear).
4. J.W.Davies, A.O.Morris, *The Schur multiplier of the generalized symmetric group*, J. London Math. Soc. (2) **8** (1974), 615-620.
5. B.Fischer, *Clifford Matrizen*, (unpublished) (1976).
6. B.Fischer, *Clifford matrices* in Representation theory of finite groups and finite dimensional algebras, G.O.Michler and C.M.Ringel(eds), Progress in Mathematics Vol. 95, Birkhäuser,Basel,(1991), 1-16.
7. P.N.Hoffman, J.F.Humphreys, *Projective representations of the symmetric group*, Oxford Mathematical Monographs, Clarendon Press Oxford, (1992).
8. I.M.Isaacs, *Character theory of finite groups*, Academic Press (1976).
9. G.D.James, A.Kerber, *The representations of the symmetric group*, Encyclopedia of Mathematics and its Applications **16**, Addison-Wesley (1981).
10. H.I.Jones, *Clifford algebras and representations of generalised symmetric groups*, Ph.D. thesis, University of Wales, Aberystwyth (1993).
11. A.Kerber, *Zur modularen Darstellungstheorie symmetrischer und alternierender Gruppen*, Mitt. Math. Sem. Univ. Giessen **68** (1966), 1-80.
12. A.Kerber, *Zur Darstellungstheorie von Kranzprodukten*, Canad. J. Math. **20** (1968), 665-672.
13. A.Kerber, *Representations of permutation groups, I and II*, Lecture Notes in Mathematics **240** (1971), **495** (1975) Springer-Verlag, Berlin.
14. A.Lascoux, B.Leclerc, J-Y.Thibon, *Ribbon tableaux, Hall-Littlewood functions, quantum affine algebras and unipotent elements*, J. Math. Physics **38** (1997), 1041-1068.
15. R.J.List, I.M.I. Mahmoud, *Fischer matrices for wreath products* $G \wr S_n$, Arch. Math. **50** (1988), 394-401.
16. M.Osima, *On the representations of the generalized symmetric group*, Math. J. Okayama Univ. **4** (1954), 39-56; II, (*ibid*) **6** (1956), 81-97.
17. E.W.Read, *The* α*-regular classes of the generalized symmetrized group*, Glasgow Math. J. **17** (1976), 144-150.
18. E.W.Read, *The projective representations of the generalized symmetric group*, J. Algebra **46** (1977), 102-132.

19. I.Schur, *Über die Darstellung der symmetrischen und der alternierenden Gruppe durch gebrochene lineare Subsitutionen*, J. Reine Ang. Math. **139** 1911, 155-250.
20. W.Specht, *Eine Verallgemeinerung der symmetrischen Gruppe*, Schriften des Math. Sem. Berlin **1** (1932), 1-32.
21. W.Specht, *Eine Verallgemeinerung der Permutationsgruppe*, Math. Zeits. **37** (1933), 321-341.
22. W.Specht, *Darstellungstheorie der Hyperoktaedergruppe*, Math. Zeits. **42** (1937), 629-640.
23. R.P.Stanley, *Enumerative Combinatorics*, Vol. I, Wadsworth Books, Monterey, California (1986).
24. J.R.Stembridge, *The projective representations of the hyperoctahedral group*, J. Algebra **145** (1992), 396-453.

New Good Codes via CQuest — A System for the Silicon Search of Linear Codes

Thomas Rehfinger, N. Suresh Babu, and Karl-Heinz Zimmermann

Technical University of Hamburg-Harburg, 21071 Hamburg, Germany

Abstract We describe a software tool called CQUEST which has been designed for the generation and manipulation of code tables. CQUEST handles one code table for each alphabet. It supports two kinds of operations for the manipulation of code tables: code modifications and code combinations. CQUEST provides an interpreter for interactive use. Some new good quaternary codes obtained by CQUEST are described.

1 Introduction

The search for good linear codes is one of the fundamental problems in coding theory. The tabulation of linear codes with block length n and dimension k which have highest possible error correction capability, i.e. minimum distance, is a cumbersome task. No general practical method is known to compute such codes. In particular, the calculation of the minimum distance is a computationally intensive task [1]. Therefore, in several approaches, upper and lower estimates for the minimum distance are calculated. Today, there is a table containing the best known upper and lower estimates for small length over small alphabets (e.g., $n \leq 255$ in the binary case) maintained by Brouwer [3].

The objective of this paper is twofold. At first, we describe a software tool called CQUEST which has been designed for the search, manipulation, and maintainance of linear codes. Secondly, we present a few new good linear codes determined by CQUEST which improve Brouwer's table.

2 Linear Codes

For the transmission of a stream of data through a noisy channel (satellite, compact disc etc.), an *encoder* is used to provide some protection against errors on the channel. The encoder divides the data stream into blocks of equal length k and adds some redundancy to each such block. We assume that the data are belonging to a finite field $\mathbb{F}_q$ of $q = p^r$ elements. The encoder is described by an injective linear mapping $\mathbb{F}_q^k \to \mathbb{F}_q^n : x \mapsto x\Gamma$, where Γ is a $k \times n$-matrix with entries from $\mathbb{F}_q$. This method of encoding produces what is called a *linear code*: The set of images

$$C = \{x\Gamma \mid x \in \mathbb{F}_q^k\}$$

forms a k-dimensional subspace of $\mathbb{F}_q^n$ and is called an $[n,k]$-*code* over $\mathbb{F}_q$; n is the *length* of the code. The elements of C are called *codewords* and the matrix Γ is called a *generator matrix* of the code. Each $[n,k]$-code C over $\mathbb{F}_q$ can also be described by the kernel of an appropriate surjective linear mapping $\mathbb{F}_q^n \to \mathbb{F}_q^{n-k}$: There is an $(n-k) \times n$-matrix Δ of rank $n-k$ such that

$$C = \{v \in \mathbb{F}_q^n \mid v\Delta^t = 0\},$$

where Δ^t is the transposed matrix of the *parity check matrix* Δ of the code.

If C is an $[n,k]$-code over $\mathbb{F}_q$, its *dual code* $C^\perp$ is the set of vectors which are orthogonal to all codewords of C (w.r.t. the standard bilinear form $\langle v,w\rangle = v_1w_1 + \ldots + v_nw_n$):

$$C^\perp = \{v \mid \forall c \in C : \langle v,c\rangle = 0\}.$$

Thus each parity check matrix of C is a generator matrix of $C^\perp$, and vice versa.

Suppose the codeword $c \in C$ is sent through the channel. Because of channel noise, the received vector $y \in \mathbb{F}_q^n$ may be different from c; the channel may distort c by adding the *error vector* $e = y - c$. The *decoder* must decide from the received vector which codeword was transmitted. His strategy is to choose the most likely error vector (under the assumption that the codewords are all equally likely and the probability that the ith component is wrong is less than the probability that its correct). In order to describe this strategy, we need some further definitions.

The *Hamming distance* between two vectors $v, w \in \mathbb{F}_q^n$ is the number of components where they differ:

$$d(v,w) = |\{i \mid v_i \neq w_i\}|.$$

It is closely related to the *Hamming weight* $\mathrm{wgt}(v)$ of a vector $v \in \mathbb{F}_q^n$ which is the number of its non-zero components:

$$\mathrm{wgt}(v) = |\{i \mid v_i \neq 0\}|.$$

Now the third important parameter of a code C, besides length and dimension, is the *minimum distance* between its codewords:

$$d(C) = \min\{d(c,c') \mid c,c' \in C, c \neq c'\}.$$

It equals the *minimum weight* of C:

$$\min\{\mathrm{wgt}(c) \mid c \in C, c \neq 0\}$$

since $d(c,c') = \mathrm{wgt}(c-c')$ and C is linear. An $[n,k]$-code with minimum distance d is also denoted as $[n,k,d]$-code. If A_i is the number of codewords of weight i, the sequence $(A_0, A_1, \ldots, A_n)$ is called the *weight distribution* of C.

Suppose the codeword $c \in C$ is transmitted and t components of c are corrupted. Thus for the received vector $y \in F_q^n$, we have $d(c, y) = t$. But every two distinct codewords differ in at least $d = d(C)$ components. So if $t \leq (d-1)/2$, then the received vector y lies in the *sphere*

$$\{x \in \mathbb{F}_q^n \mid d(c, x) \leq (d-1)/2\}$$

of radius $r = (d-1)/2$ and center c. But every two such spheres of radii r, whose centers are distinct codewords, are disjoint. Hence, the received vector can be uniquely decoded to the codeword sent: This codeword is the center of the sphere containing the received vector. This strategy of decoding is called *maximum likelihood decoding* (cf. Fig. 1). It is optimal in the sense that it minimizes the probability of the decoder making a mistake. The minimum

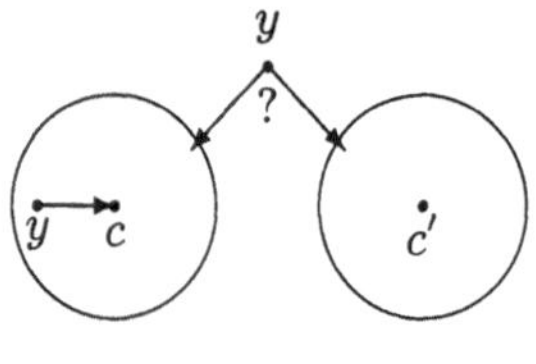

Figure1. Maximum-likelihood decoding.

distance of a linear code is thus a measure for its error correction capability [11].

Let C be a linear code of length n over $\mathbb{F}_q$. Any permutation of the n coordinates or multiplication of the coordinates by nonzero elements of $\mathbb{F}_q$ or both changes C into an *equivalent* code having many of the same properties as C (same dimension, minimum distance, etc.) [2]. As a consequence, equivalent codes have the same error correction capabilities. In particular, each $[n, k]$-code C is equivalent to a linear code C' with *systematic generator matrix* $(I_k \mid A)$, where I_k denotes the $k \times k$-unit matrix. Thus C' has parity check matrix $(-A^t \mid I_{n-k})$.

An important problem in algebraic coding is to find linear codes with the best possible minimum distance. However, no general practical method is known to compute the maximal minimum distance function (over fixed alphabet $\mathbb{F}_q$):

$$d_{max}(n, k) = \max\{d(C) \mid \exists\ [n, k]\text{-code } C \text{ over } \mathbb{F}_q\}.$$

In the binary case, Helgert and Stinaff [9] have constructed a table with lower and upper estimates for $d_{\max}(n, k)$ where $n \leq 127$. Later, Verhoeff [15] has improved this table to a large extent by an algorithm that makes the table invariant under certain simple code modification methods. In particular, his algorithm starts from a class of good linear codes (Reed-Solomon codes) for which the code parameters are known. Today, there is a table containing the best linear codes for small length over small alphabets (e.g., $n \leq 255$ in the binary case) maintained by Brouwer [3]. Important contributions to this table have been made by Wirtz [16] and, in the binary case, by Groneick and Grosse [8]; while Wirtz uses lower and upper estimates for $d_{\max}(n,k)$, Groneick and Grosse calculate the minimum distance by a simulated annealing algorithm (for binary codes of dimension $k \leq 30$).

3 The Architecture of CQUEST

CQUEST is a software tool for the generation and manipulation of code tables. Its ultimate design goal is to improve code tables. CQUEST is written in C++ and runs on machines with a UNIX operating system. The main components of CQUEST are illustrated in Fig. 2.

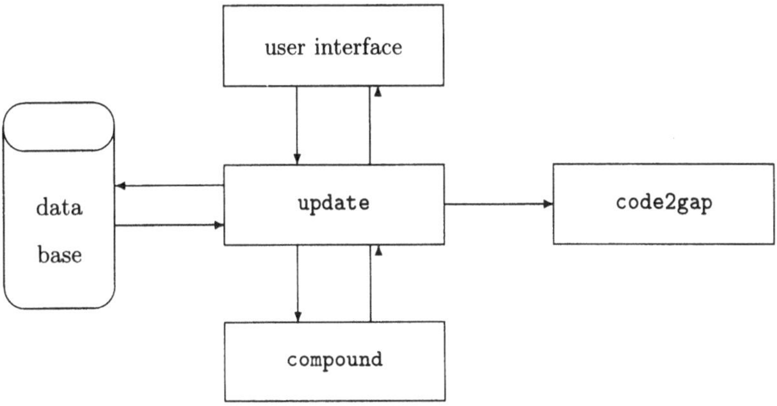

Figure2. Main components of CQUEST.

3.1 Code Tables

CQUEST handles one code table for each alphabet. Each *code table* is designed as a database generated and maintained by the GNU database manager (GDBM). Each *record* of a code table is a linked list of subrecords. Each *subrecord* specifies a linear code by the following items: block length, dimension, minimum distance, minimum weight codeword, systematic generator matrix, and a so-called *trace* (a string specifying how the code is constructed from previously inserted codes by primitive or compound operations). The trace identifies a linear code and is particularly useful for its reconstruction. The subrecords of a record specify a list of pairwise inequivalent linear codes of the same length and dimension, which all have best possible minimum distance (known to CQUEST). The *key* of each record is the pair of length and dimension.

Currently, CQUEST employs a variant of Brouwer's minimum distance algorithm [2] and uses the test described in [14] to decide whether two matrices generate equivalent linear codes. Notice that arrangements are made if the minimum distance computation exceeds a certain amount of time. Then in the subrecord corresponding to the code the minimum distance item is specified by the lower and upper estimate computed so far.

Each newly generated code table will be empty. But as soon as linear codes of length n are inserted, trivial codes up to length n will be added.

3.2 Primitive Operations

CQUEST supports two kinds of operations for the manipulation of code tables: primitive and compound operations. The *primitive operations* are code modifications [11]. The following list describes the primitive operations currently implemented and their effect when applied to an $[n, k, d]$-code C over $\mathbb{F}_q$ with generator matrix $\Gamma = (g_1 \mid g_2 \mid \dots \mid g_n)$:

3.2.1 Parity Check Extension The matrix $\Gamma' = (g_1 \mid g_2 \mid \dots \mid g_n \mid g_{n+1})$, where $g_{n+1} = -\sum_{i=1}^{n} g_i$, generates an $[n+1, k, \geq d]$-code.

3.2.2 Puncturing The matrix $\Gamma' = (g_1 \mid g_2 \mid \dots \mid g_{n-1})$ generates an $[n-1, k, \geq d-1]$-code.

3.2.3 Shortening The matrix Γ' such that

$$\Gamma = \left(\begin{array}{c|c} g_{11} & * \\ \hline 0 & \Gamma' \end{array} \right), \quad g_{11} \neq 0,$$

generates an $[n-1, k-1, \geq d]$-code.

3.2.4 A-Construction It may be assumed that the first row of Γ forms a minimum weight vector of C and that its first d components are 1's. Then the matrix Γ' such that

$$\Gamma = \left(\begin{array}{c|c} 11\dots1 & 0 \\ \hline * & \Gamma' \end{array} \right)$$

generates an $[n-d, k-1, \geq \lceil d/q \rceil]$-code.

CQUEST is designed in such a way that the user can easily add further primitive code operations.

The minimum weight vector is used in several situations to determine the minimum distance directly without any further calculation. E.g., if puncturing (resp. shortening) is applied to a component at which the given minimum weight vector has a non-zero (resp. zero) entry, then the punctured (resp. shortened) code has minimum distance $d-1$ (resp. d).

One important feature of the code tables is their *closure* under primitive operations, i.e. the codes stored in a code table cannot be improved by applying the (currently implemented) primitive operations. This closure can be realized by the following recursive `update` procedure first proposed by VERHOEFF [15]:

```
update ( R, ct ) /* R is a subrecord specifying a linear code, ct is a code table */
    if improvement of ct is possible by R then
        insert R into ct
        for each primitive operation P do
            update ( P(R), ct )
end update.
```

Moreover, each code table is invariant under duality, since whenever `update` is invoked for a linear code it will be automatically invoked for its dual.

Every improvement of an invariant code table must come from outside. This can be achieved in two ways: by insertion of new linear codes or by compound operations.

3.3 Initial Codes

Linear codes from two code classes are initially inserted into each code table, cyclic codes and codes which are modules of Hecke algebras of types A_n and B_n; those modules can be defined over any (finite) field. We have inserted Specht and James modules for the Hecke algebra of type A_n and specific homomorphic images of them [4,10,17]. It is shown in [17] that this class of codes contains the generalized Reed-Muller codes over the primes. Moreover, Specht modules for the Hecke algebra of type B_n are inserted, i.e., the duals of the modules described in [5]. It is unknown if the class of modules for the Hecke algebra of type B_n contains some interesting linear codes.

3.4 Compound Operations

The *compound operations* are code combination operations [11]. The following list describes the compound operations currently implemented:

3.4.1 $u \mid u+v$-Construction Let C_i be an $[n, k_i, d_i]$-code over $\mathbb{F}_q$ generated by Γ_i, for $i = 1, 2$. The linear code generated by

$$\Gamma = \begin{pmatrix} \Gamma_1 & \Gamma_1 \\ 0 & \Gamma_2 \end{pmatrix}$$

is a $[2n, k_1 + k_2, \min\{2d_1, d_2\}]$-code over $\mathbb{F}_q$.

3.4.2 $a+x \mid b+x \mid a+b+x$-Construction Let C_i be an $[n, k_i, d_i]$-code over $\mathbb{F}_q$ generated by Γ_i, for $i = 1, 2$. The linear code generated by

$$\Gamma = \begin{pmatrix} \Gamma_1 & 0 & \Gamma_1 \\ 0 & \Gamma_1 & \Gamma_1 \\ \Gamma_2 & \Gamma_2 & \Gamma_2 \end{pmatrix}$$

is a $[3n, 2k_1 + k_2]$-code over $\mathbb{F}_q$.

3.4.3 X-Construction Let C_1 be a $[n_1, k_1, d_1]$-code over $\mathbb{F}_q$ generated by

$$\Gamma_1 = \begin{pmatrix} \Gamma' \\ \Gamma_2 \end{pmatrix},$$

let C_2 be the $[n_1, k_2, d_2]$-subcode of C_1 generated by Γ_2, and let C_3 be a $[n_3, k_1 - k_2, d_3]$-code over $\mathbb{F}_3$ generated by Γ_3. The linear code generated by

$$\Gamma = \begin{pmatrix} \Gamma' & \Gamma_3 \\ \Gamma_2 & 0 \end{pmatrix}$$

is an $[n_1 + n_3, k_1, \geq \min\{d_2, d_1 + d_3\}]$-code over $\mathbb{F}_q$. Precautions have to be made to prevent that codewords of small weight are pasted together (cf. [8]).

3.4.4 XX-Construction Let C_i be an $[n, k_i, d_i]$-code over $\mathbb{F}_q$ generated by Γ_i, for $i = 1, \ldots, 6$, where $C_1 = C_2 + C_3$ and $C_4 = C_2 \cap C_3$. Thus

$$\Gamma_2 = \begin{pmatrix} \Gamma_2' \\ \Gamma_4 \end{pmatrix}, \quad \Gamma_3 = \begin{pmatrix} \Gamma_4 \\ \Gamma_3' \end{pmatrix}, \quad \text{and} \quad \Gamma_1 = \begin{pmatrix} \Gamma_2' \\ \Gamma_4 \\ \Gamma_3' \end{pmatrix}.$$

The linear code generated by

$$\Gamma = \begin{pmatrix} \Gamma_2' & 0 & \Gamma_6 \\ \Gamma_4 & 0 & 0 \\ \Gamma_3' & \Gamma_5 & 0 \end{pmatrix}$$

is an $[n_1 + n_5 + n_6, k_1, \geq \min\{d_4, d_2 + d_6, d_3 + d_5, d_1 + d_5 + d_6\}]$-code over $\mathbb{F}_q$.

CQuest allows the user to add further compound code operations. Also compound operations combining codes which are defined over different alphabets (as e.g. concatenation) can be easily added.

3.5 Interactive CQUEST-Interface

CQUEST provides an interpreter for the interactive generation and manipulation of code tables. At each time exactly one code table is the actual one. The following commands are currently supported:

Command	Explanation
`create` *ct q*	creates new code table *ct* for alphabet $\mathbb{F}_q$ and *ct* becomes the actual code table
`get` *ct*	*ct* becomes the actual code table
`insert` *file*	inserts linear codes from *file*
`delete`	deletes the actual code table
`compound` *n*	combines codes to codes of length *n* by compound operations
`select` *keylist*	a query selecting all records with key n, k from *keylist*
`status`	provides the status of actual code table.

The `status` command provides information on the size of the code table, the number of insertions during the last session, and the currently active primitive and compound operations. Two additional commands, `addprimitive` and `addcompound`, allow an addition of user-defined routines (primitive and compound operations) to the list of operations of the actual code table.

3.6 Parallel CQUEST

Our experience with the serial CQUEST implementation have shown that the minimum distance computation takes up to 95% of the total CPU time. Therefore, a parallel version of CQUEST has been designed to overcome this problem (in some sense). The principles upon which the parallel version of CQUEST is based include the following:

- Subsequent calls of the minimum distance routine are parallelized.
- The user is able to send commands to background processes (e.g. perform insertion or compound operations) without having to wait for them to finish, or she can select codes from a code table even if background processes are still in progress.
- Automatic recovery after a system crash, power failure, etc.

The parallel version of CQUEST makes use of the PVM (Parallel Virtual Machine) software system [6] that permits a heterogeneous collection of UNIX computers networked together to be viewed by the user's program as a single parallel computer. Our PVM-based implementation is described in [12].

3.7 The GAP-Interface

GAP (Groups, Algorithms, and Programming) is a system for computational discrete algebra with particular emphasis on computational group theory [13].

GAP includes a share library package named GUAVA (developed at Delft University of Techn.) which implements various algorithms from coding theory. GUAVA has been primarily designed for the construction and analysis of codes. We have written a GAP-program to analyze the newly found linear codes which improve Brouwer's table [3]. For each such code (of moderate length), the program determines, e.g., its weight distribution and weight histogram, whether it is cyclic and if so, computes the generator polynomial and the corresponding root set.

4 Three New Quaternary Linear Codes

At the time of writing this paper, we have filled code tables for codes over small fields $\mathbb{F}_q$ ($q = 2, 3, 4, 5, 7$) up to length $n \leq 45$ (cf. Section 3.3). Since linear codes of moderate length over $\mathbb{F}_2$ and $\mathbb{F}_3$ have been intensively searched for, no improvements of the corresponding code tables have been obtained so far. However, we have found three new linear codes over $\mathbb{F}_4$ [7]:

4.0.1 A $[36, 24, 8]$-code with systematic generator matrix $(I \mid A)$ where

$$A = \begin{pmatrix} 131330303013 \\ 122203333310 \\ 123110030323 \\ 200201102012 \\ 111310213211 \\ 120201322332 \\ 331200330200 \\ 033120033020 \\ 211202102321 \\ 233030311210 \\ 112033332133 \\ 332023131221 \\ 221312212100 \\ 022131221210 \\ 133123221133 \\ 330132120121 \\ 221103313030 \\ 301330133332 \\ 313313211300 \\ 031331321130 \\ 320313330122 \\ 220121232033 \\ 313303030113 \\ 301232321231 \end{pmatrix}.$$

The Brower table contains a $[36, 24, 7]$-code and the current upper bound for quaternary $[36, 24]$-codes is 9. The code has the trace CyP3dP2P2, i.e., it is obtained from a cyclic code by several code modifications. Notice that `P2` means parity check extension, `P3` puncturing, `P4` shortening, and `A` A-construction, while `d` indicates that the dual code is considered.

4.0.2 A [42, 31, 7]-Code with systematic generator matrix $(I \mid A)$ where

$$
\begin{pmatrix}
10233200202\\
22332020302\\
32111302123\\
13022330021\\
01202233031\\
01320223310\\
30010122211\\
21210312100\\
10212231010\\
00121223101\\
11221322102\\
12211332013\\
03121133221\\
21103213033\\
03310321312\\
32013132031\\
01301313223\\
23121231013\\
03112123132\\
12122012132\\
12121001023\\
03212100121\\
11012010232\\
22010101312\\
03202311023\\
03220231130\\
10111223332\\
32233022200\\
20332002120\\
32313222322\\
32023110023
\end{pmatrix}.
$$

The Brower table contains a [42, 31, 6]-code, the current upper bound for quaternary [42, 31]-codes is 8, and the trace of the code is `CyP3dP2P2P4P3dP2P2`.

4.0.3 A [44, 8, 27]-code with systematic generator matrix $(I \mid A)$ where

$$A = \begin{pmatrix} 132122311103233202303223201112331321 \\ 200021313311032023331013022332321003 \\ 020002131331103202333101302233232103 \\ 002000213133110320233310130223323213 \\ 132322330210122230320112212130003002 \\ 013232233021012223032011221213000302 \\ 001323223302101222303201122121300032 \\ 321223111032332023032232011123313211 \end{pmatrix}.$$

The Brower table contains a [44, 8, 26]-code, the current upper bound for quaternary [44, 8]-codes is 28, and the trace of the code `CyP3dP2P2`.

We have also found several new good linear codes over the alphabets $\mathbb{F}_5$ and $\mathbb{F}_7$, which improve Brouwer's table. However, as Brouwer has remarked, his code tables for $q = 5, 7, 8, 9$ are still weak [3]. In fact, researchers have not yet in a serious way studied bounds for linear codes over these large fields. In practice, this means that anyone who searches for such codes will come up with new results.

5 Inheritance of Decoding

We may seek for an effective decoding procedure for each of the linear codes obtained by CQUEST. However, a common property of all of these linear codes is that they are derived from an initial code by a series of primitive operations. Suppose there is a decoding procedure for each such initial code and assume there exists a *decoding scheme* for each code derived from a linear code by one of our primitive operations; such a decoding scheme will depend on a decoding procedure for the parent code and the concerned operation. Then we will be able to decode each of the linear codes constructed by CQUEST.

In order to describe such decoding schemes, let C be an $[n, k, d]$-code over $\mathbb{F}_q$ with generator matrix $\Gamma = (g_1 \mid g_2 \mid \dots \mid g_n)$ and let $\mathcal{D}$ denote a decoding procedure for C able to correct $t = \lfloor \frac{d-1}{2} \rfloor$ errors.

5.0.4 Parity Check Extension Let C' be the parity check extension of C. Suppose $y' = y_1 \dots y_{n+1}$ is the received vector. First decode $y = y_1 \dots y_n$ via $\mathcal{D}$ to obtain codeword $c \in C$ and error vector $e = e_1 \dots e_n$. This decoding is correct if at most t errors have occured at the first n positions. Now the error at the $(n+1)$-th position is $e_{n+1} = \sum_{i=1}^{n+1} y_i - \sum_{i=1}^{n} e_i$. Then $e' = e_1 \dots e_{n+1}$ is the error vector.

5.0.5 Puncturing Let C' be the $[n-1, k, d-1]$-code obtained from C by puncturing the last position. Suppose $y' = y_1 \dots y_{n-1}$ is the received vector.

For each letter $x \in \{\langle v, g_n \rangle \mid v \in \mathbb{F}_q^k\}$, consider the word $y_x = y_1 \dots y_{n-1}x$. Decode y_x via decoding scheme $\mathcal{D}$ to derive codeword $c_x = c_1 \dots c_{n-1}c_n \in C$. Notice that if d is even, i.e., $t = \lfloor \frac{d-2}{2} \rfloor$, and the n-th position is in error, then $\mathcal{D}$ may fail. Consider all the codewords of C obtained in this way. Delete the n-th position from each such codeword. Among these words of length $n-1$, the closest to y' is the codeword $c' \in C$ sent.

5.0.6 Shortening Let C' be the $[n-1, k-1, d]$-code obtained from C by shortening at the first position. Observe that $c_1 \dots c_{n-1} \in C'$ if and only if $0c_1 \dots c_{n-1} \in C$. Suppose $y' = y_1 \dots y_{n-1}$ is the received vector. Decode the word $0y_1 \dots y_{n-1}$ via decoding scheme $\mathcal{D}$ to obtain the codeword $c = 0c_1 \dots c_{n-1} \in C$. The codeword sent is thus $c_1 \dots c_{n-1} \in C'$.

5.0.7 A-Construction Let C' be the $[n-d, k-1, d']$-code obtained from C by A-construction, where $d' \geq \lceil d/q \rceil$, Suppose $y' = y_1 \dots y_{n-d}$ is the received vector. For each word $x_1 \dots x_d$, where $x_i \in \{\langle v, g_i \rangle \mid v \in \mathbb{F}_q^k\}$, consider the word $y_{x_1 \dots x_d} = x_1 \dots x_d y_1 \dots y_{n-d}$. The decoding is now similar to puncturing.

6 Concluding Remarks

We have presented CQUEST, a software tool for the search of linear codes. In the next generation of CQUEST we plan to use a parallelized version of our minimum distance routine as proposed in [12]. Furthermore, a variant of CQUEST for the search of codes over rings will be implemented.

7 Acknowledgement

We would like to thank Volker Strehl for a useful hint on Section 5.

References

1. Berlekamp E., McEliece R., van Tilborg, H.: On the inherent intractability of certain coding problems. IEEE Trans. Inform. Theory, **IT-24** (1978) 384-386.
2. Betten A., Kerber A., Fripertinger H., Wassermann A., Zimmermann, K.-H.: Codierungstheorie — Konstruktion und Anwendung linearer Codes. Springer, Berlin (1998)
3. Brouwer, A.: http://www.win.tue.nl/math/dw/personalpages/aeb/voorlincod.html
4. Dipper R., James G.: Representations of Hecke algebras of general linear groups. J. Algebra (1986) 20-52.
5. Dipper R., James G., Murphy E.: Hecke algebras of type B_n at roots of unity. Preprint (1985).

6. A. Geist et al.: *PVM: Parallel Virtual Machine. A Users' Guide and Tutorial for Networked Parallel Computing.* MIT Press, 1994; http://www.netlib.org/pvm3/book/pvm-book.html
7. Greenough P.P., Hill R.: Optimal linear codes over $GF(4)$. Discrete Math. **125** (1994) 187-199.
8. Groneick, B., Grosse, S.: Konstruktion binärer Codes. Diploma Thesis, Univ. Münster (1991).
9. Helgert H.J., Stinaff, R.D.: Minimum-distance bounds for binary linear codes. IEEE Trans. Inform. Theory, **IT-19** (1973) 344-356.
10. James G., Kerber, A.: The Representation Theory of the Symmetric Group. Addison Wesley Pub., Reading, MA (1981).
11. MacWilliams J., Sloane, N: The Theory of Error Correcting Codes. North-Holland, Amsterdam (1988).
12. Rehfinger, T.: Verteilte Suche nach guten linearen Codes. Diploma Thesis, TU Hamburg-Harburg (to appear).
13. Schönert M. et. al.: GAP – Groups, Algorithms, and Programming. Lehrstuhl D für Mathematik, RWTH Aachen, 5th Ed. (1995).
14. Suresh Babu, N., Zimmermann, K.-H.: Testing the equivalence of linear codes over finite local commutative rings. submitted (1999).
15. Verhoeff, T.: An updated table of minimum-distance bounds for binary linear codes. IEEE Trans. Inform. Theory, **IT-33** (1987) 665-680.
16. Wirtz, M.: Konstruktion und Tabellen linearer Codes. Ph.D. Thesis, Univ. Münster (1991).
17. Zimmermann, K.-H.: Beiträge zur algebraischen Codierungstheorie mittels modularer Darstellungstheorie. Bayreuther Math. Schriften, Bayreuth **48** (1994).

On Graphs with Degrees in Prescribed Intervals

Manfred Schocker

Mathematisches Seminar der Universität
Ludewig-Meyn-Str. 4
24098 Kiel

Dedicated to Adalbert Kerber on the occasion of his 60th birthday.

Abstract Necessary and sufficient conditions are given for the existence of graphs and bipartite graphs with degrees in prescribed intervals.

We consider undirected simple graphs G on n vertices $x_1, \ldots, x_n$. The *degree* $\nu_G(x_i)$ of x_i is the number of neighbours of x_i in G for all $i \leq n$.

Given nonnegative integers $r_1, \ldots, r_n, s_1, \ldots, s_n$ such that $r_i \leq s_i$ for all $i \leq n$, does there exist a graph with degrees $g_1, \ldots, g_n$ such that $r_i \leq g_i \leq s_i$ for all $i \leq n$? If $1 \leq k \leq n$, can we find such a graph which is, additionally, bipartite with respect to the decomposition $U := \{x_1, \ldots, x_k\}$, $V := \{x_{k+1}, \ldots, x_n\}$ of its vertex set?

In both cases, we give necessary and sufficient conditions for the existence of a graph of the type in question. In the case of bipartite graphs, we recover a result of Fulkerson ([Ful59]). The proof presented here makes use of the notion of a *dual dominance relation* on the set of partitions, which turns out to be equivalent to the usual dominance ordering. In the case of arbitrary graphs, our result is a natural generalization of the well-known Erdös-Gallai Criterion ([EG60]). Our (inductive) approach takes as its starting point the Gale-Ryser Theorem ([Gal57], [Rys57]) on the existence of matrices of zeros and ones with prescribed row and column sum or, equivalently, on the existence of bipartite graphs with prescribed degrees.

In [Kor76], Theorem 4.2, the problem of the existence of simple graphs with degrees in prescribed intervals which are not all trivial, is analyzed. Its essential part is an immediate consequence of our result.

1 Bipartite Graphs with Degrees in Prescribed Intervals

Let $\mathbb{N}$ ($\mathbb{N}_0$, resp.) be the set of all positive (nonnegative, resp.) integers and

$$\underline{n} := \{\, k \in \mathbb{N} \mid 1 \leq k \leq n \,\} \qquad \text{for all } n \in \mathbb{N}_0 \,.$$

If r is a mapping of $\mathbb{N}$ into $\mathbb{N}_0$ such that $r_\nu = 0$ for almost all $\nu \in \mathbb{N}$, we say that r is a *sequence of finite support* and define

$$\operatorname{sum} r := \sum_\nu r_\nu .$$

If $m \in \mathbb{N}$ such that $r_\nu = 0$ for all $\nu > m$, we sometimes use the notation $r = (r_1, \ldots, r_m)_0$.

Let r, s be sequences of finite support. We write $r \leq s$ if and only if $r_\nu \leq s_\nu$ for all $\nu \in \mathbb{N}$. Let G be a graph with vertex set $\{x_1, \ldots, x_n\}$. The sequence

$$g : \mathbb{N} \longrightarrow \mathbb{N}_0,\ i \longmapsto \begin{cases} \nu_G(x_i) & , i \leq n \\ 0 & , i > n \end{cases}$$

is called a *degree sequence* of G.

Using these definitions, the problem of this section may be stated as follows: Let p, $\tilde{p}$, q, $\tilde{q}$ be sequences of finite support such that $p \leq \tilde{p}$ and $q \leq \tilde{q}$. Does there exist a bipartite graph (with respect to the vertex sets U, V) with degree sequences g, h corresponding to U and V such that $p \leq g \leq \tilde{p}$ and $q \leq h \leq \tilde{q}$? It is easily seen that it suffices to consider the case that p and q are *partitions*, that is, nonincreasing.

Definitions 1.1. *Let r, s be sequences of finite support. r* is dominated by *s ($r \trianglelefteq s$) if and only if $\sum_{\nu=1}^k r_\nu \leq \sum_{\nu=1}^k s_\nu$ for all $k \in \mathbb{N}$.* The conjugate partition *r' of r is defined by $r'_j = |\{\nu \in \mathbb{N} \,|\, r_\nu \geq j\}|$ for all $j \in \mathbb{N}$. Let M be a $k \times l$ matrix over $\mathbb{N}_0$. We denote by $\sum_R M$ ($\sum_C M$, resp.) the sum of its rows (columns, resp.), that is,*

$$(\textstyle\sum_R M)_j = \sum_{\nu=1}^k M_{\nu j} \quad \Big((\textstyle\sum_C M)_i = \sum_{\mu=1}^l M_{i\mu},\ \textit{resp.}\Big)$$

for all $i, j \in \mathbb{N}$, where $M_{\nu\mu} := 0$ for all $\nu, \mu \in \mathbb{N}$ such that $\nu > k$ or $\mu > l$.

For example, if $M = \begin{pmatrix} 1\,3\,0 \\ 0\,2\,1 \end{pmatrix}$, we have $\sum_R M = (1,5,1)_0$ and $\sum_C M = (4,3)_0$. Note that for all sequences r of finite support we have

$$\sum_{\nu=1}^j r'_\nu = \sum_\nu \min\{r_\nu, j\}$$

for all $j \in \mathbb{N}_0$. In particular, $\operatorname{sum} r = \operatorname{sum} r'$. Furthermore, if r is a partition, then $r'' = r$, as, for all $i \in \mathbb{N}$,

$$r''_i = \Big|\Big\{ j \in \mathbb{N} \,\Big|\, |\{\nu \in \mathbb{N} \,|\, r_\nu \geq j\}| \geq i \Big\}\Big| = |\{ j \in \mathbb{N} \,|\, r_i \geq j\}| = r_i .$$

Finally, the relations $r \trianglelefteq s$ and $r \leq s$ both imply $\operatorname{sum} r \leq \operatorname{sum} s$.

As our starting point, we recall the following fundamental result.

Theorem 1.2 (Gale [Gal57], Ryser [Rys57]). *Let q be a partition and p be a sequence of finite support. Then there exists a matrix M of zeros and ones such that $\sum_C M = p$ and $\sum_R M = q$ if and only if $\operatorname{sum} p = \operatorname{sum} q$ and q is dominated by p'.*

If M is a matrix of zeros and ones, then

$$\begin{pmatrix} 0 & M^t \\ M & 0 \end{pmatrix}$$

is the adjacency matrix of a bipartite graph with degree sequences $\sum_R M$ and $\sum_C M$ (with respect to the decomposition of its vertex set). Conversely, for any bipartite graph there exists an adjacency matrix which is obtained in that way. Hence the Gale-Ryser Theorem may also be interpreted as a characterization of the pairs of degree sequences of bipartite graphs. A simple proof of the theorem can be found in [Kra96].

For any sequence $r : \mathbb{N} \longrightarrow \mathbb{N}_0$ and any $k \in \mathbb{N}_0$ we define

$$r[k]_- : \mathbb{N} \longrightarrow \mathbb{Z},\ \nu \longmapsto \begin{cases} r_\nu & , \nu \neq k \\ r_\nu - 1 & , \nu = k \end{cases}$$

and

$$r[k]_+ : \mathbb{N} \longrightarrow \mathbb{Z},\ \nu \longmapsto \begin{cases} r_\nu & , \nu \neq k \\ r_\nu + 1 & , \nu = k \end{cases},$$

where $\mathbb{Z}$ is the set of all integers. Furthermore, let the sequences $r_{\leq k}$ ($r_{>k}$, $r \ominus k$, resp.) be defined by

$$\nu \longmapsto \begin{cases} r_\nu & , \nu \leq k \\ 0 & , \nu > k \end{cases}, \quad (\nu \longmapsto r_{\nu+k}, \quad \nu \longmapsto \max\{r_\nu - k, 0\}, \text{ resp.})$$

for all $\nu \in \mathbb{N}$.

Proposition 1.3. *Let q, s be partitions such that $q \trianglelefteq s$ and $\operatorname{sum} q < \operatorname{sum} s$. Let $k \in \mathbb{N}_0$ be maximal such that $\operatorname{sum} q_{\leq k} = \operatorname{sum} s_{\leq k}$. Then, if $k > 0$, we have $q_{k+1} < q_k$.*
Particularly, if $j \in \mathbb{N}$ such that $j > k$ and $i \in \mathbb{N}$ is minimal such that $q_i = q_j$, then $q[i]_+$ is a partition dominated by s.

Proof. Let $k > 0$. The relations $\operatorname{sum} q_{\leq k-1} \leq \operatorname{sum} s_{\leq k-1}$, $\operatorname{sum} q_{\leq k} = \operatorname{sum} s_{\leq k}$ and $\operatorname{sum} q_{\leq k+1} < \operatorname{sum} s_{\leq k+1}$ imply $q_{k+1} < s_{k+1} \leq s_k \leq q_k$. In particular, $k < i$.

In order to generalize Theorem 1.2 to the case of intervals we need the following characterization of the dominance relation by means of what we call dual dominance.

Definition 1.4. Let r, s be sequences of finite support. We say that r is *dually dominated* by s ($r \trianglelefteq' s$) if and only if $\operatorname{sum} r_{>k} \leq \operatorname{sum} s_{>k}$ for all $k \in \mathbb{N}_0$.

Theorem 1.5. *Let r, s be partitions. Then the following conditions are equivalent:*

(1) $r \trianglelefteq s'$,

(2) $r' \trianglelefteq' s$.

The implication (1)$\Longrightarrow$(2) is valid even for arbitrary sequences s of finite support, and in the case of $\operatorname{sum} r = \operatorname{sum} s$ *both conditions are additionally equivalent to*

(3) $s \trianglelefteq r'$.

Proof. Assume (1). Let s be a sequence of finite support, $k \in \mathbb{N}_0$ and $m := r'_{k+1}$. Then

$$\sum_{\nu>k} r'_\nu = \left|\left\{ (i,\nu) \in \mathbb{N} \times \mathbb{N} \,\middle|\, r_i \geq \nu > k \right\}\right| = \sum_{i=1}^{m} r_i - km$$
$$\leq \sum_{\nu=1}^{m} s'_\nu - \sum_{\nu=1}^{k} \min\{s_\nu, m\} = \sum_{\nu>k} \min\{s_\nu, m\} \leq \sum_{\nu>k} s_\nu \ .$$

Now, conversely, assume (2) and let $k \in \mathbb{N}_0$. We have $\sum_{\nu>r_k} r'_\nu = \operatorname{sum} r - \sum_\nu \min\{r_\nu, r_k\} = \operatorname{sum} r_{\leq k} - k r_k$ and in the same vein $\sum_{\nu>s'_k} s_\nu = \operatorname{sum} s'_{\leq k} - k s'_k$, as $s'' = s$. In the case of $s'_k > r_k$, this implies

$$\operatorname{sum} r_{\leq k} = \sum_{\nu>r_k} r'_\nu + k r_k \leq \sum_{\nu=r_k+1}^{s'_k} r'_\nu + \sum_{\nu>s'_k} s_\nu + k s'_k - k(s'_k - r_k)$$
$$= \operatorname{sum} s'_{\leq k} + \sum_{\nu=r_k+1}^{s'_k} (r'_\nu - k) < \operatorname{sum} s'_{\leq k} \ ,$$

as, for all $\nu > r_k$, we have $\{ j \,|\, r_j \geq \nu \} \subseteq \underline{k-1}$ and therefore $r'_\nu < k$. If, on the other hand, $s'_k \leq r_k$, we have

$$\operatorname{sum} r_{\leq k} = \sum_{\nu>r_k} r'_\nu + k r_k \leq \sum_{\nu>s'_k} s_\nu + k s'_k - \sum_{\nu=s'_k+1}^{r_k} r'_\nu + k(r_k - s'_k)$$
$$= \operatorname{sum} s'_{\leq k} + \sum_{\nu=s'_k+1}^{r_k} (k - r'_\nu) \leq \operatorname{sum} s'_{\leq k} + \sum_{\nu=s'_k+1}^{r_k} (k - r'_{r_k})$$
$$\leq \operatorname{sum} s'_{\leq k} \ .$$

This shows (1). If $\operatorname{sum} r = \operatorname{sum} s$, then, for all $k \in \mathbb{N}_0$, the relation $\operatorname{sum} r'_{>k} \leq \operatorname{sum} s_{>k}$ is equivalent to $\operatorname{sum} s_{\leq k} = \operatorname{sum} s - \operatorname{sum} s_{>k} \leq \operatorname{sum} r' - \operatorname{sum} r'_{>k} = \operatorname{sum} r'_{\leq k}$, and the theorem is proved.

Theorem 1.6 (Fulkerson [Ful59]). *Let p, q be partitions and $\tilde{p}$, $\tilde{q}$ be sequences of finite support such that $p \leq \tilde{p}$ and $q \leq \tilde{q}$. Then there exists a matrix M of zeros and ones such that $p \leq \sum_C M \leq \tilde{p}$ and $q \leq \sum_R M \leq \tilde{q}$ if and only if $q \trianglelefteq \tilde{p}'$ and $p \trianglelefteq \tilde{q}'$.*

Using the correspondence between matrices of zeros and ones and bipartite graphs described above the preceding theorem gives a satisfying answer to the question mentioned in the beginning of this section. In the special case of $p = \tilde{p}$ and $q = \tilde{q}$ we recover Theorem 1.2, as, by Theorem 1.5, $q \trianglelefteq p'$ implies $p \trianglelefteq q'$ for all partitions p, q such that $\operatorname{sum} p = \operatorname{sum} q$.

Proof (of Theorem 1.6). Let M be an $l \times m$ matrix of zeros and ones such that $p \leq \sum_C M \leq \tilde{p}$ and $q \leq \sum_R M \leq \tilde{q}$. Let $j \in \underline{l}$. As $p \leq \sum_C M$, the sum of the first j rows of M is at least $\sum_{\nu=1}^{j} p_\nu$. As on the other hand $\sum_R M \leq \tilde{q}$, this number is at most $\sum_{\nu=1}^{m} \min\{\tilde{q}_\nu, j\} = \sum_{\nu=1}^{j} \tilde{q}'_\nu$. This shows $p \trianglelefteq \tilde{q}'$. Applying this argument to the transpose of M, we get the other dominance relation.

Assume now that $q \trianglelefteq \tilde{p}'$ and $p \trianglelefteq \tilde{q}'$. We will prove the existence part by induction on $\operatorname{sum} \tilde{p} - \operatorname{sum} q$. If $\operatorname{sum} \tilde{p} = \operatorname{sum} q$, there exists a Matrix M of zeros and ones such that $\sum_R M = q$ and $\sum_C M = \tilde{p}$, by Theorem 1.2.
Let $\operatorname{sum} \tilde{p} > \operatorname{sum} q$. Choose $k \in \mathbb{N}_0$ maximal such that $\operatorname{sum} \tilde{p}'_{\leq k} = \operatorname{sum} q_{\leq k}$. Assume first that there exists an index $j > k$ such that $q_j \neq \tilde{q}_j$. Choose $i \in \mathbb{N}_0$ minimal such that $q_i = q_j$ and let $r := q[i]_+$. Then, by Proposition 1.3, we have $r \trianglelefteq \tilde{p}'$. By swapping the i- and the j-component of $\tilde{q}$ we obtain a sequence $\tilde{r}$ of finite support such that $r \leq \tilde{r}$ and $p \trianglelefteq \tilde{q}' = \tilde{r}'$. Inductively, there exists a matrix $\tilde{M}$ of zeros and ones such that $r \leq \sum_R \tilde{M} \leq \tilde{r}$ and $p \leq \sum_C \tilde{M} \leq \tilde{p}$. Now swapping the i-th and j-th column of $\tilde{M}$ gives a matrix M such that $q \leq \sum_R M \leq \tilde{q}$ and $\sum_R M = \sum_R \tilde{M}$.
Assume now that $q_{>k} = \tilde{q}_{>k}$. By Theorem 1.5, p' is dually dominated by $\tilde{q}$. Hence

$$\operatorname{sum} p'_{>k} \leq \operatorname{sum} \tilde{q}_{>k} = \operatorname{sum} q - \operatorname{sum} q_{\leq k} < \operatorname{sum} \tilde{p}' - \operatorname{sum} \tilde{p}'_{\leq k} = \operatorname{sum} \tilde{p}'_{>k} \ .$$

We can find an index $\nu > k$ such that $p'_\nu < \tilde{p}'_\nu$ or, equivalently, an index $i \in \mathbb{N}$ such that $\tilde{p}_i > k$ and $\tilde{p}_i > p_i$. This implies $p \leq \tilde{p}[i]_-$ and $q \trianglelefteq (\tilde{p}[i]_-)'$ by the choice of k. Inductively, we can find a matrix M such that $q \leq \sum_R M \leq \tilde{q}$ and $p \leq \sum_C M \leq \tilde{p}[i]_-$. The theorem is proved.

2 Graphs with Degrees in Prescribed Intervals

Let r, s be sequences of finite support such that $r \leq s$. Looking for a graph with degree sequence between r and s, we may reorder the intervals given by r and s without loss of generality. Consequently, we can confine ourselves to the case that r is a partition. Then the *diagonal length* of r is defined by $d(r) := 0$ if $\operatorname{sum} r = 0$, and by

$$d(r) := \max \{ i \in \mathbb{N} \mid r_i \geq i \}$$

otherwise.

Let G be a graph with vertex set $X = \{x_1, \ldots, x_n\}$ and degree sequence g such that $r \leq g \leq s$ and $g_i = \nu_G(x_i)$ for all $i \in \underline{n}$. Let $d := d(r)$ and $k \in \underline{d}$. For any $i \in \underline{k}$, the number of neighbours of x_i in $V := \{x_{k+1}, \ldots, x_n\}$ is at least $g_\nu - (k-1)$. On the other hand, for any $i > k$, the number of neighbours of x_i in $U := \{x_1, \ldots, x_k\}$ is at most $\min\{g_\nu, k\}$. Therefore, counting the edges between U and V, we have

$$(*) \quad \sum_{\nu=1}^{k} r_\nu \leq \sum_{\nu=1}^{k} (g_\nu - (k-1)) + k(k-1)$$
$$\leq \sum_{\nu=k+1}^{n} \min\{g_\nu, k\} + k(k-1)$$
$$\leq \sum_{\nu=1}^{k} s'_\nu - k$$

or, in other words, $r_{\leq d} \trianglelefteq s' \ominus 1$.

It is surprising that, conversely, in most cases this dominance relation is already sufficient for the existence of a graph with degree sequence between r and s. It is just the simple fact that the sum of any degree sequence is even which gives rise to an additional condition for certain *critical* pairs (r, s).

Proposition 2.1. *Let r be a partition and s be a sequence of finite support such that $r \leq s$. Let $d := d(r)$ and $k \in \underline{d} \cup \{0\}$ such that*

$$\textbf{(C)} \qquad \operatorname{sum} r_{\leq k} = \operatorname{sum} (s' \ominus 1)_{\leq k} \quad \text{and} \quad r_{>k} \ominus k = s_{>k} \ominus k \ .$$

Then the only possible candidate for a degree sequence g such that $r \leq g \leq s$ is the one given by $g_{\leq d} = r_{\leq d}$ and $g_{>d} = s_{>d}$.

Proof. Let g be a degree sequence of a graph such that $r \leq g \leq s$. Then we have $r' \leq g' \leq s'$ and hence $\operatorname{sum} g'_{\leq k} - k \leq \operatorname{sum} s'_{\leq k} - k = \operatorname{sum} r_{\leq k} \leq \operatorname{sum} g_{\leq k}$, by **(C)**. Conversely, bearing $(*)$ in mind, we may deduce that

$$\operatorname{sum} g_{\leq k} \leq \sum_{\nu>k} \min\{g_\nu, k\} + k(k-1) = \sum_{\nu} \min\{g_\nu, k\} - k = \operatorname{sum} g'_{\leq k} - k \ .$$

Hence $r_{\leq k} = g_{\leq k}$ and $g'_{\leq k} = s'_{\leq k}$. As $r_{>k} \ominus k \leq g_{>k} \ominus k \leq s_{>k} \ominus k$, condition **(C)** implies $r_{>k} \ominus k = g_{>k} \ominus k = s_{>k} \ominus k$. Let $j > k$. Then we have $g_j > k$ if and only if $s_j > k$, and in this case $g_j - k = s_j - k$. As additionally

$$\sum_{\nu} \min\{g_\nu, k\} = \operatorname{sum} g'_{\leq k} = \operatorname{sum} s'_{\leq k} = \sum_{\nu} \min\{s_\nu, k\} \ ,$$

we conclude that $\sum_{g_\nu \leq k} g_\nu = \sum_{g_\nu \leq k} s_\nu$. Now $g \leq s$ leads to $g_j = s_j$ for $g_j \leq k$, too. We have proved that $g_{>k} = s_{>k}$. But for all $\nu \in \{k+1, \dots, d\}$, the equality $r_{>k} \ominus k = s_{>k} \ominus k$ implies $r_\nu = s_\nu$, and the proposition is proved.

For example, if $r = (3,2,2,2)_0$ and $s = (4,2,2,2)_0$, we see that $d(r) = 2$ and $r_{\leq 2} = (3,2)_0 \trianglelefteq (3,3)_0 = s' \ominus 1$. On the other hand, condition **(C)** holds for $k = 1$. By the preceding proposition, the only possible degree sequence of a graph between r and s is $g = (3,2,2,2)_0$, but $\operatorname{sum} g = 9$ is odd.

Definition 2.2. Let r be a partition and s be a sequence of finite support such that $r \leq s$. The pair (r, s) is called *critical* if and only if there exists an index $k \in \underline{d(r)} \cup \{0\}$ such that condition **(C)** in the preceding proposition is satisfied.

Note that, for any partition r, the pair (r, r) is critical (for $k = 0$).

Theorem 2.3. *Let r be a partition and s be a sequence of finite support such that $r \leq s$. Then there exists a graph with degree sequence g such that $r \leq g \leq s$ if and only if the following conditions hold true:*

(1) $r_{\leq d(r)} \trianglelefteq s' \ominus 1$.
(2) *If (r,s) is critical, then* $\operatorname{sum} r_{\leq d(r)} + \operatorname{sum} s_{>d(r)}$ *is even.*

More precisely, concerning the existence part, we have: If $d = d(r)$ and g is the sequence defined by $g_{\leq d} = r_{\leq d}$ and $g_\nu = \min\{s_\nu, d\}$ for all $\nu > d$, then, assuming **(1)** *and* **(2)**, *there exists a graph with degree sequence $\tilde{g}$ such that $r \leq \tilde{g} \leq s$ and*

(a) *$\tilde{g} = g$, if* $\operatorname{sum} g$ *is even,*
(b) *$\tilde{g} = g[i]_\varepsilon$ for an index $i \in \mathbb{N}$, $\varepsilon \in \{+,-\}$, if* $\operatorname{sum} g$ *is odd.*

Proof. Let $d := d(r)$. If G is a graph with degree sequence g such that $r \leq g \leq s$, (1) follows from $(*)$. Let (r,s) be critical for $k \in \underline{d} \cup \{0\}$. Then $\operatorname{sum} g = \operatorname{sum} r_{\leq d} + \operatorname{sum} s_{>d}$ is even by Proposition 2.1.
Now assume (1) and (2) and define $g : \mathbb{N} \longrightarrow \mathbb{N}_0$ by $g_{\leq d} = r_{\leq d}$ and $g_\nu = \min\{s_\nu, d\}$ for all $\nu > d$. We will show the precise existence statement by induction on $m := \operatorname{sum} s_{>d} - \operatorname{sum} r_{\leq d} + d(d-1) = \operatorname{sum}\,(s' \ominus 1)_{\leq d} - \operatorname{sum} r_{\leq d} + \operatorname{sum}\,(s_{>d} \ominus d)$. Note that $m \geq 0$, by (1). Let $m = 0$. Then, for $q := r_{\leq d} \ominus (d-1)$ and $p := g_{>d}$, we have

$$\operatorname{sum} q_{\leq j} = \operatorname{sum} r_{\leq j} - j(d-1) \leq \operatorname{sum} s'_{\leq j} - jd = \operatorname{sum} p'_{\leq j}$$

for all $j \in \underline{d}$, hence $q \trianglelefteq p'$. As $m = 0$ implies $\operatorname{sum} q = \operatorname{sum} p$, we may apply Theorem 1.2 and find a matrix M of zeros and ones such that $\sum_C M = p$ and $\sum_R M = q$. If we denote by K the adjacency matrix of the complete graph on d vertices, the adjacency matrix

$$\begin{pmatrix} K & M^t \\ M & 0 \end{pmatrix}$$

defines a graph with degree sequence g as desired. Let $m > 0$.

case 1: $\operatorname{sum} g$ is even. If $s_\nu > d$ for an index $\nu > d$, then conditions (1) and (2) are satisfied for $(r, s[\nu]_-)$: For, if $(r, s[\nu]_-)$ is critical, it follows that $(s[\nu]_-)_{>d} = g_{>d}$, and therefore $\operatorname{sum} r_{\leq d} + \operatorname{sum}(s[\nu]_-)_{>d} = \operatorname{sum} g$ is even. Hence, inductively, we are done. Let $s_\nu \leq d$ for all $\nu > d$. Then $m = \operatorname{sum} g - 2\operatorname{sum} r_{\leq d} + d(d-1)$ is even. Choose $i \in \mathbb{N}$ minimal such that $r_i = r_d$, and let $j > d$ such that $s_j = \max\{ s_\nu \,|\, \nu > d \}$. If $s_j = r_j$, we can reorder s so that $r_j > r_{j+1}$. Hence we may assume that $s_j > r_j$ or $s_j = r_j > r_{j+1}$. Let

$$\tilde{r} := \begin{cases} r[i]_+ & , s_j > r_j \\ r[i]_+[j]_- & , s_j = r_j \end{cases} \quad \text{and} \quad \tilde{s} := s[i]_+[j]_- \; .$$

Then we have $\tilde{r} \leq \tilde{s}$, and $\tilde{r}$ is a partition. By Proposition 1.3, $\tilde{r}_{\leq d} \trianglelefteq s' \ominus 1$. But $\operatorname{sum}(s' \ominus 1)_{\leq d} - \operatorname{sum} r_{\leq d} = m \geq 2$ implies $\operatorname{sum} \tilde{r}_{\leq d} < \operatorname{sum} s' \ominus 1$. Therefore we may even conclude that $\tilde{r}_{\leq d} \trianglelefteq \tilde{s}' \ominus 1$, taking into account the choice of j. Furthermore, $\operatorname{sum} \tilde{s}_{>d} - \operatorname{sum} \tilde{r}_{\leq d} + d(d-1) = m - 2$ is even, which trivially implies condition (2) for $(\tilde{r}, \tilde{s})$. Hence, by (a) and induction, there exists a graph $\tilde{G}$ with degree sequence $\tilde{g} = g[i]_+[j]_-$. As $\tilde{g}_i > r_i \geq d \geq s_j > \tilde{g}_j$ we can find a neighbour x_t ($\neq x_j$) of x_i in $\tilde{G}$, which is not a neighbour of x_j. Removing the edge between x_i and x_t in $\tilde{G}$ and adding an edge between x_t und x_j, we obtain a graph G with degree sequence g, as was claimed in (a).

case 2: $\operatorname{sum} g$ is odd. If (r, s) was critical (for $k \in \underline{d}$), we would have $r_{>k} - k = s_{>k} - k$, hence $s_{>d} = g_{>d}$ and therefore $\operatorname{sum} r_{\leq d} + \operatorname{sum} s_{>d} = \operatorname{sum} g$, which is impossible by (2). It follows that (r, s) is not critical. As $\operatorname{sum}(s' \ominus 1)_{\leq d} + \operatorname{sum} r_{\leq d} = \operatorname{sum} g + d(d-1)$ is odd, it follows that $\operatorname{sum}(s' \ominus 1)_{\leq d} > \operatorname{sum} r_{\leq d}$. Let $k \in \underline{d-1} \cup \{0\}$ be maximal such that $\operatorname{sum} r_{\leq k} = \operatorname{sum}(s' \ominus 1)_{\leq k}$. As (r, s) is not critical, we can find an index $j > k$ such that $r_j < s_j$ and $s_j > k$. Choose $i \in \underline{d}$ minimal such that $r_i = r_j$ if $j \leq d$, and $r_i = r_d$ otherwise. In the case of $j \leq d$ we may swap s_i and s_j so that $r_i < s_i$. Let

$$(\tilde{r}, \tilde{s}) := \begin{cases} (r[i]_+, s) & , j \leq d \\ (r, s[j]_-) & , j > d \text{ and } s_j \leq d \\ (r[i]_+, s[i]_+[j]_-) & , j > d \text{ and } s_j > d \end{cases} \; .$$

Then we have $\tilde{r} \leq \tilde{s}$, and, by Proposition 1.3 or the choice of j, $\tilde{r}$ is a partition such that $\tilde{r} \trianglelefteq \tilde{s}' \ominus 1$. If $j \leq d$ or $j > d$ and $s_j \leq d$, it follows that $\operatorname{sum} \tilde{s}_{>d} - \operatorname{sum} \tilde{r}_{\leq d} + d(d-1) = m - 1$. Inductively, there exists a graph with degree sequence $g[i]_+$ or $g[j]_-$ according to (b). If, on the other hand, $j > d$ and $s_j > d$, the sequence $\tilde{g} := g[i]_+$ satisfies $\tilde{g}_{\leq d} = \tilde{r}_{\leq d}$ and $\tilde{g}_\nu = \min\{\tilde{s}_\nu, d\}$ for all $\nu > d$. As $\operatorname{sum} \tilde{g}$ is even and $\operatorname{sum} \tilde{s}_{>d} - \operatorname{sum} \tilde{r}_{\leq d} + d(d-1) = m - 2$, there exists graph G with degree sequence $\tilde{g}$ by induction. We have $\tilde{g}_i > r_i \geq d \geq \tilde{g}_j$. As was described in case 1, removing a suitable edge in G and adding a suitable new one gives a graph with degree sequence $\tilde{g}[i]_-[j]_+ = g[j]_+$. This shows (b) again and completes the proof.

In the special case of $r = s$ the preceding theorem leads to the following well-known corollary:

Theorem 2.4 (Erdös, Gallai 1960). *Let g be a partition and $d := d(g)$. Then there exists a graph G with degree sequence g if and only if* $\operatorname{sum} g$ *is even and* $g_{\leq d} \trianglelefteq g' \ominus 1$.[1]

Proof. If $r = g = s$, condition (2) in Theorem 2.3 holds true if and only if $\operatorname{sum} g$ is even.

Remark (cf. [Kor76]). Let r be a partition and s be a sequence of finite support such that $r \leq s$. If the conditions (a) and (b) of Theorem (4.2) in [Kor76] are satisfied by (r, s) (instead of (ϕ, ψ)), then the conditions (1) and (2) of Theorem 2.3 are also satisfied by (r, s): Let $d := d(r)$ and $j \in \underline{d}$. For the special choice $S = \underline{j}$, condition (a) in Theorem (4.2) leads to

$$\begin{aligned}\sum_{\nu=1}^{j} r_\nu &\leq \sum_{\nu=1}^{j} \min\{s_\nu, j-1\} + \sum_{\nu>j} \min\{s_\nu, j\} = \sum_{\nu} \min\{s_\nu, j\} - j \\ &= \sum_{\nu=1}^{j} (s' \ominus 1)_\nu \ ,\end{aligned}$$

hence (1). If, additionally, (r, s) is critical for k, conditions (4.3) and (4.4) of (b) hold true for $S = \underline{k}$ and $T = \{\, i \in \mathbb{N} \mid s_i \leq k \,\}$. Consequently, condition (4.2) may not hold true, that is, $\operatorname{sum} r_{\leq d} + \operatorname{sum} s_{>d} \equiv \sum_{\nu \in T} r_\nu + \sum_{\nu \notin T} s_\nu \pmod 2$ is even, which proves (2).
Hence, the existence part of Theorem (4.2) in [Kor76] is an immediate consequence of our Theorem 2.3.

References

[EG60] P. Erdös and T. Gallai, Graphs with given degree of vertices (Hungarian). *Mat. Lapok*, 11:264–274, 1960.

[Ful59] D. R. Fulkerson, A Network-Flow Feasibility Theorem and Combinatorial Applications. *Canadian J. Math.*, 11:440–451, 1959.

[Gal57] D. Gale, A Theorem on Flows in Networks. *Pacific J. Math.*, 7:1073–1082, 1957.

[Kor76] M. Koren, Graphs with Degrees from Prescribed Intervals. *Discrete Math.*, 15:253–261, 1976.

[Kra96] M. Krause (=M. Schocker), A simple proof of the Gale-Ryser theorem. *American Mathematical Monthly*, 103, no. 4:335–337, 1996.

[Rys57] H.J. Ryser, Combinatorial properties of matrices of zeros and ones. *Canadian J. Math.*, 9:371–377, 1957.

[Sch99] M. Schocker, Über Graphen mit vorgegebenen Valenzen. *Abh. des Math. Sem. der Univ. Hamburg*, 69:265–270, 1999.

[1] In fact, Theorem 2.4 is a slightly improved version of Theorem 1.2 in [EG60], see, e.g., [Sch99].

On a Result of Cameron and Praeger on Block-transitive Point-imprimitive t-designs

Michel Sebille*

Département de Mathématiques
Campus Plaine C.P. 216
Université Libre de Bruxelles
Boulevard du Triomphe
B - 1050 Bruxelles, Belgium
msebille@cso.ulb.ac.be

Abstract In 1993, Cameron and Praeger proved that if G is a block-transitive point-imprimitive automorphism group of an $S_\lambda(t,k,c^2)$ where $c = \binom{k}{2} - 1$, $k > 5$, $k \neq 8$, $t > 1$, then there are two simple 2-transitive permutation groups T_1 and T_2 of degree c such that one of the following holds:

(*i*) G is a subgroup of the wreath product $Aut(T_1) \wr S_c$ containing T_1^c and G projects onto a 2-transitive subgroup of S_c,
(*ii*) $T_1 \times T_2 \leq G \leq Aut(T_1) \times Aut(T_2)$.

Moreover, if (*i*) or (*ii*) holds then G acts in this way on such a design.
The purpose of this paper is to construct explicit extra-examples showing that this theorem is no longer valid for $k \leq 5$ and for $k = 8$.

1 Introduction

A $t-(v,k,\lambda)$ **design** (or simply t-**design** or **design**), denoted by $S_\lambda(t,k,v)$, is a pair $(X,\mathcal{B})$ where X is a set of v elements called **points** and $\mathcal{B}$ is a collection of k-subsets of X (called **blocks**) such that every t-subset of X is contained in λ blocks of $\mathcal{B}$. If $\lambda = 1$, a t-design is a **Steiner system** (or a **linear space**) and is denoted by $S(t,k,v)$. All the designs considered in this paper will be **simple**, that is without repeated blocks. Moreover, we will always assume that $2 \leq t \leq k < v$.

In a t-design $(X,\mathcal{B})$, let I be a set of i points and λ_i the number of blocks containing I; it is well-known that $\lambda_i = \lambda\binom{v-i}{t-i}/\binom{k-i}{t-i}$ for $0 \leq i \leq t$ (note that $\lambda_t = \lambda$ and that λ_0 is the total number of blocks of the design). It follows that any $S_\lambda(t,k,v)$ is also an $S_{\lambda_i}(i,k,v)$ for $0 \leq i \leq t$.

If G is a permutation group acting on a set X of cardinality v, the **Kramer-Mesner matrix** $KM(t,k,G)$ with parameters t, k and G, is a matrix defined as follows:

(*i*) its rows are indexed by the orbit(s) of the t-subset(s) of X under G,

* This work is supported by the Fonds pour la formation à la Recherche dans l'Industrie et dans l'Agriculture.

(*ii*) its columns are indexed by the orbit(s) of the k-subset(s) of X under G,

(*iii*) $KM(t,k,G)_{i,j}$ is the number of k-subsets of the jth orbit containing a given t-subset of the ith orbit.

The following results are well-known:

Theorem 1.1 (Kramer and Mesner [13]). *G is an automorphism group of an $S_\lambda(t,k,v)$ iff there exists a vector x with entries in $\{0,1\}$ such that $KM(t,k,G) \times x = \Lambda$, where Λ is a vector all of whose entries are equal to λ.*

Corollary 1.2. *G is a block-transitive automorphism group of an $S_\lambda(t,k,v)$ iff one of the columns of the matrix $KM(t,k,G)$ has all its entries equal to λ.*

If G is a permutation group acting on a v-set X, $\mathcal{T}_G$ will denote the set of orbit(s) of the t-subset(s) of X under G and $\mathcal{K}_G$ will denote the set of orbit(s) of the k-subset(s) of X under G. If $T \in \mathcal{T}_G$ and $K \in \mathcal{K}_G$, $m(T,K,G)$ is the number of element(s) of T contained in a given element of K.

Theorem 1.3 (Alltop [1]). *G is a block-transitive automorphism group of a t-design with block size k iff there exists $K \in \mathcal{K}_G$ such that for any $T_1, T_2 \in \mathcal{T}_G$, $m(T_1,K,G) = m(T_2,K,G)$.*

We recall that a set of l points is said to be **of type** $(l_1^{n_1}, l_2^{n_2}, \ldots, l_m^{n_m})$ if it intersects n_i imprimitivity classes in l_i points ($i \in \{1,2,\ldots,m\}$); obviously, $\sum_{i=1}^{m} n_i l_i = l$.

Theorem 1.4 (Delandtsheer and Doyen [11]). *If G is a transitive permutation group of degree cs having c imprimitivity classes of size s and if G is a block-transitive automorphism group of an $S_\lambda(t,k,cs)$, then*

(1) $cs \leq (\binom{k}{2} - 1)^2$

(2) *if equality holds in (1), then $c = s = \binom{k}{2} - 1$ and every block is of type $(2^1, 1^{k-2})$.*

A permutation group G of degree c^2 is said to be **of type T4** if there are two simple 2-transitive permutation groups T_1 and T_2 of degree c such that one of the following holds:

(*i*) G is a subgroup of $Aut(T_1) \wr S_c$ containing T_1^c and G projects onto a 2-transitive subgroup of S_c,

(*ii*) $T_1 \times T_2 \leq G \leq Aut(T_1) \times Aut(T_2)$.

Cameron and Praeger [4] have proved the following theorem :

Theorem 1.5. *If G is a block-transitive point-imprimitive automorphism group of an $S_\lambda(t,k,c^2)$ with $c = \binom{k}{2} - 1$, $k > 5$ and $k \neq 8$, then G is a group of type T4.*

It is easy to deduce from the definition of a group of type T4 that such a group can never be an automorphism group of an $S(2,k,c^2)$ and so we have the following

Corollary 1.6 (Cameron and Praeger [4]). *If G is a block-transitive point-imprimitive automorphism group of an $S(t,k,c^2)$ with $c = \binom{k}{2} - 1$, then $k = 4$, 5 or 8.*

The following results solve completely the existence problem for such linear spaces:

Theorem 1.7 (Kramer, Magliveras and Mathon [12]). *There is no $S(2,4,25)$ with a block-transitive point-imprimitive automorphism group.*

Theorem 1.8 (O'keefe, Penttila and Praeger [15]). *If G is a block-transitive point-imprimitive automorphism group of an $S(2,k,(\binom{k}{2}-1)^2)$ then $k = 8$.*

Theorem 1.9 (Nickel, Niemeyer, O'Keefe, Penttila and Praeger [14]). *Up to isomorphism, there are exactly 446 block-transitive point-imprimitive $S(2,8,729)$.*

The purpose of this paper is to construct for every $k \in \{3,4,5,8\}$ at least one block-transitive point-imprimitive $S_\lambda(t,k,c^2)$ with an automorphism group which is not of type T4.

More precisely, we will prove that

(*i*) $\mathbb{Z}_4$ is a block-transitive point-imprimitive automorphism group of an $S_2(2,3,4)$,
(*ii*) $AGL(1,5) \wr AGL(1,5)$ is a block-transitive point-imprimitive automorphism group of an $S_{50}(2,4,25)$ and an $S_{100}(2,4,25)$,
(*iii*) $AGL(1,9) \wr AGL(1,9)$ is a block-transitive point-imprimitive automorphism group of an $S_{5832}(2,5,81)$,
(*iv*) $AGL(1,27) \wr AGL(1,27)$ is a block-transitive point-imprimitive automorphism group of an $S_{13\cdot 27^6}(2,8,729)$ and an $S_{26\cdot 27^6}(2,8,729)$.

2 The constructions

From Theorem 1.3, we know that the blocks of such a design are necessarily of type $(2^1, 1^{k-2})$. It follows that three points belonging the same imprimitivity class are never contained in a block, and so $t = 2$.

2.1 $k = 3$

By Theorem 1.3, we are looking for an $S_\lambda(2,3,4)$. The only possibility is obviously an $S_2(2,3,4)$. The cyclic group $\mathbb{Z}_4$ is a block-transitive point-imprimitive automorphism group of such a design and is clearly not of type T4.

2.2 $k = 4$

By Theorem 1.3, we are looking for an $S_\lambda(2,4,25)$. We will examine all the designs $S_\lambda(2,4,25)$ having $G \wr H$ as an automorphism group, where G and H are two transitive permutation groups of degree 5. The transitive groups of degree 5 are $\mathbb{Z}_5$ (generated by $(1,2,3,4,5)$), D_5 (generated by $(1,2,3,4,5)$ and $(1,5)(2,4)$), $AGL(1,5)$ (generated by $(1,2,3,4,5)$ and $(1,2,4,3)$), A_5 and S_5. If G and H are any two such groups, it follows that $\mathbb{Z}_5 \wr \mathbb{Z}_5 \leq G \wr H$.
Since all blocks are of type $(2^1, 1^2)$, we do not have to compute all the Kramer-Mesner matrix for $G \wr H$. A **truncated Kramer-Mesner matrix** with parameters t, k, G, denoted by $trKM(t,k,G)$, is a matrix whose columns are the columns of $KM(t,k,G)$ corresponding to the k-subsets of type $(2^1, 1^{k-2})$. The orbits of the 2-subsets of $\{1,2,3,4,5\}$ under $\mathbb{Z}_5$ are $\{1,2\}$ and $\{1,3\}$, and the orbits of the 3-subsets are $\{1,2,3\}$ and $\{1,2,4\}$.
The imprimitivity classes will be denoted by $C_1, C_2, C_3, \dots, C_s$. We will use the following notation : $a_1 a_2 \dots a_n, b_1^{c_1} b_2^{c_2} \dots b_m^{c_m}$ for a row (resp. column) of a truncated Kramer-Mesner matrices means that there exists an element of the orbit corresponding to this row (resp. column) and having c_j point(s) in C_{b_j} (for every $j \in \{1,2,\dots,m\}$) and for which the projection onto C_1 is $\{a_1, a_2, \dots, a_n\}$. The matrix $trKM(2,4,\mathbb{Z}_5 \wr \mathbb{Z}_5)$ turns out to be

$\mathbb{Z}_5 \wr \mathbb{Z}_5$	$12,1^2 23$	$12,12^2 3$	$12,123^2$	$12,1^2 24$	$12,12^2 4$	$12,124^2$
$12,1^2$	25	25	25	25	25	25
$13,1^2$	0	0	0	0	0	0
$1,12$	15	20	15	10	10	5
$1,13$	10	5	10	15	15	20

...

$13,1^2 23$	$13,12^2 3$	$13,123^2$	$13,1^2 24$	$13,12^2 4$	$13,124^2$
0	0	0	0	0	0
25	25	25	25	25	25
15	20	15	10	10	5
10	5	10	15	15	20

From the first row of this matrix, we deduce that $\lambda \in \{25, 50, 75, 100, 125, 150\}$. Moreover, looking at the first two rows, we see that G is 2-homogeneous, and so G is $AGL(1,5)$, A_5 or S_5. But $G \wr H$ is not of type T4, and so A_5^5 is not a subgroup of $G \wr H$. It follows that G is $AGL(1,5)$. Using the fact that $AGL(1,5) \wr \mathbb{Z}_5 \leq AGL(1,5) \wr H$, we construct $trKM(2,4,AGL(1,5) \wr \mathbb{Z}_5)$:

$AGL(1,5) \wr \mathbb{Z}_5$	$12,1^2 23$	$12,12^2 3$	$12,123^2$	$12,1^2 24$	$12,12^2 4$	$12,124^2$
$12,1^2$	25	25	25	25	25	25
$1,12$	30	40	30	30	20	10
$1,13$	20	10	20	20	30	40

From this matrix, we see that $\lambda \in \{50, 100, 150\}$. Now we construct all possible truncated Kramer-Mesner matrices:

(1) $H = D_5$

$AGL(1,5) \wr D_5$	$12,1^2 23$	$12,12^2 3$	$12,1^2 24$	$12,124^2$
$12,1^2$	50	25	50	25
$1,12$	60	40	40	10
$1,13$	40	10	60	40

(2) $H = AGL(1,5)$

$AGL(1,5) \wr AGL(1,5)$	$12,1^2 23$	$12,12^2 3$
$12,1^2$	100	50
$1,12$	100	50

(3) Since A_5 and S_5 are 3-transitive, $AGL(1,5) \wr A_5$ and $AGL(1,5) \wr S_5$ yield the same matrix, namely:

$AGL(1,5) \wr A_5$	$12,1^2 23$
$12,1$	150

Thus we have 4 examples of a block-transitive point-imprimitive design having an automorphism group which is not of type T4.

Theorem 2.1. *If $G \wr H$ is a block-transitive point-imprimitive automorphism group of an $S_\lambda(2,4,25)$ which is not of type T4, then $G = AGL(1,5)$ and one of the following holds:*

(i) $H = AGL(1,5)$ *and* $\lambda = 50$ *or* 100,
(ii) $H = A_5$ *and* $\lambda = 150$,
(iii) $H = S_5$ *and* $\lambda = 150$.

2.3 $k = 5$

We are looking for an $S_\lambda(2,5,81)$. Since there are a lot of transitive groups of degree 9, by analogy with the case $k = 4$, we will focus on the group $AGL(1,9) \wr AGL(1,9)$ which is of order 37439062426244487424. There are 144 2-subsets of type (2^1) and 1296 2-subsets of type (1^2). A block of type $(2^1,1^3)$ contains one 2-subset of type (2^1) and nine 2-subsets of type (1^2). By Theorems 1.2 and 1.3, any orbit of 5-subsets of type $(2^1,1^3)$ is a block-transitive design. Using this argument and the software DISCRETA developed at the University of Bayreuth [2], we conclude that the truncated Kramer-Mesner matrix is a 2×7 matrix all of whose entries are equal to 5832. Indeed DISCRETA gives all the Kramer-Mesner matrix but, by the above argument, the truncated Kramer-Mesner matrix consists of the columns of the complete Kramer-Mesner matrix all of whose entries are equal. Thus we have an example of a block-transitive point-imprimitive design having an automorphism group which is not of type T4.

2.4 $k = 8$

We are looking for an $S_\lambda(2,8,729)$. Like in the case $k = 5$, there are a lot of transitive groups of degree 27, and so we will focus on the group $AGL(1,27) \wr AGL(1,27)$ which is of order 702^{28}. Unfortunately, DISCRETA needs too much computational time to construct the matrix $KM(2,8,AGL(1,27) \wr AGL(1,27))$. There are 9477 pairs of type (2^1) and 255879 pairs of type (1^2). A block of type $(2^1, 1^6)$ contains one pair of type (2^1) and 27 pairs of type (1^2). Therefore, by Theorem 1.2, every orbit of blocks of type $(2^1, 1^6)$ yields an $S_\lambda(2,8,729)$. Now we have

$$\lambda_0 = \lambda \frac{729 \cdot 728}{8 \cdot 7},$$

that is

$$\lambda = \frac{\lambda_0}{9477}$$

But λ_0 is also the size of the orbit of a block of type $(2^1, 1^6)$. Using the projection onto $AGL(1,27)$, we get

$$\lambda_0 = 13 \cdot 27^7 \cdot n$$

where n is the size of the orbit of a set of six points of weight 1 and one point of weight 2 under the action of $AGL(1,27)$. But $AGL(1,27)$ is transitive, with the stabilizer of one point isomorphic to $\mathbb{Z}_{26}$, and so

$$n = 27 \cdot n'$$

where n' is the size of the orbit of a set of 6 points under $\mathbb{Z}_{26}$. Using DISCRETA, GAP or MAGMA, we see that there are 22 such orbits of size 13 and 8871 orbits of size 26. The matrix $trKM(2,8,AGL(1,27) \wr AGL(1,27))$ is then a 2×8893 matrix consisting of 22 columns with entries $13 \cdot 27^6$ and 8871 columns with entries $26 \cdot 27^6$.

3 A property of affine lines

The following question arises in a natural way : why does the above construction work only for the three affine groups $AGL(1,q)$ with $q \in \{5, 9, 27\}$?

Lemma 3.1. *If k and n are two positive integers and p a prime number, then*
$\binom{k}{2} - 1 = p^n$ if and only if $(k,p,n) \in \{(3,2,1),(4,5,1),(5,3,2),(8,3,3)\}$.

Proof : $\binom{k}{2} - 1 = p^n$ implies $k = \frac{1}{2}(1 + \sqrt{8p^n + 9})$, and so there exists a positive integer m such that $8p^n + 9 = m^2$, that is $8p^n = (m-3)(m+3)$. where $m \geq 5$.
It follows that $\{m-3, m+3\} = \{4p^s, 2p^t\}$, where $s + t = n$, $s \geq 0$ and $t \geq 0$. There are two possible cases :

(1) $4p^s - 2p^t = 6$ where $s = t + r$ $(r \geq 0)$. This implies $p^t(2p^r - 1) = 3$, and so either $p = 3, t = 1$ and $r = 0 \Longrightarrow n = 2$ and $k = 5$, or $p = 2, t = 0$ and $r = 1 \Longrightarrow n = 1$ and $k = 3$.
(2) $2p^t - 4p^s = 6$ where $t = s + r$ $(r \geq 0)$. This implies $p^s(p^r - 2) = 3$, and so either $p = 3, s = 1$ and $r = 1 \Longrightarrow n = 3$ and $k = 8$, or $p = 5, s = 0$ and $r = 1 \Longrightarrow n = 1$ and $k = 4$.

■

The following theorem is an immediate consequence of Lemma 3.1 (and also of Theorem 1.4):

Theorem 3.2. *If $AGL(1, \binom{k}{2} - 1) \wr AGL(1, \binom{k}{2} - 1)$ is a block-transitive point-imprimitive automorphism group of an $S_\lambda(2, k, (\binom{k}{2} - 1)^2)$, then $k \in \{3, 4, 5, 8\}$.*

Theorem 3.3. *If $AGL(1,q) \wr AGL(1,q)$ is a block-transitive point-imprimitive automorphism group of an $S_\lambda(2, k, q^2)$ with blocks of type $(2^1, 1^{k-2})$, then $q = \binom{k}{2} - 1$ and $(k, q) \in \{(3,2), (4,5), (5,9), (8,27)\}$*

Proof : There are $\frac{1}{2}q^2(q-1)$ pairs of type (2^1) and $\frac{1}{2}q^3(q-1)$ pairs of type (1^2).
A block of type $(2^1, 1^{k-2})$ contains one pair of type (2^1) and $\frac{1}{2}(k^2 - k - 2)$ pairs of type (1^2).
Using Theorem 1.2, we get

$$\frac{1}{2}(k^2 - k - 2) = q,$$

that is

$$\binom{k}{2} - 1 = q,$$

and Lemma 3.1 ends the proof. ■

References

1. W.O. Alltop, On the construction of block designs, J. Combin. Theory 1 (1966), 501-502.
2. A.Betten, R.Laue, A.Wassermann, http://btm2xl.mat.uni-bayreuth.de/ discreta/index.html
3. R.E. Block, On the orbits of collineation groups, Math. Z. 96 (1967), 33-49.
4. P.J.Cameron, C.E.Praeger, Block-transitive t-designs I : point-imprimitive designs, Discrete Math. 118 (1993),33-43.
5. A.R.Camina, A survey of the automorphism groups of block designs, J. Combin. Designs 2 (1994), 79-100.
6. A.R.Camina, L.Di Martino, The group of automorphisms of a transitive $2-(91,6,1)$ design, Geom. Dedicata 31 (1989), 151-164.
7. A.R.Camina, L.Di Martino, Block designs on 196 points, Arch. Math. 53 (1989), 414-416.
8. H.Davies, Flag-transitivity and primitivity, Discrete Math. 63 (1987), 91-93.
9. A.Delandtsheer, Line-primitive groups of small rank, Discrete Math. 68 (1988), 103-106.
10. A.Delandtsheer, Line-primitive automorphism groups of finite linear spaces, European J. Combin. 10 (1989), 161-169
11. A.Delandtsheer, J.Doyen, Most block-transitive t-designs are point-primitive, Geom. Dedicata 29 (1989), 307-310.
12. E.S.Kramer, S.S.Magliveras and R.Mathon, The Steiner systems $S(2,4,25)$ with non-trivial automorphism group, Discrete Math. 77 (1989), 137-157.
13. E.S.Kramer, D.M.Mesner, t-designs on hypergraphs, Discrete Math. 15 (1976), 263-296.
14. W.Nickel, A.Niemeyer, C.M.O'Keefe, T.Penttila, C.E.Praeger, The block-transitive, point imprimitive $2-(729,8,1)$ designs, Appl. Algebra Engrg. Comm. Comput. 3 (1992), 47-61.
15. C.M.O'Keefe, T.Penttila, C.E.Praeger, Block-transitive, point-imprimitive designs with $\lambda = 1$, Discrete Math. 115 (1993), 231-244.

Exact and Asymptotic Solutions for Bethe Ansatz in a Hexagon

Andrzej Wal

Institute of Physics, Pedagogical University
ul. Rejtana 16A, 35-359 Rzeszów, Poland

Abstract A complete set of solutions of the Heisenberg model for a magnet with N=6 nodes and spin s=1/2 has been described in terms of (i) immediate diagonalization, (ii) coordinate Bethe Ansatz (in terms of pseudomomenta p_α and soliton phase exchange $\Phi_{\alpha,\alpha'}$), (iii) algebraic Bethe Ansatz (in terms of spectral parameters λ_α), (iv) string hypothesis with its combinatorial content (rigged configurations of Kerov-Kirillov-Reshetikhin). It has been shown that this example transparently presents various notions and terminology associated with Bethe Ansatz, and, in particular, indicates distinction between exact and asymptotic solutions.

1 Introduction

The solution for the one-dimensional Heisenberg magnet, both for a finite number N of nodes and for the asymptotic case, was presented by Bethe [1] and is known as a Bethe Ansatz (BA) (coordinate BA). It allows us to calculate the eigenvalues and eigenfunctions for the system of N spins $s = 1/2$ which mutually interact according to the Heisenberg Hamiltonian. The mathematical form of Bethe transcendental equations, with pseudomomenta p and phase exchange φ as variables, causes difficulties when we try to solve it, even for a small number N. There is a transformation from this system of equations into the algebraic ones in which the spectral parameter λ plays the role of variable [2,3,4]. This transformation is associated with advanced mathematical methods and physical models: the monodromy matrix, quantum groups and integrable quantum chains, etc. [5,6,7,8,9,10,11]. Unfortunately, it does not lead, despite the simplest graphical form, to a simplification of the solutions. On the other hand, its advantage is, however, a classification of magnetic states by using combinatorial objects called "rigged configurations" [4]. This classification is directly connected with the string hypothesis [12] which predicts that for the asymptotic case $N \to \infty$ the spectral parameters gather together into sets with similar properties. This gathering corresponds, within the pseudoparticle picture, to the bound states of pseudoparticles.

The case of finite N described in the present paper can be solved by immediate diagonalization of Hamiltonian on the basis of all magnetic states. The solutions which are obtained in this way cannot be directly translated into "Bethe language", i.e. into pseudoparticle and phase variables. The aim of this paper is to present for the case $N = 6$ (hexagon) the exact solutions of the Heisenberg magnet eigenproblem and compare it with results obtained by coordinate and algebraic BA. The analysis of a finite magnet enables us to check how the finite effects change the asymptotic solutions, if the string hypothesis is assumed to be valid in this case.

2 A Complete Set of Solutions for the Heisenberg Magnet with $N = 6$

The symmetry adapted basis can be found by using the action of group on the set and considering the orbits of this action [13]. The natural basis for the one-dimensional eigenproblem of the Heisenberg Hamiltonian is the symmetry adapted basis, i.e. the basis adapted to the symmetric group Σ_6 and cyclic group C_6. These groups act on the set of all magnetic configurations $f : \tilde{6} \to \tilde{2}$, where

$$\tilde{6} = \{j = 1, 2, 3, \ldots 6\}$$

denotes the set of nodes, whereas

$$\tilde{2} = \{+, -\}$$

is the set of projections of spin $s = 1/2$. This action divides the set of magnetic configurations into the symmetric group orbits $Q^{(r)}$, which are in turn divided into the cyclic group orbits O_t. The orbits $Q^{(r)}$ are parameterised by the number r which denotes the number of spin deviations (from the ferromagnetic saturation). For our case $N = 6$, r can take values from the set $\{0, 1, 2, \ldots, 6\}$. Each orbit $Q^{(r)}$ spans the unitary space $\mathcal{H}^{(r)}$ [14] which is invariant under the action of the Hamiltonian

$$\hat{H} = J \sum_{j \in \tilde{6}} \left(\hat{\boldsymbol{s}}_j \cdot \hat{\boldsymbol{s}}_{(j+1) \bmod 6} - \frac{1}{4} \right)$$

that describes the dynamics of the system. J is the exchange parameter, and $\hat{\boldsymbol{s}}_j$ is the spin operator acting on the j-node. The term $1/4$ is added to establish the zero energy level. The parameter r which describes the $Q^{(r)}$ orbit determines the configurations with the same total spin projection $M = 3 - r$. This allows us to investigate the eigenvalue problem separately for each M.

Under the action of the cyclic group $C_6 \subset \Sigma_6$ which is the natural translation symmetry of the one-dim Heisenberg magnet each orbit $Q^{(r)}$ decomposes into some orbits O_t. The parameter t labels the cyclic group orbits within the symmetric orbit $Q^{(r)}$. The elements of each orbit O_t form the carrier space for the cyclic group translational representation which splits into irreducible ones. The irreducible representation is described by the formula

$$\Gamma^k(j) = \mathrm{e}^{2\pi i jk/6}, \; j \in \tilde{6}, \; k \in B \; ,$$

where k is the quantum number which characterises states within the orbit O_t, and B stands for the set of admissible k's for a given orbit.

By way of example, let us consider the case $r = 2$. This orbit contains $\binom{6}{2} = 15$ configurations $|j_1, j_2\rangle$ where the numbers j_1, j_2 denote the nodes with the spin deviations and fulfil the conditions $1 \le j_1 < j_2 \le 6$. The orbit

$$Q^{(2)} = \{|j_1, j_2\rangle, 1 \le j_1 < j_2 \le 6\}$$

is divided into 3 orbits O_t containing configurations in which the distance between reversed spins equals t:

$$\begin{aligned} O_1 &= \{12, 23, 34, 45, 56, 16\} , \\ O_2 &= \{13, 24, 35, 46, 15, 26\} , \\ O_3 &= \{14, 25, 36\} . \end{aligned}$$

For the two orbits O_1, O_2 the set B has six elements $\{0, \pm 1, \pm 2, 3\}$, whereas for the last one it has only 3 elements $B = \{0, \pm 2\}$ [15,16]. The set of quantum numbers k forms the finite analogy of the Brillouin zone B.

The orbits of the cyclic group introduce bases which are adapted to the translational symmetry of the system. The matrix of the eigenproblem has block-diagonalized form in these bases, each of the blocks is labelled by the quasimomentum k from the Brillouin zone B. The symmetry of exchanging $+ \leftrightarrow -$ restricts the number of solutions to 3 cases $r = 1, 2$ and 3.

For the aforementioned case $r = 2$ the bases read

$$\begin{aligned} |t, k\rangle &= \tfrac{1}{\sqrt{6}} e^{i\pi k t} \textstyle\sum_{j \in \tilde{6}} \Gamma_k(j) |(t, j)\rangle, \; k \in B, \; t \in \{1, 2\} , \\ |3, k\rangle &= \tfrac{1}{\sqrt{3}} \textstyle\sum_{j=1}^{3} \Gamma_k(j) |(3, j)\rangle, \; k \in \{0, \pm 2\} . \end{aligned}$$

To determine the energy spectrum for the Heisenberg magnet, the eigenproblem for $r = 0, 1, 2$ and 3 has been solved. The Table 1 contains the states for $r = 3$ described by the following quantum numbers: energy E, k and total spin S. This case ($r = 3$) has been chosen, because in this spectrum each energy level for the considered system is represented. The calculations were performed for the Hamiltonian $\hat{H}'$

$$\hat{H}' = \frac{J}{2} \hat{H} ,$$

for which matrix elements have integer values.

3 The Coordinate and Algebraic Bethe Ansatz

From the point of view of Bethe Ansatz, the integer r corresponds to a system of r pseudoparticles which move along the chain independently, except for two body scatterings, with pseudomomenta p_α, $\alpha = 1, \ldots, r$. The eigenfunctions of such a system read

$$|\Psi_r\rangle = \sum_{f \in Q^{(r)}} a_f |f\rangle ,$$

Table1. Energy spectrum, quasimomenta and the total spin for the system of $r = 3$ spin deviations

k	E	S
0	0	3
	$-5+\sqrt{5}$	1
	$-5-\sqrt{5}$	1
	-6	0
± 1	-1	2
	-4	0
	-5	1
± 2	$(-7+\sqrt{17})/2$	1
	$(-7-\sqrt{17})/2$	1
	-3	2
3	$-5+\sqrt{13}$	0
	$-5-\sqrt{13}$	0
	-2	1
	-4	2

where the coefficients a_f are obtained by the formula

$$a_f = \sum_{\pi \in \Sigma_r} A_\pi e^{i \sum_{\alpha \in \tilde{r}} p_{\pi(\alpha)} j_\alpha} ,$$

Σ_r is the Pauli group, $\tilde{r}$ denotes the set of numbers $\{1, 2, \ldots, r\}$, whereas the amplitude A_π is described by the formula

$$A_\pi = e^{i/2} \sum_{\substack{1 \le \alpha < \alpha' \le r \\ \pi(\alpha) > \pi(\alpha')}} \varphi_{\pi(\alpha), \pi(\alpha')} , \quad \pi \in \Sigma_r .$$

The phase exchange $\varphi_{\alpha\alpha'}$ arises due to the scattering of pseudoparticles with the pseudomomenta p_α and $p_{\alpha'}$. The additional Bethe equations are related to the boundary condition

$$6p_\alpha - \sum_{\alpha' \neq \alpha} \varphi_{\alpha\alpha'} = 2\pi n_\alpha , \quad \alpha \in \tilde{r} , \quad n_\alpha \text{– an integer number mod } 6$$

and to the rule that two deviations do not occupy the same node, i.e.

$$2 \cot \frac{\varphi_{\alpha\alpha'}}{2} = \cot \frac{p_\alpha}{2} - \cot \frac{p_{\alpha'}}{2} .$$

The equations (12–13) enable us to calculate the pseudomomenta p_α and the phases $\varphi_{\alpha\alpha'}$ which account for the state $|\Psi_r\rangle$ of the system.

By using the substitution

$$\lambda_\alpha = \frac{1}{2} \cot \frac{p_\alpha}{2}$$

the Bethe equation is shifted to the algebraic form

$$\left(\frac{\lambda_\alpha - \frac{i}{2}}{\lambda_\alpha + \frac{i}{2}}\right)^6 = \sum_{\substack{\alpha'=1 \\ \alpha'\neq\alpha}} \frac{\lambda_\alpha - \lambda_{\alpha'} - i}{\lambda_\alpha - \lambda_{\alpha'} + i}, \quad \alpha = 1, 2, \ldots, r .$$

The solution of the system (15) for a given r results in admissible values of the set of spectral parameters $\{\lambda_1, \lambda_2, \ldots, \lambda_r\}$ and, by virtue of Eq. (14), of pseudomomenta p_α.

The results of Bethe method in comparison with diagonalization yield two differences. The Bethe substitution presents a "particle" picture of magnet excitations (pseudomomenta of particles or spectral parameters), whereas the diagonalization determines eigenstates and quasimomenta for the system, which causes difficulties in comparison of results of those two methods. In order to compare the aforementioned results a lot of equations for a_f coefficients of wave functions $|\Psi_r\rangle$ for corresponding states have to be solved. The important property of the algebraic BA is that the solutions correspond only to the states with the highest weight, i.e. to states for which $S = M$. Other states can be obtained by using the step operator $\hat{S}_\pm$ of the total spin.

The Table 2 presents the values of the pseudomomenta, phases $\varphi_{\alpha\alpha'}$ and spectral parameters for the system of $N = 6$ spins with the three magnon excitations ($r = 3$). The number of states is equal to 5, which agrees with the number of the highest weight states in the Table 1. The divergence of the parameters describing the states of the system (character "?" in the Table 2) should be noted. The questions about the completeness of the BA method immediately arise in this connection [17].

4 The Asymptotic Solutions for BA

The values of spectral parameters can be complex. Then the corresponding pseudomomenta of Bethe particles also take complex values. Bethe has already pointed out that in this case the set of parameters λ_α forms strings. The string hypothesis predicts that for the asymptotic case $N \to \infty$ one can separate in the set of spectral parameters $\{\lambda_1, \lambda_2, \ldots, \lambda_r\}$ the subset of complex values with the following properties

$$\lambda_m^{lv} = \lambda^{lv} + im + O\left(e^{-\delta N}\right) ,$$

where λ^{lv} is the real part, constant for the whole string, l is the integer which we call the string length (the number of elements of the string), and m is the integer or half-integer, depending on the parity of l. It should be mentioned that m has the value from the set

$$m \in \tilde{s} = \{-s, -s+1, \ldots, +s\}$$

Table2. Eigenstates of Bethe for $r = 3$, $S = 0$

k	E	States characterised by		
		λ_1	λ_2	λ_3
		p_1	p_2	p_3
		φ_{12}	φ_{23}	φ_{13}
0	−6	0	−0.5i	0.5i
		3.142	i∞	−i∞
		?	?	?
1	−4	0.236 + 0.5i	0.236 − 0.5i	−0.472
		1.338 − 1.472i	1.338 + 1.472i	4.654
		−8.181i	1.745 + 0.65i	1.745 − 0.65i
−1	−4	−0.236 + 0.5i	−0.236 − 0.5i	0.472
		4.945 − 1.472i	4.945 + 1.472i	1.629
		−8.181i	4.538 + 0.65i	4.538 − 0.65i
3	$-5+\sqrt{13}$	−0.429	0.429	0
		4.46	1.723	3.142
		4.560	2.331	3.953
3	$-5-\sqrt{13}$	−1.009i	1.009i	0
		6.283 + 1.087i	- 1.087i	3.142
		1.087i	6.283 − 5.431i	5.431i

for s, which fulfils the condition $2s+1 = l$ and describes the spin of the string. The index v labels the strings with the same length l. The term $O\left(\mathrm{e}^{-\delta N}\right)$ vanishes (with $\delta > 0$) exponentially for $N \to \infty$.

Gathering of spectral parameters into a string is mathematically described by the partitions of r. It yields a classification of possible configurations of strings by the method that was proposed by Kerov, Kirillov and Reshetikhin [4], referred to as the KKR method. Each partition ν of the integer r, which is called a string configuration, is associated with the "rigging", i.e. for each row of the partition we have an additional quantum number labelling the Bethe vectors connected with this configuration. It also allows also to determining which states correspond to a bound state of pseudoparticles. For example, two configurations $\{1^r\}$ and $\{r\}$ correspond to r independent pseudoparticles and a bound state of all r magnons, respectively. The total number of states for the partition ν is calculated from the following equation

$$Z(\nu) = \prod_{l=1}^{\infty} \binom{P_l + m_l}{m_l},$$

where m_l denotes the number of strings with length l, $P_l = N - 2Q_l$ and Q_l means the number of boxes in the first l columns of the Young diagram associated with the given configuration ν.

By way of example, let us consider the case $r = 3$ for our system. The partition $\nu = \{3\}$ has the following parameters: $l = 3$, $m_3 = 1$, $Q_3 = 3$,

$P_3 = 0$ and $Z(3) = 1$. It means that this configuration corresponds to one bound state of three pseudoparticles. The partition $\nu = \{21\}$ gives $Z(21) = 3$ Bethe vectors which describe states of two bounded and one independent pseudoparticle. The last partition $\nu = \{1^3\}$ yields one state of three free particles moving along the chain.

The string hypothesis is valid for the asymptotic case $N \to \infty$. As has been mentioned the major objective of this paper is a comparison of the asymptotic solutions with the exact result for finite N. It is assumed here that the string hypothesis can be applied (in a sense of approximation) to the finite magnet.

The Table 3 presents the values of spectral parameters λ, obtained from Bethe equation and string hypothesis. On the basis of comparison it can be said that the effect of finiteness of the chain slightly changes the values of imaginary part of λ predicted by the string hypothesis; i.e. for configuration $\nu = \{2\}$ we obtain $\lambda^{\mathrm{i}} = \pm 0.512$ instead of $\lambda'^{\mathrm{i}} = \pm 0.5$, for $\nu = \{3\}$ $\lambda^{\mathrm{i}} = \pm 1.009$ instead of $\lambda'^{\mathrm{i}} = \pm 1$.

Table3. The comparison of the imaginary part of spectral parameters obtained from BA with parameters predicted by the string hypothesis (SH). The table contains parameters for partitions of type $\nu = \{r\}$ and for r=2 and 3

r	k	E	$\lambda^{\mathrm{i}}_\alpha$ from BA	$\lambda^{\mathrm{i}}_\alpha$ from SH
2	2	$(-7+\sqrt{17})/2$	$\lambda^{\mathrm{i}}_1 = -0.512$	$\lambda^{\mathrm{i}}_1 = -0.5$
			$\lambda^{\mathrm{i}}_2 = 0.512$	$\lambda^{\mathrm{i}}_2 = 0.5$
3	3	$-5-\sqrt{13}$	$\lambda^{\mathrm{i}}_1 = -1.009$	$\lambda^{\mathrm{i}}_1 = -1$
			$\lambda^{\mathrm{i}}_2 = 1.009$	$\lambda^{\mathrm{i}}_2 = 1$
			$\lambda^{\mathrm{i}}_3 = 0$	$\lambda^{\mathrm{i}}_3 = 0$

The asymptotic solution for the given configuration (given partition of r) allow us to simplify Bethe equations (15) and transform it into equations for the real part of spectral parameters λ^{lv} [2]. If we introduce the substitution $V_o(\lambda) = \frac{\lambda - \mathrm{i}}{\lambda + \mathrm{i}}$, the equations (15) read

$$V_o^N\left(\frac{\lambda^{lv}}{\frac{l-1}{2}}\right) = \prod_{l'} \prod_{\substack{v' \\ v' \neq v \text{ for } l=l'}} V_{ll'}\left(\lambda^{lv} - \lambda^{l'v'}\right) ,$$

where

$$V_{ll'}(\lambda) = \prod_{m=-s_1}^{s_1} \prod_{m'=-s_2}^{s_2} V_o(\lambda + \mathrm{i}(m+m')) .$$

If we assume that the string hypothesis can be applied to the finite case the considerable simplification of equations is expected, especially for configurations with a small number of strings. The extreme case occurs in the configuration $\nu = \{r\}$ where the right hand side of (19) is equal to 1.

5 Conclusions

Complete solution of the eigenproblem of the Heisenberg Hamiltonian for $N = 6$ nodes, each with spin 1/2 has been presented. The solution was obtained by the use of three methods: (i) immediate diagonalization, (ii) coordinate BA and (iii) algebraic BA. A comparison of the method of diagonalization with BA explicitly shows that in some cases BA does not give a complete solution. The reason is that either pseudomomenta become divergent, or the corresponding BA-like wave function vanishes (cf. the state with $k = 0$ in Table 2).

The validity of the Bethe hypothesis of strings for the finite case $N = 6$ has been assumed, and its results have been compared with the spectral parameters immediately obtained from BA equations. It follows from these considerations that the asymptotic solutions are only slightly perturbed by the finite size effects for $N = 6$.

Acknowledgements. It is a pleasure to thank to Prof. T. Lulek for innumerable valuable discussions. I am very grateful to the Local Organising Committee of the Conference "Algebraic Combinatorics and Applications" for a grant, which enabled me to take part in the meeting.

References

1. Bethe H.: Z. Physik **71** (1931), 205-26 (German). [English transl.: Mattis D.C., On the theory of metals, I. Eigenvalues and eigenfunctions of a linear chain of atoms. The Many-Body Problem. World Sci., Singapore 1993, pp. 689-716]
2. Faddeev L.D. and Takhtadzhyan L.A.: Zap. Nauch. Semin. LOMI **109** (1981), 134-78 (Russian). [English transl.: Spectrum and scattering of excitations in the one-dimensional isotropic Heisenberg model. J. Soviet Math. **24** (1984), 241-67]
3. Takhtadzhyan L.A.: The picture of low-lying excitations in the isotropic Heisenberg chain of arbitrary spins. Physics Letters **87A** 9 (1982), 479-482
4. Kerov S.V., Kirillov A.N., Reshetikhin N.Yu.: Zap. Nauch. Semin. LOMI **155** (1986), 50-64 (Russian). [English transl.: Combinatorics, Bethe Ansatz, and representations of the symmetric group. J. Soviet Math. **41** (1988), 916-24]
5. de Vega H.J.: Yang-Baxter algebras, integrable theories and Bethe Ansatz. J. Mod. Phys. **4** (1990) 735-801
6. Foerster A.: Quantum group invariant supersymmetric t - J model with periodic boundary conditions. J. Phys. A **29** (1996) 7625-7633
7. Karowski M., Zapletal A.: Quantum-group-invariant integrable n-state vertex models with periodic boundary conditions. Nucl. Phys. B **419** (1994) 567-588
8. Links J., Foerster A., Karowski M.: Bethe ansatz solution of a closed spin 1 XXZ Heisenberg chain with quantum algebra symmetry. J. Math. Phys. **40** (1999) 726-735

9. Ilakovac A., Kolanovic M., Pallua S., Prester. P.: Violation of the string hypothesis and the Heisenberg XXZ spin chain. Phys. Rev. B **60** (1999) 7271-7277
10. Faddeev L., Takhtajan L.: Hamiltonian methods in the theory of solitons. Springer, Berlin 1987
11. Michel L.: Symmetry defects and broken symmetry. Configurations. Hidden symmetry. Revs. Mod. Phys. **52** (1980) 617
12. Takahashi M.: One-dimensional Heisenberg model at finite temperature. Progr. Theor. Phys. (Kyoto) **46** (1971), 401-15
13. Kerber A.: Algebraic Combinatorics via Finite Group Actions. Wissenschaftsverlag, Manheim 1991.
14. Lulek B. and Lulek T.: Magnons as indistinguishable hard-core particles on lattices. Rep. Math. Phys. **38** (1996), 267-71
15. Lulek T.: Density of states in the reciprocal lattice for one-dimensional periodic Heisenberg magnet. J. Physique **45** (1984), 29-34
16. Florek W., Lulek T.: Symmetry properties of the density of states in the Brillouin zone for a one-dimensional periodic Heisenberg magnet. J. Phys. A **20** (1987) 1921-1940
17. Kirillov A.N.: Zap. Nauch. Semin. LOMI **131** (1983), 88-105 (Russian). [English transl.: Combinatorial identities, and completeness of eigenstates of the Heisenberg magnet. J. Soviet Math. **30** (1985), 2298-310]

50 Years of Bailey's Lemma

S. Ole Warnaar

Instituut voor Theoretische Fysica, Universiteit van Amsterdam, Valckenierstraat 65, 1018 XE Amsterdam, The Netherlands
present address: Department of Mathematics and Statistics, The University of Melbourne, Vic 3010, Australia

0 Introduction

Half a century ago, The Proceedings of the London Mathematical Society published W. N. Bailey's influential paper *Identities of the Rogers–Ramanujan type* [16]. The main result therein, which was inspired by Rogers' second proof of the Rogers–Ramanujan identities [49] (and also [48,28,15]), is what is now known as Bailey's lemma. To celebrate the occasion of the lemma's fiftiest birthday we present a history of Bailey's lemma in 5 chapters (or rather sections), covering (i) Bailey's work, (ii) the Bailey chain (iii) the Bailey lattice (iv) the Bailey lemma in statistical mechanics, and (v) conjugate Bailey pairs.

Due to size limitations of this paper the higher rank [42,40,43,41,14,60] and trinomial [11,59,19] generalizations of the Bailey lemma will be treated at the lemma's centennial in 2049. More extensive reviews of topics (i), (ii) and (iii), can be found in [5, Sec. 3], [46] and [24], respectively.

1 The Bailey lemma

In an attempt to clarify Rogers' second proof [49] of the Rogers–Ramanujan identities, Bailey [16] was led to the following simple observation.

Lemma 1.1. *If* $\alpha = \{\alpha_L\}_{L\geq 0}, \dots, \delta = \{\delta_L\}_{L\geq 0}$ *are sequences that satisfy*

$$\beta_L = \sum_{r=0}^{L} \alpha_r u_{L-r} v_{L+r} \qquad \text{and} \qquad \gamma_L = \sum_{r=L}^{\infty} \delta_r u_{r-L} v_{r+L}, \tag{1}$$

then

$$\sum_{L=0}^{\infty} \alpha_L \gamma_L = \sum_{L=0}^{\infty} \beta_L \delta_L. \tag{2}$$

The proof is straightforward and merely requires an interchange of sums. Of course, in the above suitable convergence conditions need to be imposed to make the definition of γ and the interchange of sums meaningful.

The idea behind Bailey's lemma is clear. When trying to prove a complicated identity of the form $\sum_L A_L = \sum_L B_L$ it is a considerable step in the right direction if one can find a dissection of this identity into two identities of the type (1) where $A_L = \alpha_L\gamma_L$ and $B_L = \beta_L\delta_L$. Or, as Slater put it in Bailey's obituary [57],

> *The root of the underlying idea ... is that of transforming a doubly infinite series into a simply infinite and a finite series. In a geometric sense, this involves summing over a triangle instead of over a rectangle.*

In applications of his transform, Bailey chose $u_L = 1/(q)_L$ and $v_L = 1/(aq)_L$, with the usual definition of the q-shifted factorial, $(a)_\infty = (a;q)_\infty = \prod_{k=0}^{\infty}(1-aq^k)$ and $(a)_L = (a;q)_L = (a)_\infty/(aq^L)_\infty$ for $L \in \mathbb{Z}$. (Throughout, we assume that $0 < |q| < 1$.) With this choice, equation (1) becomes

$$\beta_L = \sum_{r=0}^{L} \frac{\alpha_r}{(q)_{L-r}(aq)_{L+r}} \quad \text{and} \quad \gamma_L = \sum_{r=L}^{\infty} \frac{\delta_r}{(q)_{r-L}(aq)_{r+L}}. \tag{3}$$

A pair of sequences that satisfies the first equation of (3) is called a Bailey pair relative to a. Similarly, the second equation defines a conjugate Bailey pair relative to a.

Still following Bailey, one can employ the q-Saalschütz summation [35, Eq. (II.12)] to establish that (γ,δ) with

$$\begin{aligned}\gamma_L &= \frac{(\rho_1)_L(\rho_2)_L(aq/\rho_1\rho_2)^L}{(aq/\rho_1)_L(aq/\rho_2)_L}\frac{1}{(q)_{M-L}(aq)_{M+L}}\\ \delta_L &= \frac{(\rho_1)_L(\rho_2)_L(aq/\rho_1\rho_2)^L}{(aq/\rho_1)_M(aq/\rho_2)_M}\frac{(aq/\rho_1\rho_2)_{M-L}}{(q)_{M-L}}\end{aligned} \tag{4}$$

provides a conjugate Bailey pair.

As we shall see in the next section this conjugate Bailey pair leads to the very important concept of the Bailey chain. However, Bailey missed an opportunity here and made the (mis)judgement [16, Page 4]:

> *These values of δ_L, γ_L ... lead to ... results involving only terminating basic series. We are, however, more concerned with identities of the Rogers–Ramanujan type in this paper, as the most general formulae for basic series are too involved to be of any great interest.*

Consequently, Bailey only considered the conjugate Bailey pair (4) when the parameter M tends to infinity. Also taking the limit $\rho_1,\rho_2 \to \infty$ yields

$$\gamma_L = \frac{a^L q^{L^2}}{(aq)_\infty} \quad \text{and} \quad \delta_L = a^L q^{L^2}, \tag{5}$$

which substituted into (2) gives

$$\frac{1}{(aq)_\infty}\sum_{L=0}^{\infty} a^L q^{L^2}\alpha_L = \sum_{L=0}^{\infty} a^L q^{L^2}\beta_L. \tag{6}$$

The proof of the Rogers–Ramanujan and similar such identities requires the input of suitable Bailey pairs into (6). For example, from Rogers' work [49] one can infer the following Bailey pair relative to 1: $\alpha_0 = 1$ and

$$\alpha_L = (-1)^L q^{L(3L-1)/2}(1+q^L), \qquad \beta_L = \frac{1}{(q)_L}. \tag{7}$$

Thus one finds

$$\frac{1}{(q)_\infty}\sum_{L=-\infty}^{\infty}(-1)^L q^{L(5L-1)/2} = \sum_{n=0}^{\infty}\frac{q^{n^2}}{(q)_n}.$$

The application of the Jacobi triple product identity [35, Eq. (II.28)] yields the first Rogers–Ramanujan identity [48,49]

$$\sum_{n=0}^{\infty}\frac{q^{n^2}}{(q)_n} = \frac{1}{(q,q^4;q^5)_\infty}, \tag{8}$$

with the notation $(a_1,\ldots,a_k;q)_n = (a_1;q)_n \ldots (a_k;q)_n$. The second Rogers–Ramanujan identity

$$\sum_{n=0}^{\infty}\frac{q^{n(n+1)}}{(q)_n} = \frac{1}{(q^2,q^3;q^5)_\infty} \tag{9}$$

follows in a similar fashion using the Bailey pair [49]

$$\alpha_L = (-1)^L q^{L(3L+1)/2}(1-q^{2L+1})/(1-q), \qquad \beta_L = \frac{1}{(q)_L} \tag{10}$$

relative to q. By collecting a list of 96 Bailey pairs, and using (6) or the identity obtained from (4) and (2) by taking $M, \rho_1 \to \infty$ and $\rho_2 = -q^{k/2}$ (with k a small nonnegative integer) Slater compiled her famous list of 130 identities of the Rogers–Ramanujan type [55,56]. The next two sections deal with more systematic ways of finding Bailey pairs.

2 The Bailey chain

By dismissing the conjugate Bailey pair (4) in its finite form (i.e., with M finite, or, equivalently, with ρ_1 or ρ_2 of the form q^{-N}) Bailey missed a very effective mechanism for generating Bailey pairs. Namely, if we substitute the conjugate pair (4) into (2) the resulting equation has the same form as the defining relation (3) of a Bailey pair. This is formalized in the following theorem due to Andrews [4,5].

Theorem 2.1. *Let (α,β) form a Bailey pair relative to a. Then so does (α',β') with*

$$\begin{aligned}\alpha'_L&=\frac{(\rho_1)_L(\rho_2)_L(aq/\rho_1\rho_2)^L}{(aq/\rho_1)_L(aq/\rho_2)_L}\alpha_L\\ \beta'_L&=\sum_{r=0}^{L}\frac{(\rho_1)_r(\rho_2)_r(aq/\rho_1\rho_2)^r(aq/\rho_1\rho_2)_{L-r}}{(aq/\rho_1)_L(aq/\rho_2)_L(q)_{L-r}}\beta_r.\end{aligned}\tag{11}$$

Again letting ρ_1,ρ_2 tend to infinity leads to the important special case

$$\alpha'_L=a^Lq^{L^2}\alpha_L \qquad \text{and} \qquad \beta'_L=\sum_{r=0}^{L}\frac{a^rq^{r^2}}{(q)_{L-r}}\beta_r,\tag{12}$$

which, for $a=1$ and $a=q$, was also discovered by Paule [44]. One now finds that the Bailey pairs (7) and (10) of Rogers can be obtained from the $a=1$ and $a=q$ instances of the Bailey pair [4]

$$\alpha_L=(-1)^Lq^{\binom{L}{2}}\frac{(1-aq^{2L})(a)_L}{(1-a)(q)_L}, \qquad \beta_L=\delta_{L,0}\tag{13}$$

by application of (12). The Bailey pair (13) is an immediate consequence of the inverse Bailey transform [4]

$$\alpha_L=\frac{1-aq^{2L}}{1-a}\sum_{r=0}^{L}\frac{(-1)^{L-r}q^{\binom{L-r}{2}}(a)_{L+r}}{(q)_{L-r}}\beta_r,\tag{14}$$

which follows from (3) and [35, Eq. (II.21)] specialized to $aq=bc$. The iteration of (11) or (12) leads to what is known as the Bailey chain [4,46]:

$$(\alpha,\beta)\to(\alpha',\beta')\to(\alpha'',\beta'')\to\cdots$$

and thus, given a single Bailey pair, one immediately finds an infinite sequence of Bailey pairs. (To be compared with the 96 Bailey pairs collected by Slater!) As an example, iteration of (14) gives the Bailey pair

$$\begin{aligned}\alpha_L&=(-1)^La^{kL}q^{kL^2+\binom{L}{2}}\frac{(1-aq^{2L})(a)_L}{(1-a)(q)_L}\\ \beta_L&=\sum_{L\ge n_1\ge\cdots\ge n_{k-1}\ge 0}\frac{a^{n_1+\cdots+n_{k-1}}q^{n_1^2+\cdots+n_{k-1}^2}}{(q)_{L-n_1}(q)_{n_1-n_2}\cdots(q)_{n_{k-2}-n_{k-1}}(q)_{n_{k-1}}}.\end{aligned}$$

Substituting this into the defining relation (3) of a Bailey pair and letting L tend to infinity gives, for $a=1$ or $a=q$,

$$\sum_{n_1,\dots,n_{k-1}}\frac{a^{n_1+\cdots+n_{k-1}}q^{n_1^2+\cdots+n_{k-1}^2}}{(q)_{n_1-n_2}\cdots(q)_{n_{k-2}-n_{k-1}}(q)_{n_{k-1}}}=\frac{1}{(q)_\infty}\sum_{r=-\infty}^{\infty}(-1)^ra^{kr}q^{kr^2+\binom{r}{2}}.$$

Using Jacobi's triple product identity finally yields

$$\sum_{n_1,\ldots,n_{k-1}} \frac{q^{n_1^2+\cdots+n_{k-1}^2+n_i+\cdots+n_{k-1}}}{(q)_{n_1-n_2}\cdots(q)_{n_{k-2}-n_{k-1}}(q)_{n_{k-1}}} = \frac{(q^i,q^{2k+1-i},q^{2k+1};q^{2k+1})_\infty}{(q)_\infty}, \tag{15}$$

where $i = 1$ or $i = k$. For $k = 2$ these are the Rogers–Ramanujan identities (8) and (9), whereas for $k \geq 3$ they are Andrews' analytic counterpart of Gordon's partition theorem [3]. In fact, Andrews' identities are (15) for all $i = 1, \ldots, k$ and one concludes that the Bailey chain mechanism has failed to produce all of these. What is required is an extension of the Bailey chain known as the Bailey lattice. This will be our next topic. (Prior to the invention of the Bailey lattice Paule [44] already obtained (15) for all i using "ad hoc" Bailey lattice-like transformations.)

3 The Bailey lattice

One of the features of Theorem 2.1 is that it transforms a Bailey pair relative to a into a new Bailey pair relative to a. More generally one can of course try to transform a Bailey pair relative to a into a Bailey pair relative to b. Agarwal, Andrews and Bressoud have formulated this problem in a general setting of infinite dimensional matrices [1,24]. Here we shall only be concerned with concrete examples of such "Bailey lattice" transformations. Since the parameter a is no longer fixed we shall write $(\alpha(a), \beta(a))$ for a Bailey pair relative to a.

Theorem 3.1. *Fix N a nonnegative integer and set $b = aq^N$. Let $(\alpha(b), \beta(b))$ be a Bailey pair. Then so is $(\alpha'(a), \beta'(a))$ with*

$$\begin{aligned}\alpha'_L(a) &= (1-aq^{2L})(aq)_N \frac{(\rho_1)_L(\rho_2)_L(aq/\rho_1\rho_2)^L}{(aq/\rho_1)_L(aq/\rho_2)_L} \\ &\quad \times \sum_{j=0}^{N} (-1)^j a^j q^{2Lj-j(j+1)/2} \begin{bmatrix} N \\ j \end{bmatrix} \frac{(aq)_{2L-j-1}}{(aq)_{2L-j+N}}\, \alpha_{L-j}(b) \\ \beta'_L(a) &= \sum_{r=0}^{L} \frac{(\rho_1)_r(\rho_2)_r(aq/\rho_1\rho_2)^r(aq/\rho_1\rho_2)_{L-r}}{(aq/\rho_1)_L(aq/\rho_2)_L(q)_{L-r}}\, \beta_r(b).\end{aligned}$$

Here we have used the q-binomial coefficient defined as $\begin{bmatrix} a \\ b \end{bmatrix} = \frac{(q^{a-b+1})_b}{(q)_b}$ for $b \geq 0$ and 0 otherwise. A very similar result can be stated as follows.

Theorem 3.2. *Fix N a nonnegative integer and set $b = aq^N$. Let $(\alpha(b), \beta(b))$ be a Bailey pair. Then so is $(\alpha'(a), \beta'(a))$ with*

$$\alpha'_L(a) = (1-aq^{2L})(aq)_N \sum_{j=0}^{N} \frac{(\rho_1)_{L-j}(\rho_2)_{L-j}(bq/\rho_1\rho_2)^{L-j}}{(bq/\rho_1)_{L-j}(bq/\rho_2)_{L-j}} \times (-1)^j a^j q^{2Lj-j(j+1)/2} \begin{bmatrix} N \\ j \end{bmatrix} \frac{(aq)_{2L-j-1}}{(aq)_{2L-j+N}} \alpha_{L-j}(b)$$

$$\beta'_L(a) = \sum_{r=0}^{L} \frac{(\rho_1)_r(\rho_2)_r(bq/\rho_1\rho_2)^r(bq/\rho_1\rho_2)_{L-r}}{(bq/\rho_1)_L(bq/\rho_2)_L(q)_{L-r}} \beta_r(b).$$

The $N=0$ and $N=1$ cases of the first theorem correspond to the Bailey chain of Theorem 2.1 and the Bailey lattice of [1, Lemma 1.2], respectively. The second theorem for $N=0$ is again the Bailey chain whereas for $N=1$ it is a variation of the Bailey lattice of [51, Lemma 4.3]. Theorem 3.1 was also found by Krattenthaler and Foda [39].

First we prove Theorem 3.1. Substituting the expression for $\alpha'(a)$ in the "primed" version of (3), transforming $j \to r-j$ and then interchanging the order of summation, gives

$$\beta'_L(a) = \sum_{j=0}^{L} \frac{(\rho_1)_j(\rho_2)_j(aq/\rho_1\rho_2)^j \alpha_j(b)}{(aq/\rho_1)_j(aq/\rho_2)_j(q)_{L-j}(bq)_{2j}(aq^{2j+1})_{L-j}} \times \lim_{a_4\to\infty} {}_8W_7(aq^{2j}; a_4, a/b, \rho_1 q^j, \rho_2 q^j, q^{-L+j}; q, abq^{L+j+2}/a_4\rho_1\rho_2),$$

where we employed the conventional short-hand notation for very-well-poised basic hypergeometric series [35]. By Watson's ${}_8\phi_7$ transformation [35, Eq. (III.18)] (with $a \to aq^{2j}, b \to a_4, c \to a/b, d \to \rho_1 q^j, e \to \rho_2 q^j$ and $n \to L-j$) this can be simplified to

$$\beta'_L(a) = \sum_{j=0}^{L}\sum_{r=0}^{L-j} \frac{(\rho_1)_{j+r}(\rho_2)_{j+r}(aq/\rho_1\rho_2)^{j+r}(aq/\rho_1\rho_2)_{L-j-r}\alpha_j(b)}{(aq/\rho_1)_L(aq/\rho_2)_L(q)_{L-j-r}(q)_r(bq)_{r+2j}}.$$

Shifting $r \to r-j$, then interchanging sums and recalling the definition of $\beta_r(b)$, this indeed yields the second transformation claimed in the theorem.

The second proof proceeds in a similar manner. Substituting the expression for $\alpha'(a)$ in the "primed" version of (3), transforming $j \to r-j$ and then interchanging the order of summation, gives

$$\beta'_L(a) = \sum_{j=0}^{L} \frac{(\rho_1)_j(\rho_2)_j(bq/\rho_1\rho_2)^j \alpha_j(b)}{(bq/\rho_1)_j(bq/\rho_2)_j(q)_{L-j}(bq)_{2j}(aq^{2j+1})_{L-j}} \times \lim_{a_4\to\infty} {}_6W_5(aq^{2j}, a_4, a/b, q^{-L+j}; q, bq^{L+j+2}/a_4).$$

By Rogers' ${}_6\phi_5$ summation [35, Eq. (II.21)] (with $a \to aq^{2j}, b \to a_4, c \to a/b$ and $n \to L-j$) this can be simplified to

$$\beta'_L(a) = \sum_{j=0}^{L} \frac{(\rho_1)_j(\rho_2)_j(bq/\rho_1\rho_2)^j \alpha_j(b)}{(bq/\rho_1)_j(bq/\rho_2)_j(q)_{L-j}(bq)_{L+j}}.$$

Using the q-Saalschütz sum [35, Eq. (II.12)] (with $a \to \rho_1 q^j, b \to \rho_2 q^j, c \to bq^{2j+1}$ and $n \to L-j$) this can be rewritten as

$$\beta'_L(a) = \sum_{j=0}^{L}\sum_{r=0}^{L-j} \frac{(\rho_1)_{j+r}(\rho_2)_{j+r}(bq/\rho_1\rho_2)^{j+r}(bq/\rho_1\rho_2)_{L-j-r}\alpha_j(b)}{(bq/\rho_1)_L(bq/\rho_2)_L(q)_{L-j-r}(q)_r(bq)_{r+2j}}.$$

Shifting $r \to r-j$, interchanging sums and recalling the definition (3) yields the second expression of Theorem 3.2.

To see that we are now in the position to prove (15) for all $i = 1,\ldots,k$ we follow [1] and take the Bailey pair of equation (13) with $a = q$ as starting point. Applying (12) $k-i+1$ times, then Theorem 3.1 with $N = 1$ and $\rho_1, \rho_2 \to \infty$ once, and then again (12) $i-2$ times, one finds the Bailey pair $\alpha_0 = 1$,

$$\alpha_L = (-1)^L q^{kL^2+\binom{L}{2}+(k-i+1)L}(1+q^{(2i-2k-1)L})$$

$$\beta_L = \sum_{L\ge n_1\ge\cdots\ge n_{k-1}\ge 0} \frac{q^{n_1^2+\cdots+n_{k-1}^2+n_i+\cdots+n_{k-1}}}{(q)_{L-n_1}(q)_{n_1-n_2}\cdots(q)_{n_{k-2}-n_{k-1}}(q)_{n_{k-1}}},$$

relative to 1. Substituting this result into (3), letting L tend to infinity and using the triple product identity one arrives at the identities (15) for $i = 2,\ldots,k$.

4 The Bailey lemma in statistical mechanics

In section 1 we already mentioned Slater's famous list of 130 identities of Rogers–Ramanujan type [55,56]. She found these identities by exploiting extensive lists of Bailey pairs (grouped from A to M) extracted from Rogers' or Bailey's papers [49,16] or from known basic hypergeometric function identities. For example, the first group of Bailey pairs (all due to Rogers) reads $\alpha_0 = 1$,

	β_L	$\alpha_{3L\pm1}$	α_{3L}
A(1)	$1/(q)_{2L}$	$-q^{(2L\pm1)(3L\pm1)}$	$q^{L(6L-1)}+q^{L(6L+1)}$
A(3)	$q^L/(q)_{2L}$	$-q^{2L(3L\pm1)}$	$q^{2L(3L-1)}+q^{2L(3L+1)}$
A(5)	$q^{L^2}/(q)_{2L}$	$-q^{L(3L\pm1)}$	$q^{L(3L-1)}+q^{L(3L+1)}$
A(7)	$q^{L^2-L}/(q)_{2L}$	$-q^{(L\pm1)(3L\pm1)}$	$q^{L(3L-2)}+q^{L(3L+2)}$

with $a = 1$, and

	β_L	$\alpha_{3L-(1\mp 1)/2}$	α_{3L+1}
A(2)	$1/(q^2)_{2L}$	$q^{L(6L\pm 1)}$	$-q^{(2L+1)(3L+1)} - q^{(2L+1)(3L+2)}$
A(4)	$q^L/(q^2)_{2L}$	$q^{2L(3L\pm 2)}$	$-q^{2(L+1)(3L+1)} - q^{2L(3L+2)}$
A(6)	$q^{L^2}/(q^2)_{2L}$	$q^{L(3L\mp 1)}$	$-q^{L(3L+1)} - q^{(L+1)(3L+2)}$
A(8)	$q^{L^2+L}/(q^2)_{2L}$	$q^{L(3L\pm 2)}$	$-q^{(L+1)(3L+1)} - q^{L(3L+2)}$

with $a = q$. The Bailey pairs given in equations (7) and (10), which were used by Rogers to prove the Rogers–Ramanujan identities, are items labelled B(1) and B(3).

Remarkably, in recent work on exactly solvable lattice models in statistical mechanics identities have arisen (see [17,30,18,58,21] and references therein), which for each pair of integers (p, p'), with $1 < p < p'$ and $\gcd(p, p') = 1$ imply a family of Bailey pairs [31]. Moreover, many of the Bailey pairs of Rogers and Slater (as well as later pairs found in [4,44,1]) are included as special cases.

First we need a class of polynomials known as the one-dimensional configuration sums of the Andrews–Baxter–Forrester model [10,32]. (See [9] for a partition theoretic interpretation of the configuration sums.) For coprime integers p, p' with $1 \le p < p'$, and integers $1 \le b, s \le p' - 1$, $0 \le r \le p - 1$ and $L \ge 0$ such that $L + s + b$ is even, define

$$X_{r,s}^{(p,p')}(L, b) = \sum_{j\in\mathbb{Z}} \Big\{ q^{j(pp'j+p'r-ps)} \begin{bmatrix} L \\ \frac{L+s-b}{2} - p'j \end{bmatrix} - q^{(pj+r)(p'j+s)} \begin{bmatrix} L \\ \frac{L-s-b}{2} - p'j \end{bmatrix} \Big\}.$$

For $r = b - \lfloor (b + 1)(p' - p)/p' \rfloor$, (with $\lfloor x \rfloor$ the integer part of x) the one-dimensional configuration sums are generating functions of certain sets of restricted lattice paths, and hence are polynomials with positive coefficients. This is not at all manifest from the above definition, and the identities referred to in the above claim a different, manifestly positive representation for the configuration sums. The simplest of these identities arise when $|p'r - ps| = 1$ and $b = s$ which we assume throughout the remainder of this section.

For the moment also assume that $p < p' < 2p$, and define nonnegative integers $\nu_0, \ldots, \nu_n$ by the continued fraction expansion $p/(p' - p) = [\nu_0, \nu_1, \ldots, \nu_n]$. The integers n and ν_j, can be used to further define $t_m = \sum_{j=0}^{m-1} \nu_j$ $(1 \le m \le n)$ and $d = -2 + \sum_{j=0}^{n} \nu_j$. These latter numbers define a so-called fractional incidence matrix $\mathcal{I}$ and fractional Cartan-type matrix $B = 2I - \mathcal{I}$ (with I the d by d unit matrix) as follows

$$\mathcal{I}_{i,j} = \begin{cases} \delta_{i,j+1} + \delta_{i,j-1} & \text{for } 1 \le i < d,\ i \ne t_m, \\ \delta_{i,j+1} + \delta_{i,j} - \delta_{i,j-1} & \text{for } i = t_m,\ 1 \le m \le n - \delta_{\nu_n,2}, \\ \delta_{i,j+1} + \delta_{\nu_n,2}\delta_{i,j} & \text{for } i = d. \end{cases}$$

When $p' = p+1$ the matrix $\mathcal{I}$ has entries $\mathcal{I}_{i,j} = \delta_{|i-j|,1}$ $(i, j = 1, \ldots, p-2)$, so that B corresponds to the Cartan matrix of the Lie algebra A_{p-2}. When $p =$

$2k-1$ and $p'=2k+1$ one finds $\mathcal{I}_{i,j}=\delta_{|i-j|,1}+\delta_{i,j}\delta_{i,k-1}$ $(i,j=1,\dots,k-1)$, so that B corresponds to the Cartan-type matrix of the tadpole graph of $k-1$ nodes.

Using the above definitions we have the following result [17,30,18,58,21]:

Theorem 4.1. *Let $1<p<p'<2p$ with $\gcd(p,p')=1$ and let $r(\le p'-1)$ and $s(\le p'-1)$ satisfy $|p'r-ps|=1$. Then*

$$X^{(p,p')}_{r,s}(2L,s)=\sum_{m\in 2\mathbb{Z}^d} q^{\frac{1}{4}mBm}\prod_{j=1}^{d}\begin{bmatrix} L\delta_{j,1}+\frac{1}{2}(\mathcal{I}m)_j \\ m_j\end{bmatrix}. \tag{16}$$

Here we use the notation $mBm=\sum_{j,k}m_jB_{j,k}m_k$ and $(\mathcal{I}m)_j=\sum_k \mathcal{I}_{j,k}m_k$. The corresponding identities for $2p<p'$ follow simply from the symmetry

$$X^{(p,p')}_{r,s}(L,b;q)=q^{\frac{1}{4}(L^2-(b-s)^2)}X^{(p'-p,p')}_{b-r,s}(L,b;1/q).$$

Foda and Quano [31] used (special cases of) the above theorem and symmetry relation together with the Bailey lemma to prove conjectured q-series identities for Virasoro characters. Indeed, we can readily extract the following Bailey pairs relative to 1 [31,22]: $\alpha_0=1$,

$$\begin{aligned}\alpha_L&=\begin{cases} q^{j(jpp'+rp'-sp)}+q^{j(jpp'-rp'+sp)} & \text{for } L=jp'>0\\ -q^{(jp\pm r)(jp'\pm s)} & \text{for } L=jp'\pm s>0\\ 0 & \text{otherwise}\end{cases}\\ \beta_L&=X^{(p,p')}_{r,s}(2L,s)/(q)_{2L},\end{aligned} \tag{17}$$

where in the expression for β the representation (16) of $X^{(p,p')}_{r,s}$ is taken. We note that $(p,p')=(2,3)$ (so that $r=s=1$) corresponds to the Bailey pair A(1) and $(p,p')=(1,3)$ ($r=0$ and $s=1$) to A(5). We also remark that $(p,p')=(2,5)$ ($r=1$, $s=2$) is the Bailey pair [4, Eq. (5.3)].

5 Conjugate Bailey pairs

We have just seen that each pair of coprime integers (p,p') labels a Bailey pair. Next we discuss some recent developments which show that a similar result holds for conjugate Bailey pairs [50,51,52].

First we need to introduce the string functions associated to admissible representations of the affine algebra $\mathrm{A}_1^{(1)}$ [38]. Again fix a pair of positive, coprime integers (p,p'). Let $0\le \ell\le p'-2$ and let Λ_0,Λ_1 denote the fundamental weights of $\mathrm{A}_1^{(1)}$. Then Kac and Wakimoto showed that the $\mathrm{A}_1^{(1)}$ character of the admissible highest weight module of highest weight $(p'/p-\ell-2)\Lambda_0+\ell\Lambda_1$ is given by a generalized Weyl-Kac formula as follows

$$\chi^{(p,p')}_{\ell}(z,q)=\frac{\sum_{\sigma=\pm 1}\sigma\Theta_{\sigma(\ell+1),p'}(z,q^p)}{\sum_{\sigma=\pm 1}\sigma\Theta_{\sigma,2}(z,q)}.$$

Here $\Theta_{n,m}$ is a classical theta function, $\Theta_{n,m}(z,q)=\sum_{j\in\mathbb{Z}+n/2m}q^{mj^2}z^{-mj}$. Note that when $p>1$ we are dealing with nonintegral highest weights. The (normalized) $\mathrm{A}_1^{(1)}$ string functions of level $p'/p-2$ are defined by the expansion

$$\chi_\ell^{(p,p')}(z,q)=q^{\frac{1}{8}-\frac{(\ell+1)^2p}{4p'}}\sum_{m\in\mathbb{Z}}C_{m,\ell}^{(p,p')}(q)z^{-\frac{1}{2}m},$$

which immediately implies that $C_{m,\ell}^{(p,p')}(q)=0$ unless $m+\ell$ is even. An explicit expression for the string functions can be derived as a double sum of Hecke indefinite modular form type [37,2,52]

$$\begin{aligned}C_{m,\ell}^{(p,p')}(q)=&\frac{1}{(q)_\infty^3}\Big\{\sum_{\substack{i\geq 0\\ j\geq 0}}-\sum_{\substack{i<0\\ j<0}}\Big\}(-1)^iq^{\frac{1}{2}i(i+m)+p'j(pj+i)+\frac{1}{2}(\ell+1)(2pj+i)}\\ &-\frac{1}{(q)_\infty^3}\Big\{\sum_{\substack{i\geq 0\\ j>0}}-\sum_{\substack{i<0\\ j\leq 0}}\Big\}(-1)^iq^{\frac{1}{2}i(i+m)+p'j(pj+i)-\frac{1}{2}(\ell+1)(2pj+i)}.\end{aligned}$$

After these preliminaries let us now return to the conjugate Bailey pair of equation (5) and specialize $a=q^\eta$, with η a nonnegative integer. Let us further remark the following identities ($\ell=0,1$ and $m+\ell\equiv L+\ell\equiv 0 \pmod 2$):

$$C_{m,\ell}^{(1,3)}(q)=\frac{q^{\frac{1}{4}(m^2-\ell^2)}}{(q)_\infty}\qquad\text{and}\qquad X_{0,\ell+1}^{(1,3)}(L,1)=q^{\frac{1}{4}(L^2-\ell)}.$$

The first result is [37, Sec. 4.6, Ex. 3] whereas the second is A(5) for $\ell=0$ and A(8) for $\ell=1$. We thus infer that the conjugate Bailey pair (5) can be recast as $\gamma_L=(q)_\eta C_{2L+\eta,\ell}^{(1,3)}(q)$ and $\delta_L=X_{0,\ell+1}^{(1,3)}(2L+\eta,1)$. It now requires little imagination to conjecture the following more general result [52].

Theorem 5.1. *Fix integers $1\leq p<p'$, and let η and ℓ be nonnegative integers such that $0\leq\ell\leq p'-2$ and $\ell+\eta$ is even. Then*

$$\gamma_L=(q)_\eta C_{2L+\eta,\ell}^{(p,p')}(q)\qquad\textit{and}\qquad\delta_L=X_{0,\ell+1}^{(p,p')}(2L+\eta,1)\tag{18}$$

yields a conjugate Bailey pair relative to $a=q^\eta$.

The proof of this theorem relies on yet another class of conjugate Bailey pairs given by [52, Thm. 4.1]

$$\begin{aligned}\gamma_L&=\frac{1}{(q)_\infty^2(aq)_\infty}\sum_{i=1}^{\infty}(-1)^iq^{\frac{1}{2}i(i+2L+\eta)}\Big\{q^{\frac{1}{2}i(2j+\eta+1)}-q^{-\frac{1}{2}i(2j+\eta+1)}\Big\}\\ \delta_L&=\begin{bmatrix}2L+\eta\\ L-j\end{bmatrix}-\begin{bmatrix}2L+\eta\\ L-j-1\end{bmatrix},\end{aligned}\tag{19}$$

with $a=q^\eta$, η an nonnegative integer and j an integer. Here we remark that, incidentally, $\delta_L=K_{(2^{L-j}1^{2j+\eta}),(1^{2L+\eta})}(q)=\tilde{K}_{(L+j+\eta,L-j),(1^{2L+\eta})}(q)$, where

$K_{\lambda,\mu}(q)$ and $\tilde{K}_{\lambda,\mu}(q)$ are the Kostka and cocharge Kostka polynomial, respectively. The Bailey pair (19) can easily be derived from the summation formula [52]

$$\sum_{r=0}^{\infty} \frac{q^r (ab)_{2r}}{(q)_r (ab)_r (aq)_r (bq)_r} = \frac{1}{(q)_\infty (aq)_\infty (bq)_\infty} \sum_{i=1}^{\infty} (-1)^{i-1} q^{\binom{i}{2}} \frac{a^i - b^i}{a - b}.$$

It is again possible to give representations of the polynomials $X_{0,\ell+1}^{(p,p')}(L,1)$ that are manifestly positive [52]. Treating only the simplest cases we have the following counterpart of Theorem 4.1.

Theorem 5.2. *Let* $1 < p < p' < 2p$ *with* $\gcd(p,p') = 1$. *Then*

$$X_{0,1}^{(p,p')}(2L,1) = q^L \sum_{m \in 2\mathbb{Z}^d} q^{\frac{1}{4} mBm + \frac{1}{2} \sum_{i=1}^{n} m_{t_i}} \prod_{j=1}^{d} \begin{bmatrix} L\delta_{j,1} - \sum_{i=1}^{n} \delta_{j,t_i} + \frac{1}{2}(\mathcal{I}m)_j \\ m_j \end{bmatrix}$$

The corresponding identities for $2p < p'$ follow from

$$X_{r,s}^{(p'-p,p')}(2L,1;q) = q^{L(L+1)} X_{r,s}^{(p,p')}(2L,1;1/q).$$

Combining the Bailey pair of equation (17) with the conjugate Bailey pair of (18) and specializing some of the parameters, we find

$$\sum_{j \in \mathbb{Z}} \left\{ q^{j(jp_1p_1' + rp_1' - sp_1)} C_{2jp_1',0}^{(p_2,p_2')}(q) - q^{(jp_1+r)(jp_1'+s)} C_{2jp_1'+2s,0}^{(p_2,p_2')}(q) \right\}$$
$$= \sum_{L=0}^{\infty} X_{r,s}^{(p_1,p_1')}(2L,s) X_{0,1}^{(p_2,p_2')}(2L,1)/(q)_{2L},$$

with $1 \le p_i < p_i'$ $(i = 1,2)$ and $|p_1' r - p_1 s| = 1$. Here we have used the symmetry $C_{m,\ell}^{(p,p')} = C_{-m,\ell}^{(p,p')}$. Inserting the representations for the one-dimensional configuration sums provided by Theorems 4.1 and 5.2 this turns into a class of 'rather' nontrivial q-series identities. For $p_1 = 1$ or $p_2 = 1$ the left-hand side of the above equation can be identified as a branching function of the coset pair $(\mathrm{A}_1^{(1)} \oplus \mathrm{A}_1^{(1)}, \mathrm{A}_1^{(1)})$ at levels $N_1 = p_1/p_1 - 2$, $N_2 = p_2'/p_2 - 2$ and $N_1 + N_2$, respectively.

6 Further reading

To conclude our overview of half a century of Bailey's lemma, let us mention some further papers on (or related to) the Bailey lemma that have not been mentioned in the main text. In [45], Paule gave a short operator-type proof of the special case (12) of the Bailey chain. Riese [47] developed the Mathematica package Bailey for taking (automated) walks along the Bailey lattice.

He also shows how to apply his Mathematica package qZeil to generate Bailey pairs. The Bailey transform (3) and its inverse (14) can be formulated naturally in terms of inversion of infinite-dimensional lower-triangular matrices [1,24], making it a special case of the generalized q-Lagrange procedure of Gessel and Stanton [36]. New types of Baily lattice transformations which do not only change the value of a but also that of the base q, were very recently found and applied by Bressoud, Ismail and Stanton [25]. Bressoud [23] and Singh [54] have also applied conjugate Bailey pairs other than (4) and (5) of Bailey. For a special choice of parameters their conjugate pair can be shown to coincide with the $(p,p')=(2,3)$ case of Theorem 5.1. Andrews [6], Andrews and Hickerson [13] and Choi [27] applied the Bailey chain to prove identities for Ramanujan's mock theta functions, and Andrews [8] also used it to prove several of Ramanujan's identities for Lambert series. The Bailey lemma and its connection to $N=2$ supersymmetric conformal field theory was investigated by Berkovich, McCoy and Schilling [20]. For those left with the impression that the Bailey lemma is "merely" good for proving q-series identities we remark that Andrews utilized the Bailey machinery in [6,7] to give a proof of Gauss' theorem that every integer can be written as the sum of three triangular numbers and that Andrews, Dyson and Hickerson used Bailey's lemma in the context of algebraic number theory [12]. Finally we mention that a special case of the Bailey chain admits an extension due to Burge [26]. This was extensively applied and further developed by Foda, Lee and Welsh [29] and by Schilling and the author [53].

Acknowledgements

This work was supported by a fellowship of the Royal Netherlands Academy of Arts and Sciences.

Note added in proof

The many recent references to the Bailey lemma listed in the bibliography show that after 50 years Bailey's lemma still is a source of inspiration. This makes it quite impossible to publish an account that can claim to be complete and up to date. Indeed, after this paper was accepted for publication further advances in connection with the lemma were reported in [33,34,61].

References

1. A. K. Agarwal, G. E. Andrews and D. M. Bressoud, *The Bailey lattice*, J. Ind. Math. Soc. **51** (1987), 57–73.
2. C. Ahn, S.-W. Chung and S.-H. H. Tye, *New parafermion,* SU(2) *coset and* $N=2$ *superconformal field theories*, Nucl. Phys. B **365** (1991), 191–240.
3. G. E. Andrews, *An analytic generalization of the Rogers–Ramanujan identities for odd moduli*, Prod. Nat. Acad. Sci. USA **71** (1974), 4082–4085.

4. G. E. Andrews, *Multiple series Rogers–Ramanujan type identities*, Pacific J. Math. **114** (1984), 267–283.
5. G. E. Andrews, *q-Series: Their development and application in analysis, number theory, combinatorics, physics, and computer algebra*, in CBMS Regional Conf. Ser. in Math. **66** (AMS, Providence, Rhode Island, 1985).
6. G. E. Andrews, *The fifth and seventh order mock theta functions*, Trans. Amer. Math. Soc. **293** (1986), 113–134.
7. G. E. Andrews, *EΥPHKA! num* $= \Delta + \Delta + \Delta$, J. Number Theory **23** (1986), 285–293.
8. G. E. Andrews, *Bailey chains and generalized Lambert series: I. four identities of Ramanujan*, Illinois J. Math. **36** (1992), 251–274.
9. G. E. Andrews, R. J. Baxter, D. M. Bressoud, W. H. Burge, P. J. Forrester and G. Viennot, *Partitions with prescribed hook differences*, Europ. J. Combinatorics **8** (1987), 341–350.
10. G. E. Andrews, R. J. Baxter and P. J. Forrester, *Eight-vertex SOS model and generalized Rogers–Ramanujan-type identities*, J. Stat. Phys. **35** (1984), 193–266.
11. G. E. Andrews and A. Berkovich, *A trinomial analogue of Bailey's lemma and* $N = 2$ *superconformal invariance*, Commun. Math. Phys. **192** (1998), 245–260.
12. G. E. Andrews, F. J. Dyson and D. Hickerson, *Partitions and indefinite quadratic forms*, Invent. Math. **91** (1988), 391–407.
13. G. E. Andrews and D. Hickerson, *Ramanujan's "lost"notebook. VII. The sixth order mock theta functions*, Adv. Math. **89** (1991), 60–105.
14. G. E. Andrews, A. Schilling and S. O. Warnaar, *An* A_2 *Bailey lemma and Rogers-Ramanujan-type identities*, J. Amer. Math. Soc. **12** (1999), 677–702.
15. W. N. Bailey, *Some identities in combinatory analysis*, Proc. London Math. Soc. (2) **49** (1947), 421–435.
16. W. N. Bailey, *Identities of the Rogers–Ramanujan type*, Proc. London Math. Soc. (2) **50** (1949), 1–10.
17. A. Berkovich, *Fermionic counting of RSOS-states and Virasoro character formulas for the unitary minimal series* $M(\nu, \nu + 1)$. *Exact results*, Nucl. Phys. B **431** (1994), 315–348.
18. A. Berkovich and B. M. McCoy, *Continued fractions and fermionic representations for characters of* $M(p, p')$ *minimal models*, Lett. Math. Phys. **37** (1996), 49–66.
19. A. Berkovich, B. M. McCoy and P. A. Pearce, *The perturbations* $\phi_{2,1}$ *and* $\phi_{1,5}$ *of the minimal models* $M(p, p')$ *and the trinomial analogue of Bailey's lemma*, Nucl. Phys. B **519 [FS]** (1998), 597–625.
20. A. Berkovich, B. M. McCoy and A. Schilling, $N = 2$ *supersymmetry and Bailey pairs*, Physica A **228** (1996), 33–62.
21. A. Berkovich, B. M. McCoy and A. Schilling, *Rogers-Schur–Ramanujan type identities for the* $M(p, p')$ *minimal models of conformal field theory*, Commun. Math. Phys. **191** (1998), 325–395.
22. A. Berkovich, B. M. McCoy, A. Schilling and S. O. Warnaar, *Bailey flows and Bose–Fermi identities for the conformal coset models* $(\mathrm{A}_1^{(1)})_N \times (\mathrm{A}_1^{(1)})_{N'}/(\mathrm{A}_1^{(1)})_{N+N'}$, Nucl. Phys. B **499 [PM]** (1997), 621–649.
23. D. M. Bressoud, *Some identities for terminating q-series*, Math. Proc. Cambridge Phil. Soc. **89** (1981), 211–223.
24. D. M. Bressoud, *The Bailey lattice: An introduction*, in *Ramanujan Revisited*, pp. 57–67, G. E. Andrews *et al.* eds., (Academic Press, New York, 1988).

25. D. M. Bressoud, M. Ismail and D. Stanton, *Change of base in Bailey pairs*, preprint math.CO/9909053. To appear in The Ramanujan Journal.
26. W. H. Burge, *Restricted partition pairs*, J. Combin. Theory Ser. A **63** (1993) 210–222.
27. Y.-S. Choi, *Tenth order mock theta functions in Ramanujan's lost notebook*, Invent. Math. **136** (1999), 497–569.
28. F. J. Dyson, *Three identities in combinatory analysis*, J. London. Math. Soc. **18** (1943), 35–39.
29. O. Foda, K. S. M. Lee and T. A. Welsh, *A Burge tree of Virasoro-type polynomial identities*, Int. J. Mod. Phys. A **13** (1998) 4967–5012.
30. O. Foda and Y.-H. Quano, *Polynomial identities of the Rogers–Ramanujan type*, Int. J. Mod. Phys. A **10** (1995), 2291–2315.
31. O. Foda and Y.-H. Quano, *Virasoro character identities from the Andrews–Bailey construction*, Int. J. Mod. Phys. A **12** (1996), 1651–1675.
32. P. J. Forrester and R. J. Baxter, *Further exact solutions of the eight-vertex SOS model and generalizations of the Rogers–Ramanujan identities*, J. Stat. Phys. **38** (1985), 435–472.
33. J. Fulman, *A probabilistic proof of the Rogers–Ramanujan identities* preprint math.CO/0001078.
34. J. Fulman, *Random matrix theory over finite fields: a survey*, preprint math.GR/0003195.
35. G. Gasper and M. Rahman, *Basic Hypergeometric Series*, Encyclopedia of Mathematics and its Applications, Vol. 35, (Cambridge University Press, Cambridge, 1990).
36. I. Gessel and D. Stanton, *Applications of q-Lagrange inversion to basic hypergeometric series*, Trans. Amer. Math. Soc. **277** (1983), 173–201.
37. V. G. Kac and D. H. Peterson, *Infinite-dimensional Lie algebras, theta functions and modular forms*, Adv. Math. **53** (1984), 125–264.
38. V. G. Kac and M. Wakimoto, *Modular invariant representations of infinite-dimensional Lie algebras and superalgebras*, Proc. Nat. Acad. Sci. USA **85** (1988), 4956–4960.
39. C. Krattenthaler and O. Foda, unpublished.
40. G. M. Lilly and S. C. Milne, *The C_ℓ Bailey transform and Bailey lemma*, Constr. Approx. **9** (1993), 473–500.
41. S. C. Milne, *Balanced ${}_3\phi_2$ summation theorems for $U(n)$ basic hypergeometric series*, Adv. Math. **131** (1997), 93–187.
42. S. C. Milne and G. M. Lilly, *The A_ℓ and C_ℓ Bailey transform and lemma*, Bull. Amer. Math. Soc. (N.S.) **26** (1992), 258–263.
43. S. C. Milne and G. M. Lilly, *Consequences of the A_ℓ and C_ℓ Bailey transform and Bailey lemma*, Discrete Math. **139** (1995), 319–346.
44. P. Paule, *On identities of the Rogers–Ramanujan type*, J. Math. Anal. Appl. **107** (1985), 255–284.
45. P. Paule, *A note on Bailey's lemma*, J. Combin. Theory Ser. A **44** (1987), 164–167.
46. P. Paule, *The concept of Bailey chains*, Publ. I.R.M.A. Strasbourg 358/S-18, (1988), 53–76.
47. A. Riese, *Contributions to symbolic q-hypergeometric summation*, PhD thesis, RISC, J. Kepler University, Linz (1997), http://www.risc.uni-linz.ac.at/research/combinat/risc/publications.

48. L. J. Rogers, *Second memoir on the expansion of certain infinite products*, Proc. London Math. Soc. **25** (1894), 318–343.
49. L. J. Rogers, *On two theorems of combinatory analysis and some allied identities*, Proc. London Math. Soc. (2) **16** (1917), 315–336.
50. A. Schilling and S. O. Warnaar, *A higher-level Bailey lemma*, Int. J. Mod. Phys. B **11** (1997), 189–195.
51. A. Schilling and S. O. Warnaar, *A higher level Bailey lemma: proof and application*, The Ramanujan Journal **2** (1998), 327–349.
52. A. Schilling and S. O. Warnaar, *Conjugate Bailey pairs. From configuration sums and fractional-level string functions to Bailey's lemma*, preprint math.QA/9906092.
53. A. Schilling and S. O. Warnaar, *A generalization of the q-Saalschütz sum and the Burge transform*, in *Physical Combinatorics*, pp. 163–183, M. Kashiwara and T. Miwa eds., Progr. Math. **191**, (Birkhäuser, Boston, 2000).
54. U. B. Singh, *A note on a transformation of Bailey*, Quart. J. Math. Oxford Ser. (2) **45** (1994), 111–116.
55. L. J. Slater, *A new proof of Rogers's transformations of infinite series*, Proc. London Math. Soc. (2) **53** (1951), 460–475.
56. L. J. Slater, *Further identities of the Rogers-Ramanujan type*, Proc. London Math. Soc. (2) **54** (1952), 147–167.
57. L. J. Slater, *Wilfrid Norman Bailey*, J. London Math. Soc. **37**, (1962), 504–512.
58. S. O. Warnaar, *Fermionic solution of the Andrews-Baxter-Forrester model. II. Proof of Melzer's polynomial identities*, J. Stat. Phys. **84** (1996), 49–83.
59. S. O. Warnaar, *A note on the trinomial analogue of Bailey's lemma*, J. Combin. Theory Ser. A **81** (1998), 114–118.
60. S. O. Warnaar, *Supernomial coefficients, Bailey's lemma and Rogers-Ramanujan-type identities. A survey of results and open problems*, Sém. Lothar. Combin. **42** (1999), Art. B42n, 22 pp.
61. S. O. Warnaar, *The Bailey lemma and Kostka polynomials*, preprint.